No. 2813
$24.95

SMALL ENGINES OPERATION, MAINTENANCE AND REPAIR

American Association for
Vocational Instructional Materials

TAB BOOKS Inc.
Blue Ridge Summit, PA 17214

Part 1: Care and Operation of Small Gasoline Engines

AUTHOR

J. Howard Turner, former editor and coordinator, AAVIM

EDITORIAL AND DESIGN STAFF

First Edition

Editors

Richard M. Hylton
James E. Wren

Art and Design

James E. Wren
Marilyn A. MacMillan, Artist
Sylvia Conine, Phototypesetter

CONSULTANTS

Dan Bennett, Member Service Representative, Walton Electric Membership Corporation

Jack Carroll, Product Service Director, McCullough Corporation

Andrew B. Cochrane, Technical Writer, Briggs and Stratton Corporation

Eddie Kinnard, Area Vocational Teacher, Georgia State Department of Education

Roger Sullivan, Small Engines Specialist, Riverside Mower, Athens, Georgia

Ernie Wood, Supervisor of Education, Tecumseh Products Company

Part 2: Maintenance and Repair of Small Gasoline Engines

EDITOR
J. Howard Turner, Editor and Coordinator, AAVIM

ART DIRECTOR
George W. Smith, Jr., Art Director, AAVIM

GRAPHICS
James E. Wren, Media Development Specialist, AAVIM

ACKNOWLEDGMENTS
Stephen R. Matt, Associate Professor, Industrial Arts Department, University of Georgia

FIRST EDITION
FIRST PRINTING

Library of Congress Cataloging in Publication Data

Small engines.

Includes index.
1. Internal combustion engines, Spark ignition.
I. American Association for Vocational Instructional Materials.
TJ790.S584 1987 621.43'4 86-30142
ISBN 0-8306-0813-3
ISBN 0-8306-2813-4 (pbk.)

Questions regarding the content of this book should be addressed to:

Reader Inquiry Branch
Editorial Department
TAB BOOKS Inc.
P.O. Box 40
Blue Ridge Summit, PA 17214

Contents

Part 1

Care and Operation of Small Gasoline Engines

Preface

This section, *Care and Operation of Small Gasoline Engines,* is designed to bring the reader to an entry-level of competency for servicing and operating small gasoline engines properly and safely. This text provides information that every owner of a small engine should know.

No other reference is needed except the operator's manual for the specific engine on which the reader is working. Step-by-step procedures are given for all the specific tasks.

OBJECTIVES

Upon successful completion of this text the reader will be able to do the following:

- **Describe the features of small gasoline engines**
- **Explain how small gasoline engines work**
- **Explain the importance of proper care and operation of small gasoline engines**
- **Identify types of small gasoline engines**
- **Service small gasoline engines**
- **Operate small gasoline engines**
- **Prepare small gasoline engines for storage**

Step-by-step instructions are given for other routine service jobs such as lubrication and refueling. Also, the reader will discover what happens when small engines are not serviced properly or are neglected.

For the purpose of this text, the term *small engines* refers to all single-cylinder, air cooled, spark ignition, gasoline powered engines. These engines range in horsepower from one-half to approximately 18 horsepower. Also, the terms *four-cycle* and *two-cycle* are used interchangeably with the terms *four-stroke cycle* and *two-stroke cycle* engines.

Introduction

Several million small engines are now in use throughout the United States. It is quite certain that the amount will increase in years to come. Small gasoline engines are greatly needed, and are doing a job that no other type of equipment can do (Figure 1). Even though small engines are popular, their owners often have many complaints. These complaints are usually made because the owner doesn't understand how the small engine works. After taking the engine to a service shop, the owner often finds that a simple adjustment or minor repair will get it started.

Small engines are really tougher than most people think! When they give you trouble, it is usually caused by improper service, operation, or maintenance. There are good reasons for some engine troubles. For example:

- Many small engines operate near the ground where dirt and dust are more likely to get into them. This will cause rapid wear if the engine is not properly serviced.
- Small engines are often abused by overloading.
- Few small engines receive proper servicing or regular maintenance—as is given to truck or tractor engines.

Once you learn how your engine works and how to operate it safely and service it properly, you will be proud and satisfied that you can take care of it yourself!

After you learn how small single-cylinder engines work, you will have some understanding of how larger multi-cylinder engines (in tractors or automobiles) work. This is because the operation of all engines, regardless of size, is very similar.

Care and Operation of Small Gasoline Engines is explained under the following headings:

I. Understanding Small Gasoline Engines
II. Servicing Small Gasoline Engines
III. Starting and Stopping Small Gasoline Engines
IV. Storing Small Gasoline Engines

FIGURE 1. Each year millions of small gasoline engines are used to help make jobs easier for people around the world.

I. Understanding Small Gasoline Engines

How many small gasoline engines are used around your home, farm, or business? If you count the number of small engines in your neighborhood, you will probably be surprised at how many jobs are being done by these sturdy little power plants. Small engines do many jobs because they are:

- Compact
- Lightweight
- Easy to service and repair
- Air-cooled
- Self-contained

Some of the machines and equipment powered by small engines are:

Post-hole diggers
Irrigation pumps
Chain saws
Small tractors
Conveyors
Motorcycles
Sump pumps
Leaf blowers
Minibikes
Ice augers
Sprayers
Small feed grinders
Snow blowers
Rotary tillers
Power trimmers
Bush cutters
Mowers
Lawn edgers
Outboard motors
Air compressors
Elevators
Generators
Snow vehicles
Concrete vibrators
Concrete surfacers
All terrain vehicles (ATV)
Tillers
Golf carts

These useful engines are now offered at prices most of us can afford, thanks to advanced engineering and mass production!

The purpose of this section is to introduce you to small gasoline engines—to show you how they work and to help you understand why it is important to take care of these engines.

Understanding small gasoline engines is discussed under the following headings:

A. How Small Gasoline Engines Work
B. Importance of Proper Care and Operation

A. How Small Gasoline Engines Work

Upon successful completion of this section, you will be able to **understand how an internal-combustion engine works**. You will also be able to identify small gasoline engines by crankshaft position, type of engine, and lubrication system.

The small air-cooled gasoline engine is an internal-combustion engine. All gasoline engines are of the internal-combustion type because energy for driving the crankshaft is developed within the engine.

This is done by the fuel-and-air mixture burning inside a confined chamber, called a cylinder. Because of the heat, the mixture expands, which then forces a piston to move. The piston is connected to a crankshaft which changes linear motion into rotary motion. The crankshaft may be in the vertical or horizontal position or multi-position. All crankshafts operate at a right angle to the cylinder.

A vertical crankshaft engine has its cylinder in a horizontal position. It should be used when you are mounting a power blade directly to the shaft as in a lawnmower (Figure 2).

A horizontal crankshaft engine may have its cylinder in a vertical, a horizontal, or an intermediate position. A horizontal crankshaft engine is best used for supplying power to a horizontal transmission shaft. Such engines are often used on small tractors (Figure 2).

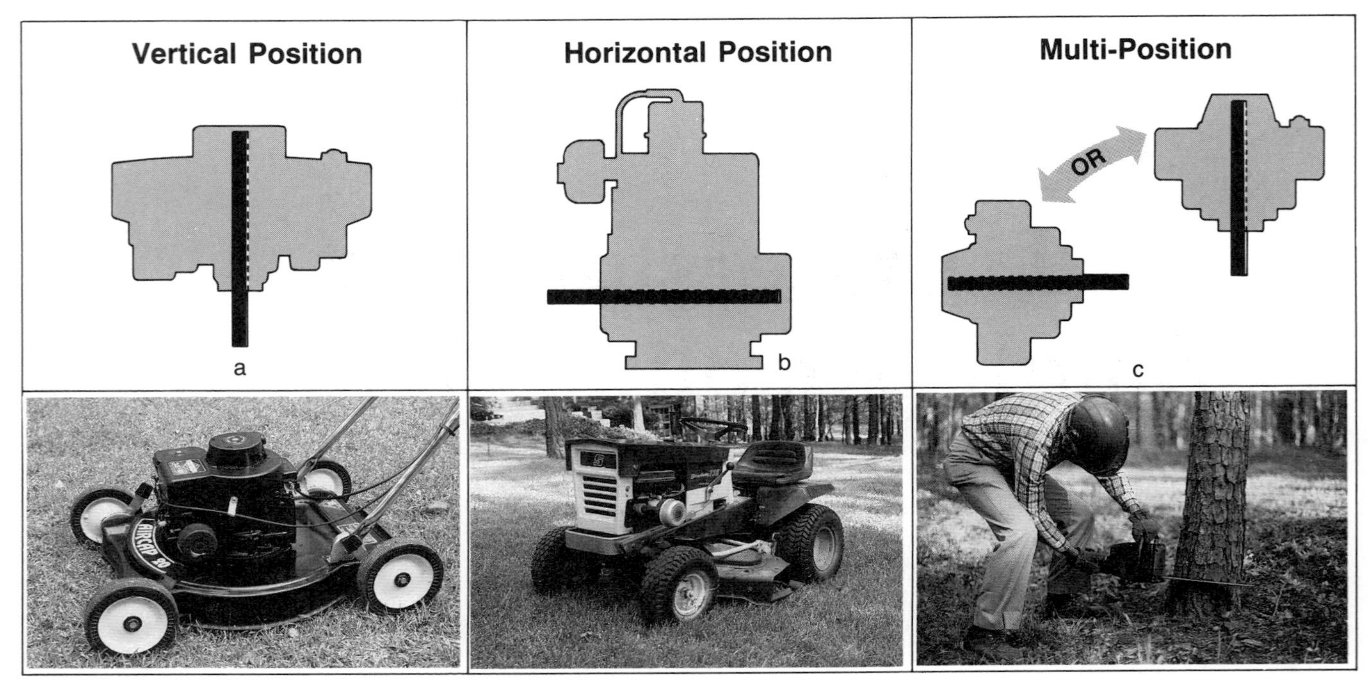

FIGURE 2. Examples of crankshaft types: (a) vertical position crankshaft engines are used on lawn mowers; (b) horizontal position crankshaft engines are commonly used on small tractors; (c) multi-position crankshaft engines are used on chain saws.

A multi-position crankshaft engine will operate in any position. Of course, the piston is always at a right angle to the position of the crankshaft. This type of engine (two-cycle) is used on chain saws (Figure 2) or on equipment where the position of operation may be at extreme angles.

To supply power (motion) to the crankshaft, a series of events must take place. The completion of this series of events is called a cycle. The events in the cycle are (Figure 3):

1. Intake of the fuel-air mixture into the cylinder
2. Compression of the fuel-air mixture
3. Power—ignition and expansion of the heated fuel-air mixture
4. Exhaust of burned gases

The full travel of the piston in one direction, either toward the crankshaft or away from the crankshaft, is called a stroke.

Some small engines are designed to complete a cycle during one revolution of the crankshaft (two strokes of the piston). Others require two revolutions of the crankshaft (four strokes of the piston). The main differences between four-stroke cycle and two-stroke cycle engines are:

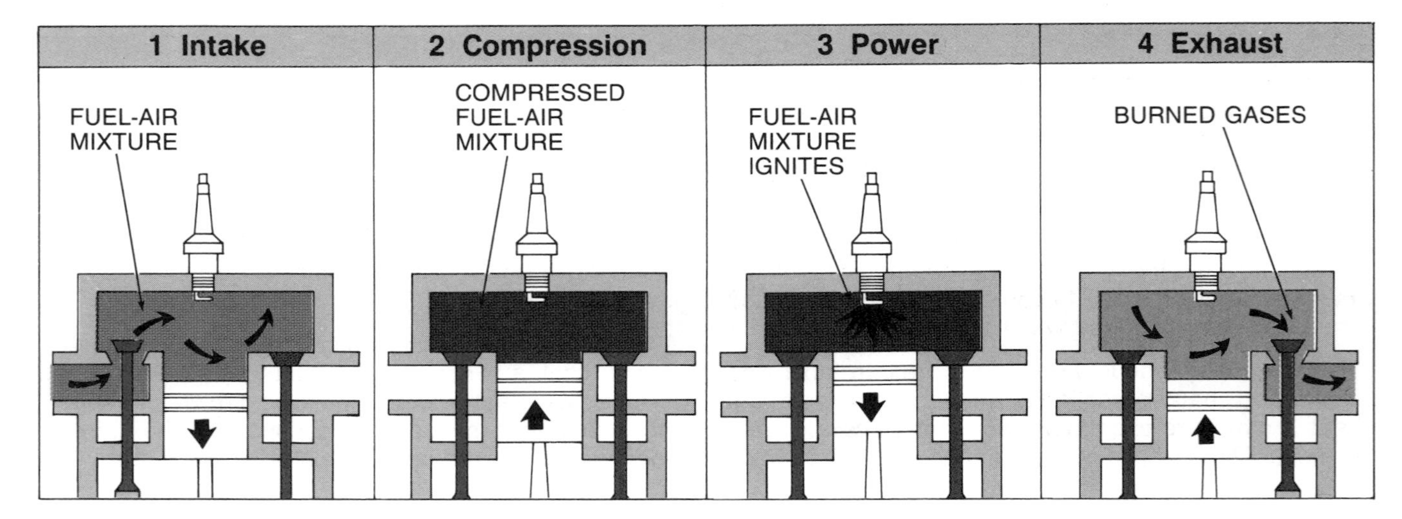

FIGURE 3. The principle events of an engine cycle.

- The number of power strokes per crankshaft revolution
- The method of getting the fuel-air mixture into the combustion chamber and the burned gases out
- The method of lubricating the internal moving parts

How these two types of engines work is explained under the following headings:

1. How a Four-stroke Cycle Engine Works
2. How a Two-stroke Cycle Engine Works

1. HOW A FOUR-STROKE CYCLE ENGINE WORKS

You can recognize a four-cycle engine by the presence of its oil sump and by the fact that it has an oil filler cap or plug where oil can be added to the crankcase. Since a four-cycle engine depends on an oil sump, the angle at which you operate it is very important. Figure 4 shows how the oil cannot be distributed properly when the engine is tilted too much.

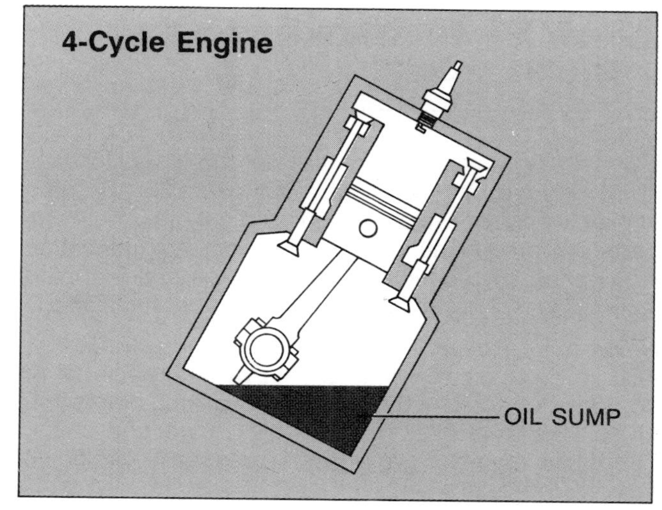

FIGURE 4. Because of its design, the four-stroke cycle engine cannot be operated at extreme angles.

Never put a mixture of oil and gasoline in the oil sump of a four-stroke cycle engine. The engine will overheat because of improper lubrication. The operating principle of a four-stroke cycle engine is shown in Figure 5. The four strokes include:

a. Stroke 1 — Intake
b. Stroke 2 — Compression
c. Stroke 3 — Power
d. Stroke 4 — Exhaust

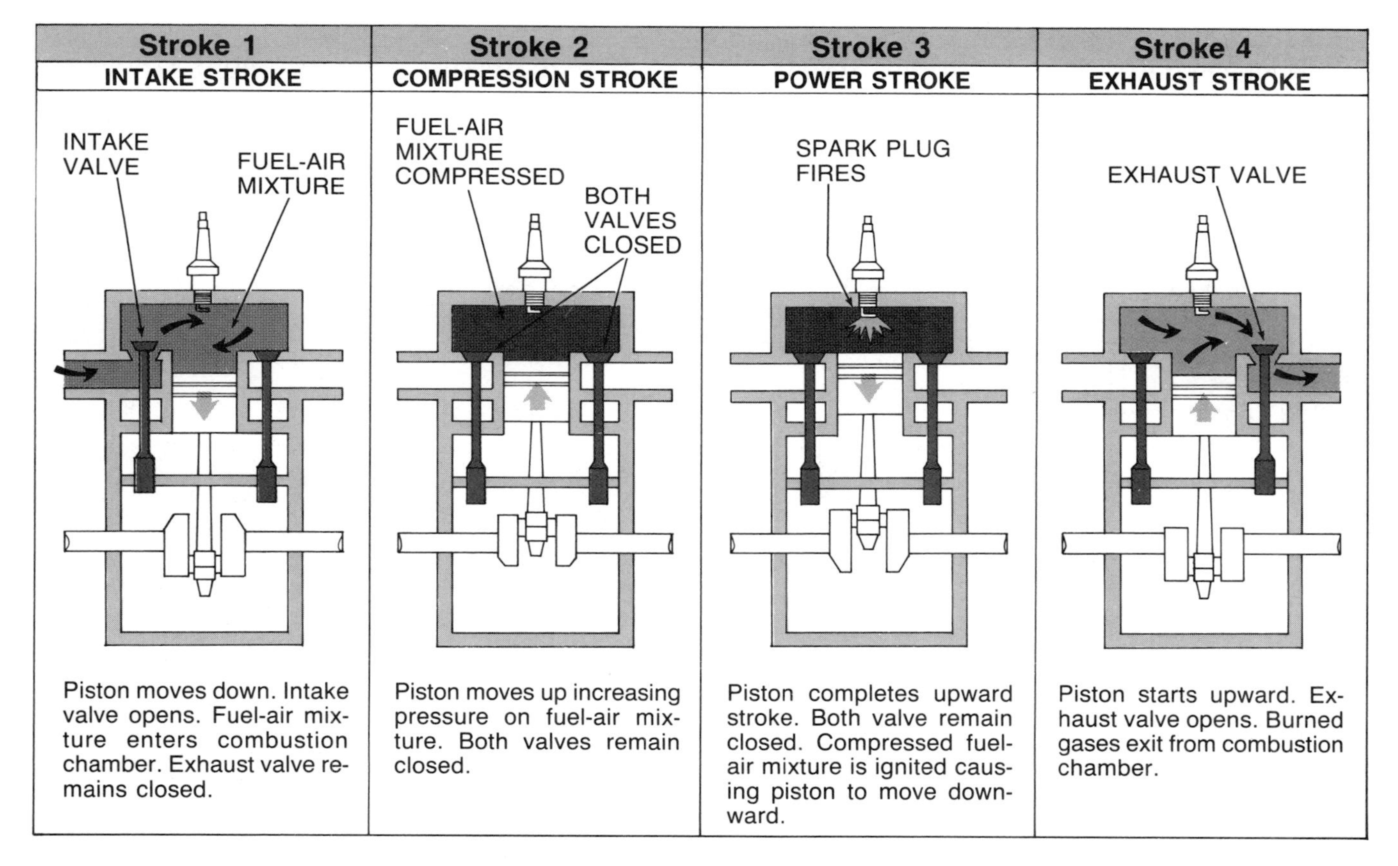

FIGURE 5. The operating principle of a four-stroke cycle engine.

2. HOW A TWO-STROKE CYCLE ENGINE WORKS

Since the two-cycle engine uses the crankcase for storing a reserve charge of fuel-air mixture for the next stroke, the crankcase cannot be used only as an oil compartment for lubricating the engine. Instead, lubrication is supplied by a specific quantity of oil that is mixed with the gasoline at the time the engine is refueled.

Never put gasoline in a two-cycle engine without first mixing it with oil, the engine will overheat because of improper lubrication. It will not run very long before the piston and bearings will overheat, score and seize. Two-cycle engines have a sealed crankcase, but no oil sump (Figure 6).

Notice that the two-cycle engine is properly lubricated at any angle (Figure 6) by small drops of oil suspended in the fuel-air mixture in the crankcase. This does not mean, however, that all two-cycle engines can be operated at any angle; it will depend on the type of carburetor used.

Here is how a two-stroke cycle engine works. It is designed to complete all the actions (a cycle) described in the four-stroke cycle engine, but it does them during one revolution of the crankshaft. The operating principle of a two-stroke cycle engine is shown in Figure 7. The two strokes include:

a. Stroke 1 — Compression and Intake
b. Stroke 2 — Power and Exhaust

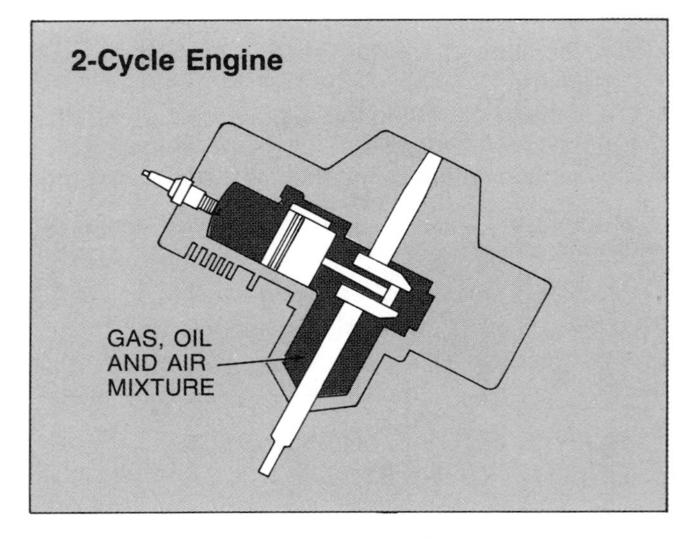

FIGURE 6. A two-stroke cycle engine can be lubricated at any position because it is lubricated by special oil mixed with the fuel.

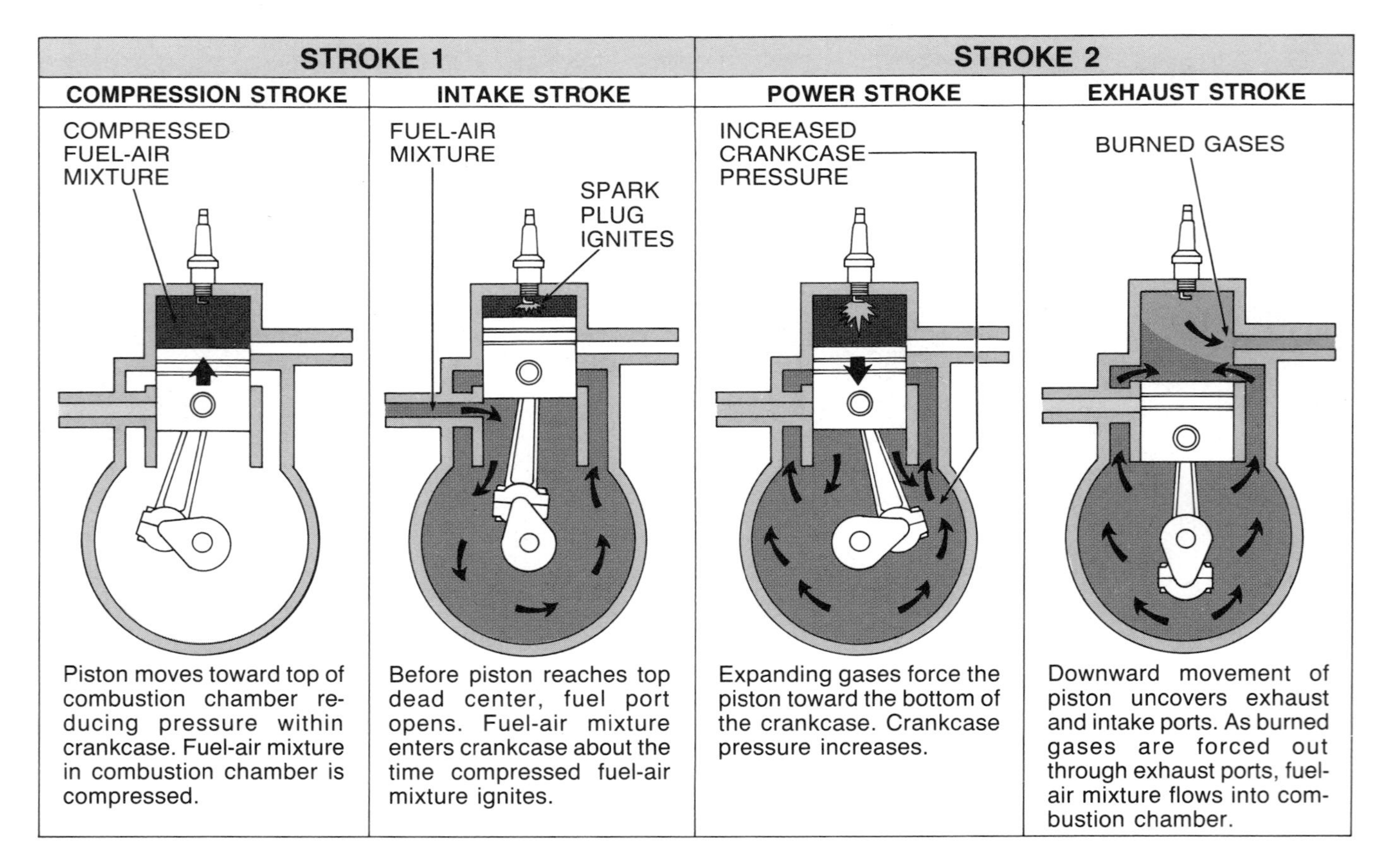

FIGURE 7. The operating principle of a two-stroke cycle engine.

STUDY QUESTIONS

1. True or False. All gasoline engines are internal combustion engines, energy for driving the crankshaft is developed within the engine.
2. List five pieces of equipment powered by small engines.
3. Which type of engine (four-cycle or two-cycle) is used in chain saws?
4. The full travel of the piston in one direction, either toward the crankshaft or away from the crankshaft, is called a ________.
5. Identify the main differences between a four-cycle and a two-cycle engine.
6. True or False. The strokes in a four-cycle engine are compaction, energy, power and input.

B. Importance of Proper Care and Operation

Upon successful completion of this section, you will be able to **prevent unnecessary wear or damage to small engines by proper care and operation**.

Some small engines have been known to run as long as 5,000 hours. One thousand hours, however, is usually considered a long life for small engines if they are not operated regularly. For example, if it takes you two hours to mow your lawn, and if you mow it every week for six months, that is only 48 hours per year. At this rate, your mower engine should last 20 years. Sometimes you hear of small engines lasting 20 years, but most of them fail much sooner.

The importance of proper care and operation is discussed under the following headings:

1. Causes of Engine Failure
2. Importance of the Operator's Manual and Engine Nameplate

1. CAUSES OF ENGINE FAILURE

If you are having operating troubles, high maintenance and repair costs and your engine doesn't last very long, your problems are probably caused by the way you service and operate your engine.

There are five major reasons recognized as causing engine failure. They are:

1. Dirt and abrasives in the engine
2. Improper lubrication
3. Improperly cleaned spark plugs
4. Blocked air cleaners and cooling fins
5. Operating engine under load for too long

2. IMPORTANCE OF THE OPERATOR'S MANUAL AND ENGINE NAMEPLATE

The operator's manual and engine nameplate are discussed under the following headings:

a. Operator's Manual
b. Engine Nameplate

a. Operator's Manual

One way to familiarize yourself with servicing and operating your engine is to save and use the operator's manual that comes with your engine. It contains important information you must have when servicing your engine.

b. Engine Nameplate

To get the proper parts for your engine, you must give the dealer the following information, which is found on the nameplate (Figure 8). If you complete a form similar to the one on page 69, you will have the necessary information.

Information you may expect to find on the nameplate includes:

1. Make of engine, or name of the manufacturer
2. Model number or name of engine
   This number usually gives a clue as to the horsepower. Some model numbers refer to other information such as the type of crankcase and accessories. Some include information about certain modifications.
3. Serial or code number
   It tells the sequence in which the engine came off the assembly line. Modifications are made as production progresses. By knowing the serial or code number, the manufacturer or dealer can tell which modifications have been made. Then the proper parts can be supplied.
4. Type number
   The type number identifies the engines that require certain parts and accessories which are different from the parts originally designed for the engine.
5. Specification number
   This refers to different designs, requested by the equipment manufacturers, such as the length of power take-off end of the crankshaft.

The type and specifications information is included in the model and serial number on some engines. When recording the numbers, be careful not to confuse the

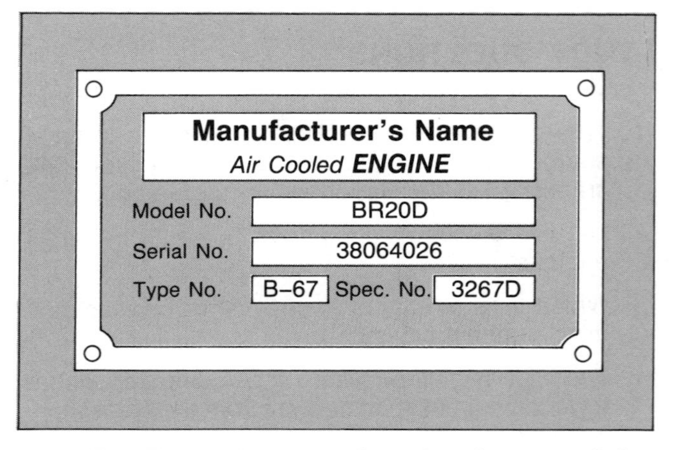

FIGURE 8. The engine nameplate gives important information for ordering new parts.

engine number with the equipment number. They are very similar. Look for the engine number on the engine proper. Sometimes it is necessary on an old engine to rub the numbers with chalk to bring out the raised, or indented numbers.

STUDY QUESTIONS

1. List five major reasons for small engine failure.
2. When caring for and operating small engines, which manual should be saved and consulted regularly?
3. Identify types of information you might find on an engine nameplate.
4. On older engines, ________ may be used to bring out raised or indented numbers on engine nameplates so that they can be easily and accurately read.

II. Servicing Small Gasoline Engines

The purpose of this section is to give you the proper procedures and time intervals for servicing small engines.

If you expect trouble-free service from your engine, you must take time to service it regularly. When this should be done will vary with the different manufacturers' recommendations, with the operating conditions, and with the type of service to be done.

Most manufacturers agree on the minimum time interval for doing most jobs. Before each operation, you should (1) check the crankcase oil level in four-cycle engines, and (2) fill the fuel tank with clean, fresh gasoline. If yours is a two-cycle engine, be sure the proper oil is thoroughly mixed with the gasoline. This can be done by agitating the fuel container prior to filling the fuel tank.

Two jobs should be done at least every 25 hours of operation. They are (1) servicing carburetor air cleaners, and (2) changing the crankcase oil (four-cycle engines). If you are operating your engine under very dirty and dusty conditions, you should perform these jobs more often.

As a rule, an annual cleaning and general inspection is recommended. At this time the fuel strainers and crankcase breather are cleaned and serviced.

The following service jobs are discussed in the order you would perform them from the start of an operating season. Of course, most of them will need to be repeated throughout the season.

The proper procedures for doing these and other service jobs are given under the following headings:

A. Identifying Types of Small Engines
B. Cleaning Small Engines
C. Servicing Carburetor Air Cleaners
D. Servicing Fuel Strainers
E. Servicing Crankcase Breathers (Four-cycle Engines)
F. Lubricating Small Engines
G. Refueling Small Engines
H. Servicing Spark Plugs
I. Checking and Adjusting Carburetors
J. Checking Compression
K. Checking and Servicing Batteries

A. Identifying Types of Small Engines

Upon successful completion of this section, you will be able to **use the four methods of recognizing a four-cycle from a two-cycle engine**.

Before you start servicing your small engine, it is important that you know what type of engine it is. This is because servicing procedures for four-cycle engines are not always the same as for two-cycle engines.

It is difficult to recognize a four-cycle from a two-cycle engine unless you understand the main differences.

Four methods can be used to learn the differences. They are discussed under the following headings:

1. Checking for Oil Sump and Filler Cap
2. Locating the Exhaust Ports or Muffler
3. Reading the Nameplate
4. Identifying Engines by Compression

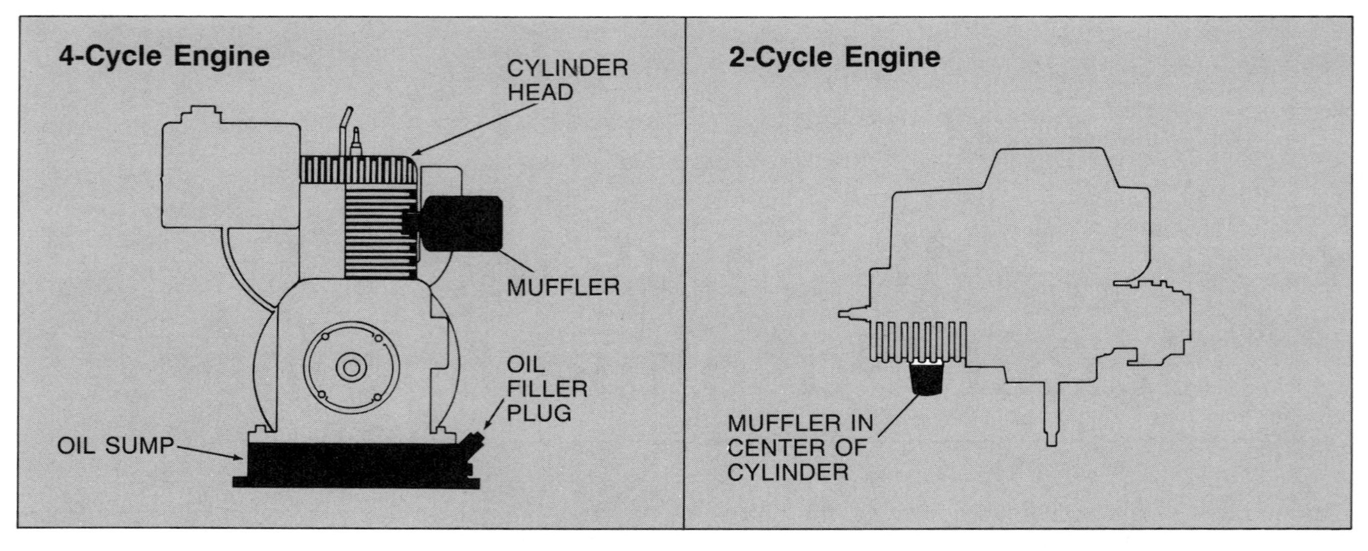

FIGURE 9. Four-cycle engines can be distinguished by their oil sump, oil filler plug and muffler located at the end of the cylinder head. Two-cycle engines have no oil sump and the muffler is located between the cylinder head and the crankcase.

1. CHECKING FOR OIL SUMP AND FILLER CAP

Check for an oil sump and oil filler plug or cap. If your engine has a sump and filler plug or cap, it is a four-cycle engine. There is no oil sump on two-cycle engines. Chainsaws (two-cycle) have a chain oil reservoir. This should not be confused as an oil sump.

2. LOCATING THE EXHAUST PORTS OR MUFFLER

Check for the location of the exhaust ports or muffler (Figure 9). On a four-cycle engine the exhaust muffler connects at the cylinder-head end of the engine cyclinder. The two-cycle engine has an exhaust port about midpoint on the cylinder.

3. READING THE NAMEPLATE

Check the information on the nameplate or check your operator's instructions. One or both of them should mention the oil specifications, or fuel-and-oil specifications. If either one gives the crankcase capacity or a kind of crankcase oil, this applies only to four-cycle engines. If mixing oil and gasoline is mentioned, this would identify it as a two-cycle engine.

4. IDENTIFYING ENGINES BY COMPRESSION

If after using methods 1, 2, or 3, you are still not sure about the type of your engine, use method 4, the compression method. Proceed as follows:

1. *Disconnect the spark plug to prevent the engine from starting.*
   Make sure the connector is not touching the spark-plug terminal.
2. *Put a chalk mark on the starter flange or pulley.*
3. *Crank the engine slowly by hand by pulling on the crank rope.*
   If resistance—caused by compression—is felt at only every other revolution, it is a four-cycle engine.
   If the resistance—caused by compression—is felt at each revolution, it is a two-cycle engine.

SECTION QUESTIONS

1. Two jobs should be completed on small engines every 25 hours of operation. Identify these jobs.
2. True or False. Servicing procedures for small engines are exactly the same for two-cycle and four-cycle engines.
3. List four methods of recognizing a four-cycle from a two-cycle engine.

B. Cleaning Small Engines

Upon successful completion of this section, you will be able to do the following:

- **Tell what can happen to an engine if it is not cleaned properly and regularly**
- **Select the tools and materials needed to clean an engine**
- **Give the step-by-step procedures for cleaning and inspecting an engine's crankcase and accessories**
- **Follow the step-by-step procedures for cleaning and inspecting an engine's cooling system**

The term *cleaning the engine* in this section refers to cleaning the outside of the crankcase and accessories, the cooling system, and the exhaust system.

The importance of keeping your small engine clean and how to clean and inspect it are discussed under the following headings:

1. Importance of Cleaning the Engine
2. Tools and Materials Needed
3. Cleaning and Inspecting the Crankcase and Accessories
4. Cleaning and Inspecting the Cooling System

1. IMPORTANCE OF CLEANING THE ENGINE

If you fail to clean your engine regularly, it will not run properly and its service life will be shortened. For example, the following problems may occur:

- The engine overheats
- Dirt gets inside the engine
- Rubber parts soften and break down
- Loose nuts and cracks go unnoticed
- Neglect

These problems are discussed under the following headings:

a. Overheating
b. Dirt
c. Neglect

a. Overheating

It will be easier for you to understand why it is important to keep the cooling system clean if you first understand how the cooling system works.

The average temperature of burned gases inside the cylinder is about 1980°C (3600°F). About one-third of the heat is given off through the cooling system. Another one-third is lost through the exhaust system. The remaining one-third is used to develop power.

Heat travels through the cooling system in the following way: The heat moves from the cylinder through the cylinder walls by conduction*. When it reaches the outer surfaces of the cylinder, it is taken up by the air through forced convection**. Forced convection is accomplished by a flywheel blower forcing air past the cylinder wall and the cooling fins. It is directed around the cylinder by the cylinder baffles (Figure 10). This is why you should never operate an engine with the baffles removed.

The cooling fins are used to increase the surface around the outside of the cylinder for additional cooling area. This is necessary on air-cooled engines, but not on water-cooled engines because air is about one twenty-fifth as effective as water when used for cooling engines.

The intake area is screened to prevent large particles of dirt and trash from entering (Figure 10). Small particles, however, enter and collect on the fins and

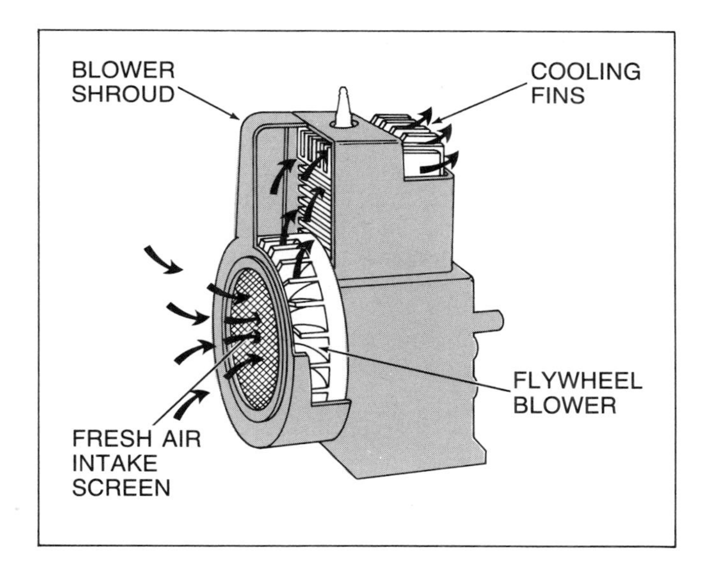

FIGURE 10. Small engines are cooled by air directed over the cooling fins of the engine by a shroud or baffle.

* Conduction is the term used to describe heat transfer through a solid conductor.

** Convection is the term used to describe heat transfer through movement of a gas—in this case air. When air is heated, it becomes less dense and rises. Cool air moves in to replace it. This is natural convection. Forced convection is accomplished by using a fan to help speed the convection action. In this way the hot air on the cooling fins is replaced faster by cooler air being forced past them by the flywheel blower.

throughout the cooling system. This slows the flow of air and decreases the fins' effectiveness for giving off heat unless the accumulation is removed regularly.

Overheating can be caused by dirt and grime on the outside of the engine and by a clogged or restricted exhaust system.

When dirt and grime collect on the fins, they tend to insulate the fin area. When this happens, heat does not move as easily as it should from the fins into the airstream. Then when the engine runs, it overheats.

A clogged muffler will also cause an engine to overheat. This most often happens with two-stroke cycle engines. It causes a back pressure in the cylinder and prevents the hot gases from escaping easily. Remember that one-third of the combusion heat is given off through the exhaust system.

Overheating, in turn, may cause valve-guide distortions, cylinder warping, scuffing and scoring of cylinder walls, sticking valves, loss of power and eventually engine failure.

b. Dirt

Another reason for keeping your engine clean is to prevent dirt from getting inside the engine. Much of the dirt that enters the engine comes from deposits on the outside. When dirt mixes with oil and gets into the lubrication area of an engine, a harmful grinding mixture develops. This will cause rapid wear on bearings and other surfaces which slide against one another.

Unless you are very careful, dirt will enter the engine when you check the crankcase oil level or refuel the engine. Dirt may also get into the air cleaner assembly when the filter is removed. This restricts air flow into the carburetor; then some of the dirt will enter the cylinder during engine operation. With two-cycle engines dirt and other foreign materials may enter the exhaust ports and lodge on the cylinder walls (Figure 11).

c. Neglect

Oil or gasoline which collect on rubber parts, such as the spark plug wire, V-belt, rubber hoses or vibration dampening mount, causes them to soften and wear out quickly.

An important reason for cleaning an engine is to be able to see any parts that might have become defective, loose or broken. These parts can be repaired before any real damage is done. Therefore it is important not to neglect your engine.

FIGURE 11. Muffler of a two-cycle engine removed to show exhaust ports.

It is important to set aside a definite time each year—at the start of the season or at regular times during the season—for completely cleaning the outside of the engine and the cooling system. You should also clean the engine before it is to be serviced or repaired.

2. TOOLS AND MATERIALS NEEDED

1. Slot-head screwdrivers — 4″ and 6″
   Phillips-head screwdrivers — 4″ and 6″
2. Socket set — including 3/8″, 7/16″, 1/2″, 5/8″, and 9/16″ sockets and 3/8″ ratchet handle
3. Open-end wrenches of the same sizes as indicated for socket wrenches
4. Nut drivers — 1/4″, 3/8″
5. Wire brush
6. Pail — approximately 10 quart capacity
7. Paint brush
8. Water hose equipped with nozzle
9. Small wooden scraper and/or a small putty knife
10. Old toothbrush
11. Commercial degreaser, petroleum solvent (mineral spirits, kerosene, or diesel fuel), or steam cleaning equipment
12. Hand sprayer

A fire extinguisher is also important. It should be of the dry-chemical, carbon-dioxide, or foam type to be effective for gasoline or other petroleum-type fires.

An air compressor, set for a pressure of approximately 100 pounds (per square inch) is very helpful in removing dirt from hard-to-reach places. Goggles and gloves should be worn when compressed air is used.

3. CLEANING AND INSPECTING THE CRANKCASE AND ACCESSORIES

There are various methods for cleaning an engine. Most recognize using a good cleaning solvent along with a brush and cloth. Be careful when selecting solvents. Some may remove the engine paint leaving the engine unsightly and without protection from rust formation.

A degreaser is usually faster acting than other solvents. It does a good job of cleaning hard to reach places and is available in spray cans or in a concentrate form. Common petroleum solvents can be used but are not adequate for removing substances which do not contain oil or grease.

Cleaning with a strong, forceful stream of water is not recommended because water could contaminate the fuel system.

CAUTION! Do not use gasoline for cleaning your engine. It is highly flammable and an extreme fire hazard. If you use a degreaser, check the instructions on the can. Some are flammable and dangerous if used in a closed building or near a flame. When using solvents be careful not to touch plastic engine parts. Solvents may damage some plastics.

Procedures for cleaning your engine are as follows:

1. *Allow the engine to cool if it has been running.*
   A hot engine will evaporate the cleaning solution before it has had time to become effective.
2. *Remove the blower shroud and cylinder baffles.*
3. *Inspect for oil leaks and fuel leaks.*
   Where there is an oil leak, there is often a large accumulation of dirt which helps you to locate the leak more quickly than after the engine is cleaned.
   Where there is a fuel leak, the area may be clean, except for a reddish deposit left from the evaporation of the fuel, and a tendency for the paint to blister.
4. *Remove air cleaner and cover air cleaner opening* (Figure 12).
   It is best to remove the air cleaner and to cover the carburetor intake with a piece of plastic sheeting. Use a rubber band to hold the plastic sheet in place. This prevents the cleaning solution from entering the carburetor.
5. *Clean the exhaust system.*
   a. If you have a four-cycle engine, you will not likely have trouble with the muffler clogging. Proceed to step 6.
   b. If your engine is the two-cycle type, it is important that you clean the exhaust ports. Proceed as follows:

FIGURE 12. Remove the air-cleaner and cover the opening with a piece of plastic.

FIGURE 13. Remove the muffler and check exhaust ports.

(1) Remove the muffler (Figure 13). Many engines are equipped with spark arrestor screens as required by the U.S. Forestry Service. These screens plug onto two- and four-cycle engine mufflers.
(2) Check cylinder exhaust ports for carbonizing.
(3) Rotate crankshaft until the piston covers the exhaust ports.
This avoids the possibility of scraping carbon and dirt into the piston rings and causing more wear.
(4) Clean exhaust ports (Figure 14).
Use a wood scraper so you won't damage the chamfered edges. Scrape so that carbon will not enter cylinder.
(5) Clean the muffler in solvent.
(6) Replace the muffler before cleaning the engine.

FIGURE 14. Clean exhaust ports using a wooden scraper.

This is to prevent the cleaning solvent from getting into the cylinder. It can cause corrosion and interfere with lubrication. Install a new gasket when installing the muffler. A leaking gasket will cause noisy operation. Mufflers vary in shape and size, but they are usually attached to the cylinder by two capscrews and a gasket.

6. *Apply solvent on areas that need cleaning* (Figure 15).

   Apply a thin, well-distributed film. Apply until the surface has a moist appearance.

   If you are using a petroleum solvent without a degreaser, apply it with a paint brush to help get better penetration.

7. *Let the solvent set about 5 minutes.*

   This gives the solution time to do a good job of penetrating and loosening the grease and oil particles.

8. *Remove solvent from the engine surface.*

FIGURE 15. Apply a cleaning solvent in a thin, even film until surface is moist.

FIGURE 16. Remove the cleaning solution by flushing with clean water.

If you used a degreaser, flush with water and continue to wash until all milky substance is removed from engine (Figure 16).

If you used a petroleum solvent, remove it with a strong soap solution, then flush it off with water.

To avoid possible shorting of the ignition circuit, do not apply water in the area of the breaker points.

9. *Check for areas that have been missed.*

   Use a scraper and a wire or bristle brush to remove accumulations that did not flush off. Reapply degreaser if needed to remove remaining accumulations.

10. *Replace protective covers.*

11. *Replace the carburetor air cleaner.*

    The air cleaner should be serviced before installation. See procedures under "Servicing Carburetor Air Cleaners."

12. *Operate the engine immediately for three to five minutes.*

    Heat from the engine operation will help it to dry and help to remove water from vital parts.

4. CLEANING AND INSPECTING THE COOLING SYSTEM

The steps for cleaning and inspecting the cooling system are as follows:

1. *Close the fuel shut-off valve* (Figure 17).

   Not all engines have shut-off valves. If yours has one, it may be either a part of the sediment bowl or a single unit.

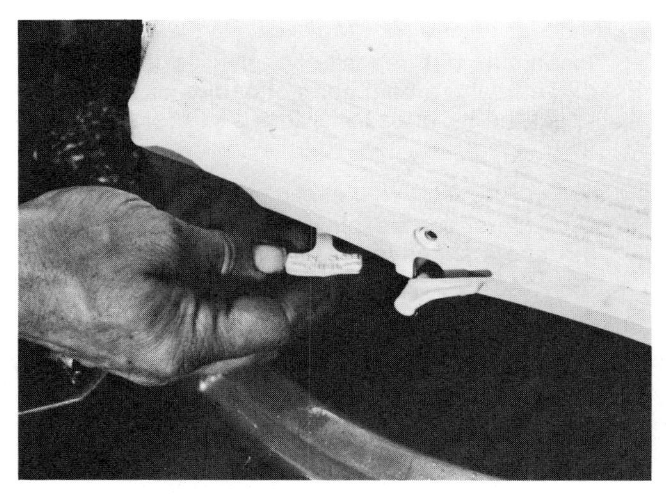

FIGURE 17. Close the fuel shut-off valve before cleaning the cooling system.

FIGURE 19. Remove the blower shroud.

If your engine has no shut-off valve, disconnect the supply tube to drain the tank.

CAUTION! Do not remove fuel in an unvented area. Store the drained fuel in proper containers in vented storage areas.

2. *Remove fuel tank, if necessary, to remove the blower shroud* (Figure 18).

   This will require disconnecting a flexible hose or a metal tube. Be sure to cover the end of the fuel lines so dirt and cleaner will not enter while you are cleaning the engine. Use a piece of plastic sheeting to cover the opening and a rubber band to hold it in place.

3. *Remove the blower shroud* (Figure 19).

   Notice the positions of cylinder head bolts. With some engines you will have to remove the recoil starter or screened sheave if equipped with a self-starter. You may also have to remove some of the accessories such as the air cleaner, muffler, governor spring, or sparkplug wire before removing the blower shroud.

FIGURE 18. Remove the fuel tank and cover the end of the fuel line with plastic.

4. *Remove remaining baffles and deflectors that divert air from shroud around and over end area of cylinder.*

   Some engines have additional baffles welded to the blower shroud. Others use one or more separately mounted baffles as needed to direct the air over the finned area.

   WARNING! Do not operate your engine with the shroud and baffles removed. It will overheat. Those equipped with an air vane governor are sensitive to air flow through the cooling system, so there is no speed control, and the engine will overspeed.

5. *Clean inside of shroud and baffles.*

   Use a small bristle brush. A putty knife makes a good scraper for removing heavy accumulations of dirt (Figure 20).

6. *Clean intake screen* (Figure 21).

   Wash it in solvent.

7. *Clean dirt from cylinder fins* (Figure 22).

   Use a wooden scraper. Steel scrapers scratch the finned surfaces, especially if they are made of aluminum. The scratches will collect dirt. Blow out with compressed air. Take care not to blow dirt into engine while cleaning. Install cylinder head bolts in correct order and torque according to specifications.

8. *Clean dirt from blower flywheel fins.*

   Use same procedure as for cleaning dirt from cylinder head and fins.

FIGURE 20. Clean shroud with a putty knife or brush.

FIGURE 21. Clean intake screen with a bristle brush.

FIGURE 22. Clean cylinder fins with a wooden scraper.

9. *Inspect engine for cracks and broken parts.*
   Inspect for dirt deposits on other parts of engine, and straighten bent parts of baffles or shroud. Inspect the cylinder for cracks and broken cooling fins.
10. *Repeat the spraying and cleaning process.*
    Procedures were explained in the section "Cleaning and Inspecting the Crankcase and Accessories."
11. *Replace broken or damaged parts.*
12. *Reassemble parts in reverse order. Replace all old gaskets with new gaskets.*
13. *Check nuts, screws and bolts (including engine mounting bolts) to make sure they are properly torqued.*
14. *Repeat steps 10, 11, and 12 of* "Cleaning and Inspecting the Crankcase and Accessories."

STUDY QUESTIONS

1. List problems that may occur if small engines are not cleaned regularly.
2. About ________ of the heat developed in an engine is used to develop power.
3. Dirt should never be allowed to enter the small engine when checking the ________ or filling the fuel ________.
4. Individuals who work with small engines should always have a ________ ________ available for possible fires.
5. True or False. Cleaning engines with a strong forceful stream of water is not recommended.
6. True or False. Gasoline is highly recommended for cleaning small engines.
7. Describe why it is necessary to replace all gaskets when cleaning a small engine.

C. Servicing Carburetor Air Cleaners

Upon successful completion of this section, you will be able to do the following:

- **Identify the type of air cleaner for your engine and know how the air cleaner works**
- **Tell why it is important to service air cleaners and when it should be done**
- **Select the tools and materials needed for servicing air cleaners**
- **Follow procedures for servicing oiled-filter type and dry-filter type air cleaners**

Carburetor air cleaners filter the air going into your engine through the carburetor. Without them, dirt would soon ruin your engine.

The carburetor air cleaner is one of the most important parts of your engine. One manufacturer who conducted an experiment to learn the effects of operating an engine under dirty conditions without an air cleaner found that the engine failed after only three and one-half hours of operation. The cylinders, pistons, rings and bearings were badly worn.

For the air cleaner to protect your engine from dirt, it must be serviced properly. Servicing consists mainly of cleaning the filter element. Why it is important and how to service the air cleaner on your engine are explained under the following headings:

1. Types of Air Cleaners and How They Work
2. When to Service the Air Cleaner
3. Tools and Materials Needed
4. Servicing the Oiled-Filter Type of Air Cleaner
5. Servicing the Dry-Filter Type of Air Cleaner

1. TYPES OF AIR CLEANERS AND HOW THEY WORK

There are two common types of air cleaners used on small gasoline engines. When serviced properly, both of them do a good job of removing harmful dirt from the air entering the engine.

In order to service an air cleaner properly, you must first know the type of air cleaner in your engine and how it works.

The common types of air cleaners are described as follows:

a. Oiled-Filter Type
b. Dry-Filter Type

a. Oiled-Filter Type

The oiled-filter type air cleaner of today is commonly made of a sponge-like filtering material called polyurethane (Figure 23). It is relatively inexpensive and comes in many shapes and sizes. To be an effective filter, polyurethane must be coated with oil before being installed.

The filter is designed to allow air to pass over a large area of oiled surface. The oil on the filter material picks up dust and dirt particles and prevents them from entering the engine.

FIGURE 23. (left) Typical polyurethane filters. (right) A combination polyurethane-paper filter.

FIGURE 24. Typical dry-filter type air cleaners.

b. Dry-Filter Type

The dry-filter type air cleaner consists of a porous filtering element usually made of paper, felt or fiber (Figure 24).

The dry-filter type air cleaner has a filter with very small openings that keep harmful particles from passing through. This type of filter does not depend on oil for catching the dust and dirt.

Dry-filter type air cleaners have the following advantages:

- They are easier to service.
- Trapped particles in this type of filter cause less restriction to air passage.
- They are more efficient at a wide range of engine speeds
- When the air cleaner needs servicing, the engine warns you by failing to run properly. Dirt builds up on the filter, and the engine starves for air. This results in a choking effect.

Dry-filter type air cleaners cost more to maintain than other cleaners because the filter element must be replaced quite often. How often depends on the operating conditions.

2. WHEN TO SERVICE THE AIR CLEANER

If you fail to service your air cleaner properly, dirt and dust, which should be collected in the air cleaner, will either (1) pass into the engine and mix with the lubrication oil; or (2) it will build up on the air cleaner filter, choke the engine, and cause an excessively rich fuel-air mixture. Either condition will shorten the life of your engine.

If you allow the air cleaner to become severely clogged, you will get far too much fuel for the amount of air going into the cylinder. All the gasoline does not burn. The unburned gasoline washes the oil from the cylinder walls. Then the piston and cylinder walls become scuffed and scored from lack of lubrication. Fuel also enters the crankcase (four-cycle engines) and dilutes the lubricating oil.

It is difficult for manufacturers to say when an air cleaner should be serviced. Most of them recommend servicing the air cleaner every 25 hours if the engine is being operated under the best of conditions.

When the engine is operated continuously in extremely dirty or dusty conditions, there may be times when you will need to service the air cleaner two or three times per day. To be safe, inspect the air cleaner at least once each day or before each use.

When you are operating a chain saw continuously, carry an extra clean filter for replacement. It is often not possible to clean the filter on the job; as it usually requires cleaning twice each day.

Procedures for servicing air cleaners vary with the different types. They are given under headings 5 and 6 that follow. You will also find instructions in your operator's manual or on a label stamped on the air cleaner.

3. TOOLS AND MATERIALS NEEDED

1. Slot-head screwdrivers — 4″ and 6″
2. Phillips-head screwdrivers — 4″ and 6″
3. Container for washing parts
4. Crankcase oil or special filter lubricant for polyurethane filters
5. Clean rags
6. Wooden scraper
7. Paint brush
8. Petroleum solvent (mineral spirits, kerosene, or diesel fuel)

CAUTION: Do not use gasoline, naptha, or benzine. They are extremely flammable.

4. SERVICING THE OILED-FILTER TYPE AIR CLEANER

The procedures for servicing are as follows:

1. *Disconnect the spark plug wire* (Figure 25).

   This is to prevent the engine from starting accidentally.

FIGURE 25. Disconnect the spark plug to prevent accidentally starting the engine.

2. *Clean area around the air cleaner.*

   This is to prevent loose dirt on the engine from getting into the carburetor air duct.

3. *Remove filter element cover* (Figure 26).

4. *Remove filter element* (Figure 26).

   Be sure to note the order in which the parts are disassembled so you will be able to reassemble them in the proper order.

5. *Check condition of the filter element and other parts of the air cleaner.*

   Some polyurethane filters become brittle and crumble, and should be replaced. Some manufacturers are now installing oiled-filters which are not servicable and must be replaced. Replace all other damaged or broken parts.

6. *Cover carburetor air intake.*

   This is to prevent dirt from blowing or falling into the carburetor while the air cleaner filter is removed.

FIGURE 26. Remove the filter cover and element.

FIGURE 27. Clean the filter element with a household detergent and hot water.

7. *Clean filter* (Figure 27).

   A strong solution of household detergent and hot water is best for cleaning filters. Do not use gasoline for cleaning your air filter, it is a serious fire hazard.

   Never use carbon tetrachloride or paint thinner. It dissolves certain glues in polyurethane filters. Also, when it is inhaled, carbon tetrachloride will make you sick. Some manufacturers recommend using petroleum solvent, which is fairly satisfactory for both types of filter elements. Clean the housing and cover with a stiff bristled brush; do not leave any loose bristles in the air cleaner.

8. *Dry the filter element.*

   If your filter is the polyurethane sponge type, dry it by squeezing or by using compressed air.

   If your filter is the metal mesh type, allow it to air dry or use compressed air.

9. *Remove the protective cover from the carburetor intake.*

10. *Clean the carburetor intake.*

    Use a clean cloth dampened with solvent.

11. *Oil the filter element* (Figure 28).

    Oiling the filter element is important. Many people neglect this because it is messy and because they think it is not needed. The oiled surfaces, however, are what catch the dirt. Use the same grade and type of oil used in your engine crankcase, or use special oil provided by your manufacturer. Oil must be well distributed throughout the element for proper filtering action.

    For polyurethane filters, manufacturers recommend using one, two, or three tablespoons of oil. This depends on the size of the filter element. Squeeze it several times to be sure oil is distributed evenly.

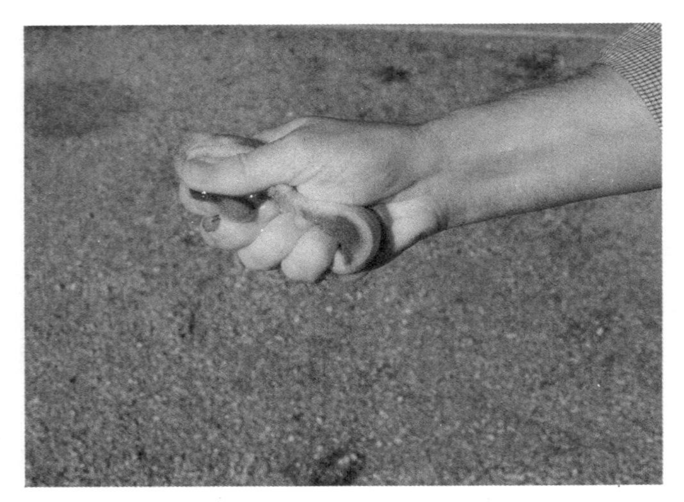

FIGURE 28. Make sure the oil is evenly distributed when oiling the filter element.

FIGURE 29. Clean a paper filter element by tapping it on a flat surface.

If you have a metal mesh filter, dip it into oil and drain off the excess.

12. *Install filter element.*

    Some polyurethane filters come in two sections, coarse filter on the outside and a fine filter next to the carburetor. The coarse filter stops larger particles and prevents early clogging. The fine filter removes most of the smaller particles.

13. *Install cover.*

14. *Reconnect spark plug wire.*

5. SERVICING THE DRY-FILTER TYPE OF AIR CLEANER

Servicing the dry-filter type of air cleaner is similar to the oiled-filter type. Follow the same procedures as in servicing the oiled-filter type except in cleaning the filter. Some dry-filter type air cleaners are made in one piece (self-contained). There may be no removable cover.

To clean the dry-filter air cleaner, proceed as follows:

1. *Clean air filter element.*

   If your filter element is paper, clean by tapping it on a flat surface (Figure 29). Some manufacturers recommend replacing dry-filter type air cleaners rather than cleaning them.

   Do not wash it unless instructed to do so by the manufacturer. The open pores in some filter elements will close. Replace paper elements if the dust does not drop off easily during or after tapping—or if it is bent, crushed, or damaged. Handle paper element with care. One small hole in the filter may be as bad as no filter at all. Do not use compressed air. It may rupture the filter.

   If your filter element is felt or fiber, direct compressed air from the inside out (Figure 30). Wash it in soap and water. Petroleum solvents are too oily. Do not put oil on a dry type filter. It clogs the pores in the paper, and air will not be able to pass through.

2. *Replace filter element and cover.*

   Be sure the filter fits snugly around the air cleaner base to prevent unfiltered air from bypassing the filter element. Install gasket if one is used.

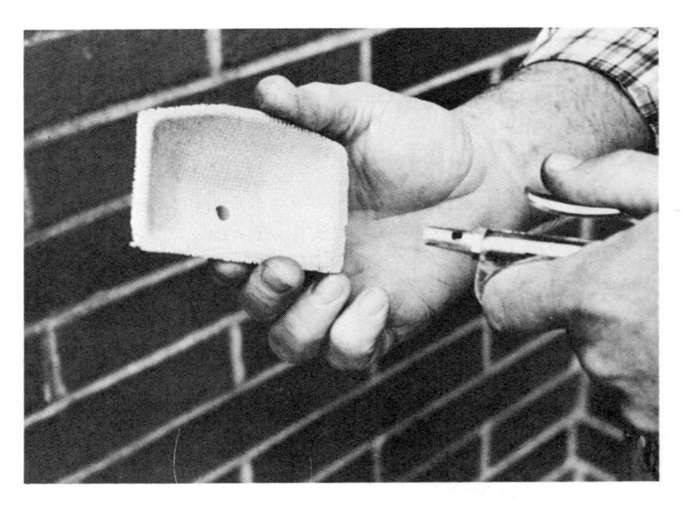

FIGURE 30. Clean a felt or fiber filter element by directing compressed air from the inside out.

STUDY QUESTIONS

1. List the two common types of air cleaners found on small engines.
2. ________ is a common material used for oiled-filter type air cleaners in small engines.
3. Most dry-filter type air cleaners are made of ________.
4. True or False. Paint thinner is an ideal cleaner to use in cleaning oiied-filter type air cleaners.
5. True or False. A solution of household detergent and hot water is used when cleaning oiled-filter type air cleaners.
6. Some manufacturers recommend ________ the dry-filter type air cleaner rather than attempting to clean them.

D. Servicing Fuel Strainers

Upon successful completion of this section, you will be able to do the following:

- **Recognize the most common types of fuel strainers in small engines**
- **Understand the importance of keeping fuel strainers clean**
- **Know the tools and materials needed for servicing fuel strainers**
- **Use the proper procedures for servicing screen-type fuel strainers**

The fuel strainer may be located between the fuel intake and the carburetor, or inside the fuel tank. Its purpose is to keep dirt and trash from getting into the engine through the fuel system. If the fuel strainer gets clogged, the engine does not get enough fuel. Check the fuel strainer after about 25 hours of operation. Fuel strainers on chain saws (if used often) should be cleaned about once a week.

The steps for servicing fuel strainers are explained under the following headings:

1. Types of Fuel Strainers
2. Importance of Servicing the Fuel Strainer
3. Tools and Materials Needed
4. Servicing the Screen-Type Fuel Strainer

1. TYPES OF FUEL STRAINERS

The most common types of fuel strainers used in small gasoline engines are the following:

- A screen in the fuel tank attached either to the fuel shut-off valve, the fuel-line tank fitting, or on the bottom on the fuel pickup tube (if there is no shut-off valve)
- A strainer attached to the end of a fuel pickup hose in the fuel tank
- An inline filter screen in the carburetor

They are described under the following headings:

a. Screen in Fuel Tank
b. Strainer Attached to Flexible Hose
c. Screen in Carburetor

a. Screen in Fuel Tank

One type of strainer has a screen in the fuel tank (Figure 31a). The most common is a fine wire screen attached to the fuel shut-off valve, the fuel inlet line, or on the bottom of the fuel pickup tube. Many fuel filters are now molded into the polyurethane fuel tank.

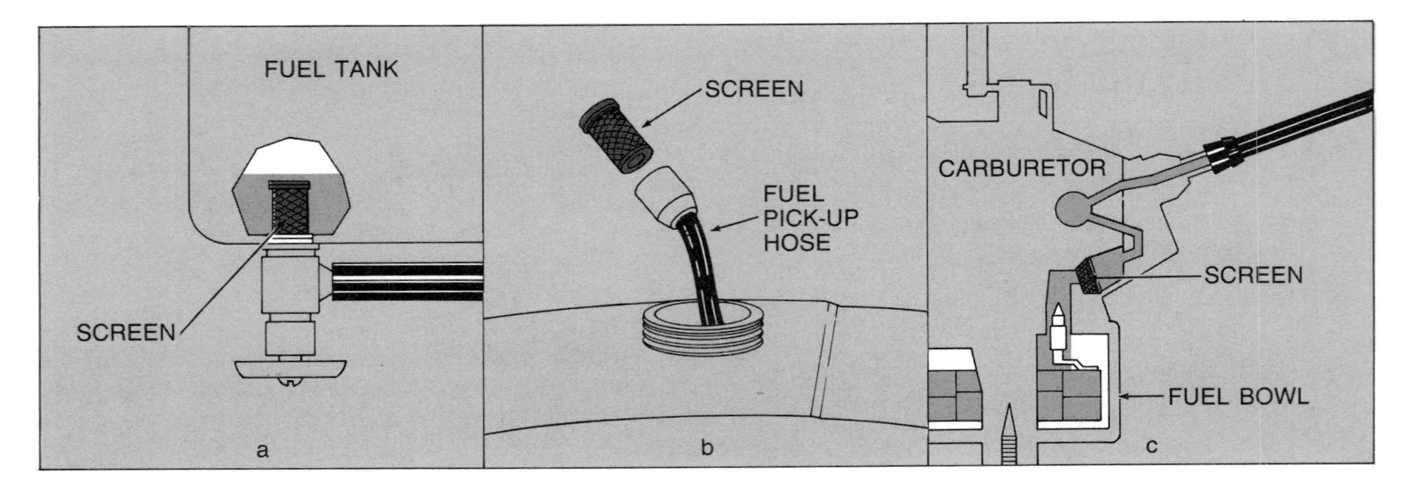

FIGURE 31. Types of fuel strainers.

b. Strainer Attached to Flexible Hose

On engines that operate in any position, such as chain saws, the strainer may be attached to the end of a flexible hose (Figure 31b). The strainer end is weighted so that the pick-up end will move to the lowest point in the fuel tank. In this way, the end of the hose stays in the fuel tank, regardless of how the engine is tilted or turned.

c. Screen in the Carburetor

The carburetor may also have a screen for filtering dirt from the fuel (Figure 31c). This screen in the carburetor may be used in addition to one of the other types of fuel strainers. Before servicing the fuel strainer on your engine, you will need to know which of these types is used.

2. IMPORTANCE OF SERVICING THE FUEL STRAINER

If the fuel strainer is not kept clean, it will clog and the engine will not get enough fuel. When this happens, your engine may not start or, if it does, it will run rough and uneven, will lack power, or may stall.

An engine will not run properly if there is dirt or trash in the gasoline. Gasoline is metered at the carburetor through small needle-valve jets. These jets let just enough gasoline flow to make the right fuel-air mixture. And even a tiny bit of fine dirt can clog the carburetor jets and cause the engine to skip or stall.

3. TOOLS AND MATERIALS NEEDED

1. Open-end wrenches — 7/16″ and 1/2″
2. Combination pliers — 7″
3. Petroleum solvent (mineral spirits, kerosene, or diesel fuel)
4. Pan for cleaning parts in solvent
5. Cleaning rags

4. SERVICING THE SCREEN-TYPE FUEL STRAINER

If your engine does not have a sediment-bowl type fuel strainer, it will have a wire mesh or a fiber strainer in the fuel tank to filter dust and dirt from the gasoline. To service a screen-type strainer, use the following steps:

1. *Disconnect the spark plug wire.*
   This prevents the engine from starting accidentally.
2. *Close fuel shut-off valve (if one is installed).*
3. *Remove fuel line from fuel tank* (Figure 32).
   If you have no fuel shut-off valve, drain the fuel tank by disconnecting the fuel line and/or removing the fuel tank, or by removing the filler cap and tipping the engine until the fuel drains through the filler hole. Remember to always store the drained fuel in a well vented area.
4. *Remove fuel shut-off valve and/or fuel strainer from fuel tank, if possible.*
   On most engines, this is done by unscrewing the fuel shut-off valve or the tank fitting on which the screen is mounted. On many engines the screen is

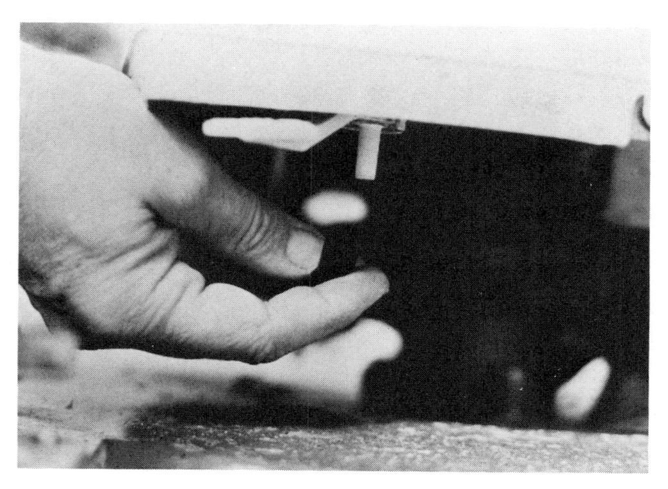

FIGURE 32. Remove the fuel line from the fuel tank.

made on the end of the fuel pickup tube and the carburetor must be removed from the gasoline tank to clean the screen. Be careful not to damage the screen, gasket or diaphragm when servicing. If there is a flexible hose and strainer and the strainer is inside the fuel tank, carefully remove the strainer with a wire hook.

Some engines have strainers built into a plastic fuel tank. If yours is built into the fuel tank, remove the tank from the engine and clean the tank and strainer without removing the strainer.

5. *Clean fuel strainer with cleaning solvent and allow to dry.*
6. *Replace fuel strainer assembly.*

   When installing fuel hose, check for cracks and signs of wear. If worn, replace the fuel line.
7. *Reconnect spark plug wire to spark plug.*

STUDY QUESTIONS

1. What is the purpose of a fuel strainer on a small engine?
2. List three types of fuel strainers.
3. True or False. Small particles of dirt getting into the carburetor through the fuel system may cause the small engine to skip or stall.
4. True or False. When cleaning a fuel strainer it is advisable to drain the fuel tank.
5. True or False. Fuel hoses should be checked for cracks and signs of wear when cleaning the fuel strainer.
6. The ________ ________ wire should always be disconnected when cleaning the fuel strainer.

E. Servicing Crankcase Breathers

Upon successful completion of this section, you will be able to do the following:

- **Recognize the different types of crankcase breathers**
- **Understand how crankcase breathers work**
- **Understand why it is important to keep crankcase breathers clean**
- **Know the tools and materials needed for servicing crankcase breathers**
- **Use the proper steps for cleaning crankcase breathers**

Four-cycle engines have a "breather vent" in the crankcase. Crankcase breathers have three purposes:

1. Remove harmful gases and vapors from the crankcase
2. Keep a partial vacuum in the crankcase
3. Keep out dust and dirt

Two-cycle engine crankcases are not vented because the fuel and oil enter the combustion chamber through the crankcase. The crankcase is sealed and operates under pressure and vacuum.

Crankcase breathers which are servicable should be cleaned at least once each year. To service the breather, clean the filter element and check the breather valve for proper operation.

The types of crankcase breathers used on four-cycle engines, how they work, and how to service them are explained under the following headings:

1. Types of Crankcase Breathers and How They Work
2. Importance of Proper Servicing
3. Tools and Materials Needed
4. Cleaning the Crankcase Breather

1. TYPES OF CRANKCASE BREATHERS AND HOW THEY WORK

The breather is usually located in the valve-tappet access well (Figure 33) or in the valve-tappet access-well cover.

How does the crankcase breather prevent too much pressure build-up in the crankcase? During combustion a certain amount of blow-by (blow-by is the term used to describe gases under pressure leaking by the piston rings during combustion and the power stroke) from the combustion chamber passes into the crankcase.

If your engine had no breather, this pressure would build up to the point where oil seals and gaskets would break, and there would be oil leaks. So if the breather does not work properly, or if it gets clogged, it would be the same as if the engine had no breather. The crankcase breather valve helps prevent too much pressure by letting air escape when the pressure in the crankcase is greater than the atmospheric pressure.

At the same time, harmful gases and vapors are removed from the crankcase. These gases and vapors get into the crankcase as a result of blow-by past the piston. If not removed, they use up the additives in the oil and cause it to oxidize. This causes corrosion of engine parts and the oil will not lubricate properly.

On some four-cycle engines the breather provides a partial vacuum in the crankcase. The breather allows the gases to escape from the crankcase, but it also allows a small amount of air to enter. As a result, a partial vacuum is maintained in the crankcase. This vacuum helps prevent oil leaks through the oil seals

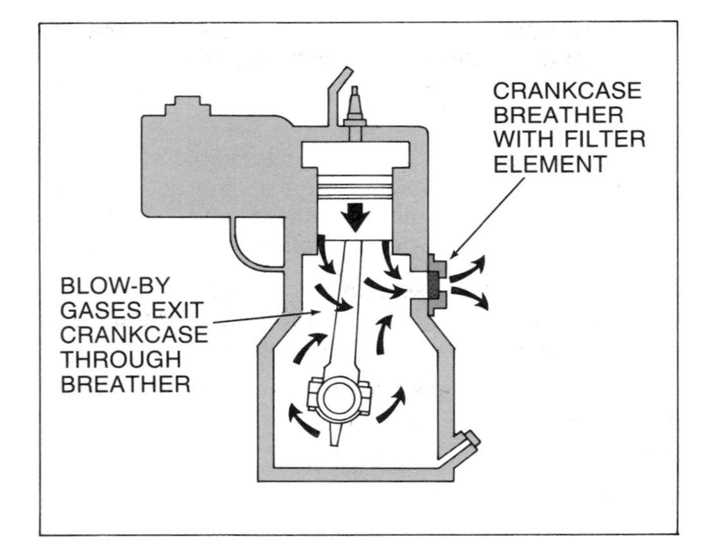

FIGURE 33. A four-stroke cycle engine is vented by a crankcase breather usually located in the valve-tappet access well.

FIGURE 34. A floating-disk type crankcase breather.

and gaskets. This vacuum is accomplished by two different methods: (1) some valves do not close completely, or (2) some valves have a small opening to allow air to enter. An air filter is required on such engines.

There are two types of valves used in crankcase breathers. Crankcase breathers are discussed under the following headings:

a. Floating-Disk Type
b. Air Filter

a. Floating-Disk Type

The floating-disk valve floats freely (Figure 34). It is opened by pressure in the crankcase, and partly closed by atmospheric pressure.

b. Air Filter

Air from the crankcase may be vented either to the outside air or to the carburetor (Figure 35). If the breather is vented to the outside air, a filter is necessary. Its job is to remove dust and dirt from the air entering the crankcase. It is a part of the crankcase breather. Filters are made of metal maze, animal hair, or polyurethane.

2. IMPORTANCE OF PROPER SERVICING

Failure to clean the crankcase breather filter regularly may cause the breather to become clogged. The result is too much pressure or vacuum buildup in the crankcase—breaking oil seals and gaskets which may cause oil leaks to develop.

FIGURE 35. An air-filter type crankcase breather vented to the carburetor.

If the inlet passage clogs, air can get out but not in. Too much negative pressure (vacuum) will develop and dirt will be drawn in through the oil seals. Portions of gaskets will have a tendency to be drawn into the engine.

If oil leaks out through the breather, it is a sign of a clogged breather valve, or worn or damaged parts. It may also mean one of the following: that the breather has not been assembled properly; that the engine is running too fast; that the engine is being operated at too much of an angle; that the piston rings are worn; or that the breather was installed upside down.

3. TOOLS AND MATERIALS NEEDED

1. Open-end wrenches — 7⁄16″ and 1⁄2″
2. Slot-head screwdriver — 8″
3. Combination pliers — 7″
4. Clean rags
5. Petroleum solvent (mineral spirits, kerosene or diesel fuel)
6. Container for cleaning parts
7. New gasket

4. CLEANING THE CRANKCASE BREATHER

Cleaning procedures are as follows:

1. *Disconnect spark-plug wire.*
2. *Check breather for proper operation.*

   Unusual oil leaks anywhere on the crankcase may be a sign of a clogged or inoperative breather.

   Breathers that are vented to the carburetor can be checked by removing the hose and placing your finger over the breather opening. You can feel the pressure and suction when the engine is turned over.
3. *Remove crankcase breather cover.*

   If there is a breather tube, you must first release hose clamp and remove the tube.

   The valve cover may be held by a stud bolt and nut, or by capscrews. It may be necessary to remove the carburetor or other accessories to get to the breather.
4. *Remove the crankcase breather.*

   The breather may be located in the valve cover or valve-tappet access well.
5. *Check breather valve for clearance* (Figure 36).

   Valves should never close completely. Check floating disk valves for freedom of action and some clearance. Refer to the service manual for specific clearance. Some engine manufacturers do not give a specific clearance because it is difficult to measure.
6. *Disassemble crankcase breather.*

   Be sure to remember the order in which the breather is assembled. Make a drawing if necessary.
7. *Clean parts in petroleum solvent.*

   Refer to "Servicing Carburetor Air Cleaners" for directions for cleaning the filter element.

   Let the parts dry.
8. *Reassemble and install breather.*

   Be sure to reassemble breather in proper order. If you are not certain, refer to your drawing.

   Some breathers have oil drain holes. Install the breather with the drain hole toward the base of the engine. If you don't, the oil will be pumped out through the breather. Replace all gaskets with new gaskets. Old gaskets will not provide a good seal.

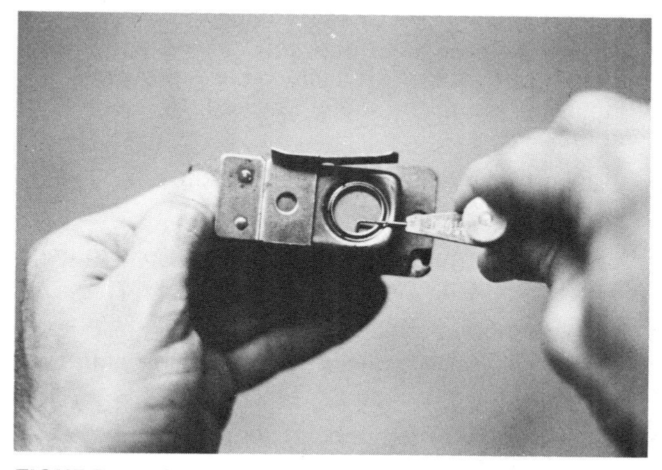

FIGURE 36. Check the breather valve for clearance.

STUDY QUESTIONS

1. What are the three purposes of crankcase breathers on small engines?
2. On some four-cycle engines the crankcase breather provides a partial ________ in the crankcase.
3. If oil leaks out through the crankcase breather it is a sign of a ________ breather valve or worn or damaged parts.

F. Lubricating Small Engines

Upon successful completion of this section, you will be able to do the following:

- **Identify the different types of lubrication systems and explain how they work**
- **State the importance of lubricating small engines properly**
- **Select the proper oil for lubricating your small engine**
- **Select the tools and materials needed for lubricating your engine**
- **Check the oil level in four-cycle engines**
- **Change the oil in four-cycle engines**
- **Protect unused oil by storing it properly**

Oil is needed for the operation of your engine. Proper lubrication is very important. Many engines wear out before they should due to operating them without enough oil or with the wrong kind of oil.

Without lubrication the pistons would seize and bearings would burn out. Engine life also depends on how well the engine is lubricated. Engine oil is vital to engine life. The reasons are as follows:

1. Oil reduces friction between moving parts.
   It provides a cushion between moving parts, and keeps them apart. This cushion helps prevent scoring and fusing of metal caused by heat and abrasion.
2. Oil reduces heat by reducing friction.
3. Oil cleans.
   Special detergents are added to most oils to help keep the engine clean.
4. Oil prevents corrosion.
   Some oils have special rust inhibitors for this purpose.
5. Oil helps seal piston rings to help prevent blow-by.
6. Oil helps increase power output by reducing friction.

The importance of and how to lubricate your engine properly are discussed under the following headings:

1. Types of Lubrication Systems and How They Work
2. Importance of Proper Lubrication
3. Selecting the Crankcase Oil
4. Tools and Materials Needed
5. Checking the Crankcase Oil Level (Four-Cycle Engines)
6. Changing the Crankcase Oil (Four-Cycle Engines)
7. Storing Crankcase Oil

1. TYPES OF LUBRICATION SYSTEMS AND HOW THEY WORK

The lubrication systems of four-cycle and two-cycle engines are not the same.

These differences are discussed under the following headings:

a. Lubrication of Four-Cycle Engines
b. Lubrication of Two-Cycle Engines

a. Lubrication of Four-Cycle Engines

All four-cycle engines are lubricated from an oil reservoir. However, there are several methods used to get oil to the bearing surfaces. Four common methods of lubrication are:

1. Dipper and Sump (Figure 37a)
   As the crankshaft turns, oil is picked up from the sump by a dipper attached to the connecting rod bearing cap, and splashed about inside the crankcase.

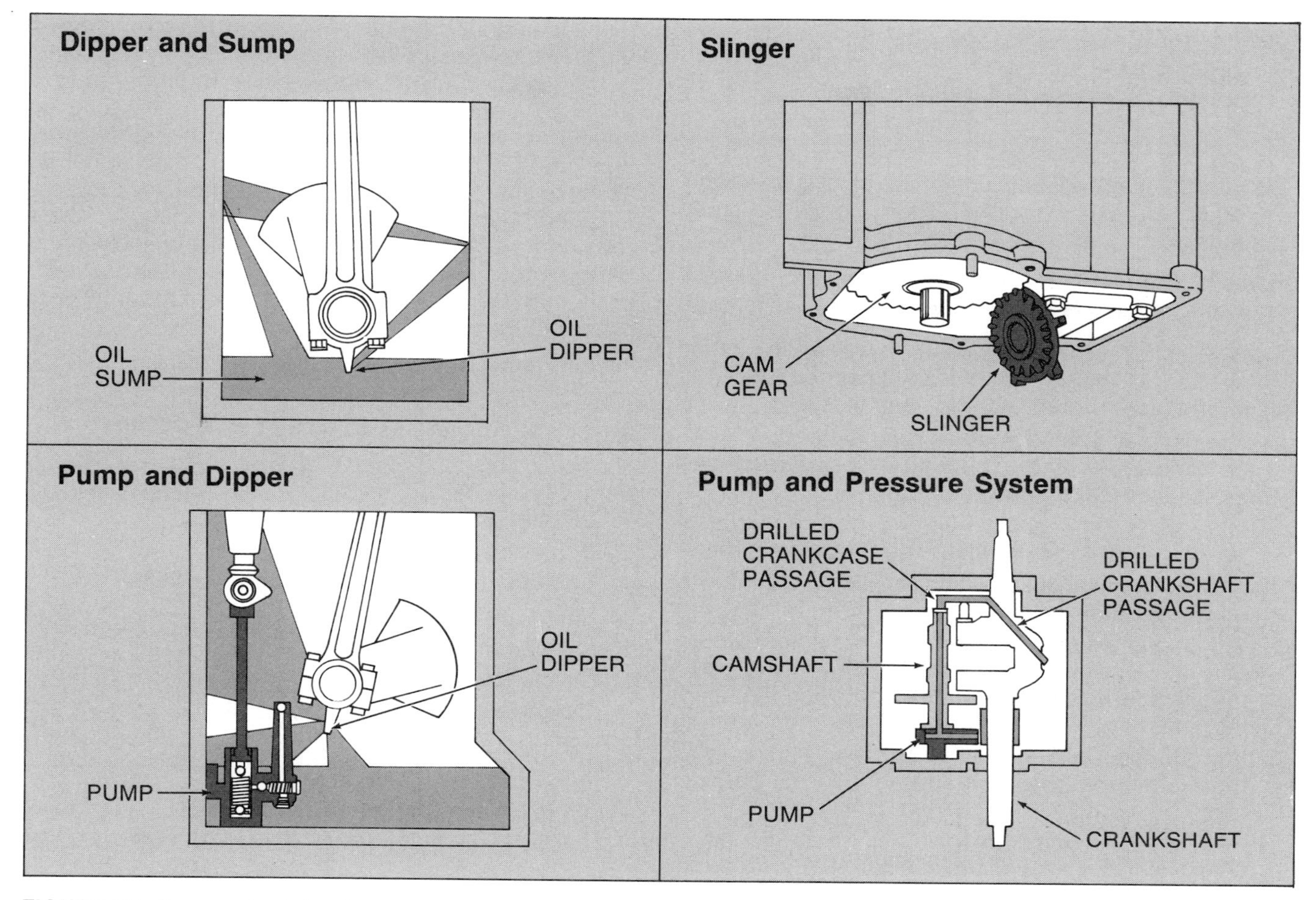

FIGURE 37. Types of lubrication systems for four-cycle engines.

2. Slinger (Figure 37b)

   Oil is picked up from the oil sump by a rotating slinger, and it is splashed around inside the crankcase. The slinger is driven by the cam gear.

3. Pump and dipper (Figure 37c)

   Oil is pumped from the oil sump and sprayed onto a dipper; then it is splashed around inside the crankcase. Or oil is pumped into a constant level sump. Then it is picked up by a dipper and splashed around inside the crankcase.

4. Pump and pressure system (Figure 37d)

   Oil is pumped from the oil sump through drilled passageways to the bearings. The pressure is kept constant by a relief valve.

Larger multi-cylinder engines have crankcase lubrication systems that are similar to small four-cycle engines.

b. Lubrication of Two-Cycle Engines

Two-cycle engines are lubricated by specific types of oil in the fuel. The mixture passes through the crankcase. When the piston moves toward the crankshaft, the oil and fuel are pressurized in the crankcase for a few seconds. This causes some of the oil to condense (thicken) on the bearing surfaces.

As the piston moves upward, oil condenses from the mixture in the combustion chamber to lubricate the piston, rings, and cylinder walls. The remainder is burned during combustion. Mixing oil and gasoline for two-cycle engines is discussed under "Refueling Small Engines."

2. IMPORTANCE OF PROPER LUBRICATION

Lubrication of all engines is important but it is especially important for small engines. This is because small engines are not as well protected as large engines.

Some of the many reasons why you should be sure your small engine is lubricated properly are as follows:

1. Few small gasoline engines have an oil filter. Therefore, the oil should be changed often to remove metal particles, dirt, and sludge.
2. The oil in air-cooled engines runs hotter than oil in water-cooled engines. This high temperature causes the oil to oxidize and break down.
3. Most small engines have no oil-pressure gauge or warning light to show when the pressure is low.
4. The amount of oil available in small engines is relatively small.
5. Small engines usually operate at maximum power output. Extreme pressures between moving parts tend to squeeze the oil out from between the bearing surfaces.
6. Small engines are lightweight and may vibrate more than large engines. Vibration adds to the bearing load.
7. Few small engines are given a warm-up period before a load is applied. Most small engines are run for only short periods of time.

Oil may not reach the piston rings until after the engine has turned over several times. This lack of lubrication can cause two conditions: (1) fast wear of the piston rings and cylinder walls, and (2) failure of the rings to seal off compression. Blow-by is increased. Blow-by sends harmful chemicals by the piston rings to the crankcase. During choking, raw gasoline in the combustion chamber washes down the cylinder walls and makes the conditions even worse. If this continues, oil in the crankcase is diluted and loses its effectiveness. After the engine warms, however, the gasoline is evaporated by heat.

3. SELECTING THE CRANKCASE OIL

Even though oil is the smallest expense in operating your engine, selecting the proper oil is one of your most important decisions. The proper oil for lubricating your small engine will be explained on the engine nameplate, in the operator's manual, or in lubrication guides from the major oil manufacturers. Specifications may be given in two ratings: (1) API (American Petroleum Institute) Service classification often called type of oil, and (2) SAE (Society of Automotive Engineers) Viscosity often called viscosity grade. Some two-cycle, and many four-cycle, engine manufacturers recommend a single viscosity grade oil, usually SAE 30, service class, SC, SD, SE or MS. Most two-cycle engine manufacturers now recommend SAE 30 oil (two-cycle rating) as an alternative to special two-cycle oils. These oils contain characteristics such as anti-rust, anti-oxidants, ability to stay suspended in the fuel and dyes for identification. Also, most four-cycle engine manufacturers recommend detergent SAE 30, service classification SE or better. Detergent oil **should not** be used for two-cycle engines. Refer to owner's manual for additional manufacturers' recommendations.

4. TOOLS AND MATERIALS NEEDED

1. Slot-head screwdriver — 8″
2. Combination pliers — 7″
3. End wrenches — 7/16″, 1/2″, 9/16″, and 5/8″
4. Funnel with a flexible spout
5. Oil as recommended
6. Container for used oil
7. Clean rags

5. CHECKING THE CRANKCASE OIL LEVEL (FOUR-CYCLE ENGINES)

Check the oil after every two to four hours of operation, or according to the recommendations of the manufacturer of your engine.

A good policy is to check the oil level each time you refuel the engine. The steps are as follows:

1. *Locate the oil-filler plug or cap* (Figure 38).

FIGURE 38. Locate the oil-filler plug on your engine.

FIGURE 39. Be sure your engine is level when checking the oil level.

On small tractors, you may have to raise the hood to locate the oil-filler plug. A dip stick may be attached to the filler plug on the crankcase or to a cap on the filler neck. Some small engines have no dip stick. They have a filler plug on the top of the oil sump.

2. *Clean dust and dirt from the filler plug or cap.*
   This prevents dust from falling into the crankcase when the plug or cap is removed.
3. *Remove filler plug or cap.*
4. *Check oil level* (Figure 39). Be sure your engine is level.
   If your engine has a dipstick, proceed as follows:
   a. Remove dipstick.
   b. Wipe dipstick with a clean rag.
   c. Reinsert dipstick.
      Follow manufacturer's recommendations. Some must be screwed in for accurate reading.
   d. Remove dipstick and check oil level.
      Oil should be between the marks shown on the dipstick, never above the FULL mark or below the ADD mark.

   If your engine has no dipstick, check to see if the oil comes to the top of the filler hole, or to a marking or slot that indicates the FULL oil level.
5. *Add or change oil.*
   If the oil is low and it is time for a change, do not add oil; change it. Follow instructions under the heading, "Changing the Crankcase Oil." If you add oil, do not overfill.
6. *Replace plug or cap.*
   Be sure the plug or cap fits tight. If your crankcase operates on a partial vacuum, dirt will get in through any small opening that results from a poor seal.

6. CHANGING THE CRANKCASE OIL (FOUR-CYCLE ENGINES)

Most manufacturers recommend changing oil every 25 hours of operation in ideal conditions. When operating in dusty conditions, change oil more often. Steps for changing oil in your engine are as follows:

1. *Operate engine until it is completely heated.*
   Oil will drain better when hot. Also more contaminants are removed while the oil is still being moved around.
   If the crankcase is drained while the oil is cold, some of the more contaminated oil stays in the engine. Never drain oil while engine is running.
2. *Stop the engine and disconnect the spark plug wire.*
3. *Locate the drain plug.*
   Most drain plugs are located on the outside edge of the bottom of the oil sump while others are located under the oil sump (Figure 40).
   Some engines have no drain plug and are drained through the filler neck.
4. *Clean dirt from around plug before removing it.*
5. *Remove drain plug.*
   Use a wrench that fits the drain plug, not a pair of pliers. Pliers will gradually round off the corners of the plug until it will be difficult to remove or to tighten the plug.
6. *Let crankcase drain.*
   It's important to let it drain long enough for oil to drain from other parts of the engine. If the drain is in the side or top of the sump, tilt the engine towards the drain. If your engine is drained from the filler neck, tilt it over on the filler-neck side until the oil drains.

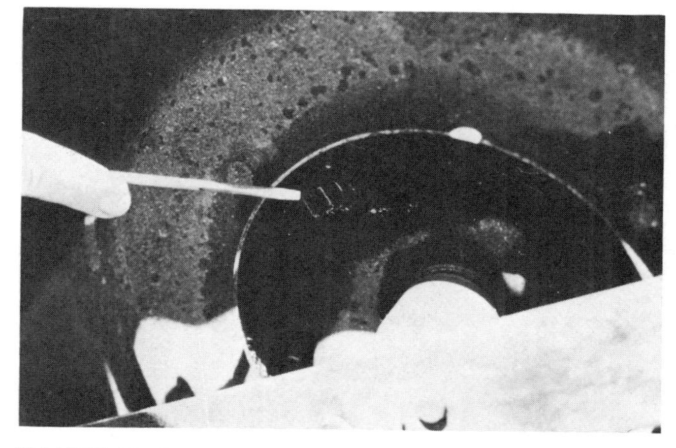

FIGURE 40. Some drain plugs are located under the oil sump.

7. *Replace drain plug.*
   If drain plug is equipped with a gasket, be sure to install it.
8. *Refill crankcase with new oil.*
   Use the viscosity and type of oil recommended in your operator's manual. Be sure the tops of the oil cans are very clean. Check the funnel, or any container you use, to make sure it's free of dirt. Use the amount of oil recommended for your engine. Wipe away excess oil around filler plug. Replace filler cap.
9. *Connect spark plug wire and start engine.*
   Operate it a few minutes. This gives the oil an opportunity to establish a true level on the dipstick.
10. *Check for oil leaks.*
11. *Stop engine.*
12. *Recheck oil level.*
    If oil is not to the full line, add more until it reaches that level. Do not overfill.
13. *Clean or destroy oily rags.*
    Do not store them. They may cause spontaneous (self-ignited) combustion.

7. STORING CRANKCASE OIL

After adding or changing oil in your engine, you often have part of a can of oil left over. If you plan to use it later, it should be protected from dirt, moisture and heat. The plastic tops used on vacuum packed food cans are good covers for quart oil cans (Figure 41).

If you keep oil in large cans with screw caps, be sure to keep the tops on and tight.

Store oil in a clean, dry place.

FIGURE 41. Plastic tops from vacuum packed food cans provide excellent covers for oil cans. Manufacturers now package oil in plastic resealable bottles.

STUDY QUESTIONS

1. Engine oil is vital to engine life. Oil serves what purpose in a small engine?
2. List four common methods of getting oil to bearing surfaces in a four-cycle engine.
3. Two-cycle engines are lubricated by specific types of oil in the ________.
4. True or False. Most small engines have an oil filter.
5. True or False. The amount of oil available in small engines is relatively small.
6. Check the oil in the small engine after every ________ hours of operation.
7. Oil should be changed when the engine oil is (hot or cold).
8. True or False. Engine oil, when changed, should be stored and reused in other small engines.
9. Describe why engine oil should be hot when changed and drained from the engine.
10. True or False. Pliers should be used when removing the oil drain plug on small engines.

G. Refueling Small Engines

Upon successful completion of this section, you will be able to do the following:

- **Select the proper fuel for your engine**
- **Select the tools and materials needed for refueling**
- **Mix gasoline and oil for two-cycle engines**
- **Fill the fuel tank for small engines**
- **Store fuel safely**

Gasoline is well suited for spark-ignition engines for several reasons. Some of these reasons are:

1. Gasoline vaporizes quickly.

   Petroleum products will not burn when in a liquid state. They must first be evaporated and mixed with air. Since gasoline vaporizes quickly, it makes for easy starting and complete combustion during operation.

2. Gasoline does not burn too fast.

   The fuel in the combustion chamber must not explode. It must burn at a smooth, even rate.

Follow manufacturer's recommendations for fuel selection. Many manufacturers recommend regular or unleaded gas for their engines.

You might think that anyone can put gasoline in an engine properly. There are several facts about preparing fuel for two-cycle engines, and about the method you use when filling the tank that are important for you to know. These are discussed under the following headings:

1. Preparing Fuel for Two-Cycle Engines
2. Filling the Fuel Tank

1. PREPARING FUEL FOR TWO-CYCLE ENGINES

Use clean and fresh regular or unleaded gasoline and the oil recommended in your operator's manual or on the engine nameplate. Avoid using gasoline containing more than 10% alcohol or gasahol.

The amount of oil you mix with gasoline is very important. It depends on the engine design, horsepower, size and engine speed. Recommended mixes range from 16 parts gasoline to one part oil, to 50 parts gasoline to one part oil. Recommended fuel mixes also vary during the engine break-in period. Follow the directions in your operator's manual.

The addition of more oil than is recommended or using the wrong type oil will cause poor combustion and will form gum, varnishes, and carbon deposits. The exhaust ports can become clogged, pre-ignition can occur, and the engine can lose power and eventually stop running.

Too little oil may cause poor lubrication, which greatly increases wear on moving parts.

The method of mixing oil and gasoline is also important. Oil and gasoline are not easily mixed, but once they are mixed, they do not separate easily.

To be sure of a proper mixture of fuel for your two-cycle engine proceed as follows:

1. *Use an approved gasoline container that is large enough for mixing.*

   The container should have twice the capacity of the amount of fuel you intend to mix.

   Make sure it is clean.

2. *Fill container ¼ full of gasoline.*

   To remove any trash or dirt, strain the gasoline through a clean 120-mesh strainer.

3. *Add oil to the gasoline* (Figure 42).

   Use the oil recommended by your manufacturer. Make sure the oil is rated for two-cycle engines.

4. *Shake the can vigorously.*
5. *Add more gasoline until container is ½ full.*
6. *Shake the can again.*
7. *When not in use, keep can tightly closed and well marked.*

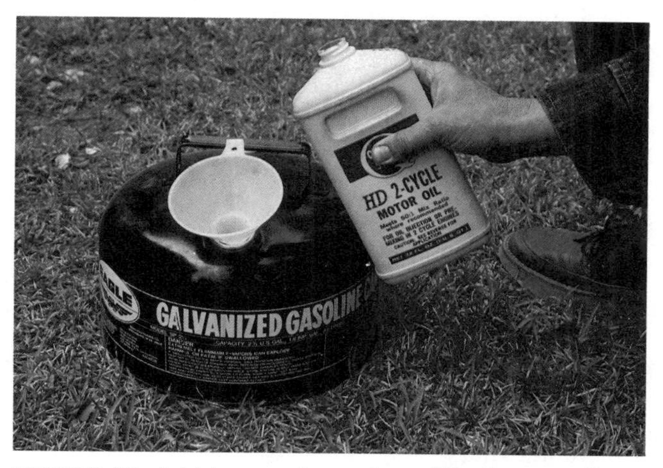

FIGURE 42. Add two-cycle engine oil to the gasoline.

2. FILLING THE FUEL TANK

Steps for filling the fuel tank are as follows:

1. *Make sure the engine is off, and locate the fuel-tank filler cap.* Remember, never refill a hot engine.

   On some small tractors, you will have to raise the hood. Check gasket in tank cap. Replace it if it does not seal.
2. *Wipe dust and dirt from around the filler cap.*
3. *Remove filler cap.*

   Never smoke or light a match near the open fuel tank.
4. *Visually check the fuel level in the tank.*
5. *Refill the tank.*

   NOTE: Never use the last few drops from a gasoline can; water and trash are likely to be present.

   If possible, use a funnel with a filter (Figure 43). When pouring fuel, keep the nozzle in contact with the metal fuel tank. Fuel movement can cause static electricity. This contact grounds the electrical charge and avoids any risk of igniting the gasoline vapors. If fuel is accidentally spilled, wipe it away with a rag. It is a good idea to move the engine away from the area where the tank is filled before starting. This will prevent a possible ignition spark from igniting spilled fuel.
6. *Replace fuel tank filler cap tightly.*
7. *Fasten the hood if it was opened.*
8. *Store remaining gasoline.*

FIGURE 43. If possible, use a funnel with a filter when refilling the tank.

STUDY QUESTIONS

1. Why is gasoline an excellent fuel for spark-ignition engines?
2. Avoid using gasoline containing more than ________ percent alcohol or gasohol in small engines.
3. In two-cycle engines, oil (rated two-cycle) must be mixed with ________ for proper engine lubrication.
4. True or False. Never refuel a hot engine.
5. True or False. If possible, use a fuel strainer when refueling a small engine.

H. Servicing Spark Plugs

Upon successful completion of this section, you will be able to do the following:

- **Describe what happens when spark plugs wear out**
- **Relate importance of proper spark plug maintenance**
- **Select the tools and materials needed to service spark plugs**
- **Give the steps for checking spark plugs for proper operation**
- **Remove, inspect, clean, and regap spark plugs**
- **Install spark plugs**

The spark plug makes it possible for a spark to occur inside the combustion chamber (Figure 44).

The spark plug works in the following way:

- Electric current from the magneto or ignition coil travels at regular intervals through the high-tension lead wire to the spark plug terminal.
- The current is conducted through the center electrode to the spark gap.
- High voltage (10,000 to 20,000 volts) pushes the current across the spark gap to the ground electrode.
- As the current jumps the gap, a spark is formed.
- The fuel-air mixture in the combustion chamber is ignited by the spark.

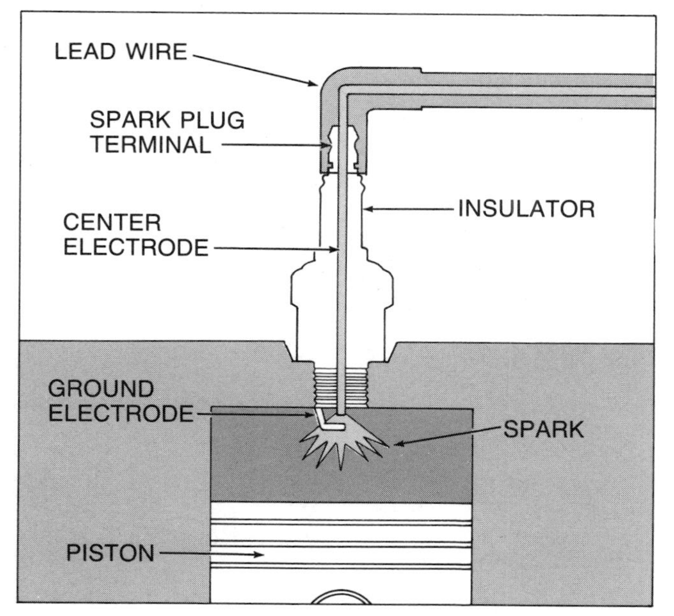

FIGURE 44. The function of a spark plug is to provide a spark to the combustion chamber.

A fouled spark plug is a common ailment with small engines, one that you should be able to detect and correct. How to detect and how to correct spark plug troubles are discussed under the following headings:

1. Importance of Proper Maintenance
2. Tools and Materials Needed
3. Inspecting and Servicing the Spark Plug
4. Installing the Spark Plug

1. IMPORTANCE OF PROPER MAINTENANCE

Anything that inhibits the flow of current through the spark plug and across the spark gap will affect the operation of the engine. The plug may become fouled with oil or lead deposits. These deposits make a path for part of the electric current to by-pass the spark gap. This may result in a weak spark or no spark. A cracked porcelain insulator will also cause a weak spark.

Here is what happens when a spark plug wears out. When a plug is new, the edges of the electrodes are sharp. A spark will jump the gap under cylinder pressure with as little as 10,000 volts. But the spark gap widens as the engine is used.

Hot combustion gases and continuous electrical discharge both erode and corrode (wear away) electrodes until the edges become rounded and the spark gap widens. The gap increases about .025 mm (.001 in.) each 20 hours of engine operation. After 100 hours of operation, the gap may be wide enough to require 15,000 volts, or more, to jump the gap.

If your engine is new and the ignition system is in good condition, it will have little trouble producing 10,000 volts, or even up to 20,000 volts. As the ignition system gets older, however, it may have trouble developing enough voltage to fire a worn plug. This makes spark-plug maintenance important in older engines.

A plug may reach a point where it cannot be cleaned and regapped properly. It should then be replaced with a new one.

A spark plug that does not work properly will increase fuel consumption, cause crankcase oil dilution, cause excessive deposits in the combustion chamber, cause hard starting, and may cause the engine to skip.

2. TOOLS AND MATERIALS NEEDED

1. Spark plug deep sockets — 13⁄16″ × 3⁄8″ drive and 3⁄4″ × 3⁄8″
2. T-handle — 3⁄8″
3. Torque wrench handle — 3⁄8″
4. Wire feeler gauge
5. Ignition file
6. Pen knife
7. Wire brush
8. Small paint brush
9. Pan of petroleum solvent (mineral spirits, kerosene, or diesel fuel)
10. Goggles
11. Safety glasses

3. INSPECTING AND SERVICING THE SPARK PLUG

Most manufacturers recommend removing, inspecting, cleaning, and regapping a spark plug after every 100 hours of operation. A good time for servicing the spark plug is at the beginning of each season. This may prevent trouble later when you have less time to work on the engine.

Proceed as follows:

1. *Remove the spark plug wire from the spark plug* (Figure 45).
2. *Remove the spark plug and reconnect the spark plug wire.*

FIGURE 45. Remove the spark plug wire from the spark plug.

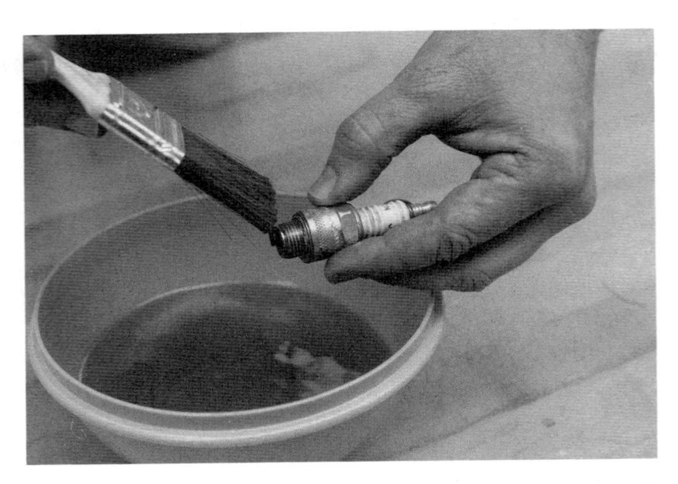

FIGURE 46. Use a cleaning solvent to remove oily deposits from the spark plug.

3. *Check the condition of the plug.*
   Plugs that are worn should be replaced with new ones of the type and size recommended for your engine. Check your operator's manual or see your dealer for proper selection.
4. *Remove oily deposits from plugs* (Figure 46).
   Put plugs in a pan of solvent (kerosene, distillate, or diesel fuel). Remove oily film from porcelain body of plug with a small paint brush. Then wipe solvent off plug with a clean cloth.
5. *Clean threads with a wire brush* (Figure 47).
   This cleaning is important for removing dirt so the plug will not bind when reinstalled. Wear safety glasses or goggles.
   Do not brush the insulator. It will leave a metallic film which may provide an electrical short to ground.

FIGURE 47. Use a wire brush to clean the threads of the plug.

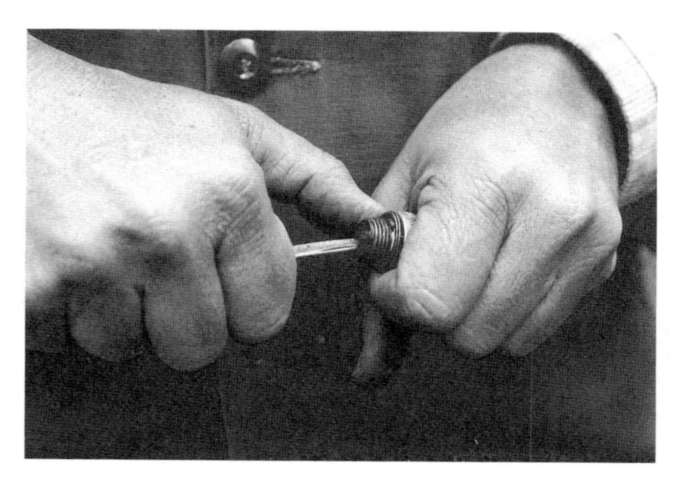

FIGURE 48. A small knife is good for removing hard deposits on the plug.

6. *Remove deposits from plugs.*
   Use a small bladed knife for removing hard deposits (Figure 48).
7. *Blow loose particles from plug with compressed air* (Figure 49). Wear safety glasses or goggles.
   This is important to get rid of any remaining particles.
8. *Determine proper spark gap spacing for your engine.*
   Check your operator's manual or one of the charts issued by spark plug companies.
   Most plugs are set on 0.6 mm (.025 in.). The range is from .5 mm to 1.0 mm (.020 in. to .040 in.).
9. *Regap plug.*
   Make all adjustments by bending the ground electrode. This is done by a trial check with a feeler gauge for proper thickness until you get the proper gap (Figure 50).

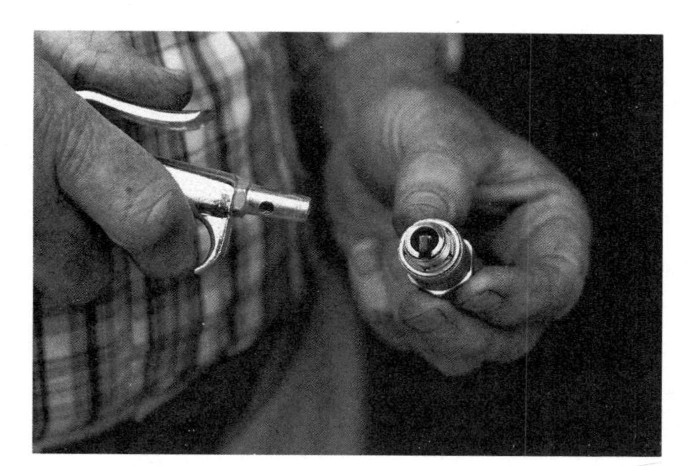

FIGURE 49. Use compressed air to clean the plug of any remaining particles.

FIGURE 50. Regap the plug using a feeler gauge.

A flat feeler gauge can be used for new plugs, but for used plugs it is difficult to completely remove the 'cup' in the ground electrode. A flat feeler gauge tends to bridge over the cup—if not filed away—which results in your setting the gap too wide (Figure 51). Use a wire (round) feeler gauge to avoid this difficulty.

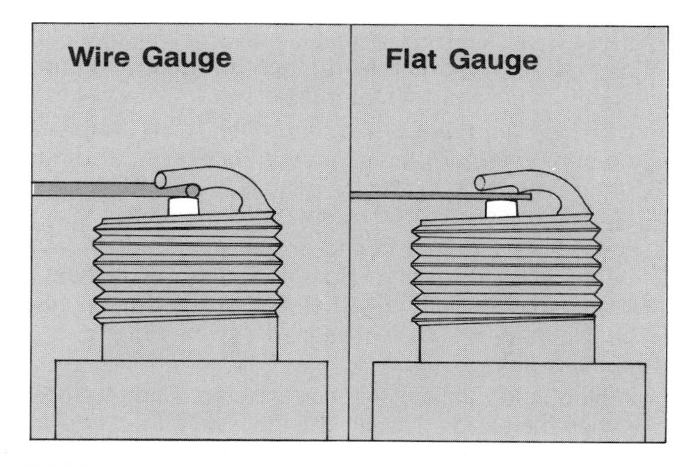

FIGURE 51. Use a wire (round) feeler gauge when regapping a used plug after cleaning.

4. INSTALLING THE SPARK PLUG

The steps for installing spark plugs are as follows:

1. *Replace plug and tighten with fingers.*
   Always be sure to check the gasket before installing the plug (Figure 52). The only plugs that do not require gaskets are ones with tapered seats. Without a gasket, a spark plug may extend into the combustion chamber far enough to become damaged.

FIGURE 52. Before installing the plug, make sure the gasket has not been damaged.

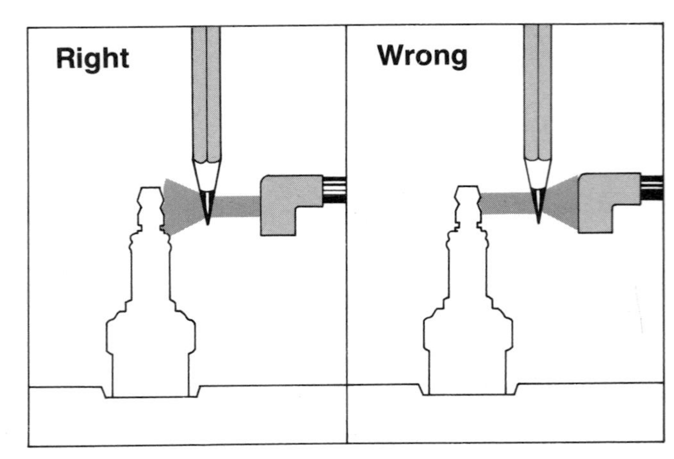

FIGURE 53. Testing the polarity of a spark plug.

NOTE: Some spark plugs are designed for use without gaskets. Check the manufacturer's instructions.

If you cannot seat the plug in the cylinder head by hand, remove it and wipe the threads in the cylinder head with a clean cloth.

2. *Completely tighten the plug with a spark plug socket wrench.*

   If you use a torque wrench, check a mechanic's manual to determine how much torque should be applied; or if it is not stated, see Table I for a guide.

   If you do not have a torque wrench, give the plug an additional turn, as indicated under the heading of "Turns with Wrench," Table I, which should tighten the plug adequately on the gasket.

   If the plug is not tightened properly on the gasket, heat will build up and cause the plug to overheat. If properly seated, the gasket is compressed enough so that heat will travel readily from the plug to the cooling fins.

   If plug is too tight, it will be distorted. The gap between the electrodes will change. The added stress may also break the porcelain.

3. *Check condition of connections and insulation on spark plug wire when reattaching wire to spark plug.*

4. *Check polarity of spark at spark plug on a battery ignition system.*

   If ground electrode on spark plug is cupped, the polarity is wrong—from five to 45 percent more voltage may be required to fire the plug.

   Electrons (a spark) jump quicker from a hot surface to a cool one than from cool to hot. Because the center electrode of the spark plug is hotter than the ground electrode, it is important that the primary circuit to the ignition coil be connected so the spark jumps from the center terminal.

To check polarity, hold the metal connector about 6 mm (¼ in.) from the spark plug terminal (Figure 53). Insert pencil as shown.

If the spark feathers and has a slight orange tinge on the plug side, polarity is correct.

If the spark feathers on the connector side, polarity is reversed. Correct by interchanging primary wire connections on the ignition coil.

TABLE I. TORQUE RECOMMENDED FOR TIGHTENING SPARK PLUGS

SPARK PLUG THREAD SIZES	TORQUE WRENCH* (Cast Iron Head)		TORQUE WRENCH* (Aluminum Head)		WITHOUT TORQUE WRENCH Engine Cool
Approximate Size (in Inches)	J	ft • lbf	J	ft • lbf	Turns with Wrench (beyond finger tight)
10 mm (3/8 +)	16–20	(12–15)	13–15	(10–15)	¾ to 1
14 mm (9/16 −)	34–40	(25–30)	30–37	(22–27)	½
18 mm (a11/16 +)	40–54	(30–40)	34–47	(25–35)	½ to ¾

* Combined recommendations of two spark plug manufacturers

STUDY QUESTIONS

1. Describe the role a spark plug plays in igniting the fuel in the combustion chamber of a small engine.
2. When a spark plug reaches a point where it cannot be cleaned and regapped properly it should be ________.
3. True or False. Spark plugs should be cleaned with a shop sand blaster.
4. Most manufacturers recommend removing, inspecting, cleaning and regapping a spark plug after every ________ hours of operation.
5. True or False. All spark plugs have the same gap spacing measurement.

I. Checking and Adjusting Carburetors

Upon successful completion of this section, you will be able to do the following:

- **Tell how a carburetor operates**
- **Select tools and materials needed for adjusting carburetors**
- **Check a carburetor to learn if it is operating properly**
- **Tell what jobs must be done before you can adjust a carburetor**
- **Adjust the carburetor choke valve, the high-speed load valve, the idling valve, and the idle-speed stop screw**

The carburetor performs three functions:

1. It breaks up (atomizes) the fuel into a fine spray and mixes it with air to make a mixture that will easily burn.
2. It regulates the ratio of fuel to air.
3. It regulates the amount of the fuel-air mixture going into the combustion chamber.

The ratio of fuel to air is controlled by adjustable needle valves. Adjustments help you select the right mixture for the conditions under which you operate your engine. They are usually set at the factory and rechecked by your dealer when you buy the engine. Some carburetors are made without operator adjustments.

As you use the engine, normal wear and vibration cause the carburetor valve openings to change. The carburetor is then out of adjustment. Operating your engine with the carburetor out of adjustment may result in high fuel consumption and loss of power.

Too much fuel consumption is not usually important when operating a small gasoline engine. The loss of 10 to 20 percent of the engine horsepower is not noticeable on many applications. The harmful results come from operating the engine with the carburetor out of adjustment.

Too lean a fuel-air mixture (too much air - not enough fuel) may result in hard starting, overheating, pre-ignition and valve burning. Too rich a fuel-air mixture (too much fuel - not enough air) may result in high fuel consumption, carbon build-up in the cylinder and cause pre-ignition.

To avoid such problems, you need to know how a carburetor operates, and how to adjust it for smooth engine operation. This information is given under the following headings:

1. Principles of Operation
2. Types of Carburetors
3. Tools and Materials Needed
4. Checking the Carburetor for Proper Operation
5. Preparing for Carburetor Adjustment
6. Adjusting the Carburetor Choke Valve
7. Adjusting the High Speed Load Valve
8. Adjusting the Idling Valve
9. Adjusting the Idle-Speed Stop Screw

1. PRINCIPLES OF OPERATION

The purpose of the carburetor is to mix the fuel with air and to control the amount of the mixture going into the cylinder. These functions are described under the following headings:

a. Atomizing the Fuel
b. Regulating the Air-Fuel Mixture
c. Controlling the Engine Speed

a. Atomizing the Fuel

The carburetor atomizes (mixes) the fuel and air in the following manner:

1. As the piston moves down (on the intake stroke) a partial vacuum develops in the cylinder.

   On four-cycle engines, during the intake stroke the intake valve is open and the exhaust valve closed. On two-cycle engines, the low pressure is created in the crankcase on each upward motion of the piston.

2. Atmospheric pressure pushes air through the carburetor air intake to equalize the pressure (Figure 54).

3. The air speed increases in the carburetor by air passing through narrow passages (venturi) thereby increasing air speed. As air speed increases, its pressure is lowered.

4. When the pressure lowers in the carburetor, atmospheric pressure in the fuel bowl pushes fuel in the carburetor into the air stream.

   The speed of the air in the carburetor and the air turbulence past the narrow passages atomize the fuel and mix the tiny droplets with the air.

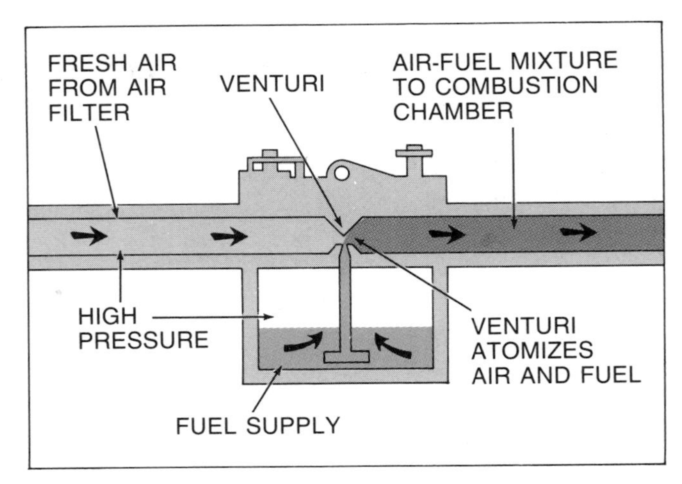

FIGURE 54. The basic operating principle of a carburetor.

b. Regulating the Air-Fuel Mixture

Certain adjustments to the carburetor must be made to make your engine operate properly in the different conditions. Therefore, when the engine is operating different ratios of fuel-to-air are regulated by adjusting valves. Different atmospheric conditions may also affect the ratio of fuel-to-air needed for proper engine operation.

Two adjustable needle valves and a choke control valve are used to make the air-fuel mixture either leaner or richer. Not all carburetors have two needle valves. Some chain saws and smaller lawnmower engines have only one valve. They are described as follows:

1. The Idle-Mixture Valve (Figure 55)

   This valve is used mainly for supplying fuel to the engine when it is idling.

2. The High-Speed, Load Valve (Figure 55)

   This valve supplies the fuel for operations other than idling.

   On most carburetors, both valves are used to adjust the amount of liquid fuel being fed into the air stream. This is what determines the fuel-air ratio. The idling jet is adjusted while the engine is idling, and the load jet while the engine is under load or at least above idling speed.

3. The Choke Valve (Figure 55)

   A choke valve is used to help in cold starting. The choke valve provides for better vaporization of the fuel and also provides for more fuel-to-air (a richer mixture). Only vaporized gasoline will burn, and the richer mixture provides for more gasoline to be vaporized. When the choke is closed, the air going into the carburetor is restricted. And the pressure inside the carburetor and cylinder is reduced even more than it was by the downward stroke of the piston. Remember, the lower pressure increases vaporization of fuel.

c. Controlling the Engine Speed

Once the air and fuel are mixed, the next job of the carburetor is to control the amount of mixture that enters the engine cylinder.

A throttle (butterfly) valve in the manifold controls the fuel and air (Figure 56). If you want your engine to run fast, open the throttle (butterfly) valve. The more fuel and air going into the engine, the faster it runs.

If you want your engine to run slowly, close the throttle valve. The throttle valve does not change the ratio of fuel to air. It simply regulates the amount of fuel-air mixture going into the engine.

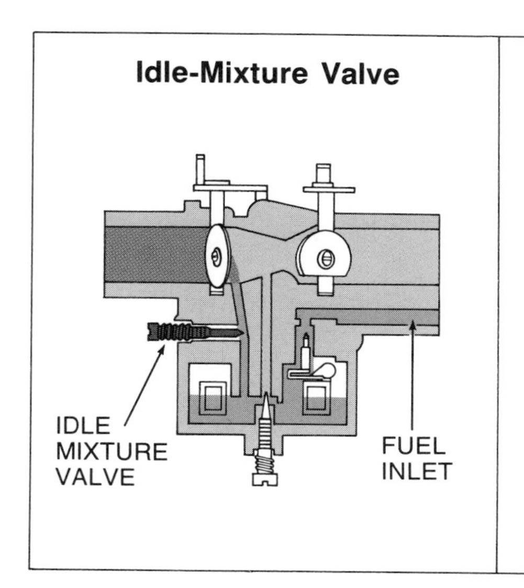

High-Speed Load Valve

HIGH-SPEED
LOAD VALVE

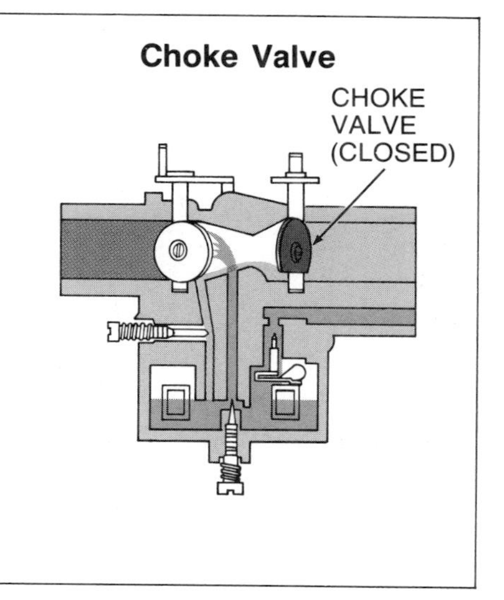

FIGURE 55. The basic adjustments on a carburetor.

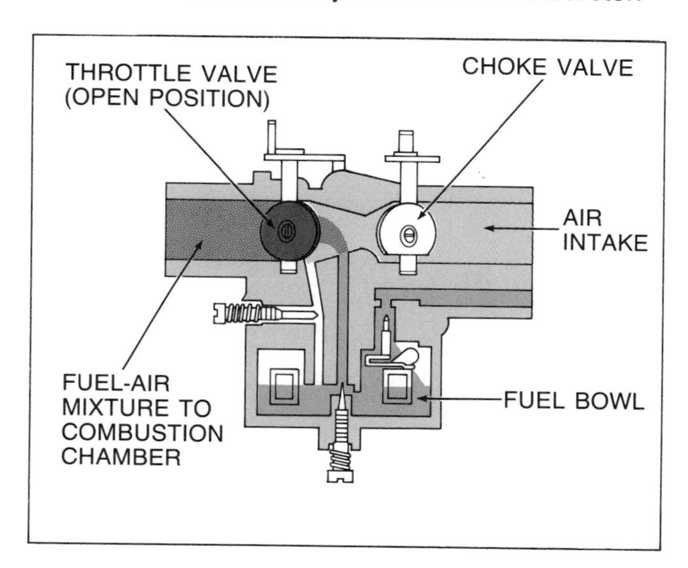

FIGURE 56. A throttle (butterfly) valve controls the engine speed.

2. TYPES OF CARBURETORS

There are many different types of carburetors used on small gasoline engines. These carburetors differ in design and modifications by manufacturer. Consult your operator's manual or maintenance manual to determine the type carburetor used on your small engine.

3. TOOLS AND MATERIALS NEEDED

1. Slot-head screwdriver — 6″
2. Phillips head screwdriver — 6″
3. Open-end wrenches — ¼″ through ½″
4. Needle nose pliers

4. CHECKING THE CARBURETOR FOR PROPER OPERATION

There are three main conditions that tell you there is carburetor trouble. They are:

1. No fuel is going to the combustion chamber from the carburetor
2. Not enough fuel is going to the combustion chamber from the carburetor
3. Too much fuel is going to the combustion chamber from the carburetor

To find out if the carburetor is working properly, follow these steps:

1. *Start the engine.*

   If the engine does not start and you have a strong spark at the spark plug, check how much, if any, fuel is going to the combustion chamber from the carburetor.

FIGURE 57. Add fuel to the combustion chamber using a soda straw.

Proceed as follows:

a. Remove the spark plug.

b. Put one-half teaspoon of gasoline into the combustion chamber (Figure 57).

You can do this by putting a soda straw or some small tubing into gasoline.

When it fills with gasoline, hold your finger over the top to hold the gasoline until you can put the straw or tubing into the combustion chamber through the spark plug hole.

Then take your finger off the straw or tube. Do not put too much fuel into the combustion chamber or you will flood the engine.

c. Reinstall spark plug.

d. Start the engine.

If engine starts, runs for a short time and then stops, this means that there isn't enough fuel getting through the carburetor. Then you will know the trouble is most likely in the carburetor. See instructions for adjusting the carburetor. If your engine does not start at all, this means that the trouble is not in the carburetor (unless it is flooding). Check other accessories for problems.

2. *Run the engine at idling speed for warm-up.*

3. *Advance the throttle suddenly.*

The engine should accelerate smoothly and evenly. If it does not, this means the fuel mixture is too lean. Adjust the carburetor.

4. *Check the exhaust smoke.*

If the engine gives off black smoke, burns a lot of fuel and idles rough, the fuel mixture is too rich (too much fuel is going to the combustion chamber). Adjust the carburetor, procedures follow.

Two-cycle engines may smoke slightly because of the oil in the fuel. Some higher quality oils will not smoke when mixed properly.

5. PREPARING FOR CARBURETOR ADJUSTMENT

In preparing for carburetor adjustment follow these steps:

1. *Fill the fuel tank with clean, fresh fuel (or fill to the level recommended by the service manual).* Some manufacturers recommend only a half-full fuel level.

2. *Check the throttle for mechanical condition and freedom of action.*

3. *Service the engine.*

The following service jobs should be done before you adjust the carburetor (otherwise, the adjustment will not be accurate):

a. Check the crankcase oil level

b. Check the fuel tank vent

c. Service the fuel strainer(s)

d. Service the air cleaner

4. *Check for air leaks in the carburetor manifold.*

This also includes checking for air leaks in the crankcase of two-cycle engines. If you suspect a leak at any point, put a drop of oil on the area, crank the engine, and see if the oil disappears. If so, there is an air leak.

If you have an air leak, tighten the flanges and recheck. It may be necessary to replace the gasket(s).

5. *Check the spark plug.*

6. *Find the idle-speed stop screw* (Figure 58).

FIGURE 58. Locate the idle-speed stop screw.

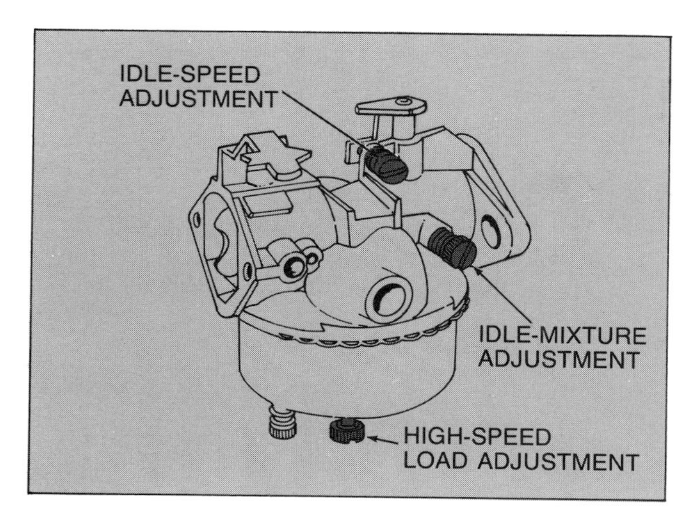

FIGURE 59. Locate the fuel-air mixture adjusting screws.

If you can't find the stop screw, operate the throttle control and notice where the linkage is attached to the carburetor or check your operator's manual for the location on your engine. The stop screw limits the travel of the throttle valve toward the closed position.

7. *Find the fuel-air mixture adjusting screws* (Figure 59).

   There are two of these screws on most carburetors—an idle-mixture adjustment and a high-speed load adjustment screw. On most carburetors the idle-mixture adjustment is the screw nearer the engine.

   If you are not sure, start the engine and operate it at idle speed. Turn the most likely adjusting screw to closed position (clockwise in most cases). If the engine slows down or stops, you have closed the idle-mixture adjusting screw.

   If there is very little difference in the engine speed, try changing the engine speed to ½ to ¾ throttle.

   If you get a difference in operation at this speed, then you have closed the high-speed load adjusting screw.

   A few carburetors have no idle-mixture adjustment. Also, some engines have no high-speed load adjustment.

8. *Begin carburetor adjustments.*

   Turn each of the adjusting needles all the way in by hand (usually clockwise). **Never tighten the adjusting screws more than finger tight.** Then open the idle speed about one turn. This amount varies from ¾ to 1½ turns on different carburetors. Open the high speed load to 1 to 2½ turns.

NOTE: The best advice for adjusting your carburetor comes from the manufacturer of your engine. So be sure to check your operator's manual. Adjustments for smooth operation will be slightly different with different engines.

FIGURE 60. Remove the air cleaner and check the choke valve position.

6. ADJUSTING THE CARBURETOR CHOKE VALVE

With manually operated choke valves, usually the choke valve linkage is adjusted; with spring operated automatic choke valves, the spring tension may or may not be adjustable, depending on the manufacturer.

The three types of chokes are as follows:

a. Manual (separate valve)
b. Automatic (heat sensitive spring)
c. Automatic (thermostat and solenoid)

a. Manual (Separate Valve)

If your carburetor has an adjustable manual choke valve, the adjustment will be in the control linkage. Adjust it as follows:

1. *Remove carburetor air cleaner.*

   Check the choke valve position (Figure 60).

2. *Close choke valve with the control.*

   Check for freedom of operation.

3. *Notice position of the valve, after closing the control.*

   The valve should be completely closed.

   If it is not closed, adjust the control linkage.

   Loosen the set screw (Figure 61), close the choke valve; then tighten the set screw.

4. *Open choke valve with the control.*

   It should open all the way.

   If it does not open, recheck your adjustment in step 3.

FIGURE 61. Adjust the control linkage.

b. Automatic (Heat Sensitive Spring)

If your carburetor has an automatic choke which is operated by the heat sensitive (thermostat) spring, (Figure 62), adjust it as follows:

1. *Remove air cleaner.*
2. *See position of the choke valve.*

   Adjustment must be made while the engine is cold. Valve should be closed at room temperature. If you make adjustments while the engine is warm, the spring will expand and have a tendency to open the valve.

   If valve is not closed, go to step 3.
3. *Loosen adjusting lock screw* (Figure 63).

   Some lock screws are on the spring bracket, and some have a set screw on the adjusting arm.
4. *Move adjusting link until choke is closed.*
5. *Tighten adjusting lock screw.*

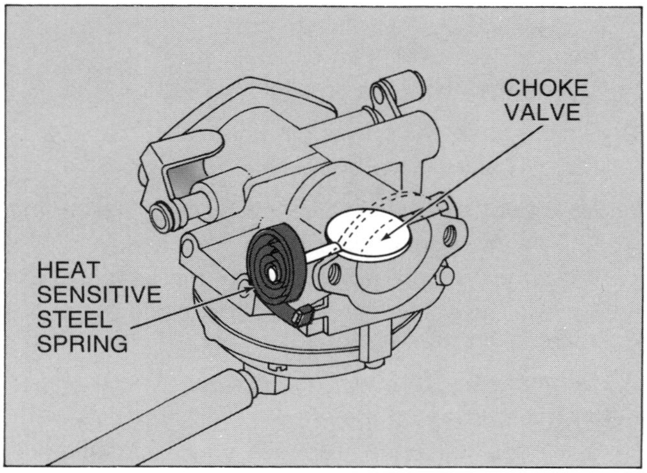

FIGURE 62. An automatic choke operated by a heat sensitive spring.

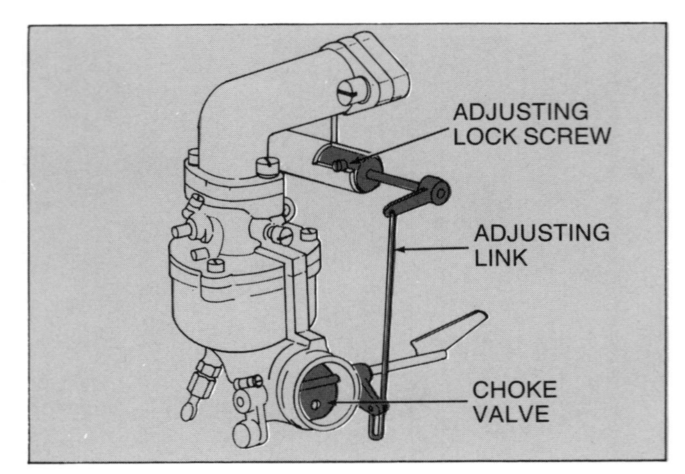

FIGURE 63. Loosen the adjusting lock screw.

c. Automatic (Thermostat and Solenoid)

If your automatic choke is controlled by a combination electric thermostat and solenoid. Adjust as follows:

1. *Turn ignition switch off if your engine is so equipped.*
2. *Close fuel shut-off valve.*
3. *Remove carburetor air cleaner.*
4. *Operate starter and notice the extreme position of the solenoid lever arm and choke valve* (Figure 64).

   This should be done with the engine cold. The choke valve should be closed. If the choke valve is not closed, go to step 5.
5. *Lock the solenoid lever in extreme closed position.* Some chokes have special steps for doing this. See your service manual.
6. *Loosen clamp on solenoid arm.*
7. *Adjust choke valve to close.*

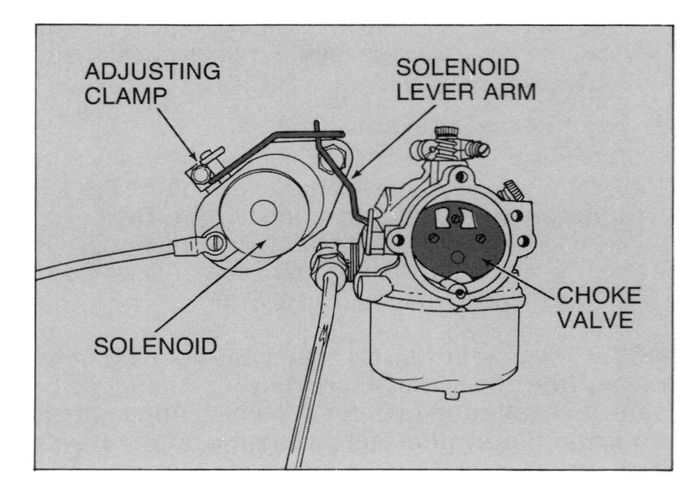

FIGURE 64. Check the position of the solenoid lever arm and choke valve.

FIGURE 65. The high-speed load valve adjustment.

8. *Tighten clamp.*
9. *Check for proper operation.*

7. ADJUSTING THE HIGH-SPEED LOAD VALVE

The valve that controls the fuel flow for full load at high speed is known by such names as main adjusting needle, power adjusting needle and high speed load adjustment. All refer to the same adjustment (Figure 65).

It is very important that the engine be up to full operating temperature when this adjustment is made. If it is not, your adjustment will not be right when the engine reaches normal operating temperature.

To adjust the high-speed load valve, follow these steps:

1. *Run engine at full throttle load, if possible.*

   You will get a better adjustment if there is a constant load without having your equipment in motion.

   If you make adjustments without load, check to make sure the adjustment is satisfactory under load. Chain saws should always be checked under normal sawing conditions (Figure 66).

2. *Add load until speed decreases slightly from its high idle speed.*
3. *Turn load adjusting screw clockwise until engine begins to lose power.*

   This shows that you have reached the border line on the lean side of the mixture.

4. *Turn adjusting screw counterclockwise until engine gives off black smoke from exhaust.*
5. *Turn screw clockwise in 1/8 increments until engine runs smoothly and at full speed.*

   Make 1/8 turn of screw and wait for engine to adjust.

FIGURE 66. Always check chain saws under normal sawing conditions.

6. *Check carburetor adjustment by operating the engine under load.*

   If the carburetor is properly adjusted, the engine should accelerate smoothly when you adjust the speed control lever suddenly.

   Backfiring indicates too lean a mixture, and dark colored smoke indicates too rich a mixture.

   If you operate your engine under a heavy load or during cold weather or both, adjust the fuel-air mixture so it is slightly on the rich side—about 1/16 turn. Do not overdo it. This adjustment is not to take the place of your choke for cold weather starting.

8. ADJUSTING THE IDLING VALVE

Some operating manuals suggest a first setting of 3/4 to 1 1/2 turns open for the idling valve, but be sure to use the setting that applies to your particular engine.

Use this setting when your engine will not start because of carburetor trouble or when you are repairing the carburetor.

If you want a more exact adjustment, follow these steps:

1. *Warm the engine.*
2. *Loosen idle-speed stop screw.*
3. *Slowly tighten idle valve* (Figure 67) *until engine begins to slow down.*

   Use 1/8 turn increments when adjusting the needle valve.

   After each 1/8 turn, wait a few seconds until the engine has a chance to adjust to the new fuel mixture.

4. *Turn idle screw back slowly until engine runs smoothly.*

FIGURE 67. Tighten idle valve until engine begins to slow down.

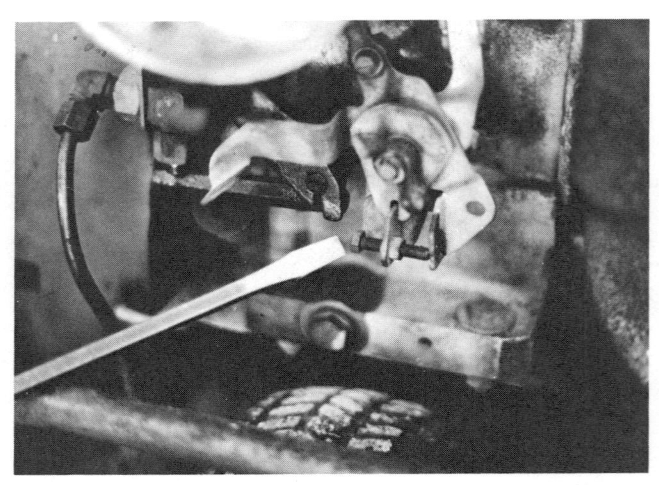

FIGURE 68. A set screw acts as a stop to the throttle stop lever.

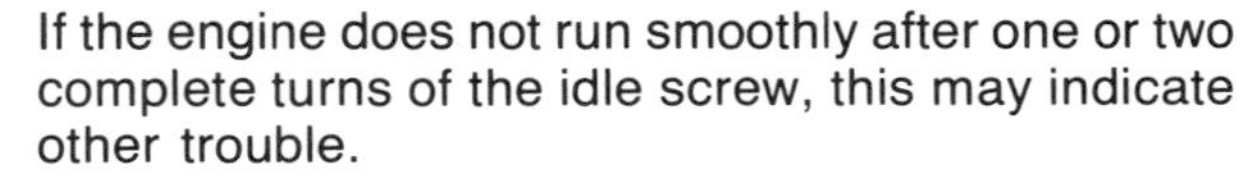

If the engine does not run smoothly after one or two complete turns of the idle screw, this may indicate other trouble.

5. *Recheck high speed load adjustment.*

   Adjusting the idle speed may change the fuel mixture enough to affect the original adjustment.

6. *Recheck idle mixture adjustment.*

7. *Check for proper operation.*

   Operate the throttle back and forth a couple of times to see if your engine will accelerate properly from the idle position. If a chainsaw does not accelerate smoothly, open idling valve an additional ⅛ turn and check again.

8. *Adjust the idle-speed stop screw.*

   See procedures under next heading.

9. ADJUSTING THE IDLE-SPEED STOP SCREW

Adjusting the idle speed is really a throttle stop adjustment. A set screw acts as a stop to the throttle stop lever (Figure 68). It can be adjusted to hold the throttle at the desired engine speed. A spring is usually installed on the screw to hold it in place.

When you change from cold weather operation to warm weather operation and vice versa, the idle speed may need adjustment. A difference in altitude will also affect the carburetor adjustment.

Never get the idea that you are taking care of your small engine by operating it at a slow idle speed. Small engines are meant to operate at full throttle. The fuel-air mixture is too rich for slow speed. Unburned fuel will foul the spark plugs and will contaminate lubrication oil.

To adjust the idle-speed stop screw, follow these steps:

1. *Run the engine at about half throttle for two minutes for warm-up.*

2. *Set speed control lever at idle position.*

3. *Adjust to normal idling speed.*

   Turning the adjusting screw clockwise usually increases engine speed. Turning it counterclockwise decreases it.

   Check your operator's manual for proper idle speed.

STUDY QUESTIONS

1. Describe the three functions of a small engine carburetor.

2. Narrow passages in the carburetor that increase air speed which help atomize the fuel are called the ________.

3. Identify two valves on small engines which regulate the air-fuel mixture in the carburetor.

4. List three types of choke valves.

5. True or False. When adjusting the high-speed load valve, the engine should be under a work load.

J. Checking Compression

Upon successful completion of this section, you will be able to do the following:

- **Explain the importance of checking the compression in your engine**
- **Tell how good compression helps in starting your engine**
- **Give the causes of poor compression**
- **Determine how often compression should be checked**
- **Follow the procedures for checking the compression of your engine**

Compression should be checked every 50 hours of operation, or more often if your engine is hard to start, won't accelerate or if it seems to be losing power.

There is a simple test for checking compresion. This test can be performed on most small engines.

It is important to check the compression in your engine because if you continue to operate your engine when the compression is low, the engine will lack power and damage to the engine parts will increase rapidly. Fuel and oil consumption will increase, and the engine will become more difficult to start.

Good compression is needed for easy starting and for proper engine operation. It helps starting in the following four ways:

1. It gives high pressure and temperatures that are near the ignition point. (The spark plug provides the final spark.)
2. It keeps the fuel mixture in a more restricted space, making it easier to ignite.
3. It helps to distribute the fuel particles in the cylinder for better burning.
4. The higher the compression, the greater the expansion of gases after the fuel burns in the cylinder. This expansion pressure gives more power to the piston.

Any condition that allows the gases to escape from the cylinder during the compression stroke will cause poor compression. You can easily correct conditions, such as a loose spark plug, loose cylinder head bolts, low cranking speed or dry cylinder walls.

Other causes of poor compression are more difficult to correct, such as stuck or burned valves or worn piston rings and cylinders.

Simple steps for checking the compression are as follows:

1. *Disconnect the spark plug wire to prevent the engine from starting.*
2. *Turn the flywheel by hand until it comes up to compression stroke* (Figure 69).

FIGURE 69. Turn the flywheel by hand.

   If your engine is the four-cycle type, you should notice a definite resistance at every second revolution of the crankshaft.

   If your engine is the two-cycle type, you should notice a definite resistance on each revolution of the crankshaft.

3. *Give the flywheel a quick twist by hand or with the manual starter.*

   On four-cycle engines, turn the flywheel opposite the direction of hand rotation. This is done because in small engines there are provisions for reducing compression at low speeds for easy cranking. The compression reducers are not effective when the crankshaft is turning backward.

   If the compression is good, the flywheel will spring back from the force of the compressed air in the cylinder.

   If the compression is poor, there will be little or no resistance to the flywheel turning; and it will not spring back. Go to step 4.

4. *Check for cylinder head trouble.*

   Crank the engine and listen for air escaping at the cylinder head. Look for burned spots that show possible leaks.

5. *If your engine has not been operated for several weeks, check for dry piston rings and cylinder walls.*

   Sometimes oil will drain from the piston rings while the engine is setting. If the rings are not oiled, air will escape by the rings.

   Remove the spark plug and pour a tablespoon of engine oil in the cylinder. Recheck compression.

   If compression is good, the piston rings are in good condition. The oil made a seal between the rings and the cylinder wall.

If compression is still low, check for valve (four-cycle engines), cylinder and piston ring troubles.

STUDY QUESTIONS

1. Proper engine compression is essential for starting. In what ways does compression aid engine starting?
2. Compression should be checked every ________ hours of operation.

K. Checking and Servicing Batteries

Upon successful completion of this section, you will be able to do the following:

- **Explain the importance of properly servicing a battery**
- **Select the tools and materials needed to service a battery**
- **Check and refill the liquid in your battery**
- **Check battery frame and cable connections**
- **Test battery charge**
- **Clean a battery**

The battery on your small engine supplies electrical energy for (1) engines with electric starters and (2) for igniting the fuel-air mixture in the combustion chamber on engines with battery ignition and (3) for lights and other electrical accessories.

Servicing the battery is a simple job for the small engine operator. Perhaps this is the reason it is often neglected.

Your battery will last longer and give better service if you check and service it regularly. Operators' manuals give different recommendations on how often to check and add water. These differences may be due to the liquid capacity of the battery or the temperature at which the battery operates. If the battery is near the engine and is operating in a warm climate, more water will evaporate. Be sure to check the liquid level every week. Also clean the battery top and connections when they become coated.

How a battery supplies power and the procedures for checking and servicing the battery are explained under the following headings:

1. Functions of the Battery and How It Works
2. Importance of Proper Servicing
3. Tools and Materials Needed
4. Checking and Replacing the Battery Liquid
5. Checking the Battery Frame and Cable Connections
6. Checking the Battery Charge
7. Cleaning the Battery

1. FUNCTIONS OF THE BATTERY AND HOW IT WORKS

To care for a battery, you need to understand how it is made and how it works. The parts of a battery and how they are assembled are shown in Figure 70.

A storage (lead acid) battery contains metallic lead grids to which lead oxide is added.

The plates are electro-chemically treated in a diluted solution of sulfuric acid. This treatment causes the lead oxide (PbO) in the positive plates to change to lead peroxide (PbO_2). The lead oxide in the negative plates is changed to metallic spongy lead (Pb).

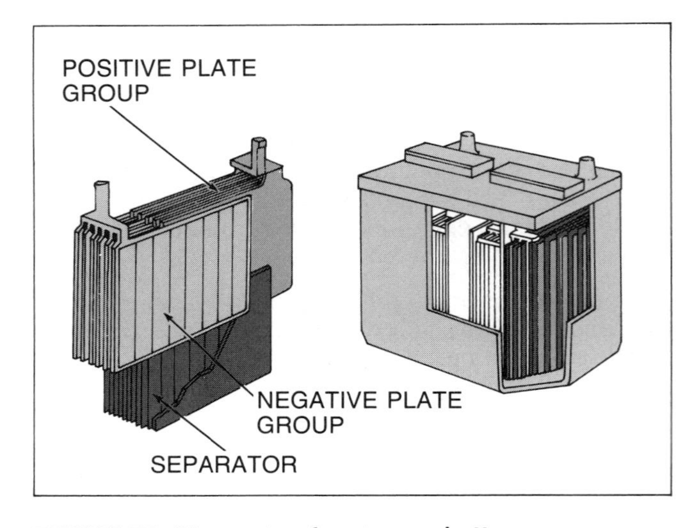

FIGURE 70. The parts of a storage battery.

Groups of positive and negative plates are assembled alternately with separators between each two plates. The separators are made of a plastic or cellulose and are not usually affected by the acid. Each grouping of positive and negative plates forms a two-volt element when immersed in sulfuric acid.

Each element is assembled in a hard rubber or plastic cell. The elements are connected in series (positive to negative plates) by means of welded-on lead connectors. A battery of six such cells give a nominal voltage of 12.

The chemical action inside a battery is shown in the following chemical equation:

$$Pb + PbO_2 + 2H_2SO_4 \underset{\text{charging}}{\overset{\text{discharging}}{\rightleftarrows}} 2PbSO_4 + 2H_2O$$

The left side of the equation shows a battery cell in a charged condition. The right side shows a cell in a discharged condition. In the charged condition, the positive plate contains lead peroxide (PbO_2), and the negative plate contains sponge lead (Pb). The liquid in the battery is called electrolyte. It contains about 36 percent sulfuric acid (H_2SO_4) and 64 percent water.

When you use energy for lights or for engine starting, the electrolyte reacts with the sponge lead on the negative plate and with the lead peroxide on the positive plate. Lead sulfate forms on both the positive and negative plates.

The acid content of the electrolyte becomes less and less because it is used in forming lead sulfate ($PbSO_4$). The specific gravity of the electrolyte decreases from about 1.275 when fully charged to 1.150 when discharged.

If you do not recharge the battery, the active material will change into lead sulfate. Then the cells can't produce enough current. At this point the cells are said to be discharged.

In the discharged condition, both plates contain lead sulfate ($PbSO_4$), and the electrolyte is mostly water. Recharging is done by forcing electric current through the battery in the opposite direction from normal battery current flow. This causes the lead sulfate and water to change back to lead, lead peroxide and sulfuric acid (as shown on the left side of the chemical equation).

The sulfuric acid does not need to be replaced unless it is lost through leakage or is spilled from the battery.

The water is lost partly from evaporation, but mostly from the chemical action within the battery while it is being charged. It is then that some of the water in the electrolyte is changed to hydrogen and oxygen gases. These pass out the vent holes in the battery caps. If water is not added to replace the amount that is lost, the tops of the plates become exposed.

2. IMPORTANCE OF PROPER SERVICING

Operator's manuals state that you should keep the battery liquid above the level of the plates for several reasons.

When parts of the plates and separators are exposed to air, they dry out. The parts that dry can't function and may become permanently damaged. This means less battery power for cranking.

It also means shorter battery life since water is lost only from evaporation and forming of gas. The electrolyte that remains becomes strongly concentrated and may break down the separators and plates during the time the liquid level is low.

Your battery requires more water (a) when it is being overcharged or (b) when the weather becomes hotter.

Some batteries are advertised as requiring water only a few times a year. They do not use less water than other batteries, but they provide more space above the plates for reserve electrolyte. Any battery will last longer, however, if the liquid level is kept near the top.

It is important that you keep your battery at or near full charge. There are two reasons:

(1) When a battery remains discharged for several weeks, the lead sulfate on the positive and negative plates becomes hard. This is called harmful sulfation. When recharging your battery, the hardened lead sulfate remains and prevents the plates from taking a full charge. This lowers the overall electrical capacity of your battery and shortens its life.

(2) The capacity of your battery for cold weather starting is greatly reduced. Even a fully charged battery at −18°C (0°F) has only 40 percent of the capacity it has at 20°C (70°F). That is the reason a weak battery may give fair service until the weather turns cold. Then it appears to go bad all at once.

Checking the specific gravity, or state of charge, of the battery is recommended for every 50 hours of use. Keep liquid level above the plates and separators. Charge the battery after each water addition and keep it charged. A fully charged battery will not freeze. This is particularly important if you are using a 12-volt battery. It may have a high enough charge to crank the engine but still have low enough charge (weak acid) to freeze.

Cleaning the battery is important. Dirt, moisture and acid gradually accumulate on top of the battery. While the battery is being charged, acid spray is carried out of the battery along with the gas. It settles on the battery top and makes a damp surface where dust and dirt accumulate. This also causes corrosion on the battery posts, terminals and equipment (Figure 71).

FIGURE 71. Dirt, moisture and acid accumulate on a battery causing corrosion on the battery posts, terminals and parts of the equipment. This corrosion can cause the battery to slowly discharge.

The amount of acid present is very small, but it is enough to provide an electrical path to the metal frame or to the opposite terminal. The battery will then slowly discharge itself. With 12-volt batteries the discharge is much faster than with six-volt batteries.

A battery, even if charged but not being used, has a natural tendency to discharge slowly. At −18°C (0°F) a fully charged battery may last a year before it becomes completely discharged. At 50°C (125°F) the discharge period may be shortened to a month. If acid and corrosion have collected on top, it will discharge much faster.

3. TOOLS AND MATERIALS NEEDED

1. Battery syringe
2. Slot-head screwdriver — 8″
3. Box-end wrenches — 7/16″, 1/2″, and 9/16″
4. Wrenches that fit nuts on battery hold-down clamp and nuts on battery posts
5. 0–20 direct current voltmeter
6. Wire brush and sandpaper
7. Small paint brush
8. Battery clamp puller
9. Clean water
10. Battery hydrometer
11. Baking soda
12. Light grease (petroleum jelly)
13. Bristle brush
14. Goggles
15. Safety glasses

4. CHECKING AND REPLACING THE BATTERY LIQUID

If you have reason to believe that your battery has a low charge, follow procedures under "Checking the Battery Charge." If your battery has been giving satisfactory service, proceed as follows:

1. *Remove caps from battery cells.*

   Most caps are threaded. Some are pressed into position. Turn them upside down and lay them on the side of the battery case. This keeps acid off the battery top.

   CAUTION! Do not smoke or light a match while the caps are removed. If the battery has been charging, hydrogen gas is present. A concentration as low as seven percent may burn or explode in the presence of a spark or flame and spatter acid all over you. Be sure to wear goggles or safety glasses.

2. *Fill each cell to fill level with clean water.* A battery syringe is best for filling (Figure 72).

   Use distilled water whenever possible; otherwise, clean tap water may be added.

   In many batteries the vent well goes into the battery far enough for the lower end to be used as a fill level indicator. Batteries without this are usually filled until the level of the electrolyte is about 3/8 in. above the level of the separators.

   Avoid overfilling. If the level is too high, the electrolyte will overflow through the caps and onto the top of the battery. As the electrolyte spreads, it may reach the battery terminals and the battery frame.

FIGURE 72. Use a battery syringe for filling the battery with clean water.

It then provides a path for leakage and causes corrosion and deterioration (damage, wear) of the metal parts.

CAUTION! Wear safety glasses to protect your eyes from acid when servicing the battery. If acid does splash on you, wash it off with lots of cold water as soon as possible. If it gets into your eyes, splash them with cold water. See a doctor.

3. *Replace caps on battery.*

   If operating engine under dusty conditions, check the vent hole in the cap to make certain it is not filled with dirt. If clogged, clean with water and replace caps on battery. If the vent is plugged, enough pressure can develop in the battery to break the seal or case.

5. CHECKING THE BATTERY FRAME AND CABLE CONNECTIONS

Check for loose terminal connections on the battery posts and for loose hold-down clamps on the battery (Figure 73).

If the terminals are loose, there is resistance to the flow of current so that equipment supplied by the battery does not get the full benefit of the battery voltage.

If the hold-down clamp is loose, the battery is free to bounce, which may damage the plates and cause internal short circuiting. It is also possible to tighten the hold-down clamps too much, which can result in breaking the case. Tighten just enough to prevent movement of the battery.

If there is corrosion on the battery terminals, proceed to " Cleaning the Battery."

6. CHECKING THE BATTERY CHARGE

There are two methods of testing your battery. They are discussed under the following headings:

a. Specific Gravity Method
b. Load Capacity Test

a. Specific Gravity Method

The specific gravity method (hydrometer) is used most often. This is partly because a hydrometer is economical and does a good job of testing.

Proceed as follows:

1. *Remove battery caps.*

   Notice the level of the electrolyte in the cells. If it does not cover the separators, add water and delay checking until you have operated your engine for about four hours so that the water has time to mix into the electrolyte.

2. *Insert hydrometer nozzle, squeeze bulb, then slowly release to draw electrolyte into barrel* (Figure 74).

3. *Adjust electrolyte level until float rides freely.*

   If the float tends to stick to the side of the hydrometer barrel, shake it sideways gently. If it still sticks, flush the inside of the hydrometer with soap and water and rinse with clean water before further testing.

   If the paper scale inside the float is wet, the float probably leaks and your reading will not be accurate.

4. *Hold hydrometer vertically while taking reading.*

FIGURE 73. Check the battery for loose terminal connections and hold-down clamps.

FIGURE 74. Using a hydrometer to check the specific gravity of the electrolyte in a battery.

Leave the hydrometer nozzle in the battery cell to avoid dropping acid on the battery and metal parts of the engine or equipment.

While reading the scale, adjust your position so that your eye is level with the liquid. This will improve the accuracy of your reading.

5. *Return electrolyte to cell from where it was removed.*

   CAUTION! If you accidentally spill some of the electrolyte on your skin or clothing, rinse with water immediately. Apply baking soda or lime to neutralize the acid.

6. *Check remaining cells in the same way.*
7. *Flush hydrometer with clear water.*
8. *To interpret results of readings from each of the battery cells, see Table II.*

b. Load Capacity Test

To perform this test, use a voltmeter (0–10 for six-volt and 0–20 for 12-volt) with a 1/10 scale. Or, use a special battery load tester that includes a voltmeter, ammeter, and a variable resistor. The temperature should be 16°C to 25°C (60°F to 80°F). If you have a special battery tester, follow instructions by the manufacturer. Proceed as follows:

1. *Connect the instrument to the battery* (Figure 75).

   Follow instructions by the manufacturer. Be sure to connect the positive lead of the voltmeter to the positive post of the battery and the negative lead to the negative post of the battery.

TABLE II. INTERPRETING THE BATTERY SPECIFIC GRAVITY READING

Specific Gravity Reading	What It Means
1.300	Electrolyte level is low. Battery is being **overcharged** or **electrolyte** has been added instead of water.
Between 1.225 and 1.280	Battery is in good condition. Some batteries are fully charged with a reading of 1.280 while others are fully charged at 1.250 or less. Batteries intended for use in warmer climates may be fully charged with the specific gravity as 1.225.
Under 1.225	Battery charge is **too low**. Have it recharged. The electrolyte in various stages of charge will start to freeze at the following temperatures: Specific Gravity Reading (Corrected to 1 + 25°C (80°F) — Electrolyte Freezing Temperatures 1.250 .... −52°C (−62°F) 1.200 .... −27°C (−16°F) 1.150 .... −15°C (+ 5°F) 1.100 .... − 7°C (+19°F)
If, after charging, there are more than 50 gravity points difference between the highest and the lowest cell readings (example: 1.260 high and 1.200 low).	Battery may need to be replaced. Check with your dealer. The condition may indicate: short circuits in the battery; unequal losses of acid from the cells; that the battery is worn out due to loss of active plate materials, breakdown of the separators, or accumulation of impurities. Battery additives are sometimes offered for sale. They are supposed to rejuvenate worn out or "dead" batteries. Extensive tests by the National Bureau of Standards have shown that none of those tested are of value.

FIGURE 75. Connect the load capacity tester to the battery.

FIGURE 76. Use a wire brush to clean the outside surfaces of the battery posts.

2. *Remove the coil wire and ground it to the engine.*

   This will prevent the engine from starting and from damaging the diodes in the alternator (if installed).

3. *Apply load to the battery.*

   If you are using a voltmeter and the starter for loading, turn engine with the starter for 15 seconds: watch the reading while starter is turning.

   Do not crank the starter more than 15 seconds at a time, or it may be damaged from overheat.

   If you are using a variable resistor, use three times the amperage rating of the battery (300 amps for 12-volt batteries; 180 amps for six-volt batteries).

4. *Read the voltage.*

   If the voltage is 9.6 (12-volt) or 4.8 (six-volt) or more, this means the battery is good.

   If the voltage is less than 9.6 (12-volt) or 4.8 (six-volt), the battery is either bad or it is discharged. Charge it and retest, or do the three-minute charge test.

   NOTE: These results are for a temperature of 21.1°C (70°F).

7. CLEANING THE BATTERY

The steps for cleaning a battery are as follows:

1. *Disconnect cable and ground strap from battery terminals.*

   Loosen and remove ground strap first.

   This prevents short circuits in case you accidentally (a) touch metal with the wrench used to loosen ground strap, or (b) make contact with the opposite battery terminal or some part of frame with wrench or screwdriver you have set down on the battery.

   If you remove the battery, tag the terminal connections so you will know how to reinstall it. If you connect it wrong, there is a danger of burning out the voltage regulator (on generator systems that use an alternator). A wrong connection will only take a second to burn out the alternator.

2. *Clean cable clamps and battery posts* (Figure 76).

   A wire brush may be used on outside surfaces. Sandpaper is satisfactory for cleaning inside of clamps. Pay special attention to inside of cable clamps and outside of battery posts. Special battery connection cleaning tools are available.

   CAUTION! Use safety glasses to protect your eyes when using a wire brush.

3. *Remove loose dirt and corrosion particles from top of battery.*

4. *Brush soda and water mixture on top of battery, on posts, and on clamps* (Figure 77).

FIGURE 77. A solution of soda and water will neutralize battery acid.

Use about two tablespoons of baking soda in a pint of water. Mix well and spread on the battery. This will react with the acid and cause a foaming action. Keep spreading until the foaming stops.

Some manufacturers recommend hot water (temperature of 125 to 150 degrees) in place of soda and water.

WARNING! Keep water or soda and water from entering the battery through the breather hole in each cap; it will weaken the acid in the electrolyte.

5. *Wash away residue with clean water.*

   Remove residue stuck on battery frame or parts of engine and equipment.

   Breather caps may be plugged with toothpicks to prevent water and dirt from going into the battery.

6. *Dry top of battery with a clean cloth.*

7. *Reconnect power cable and ground strap.*

   Connect power cable first to help avoid grounding the battery with your tools.

   Do not hammer the clamps into place on the battery posts. You can break the seal around the battery terminal or cause a crack in the cell cover. In either instance, the electrolyte will work out on top of the battery and speed up battery failure.

   If the jaws of the clamp meet before the clamp tightens on the battery post, take your knife and cut a small amount of metal from between the jaws.

8. *Apply a coating of petroleum jelly or silicone spray on post and cable-clamp connections* (Figure 78). This helps protect against corrosion.

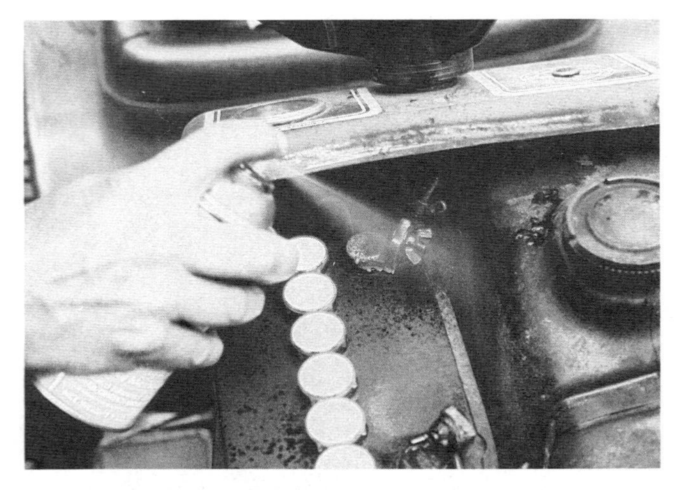

FIGURE 78. Apply silicone spray to the battery posts and cable-clamp connections.

STUDY QUESTIONS

1. True or False. Batteries in which a liquid level must be checked should be checked weekly.
2. True or False. Battery terminals never have to be cleaned.
3. The liquid in a battery is called ________.
4. The water in a battery is lost partly due to ________, but mostly due to the ________ action within the battery while it is is being charged.
5. Battery charge should be checked every ________ hours of operation.
6. True or False. A fully charged battery tends to discharge slowly when not in use.
7. ________ gas is present when a battery has been charging. Avoid a spark or flame because the gas may burn or explode.
8. ________ water is recommended for use in a battery.
9. A ________ is used when checking the battery charge using the specific gravity method.
10. True or False. The battery cables should be connected when cleaning a battery.

III. Starting and Stopping Small Gasoline Engines

Every owner or operator of a small engine should be familiar with, and follow the proper procedures for starting and stopping the engine. If you follow the proper procedures, you will save time and irritation. Also, your engine will last longer and give better service.

To help you become a better operator and a better judge of the cause of starting and operating troubles, review the information given under the following headings:

A. Starting Engines
B. Stopping Engines

A. Starting Engines

Upon successful completion of this section, you will be able to do the following:

- **Explain the importance of following the proper steps in preparing to start small engines**
- **Follow the proper procedures for starting different types of small engines**

Most operators have more trouble starting a small engine than they do with any other phase of operation. Procedures for starting an engine are discussed under the following headings:

1. Preparing to Start the Engine
2. Starting the Engine

1. PREPARING TO START THE ENGINE

When preparing to start a small engine, read your operator's manual carefully (Figure 79). Understand how to operate the engine and equipment controls. Learn how to stop the engine quickly in case of emergency.

Always dress properly. Never wear loose-fitting clothing which may catch in the moving parts of the engine or equipment.

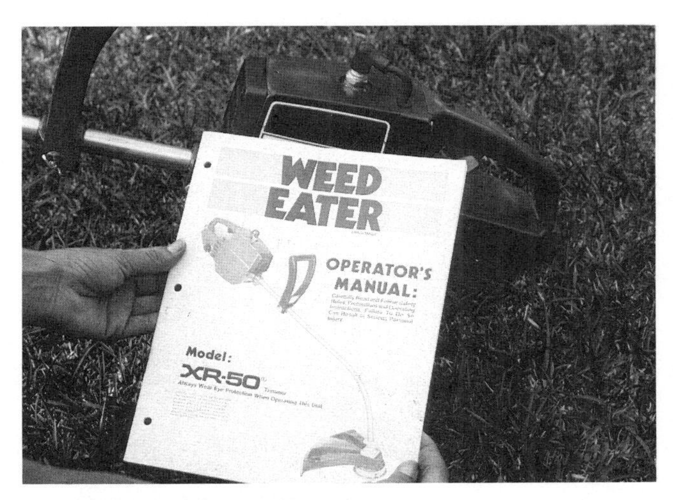

FIGURE 79. Before starting your engine, read your operator's manual carefully.

Follow these steps in preparing to start your engine:

1. *Be sure that your engine has been serviced properly.*
2. *Move the engine to an outside area.*

   Do not operate the engine in an enclosed area because exhaust gases contain carbon monoxide, an odorless deadly poison.
3. *Make sure the area is clean.*

   Gasoline spilled on the floor is a fire hazard. Wash it away with water.
4. *Be sure the engine is level and well balanced.*

   There is a danger of pulling the equipment over when using a pull-type starter.
5. *Apply the brakes (if the equipment has them).*
6. *Disengage clutch on all power-driven equipment.*

   This takes the load off the engine during cranking and warm-up. It is also a safety measure.

   Some small tractors will not start unless the power take-off clutch is disengaged.
7. *Crank the engine.*
8. *Adjust carburetor as needed.*

   Most carburetors are preset at the factory. But some manufacturers recommend checking the adjustments before starting an engine for the first time.

---

2. STARTING THE ENGINE

The steps for starting the engine are as follows:

1. *Open the fuel shut-off valve, if your engine is equipped with one, all the way to prevent leakage around the valve stem.*
2. *If your engine is equipped with a choke it may be necessary to close the choke valve* (Figure 80). The choke helps increase vaporized fuel therefore increasing the chances for ignition.

FIGURE 81. The throttle position on most engines should be ¼ to ½ open for starting.

   Some engines need little or no choking during warm weather. Warm engines usually don't need choking.

   CAUTION! Operate the controls from the operator's position if possible. You are in position to stop the engine if necessary and to control the equipment.
3. *Set throttle at position recommended for your engine* (Figure 81).

   Usually ¼ to ½ open is sufficient.

   If your choke and throttle lever are combined, move lever to choke position for starting. Then when engine starts, move it to the run position.

   This operation is not necessary on engines with automatic chokes. The valve opens automatically when the engine starts.

   If your engine has no throttle control lever, the throttle is adjusted at the carburetor for normal operations. There is no manual control. Do not adjust this type throttle for starting.

   Turn on the ignition switch (Figure 82) if engine has one.

   Ignition switches are usually marked on or off.

FIGURE 80. Close the choke valve.

FIGURE 82. Turn on the ignition switch.

FIGURE 83. A typical grounding switch for a magneto type ignition.

Engines equipped with magneto ignitions have some type of grounding switch. As shown in Figure 83, the switch is on when the grounding switch is in the open position. To stop the engine, the grounding switch is closed. This gives another path for the spark plug current to follow. The engine stops.

On some engines this switch is part of the throttle control. When the throttle is opened, the grounding switch is open.

4. *Crank the engine.*

   There are several different types of starters used on small engines. The procedures you use for cranking will depend upon the type starter you have.

Procedures for four common types of starters are explained as follows:

- Rope-Wind Starter
- Rope-Rewind Starter
- Wind-Up Starter
- Electric Starter

If your engine has a rope-wind starter (Figure 84), follow these steps:

FIGURE 84. A rope-wind starter.

FIGURE 85. A rope-rewind starter.

1. *Wind rope around flywheel pulley.*
2. *Crank engine slowly until it reaches the compression stroke.*
3. *Rewind rope until you have a short section left.*
4. *Pull rope all the way.*

   Brace yourself against the engine. Pull rope straight and evenly.

   CAUTION! Keep hands away from pulley. When cranking an engine, steady the engine by bracing your hand or foot against it to keep it from tipping over.

If your engine is equipped with a rope-rewind starter (Figure 85), follow these steps:

1. *Pull rope to crank engine.*

   If you are starting a chain saw, place it on the ground and place your foot in the rear handle. While holding down on the handle bar, pull the starter rope (Figure 86).

   CAUTION! Never try to crank a chain saw without bracing the saw securely.

FIGURE 86. The proper method for cranking a chain saw.

FIGURE 87. A wind-up starter.

2. *Allow the rope to rewind.*
3. *Pull rope briskly and firmly—but not all the way.*
   The end of the rope is anchored to the starter pulley, and there is a danger of breaking the rope.
4. *Hold on to the rope handle and allow the rope to rewind.*
   This action prevents damage to the starter spring and avoids binding of the rope.

If your engine is equipped with a wind-up starter (Figure 87), follow these steps:

1. *Place release lever in wind-up position.*
2. *Lift handle to cranking position.*
3. *Wind starter.*
   Turn handle in direction indicated on your engine.
4. *Fold wind-up handle to retracted position.*
5. *Move release lever to run position.*
   CAUTION! Never work on engine or equipment when a wind-up starter is wound.
   NOTE: While you are operating your engine, leave release lever in run position to prevent damage to starter.

If your engine is equipped with an electric starter, proceed as follows (Figure 88):

1. *Crank the starter switch for not more than 10 seconds and release.* Electric starters may be operated by means of the ignition key, by a push button or by a pull knob. The starter switch automatically returns to the off position when released. This prevents the starter from overheating.
2. *Open the choke valve (part or all the way) when the engine starts.*

FIGURE 88. An engine equipped with an electric starter.

   Some choke valves snap completely open. Others are designed for partial opening until the engine warms up.
3. *Adjust the throttle for warm-up.*
   Set for fast idle.
   Never open the throttle completely, gun the engine or apply load until the engine is warmed up. Let it run for at least one minute. This gives time for oil to get to the bearing surfaces.
   CAUTION! To prevent injury stay clear of moving parts any time the engine is running.
4. *After warm-up, readjust the throttle for best operation.*

STUDY QUESTIONS

1. Clutches on all power driven equipment should ________ be engaged when starting an engine.
2. Do not operate small engines in an enclosed area because exhaust gases contain ________, an odorless deadly poison.
3. True or False. Throttles should always be set completely open when starting small engines.
4. List four types of starters commonly found on small engines.
5. True or False. Never work on a small engine equipped with a wind-up starter when the starter has been wound.

B. Stopping Engines

Upon successful completion of this section, you will be able to **properly stop small engines**.

The steps for stopping the engine are as follows:

1. *Remove load from the engine.*

   If you suddenly stop an engine under load, it suffers from shock. Instead of coasting to a standstill, it is actually stalled. This causes shock on the bearings and increases wear.

2. *Reduce engine speed to idle.*

   Allow engine to cool for one or two minutes at ⅓ throttle speed and no load.

   The engine temperature may be three times higher at operating speeds than at idle speeds. By stopping the engine at a high temperature, the sudden cooling causes stress on all the engine parts.

3. *Turn off the engine.*

   On battery ignition engines, leaving the ignition switch on will cause the battery to discharge.

   CAUTION! If your engine is controlled by an ignition lock and key, remove it before leaving the engine. This is a safeguard against children starting the engine, operating the equipment, and getting hurt. If the engine has no ignition key, remove the spark plug wire from the spark plug.

4. *Close the fuel tank shut-off valve (if the engine has one).*

   This care takes the pressure off the carburetor diaphragms and/or float and prevents fuel leaks.

5. *Store the engine in a dry, protected area.*

   If you do not plan to use your engine within a period of 30 days, drain the tank and carburetor. Then, just before using it, fill the tank with clean, fresh gasoline. This prevents gum from forming in the fuel system.

   If you do not plan to operate your engine for three months or longer, follow procedures given under "Storing Small Gasoline Engines."

STUDY QUESTIONS

1. True or False. Engines should not be stopped suddenly, especially if they are under load.
2. Engines should be allowed to cool at ________ speed before completely stopping.

NOTES

IV. Storing Small Gasoline Engines

Upon successful completion of this section, you will be able to do the following:

- **Tell what can happen to small engines if they are not properly prepared for storage**
- **Give the steps that must be taken to protect small engines from corrosion, moisture, and gum deposits**
- **Select the tools and materials needed to properly store an engine**

Few people bother to prepare their engines for storage. But proper storage of your engine will pay off in convenience, easy starting, good performance and a longer engine life.

Why you should prepare your engine for storage, and how to do it are discussed under the following headings:

1. Importance of Proper Storage
2. Tools and Materials Needed
3. Protecting the Engine from Corrosion, Moisture and Gum Deposits
4. Protecting the Engine from Dirt and Physical Damage

1. IMPORTANCE OF PROPER STORAGE

When engines are properly prepared for storage, they have been known to start and run perfectly after being stored for five years or longer.

The end of the season is a good time to check and clean your engine for two reasons:

1. If you find repairs are needed, you can perform them right away, or make a note to perform them at a future date—but before the next operating season.
2. To remove dirt and grime. When dirt is left on an engine, it tends to collect more dirt during storage.

Furthermore, you will enjoy taking a clean engine, one that is ready for service, out of storage. It will be almost like having a new engine.

Even if a small engine is stored for only six months, you should protect it from the following:

- Corrosion of any metal and iron parts such as the valves, cylinder, piston and piston rings
- Gum which develops from leftover fuel in the fuel system
- Moisture that collects on the outside of the engine and penetrates into electrical system and the inside of the engine
- Dust and dirt which accumulate on the outside of the engine
- Physical damage

These are discussed under the following headings:

a. Corrosion
b. Gum Deposits
c. Moisture
d. Dust and Dirt
e. Physical Damage

a. Corrosion

Corrosion during storage is not so great a problem now as it once was because crankcase oils now contain rust inhibitors which help protect the steel parts. Also many small engine cylinder blocks, cylinders and pistons are now made of aluminum, which is not affected by rust.

In addition, rust flakes accumulating in the fuel tank have been eliminated by several manufacturers. They are now making tanks with plastic and non-corrosive metals.

There is, however, still a chance of steel pistons, piston rings, valves and other steel parts becoming corroded during long periods of storage. Manufacturers continue to recommend that you protect your engine with a rust preventive before storing it.

b. Gum Deposits

Gum deposits clog the carburetor and other parts of the fuel system. This is caused from stale gasoline being left in the engine during storage.

Although today's gasolines have gum inhibitors which slow down the formation of gum deposits, experience has shown that when fuel is drained from the fuel system of a stored engine, it usually starts and runs without difficulty.

This is not always true with engines which have not been properly drained. Many times it is necessary to disassemble and clean the carburetor to remove gum deposits.

Draining the fuel from the fuel system also eliminates possible fire hazards.

c. Moisture

Moisture condenses on the outside and inside of the engine while it is not being used. This can be harmful to the electrical system even when other parts of the engine have been protected by a rust preventive.

Moisture helps to oxidize the breaker points and electrical connections which insulates them. This prevents the proper flow of electricity. It may also cause the condenser to become shorted; then the breaker points arc and become pitted.

Moisture will also cause corrosion of steel parts that are not protected.

d. Dust and Dirt

Dust and dirt, which may collect on the engine during storage, stick to the cooling fins and will interfere with cooling when the engine is put back into operation.

There is a danger of the dirt getting inside the engine when you service it for operation and/or when you remove some part for maintenance or repair. Dirt inside the engine causes rapid wear.

e. Physical Damage

Always store your engine in a safe place away from physical damage.

When stored in a machinery shed or garage, make sure it is out of the way of vehicles and falling objects which might damage the cooling fins and cylinder baffles.

Preparing your engine properly for storage and keeping it in a safe place may also prevent injury to someone who may trip over it.

Proper storage will also prevent the danger of small children starting small engines or equipment which might injure them. This is especially true with engines on such machines as chain saws, portable generators, small tractors, mowers and air compressors.

2. TOOLS AND MATERIALS NEEDED

1. Tools and materials for servicing air cleaners
2. Tools and materials for cleaning engines
3. Tools and materials for maintaining spark plugs
4. Tools and materials for changing crankcase oil
5. Crankcase oil
6. Kerosene
7. Oil can (squirt)
8. Rust preventive, paint and brush
9. Paint, heat resistant enamel
10. Paint brush
11. Plastic sheet for covering engine

3. PROTECTING THE ENGINE FROM CORROSION, MOISTURE AND GUM DEPOSITS

Follow these steps:

1. *Operate the engine until it runs out of fuel.*
   This empties the fuel tank and the carburetor.
2. *Check and clean the engine.*
   Refer to "Cleaning Small Gasoline Engines."
3. *Check and clean the carburetor air cleaner.*
   Refer to "Servicing Carburetor Air Cleaners."

4. *Coat the inside of the engine with a rust preventive.*

   There are two methods recommended for coating the inside of an engine with rust preventive: (a) introducing the rust preventive through the carburetor intake while the engine is running or (b) pouring the rust preventive into the cylinder through the spark plug hole and turning the engine by hand.

   When done properly, the first method (a) distributes the rust preventive better. Most manufacturers, however, recommend the second method (b) because it is simpler and more likely to be done.

   a. If you introduce the rust preventive into the carburetor intake, follow these steps:

      (1) Prepare a mixture of ¼ cup of crankcase oil and ¼ cup kerosene.

      Use SD or SE oil. They have more rust preventive additives than SA or SB types.

      (2) Pour mixture into a squirt can.

      (3) Remove the carburetor air cleaner.

      (4) Start the engine.

      This should be done in a clean area to avoid dirt and dust being drawn into the engine while the air cleaner is removed.

      (5) Put rust preventive mixture into the carburetor air intake with the squirt can (Figure 89).

      (6) Stop the engine.

      If your engine is the four-cycle type, stop it as soon as you see heavy blue smoke coming from the exhaust.

      If your engine is the two-cycle type, close both the fuel shut-off valve and the choke valve. Allow engine to run until it stops. This helps lubricate the internal parts of the engine and also drains the carburetor.

      (7) Drain and replace the oil in the crankcase (four-cycle engines) while the engine is warm.

      Any sludge that may have settled out because of your failure to change the oil at the proper time will drain with the oil.

      (8) Replace the air cleaner.

   b. If you pour the rust preventive through the spark plug hole, follow these steps:

      (1) Fill a squirt can with SD or SE type crankcase oil.

      Do not dilute with kerosene.

      (2) Allow the engine to cool (if it has been running).

      (3) Drain the fuel tank and carburetor. This prevents fuel from diluting the oil in the cylinder.

      (4) Remove the spark plug.

      (5) Squirt a tablespoon of oil into the cylinder through the spark plug hole (Figure 90).

      (6) Turn the crankshaft two or three times by hand.

      This distributes the oil on the cylinder wall, around the valves and on the piston rings. Raise the piston to top position, compression stroke.

5. *Clean, regap and replace the spark plug.*

   For procedures refer to "Servicing Spark Plugs."

6. *Drain the carburetor, fuel strainer and fuel tank.*

   (Gasoline forms gum deposits if left in the engine.)

7. *Place piston in top position on compression stroke.*

   At this point, both valves are closed—protecting the combustion chamber, and most of the cylinder is exposed to the crankcase, which is well oiled.

FIGURE 89. Add a rust preventive to the engine through the carburetor air intake.

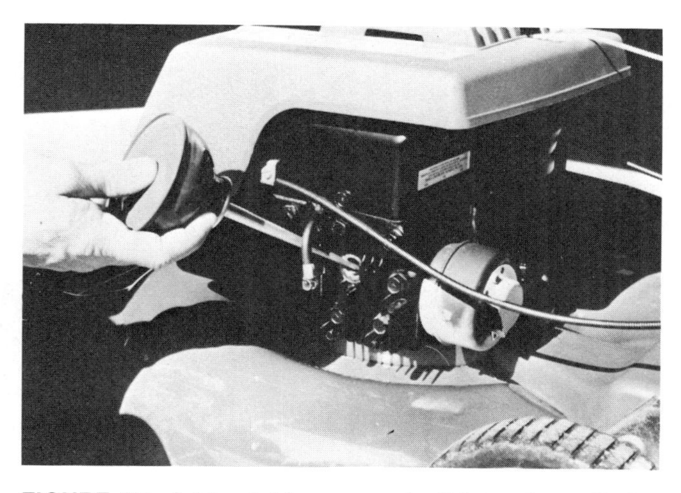

FIGURE 90. Add a tablespoon of oil into the cylinder through the spark plug opening.

Turn the crankshaft until you feel the compression pressure; continue until the pressure is released, then stop.

4. PROTECTING THE ENGINE FROM DIRT AND PHYSICAL DAMAGE

Proceed as follows:

1. *Store engine in a dry, safe place away from vehicles and falling objects that might damage it.*
2. *Cover the engine with a sheet of plastic to protect it from dirt and dust while it is stored* (Figure 91). Make sure not to seal the plastic sheet around the engine. Allow enough air circulation to keep out condensation which might cause harm to the engine.

FIGURE 91. A plastic covering will protect the engine from dirt and dust while it is stored.

STUDY QUESTIONS

1. Why should small engines be properly stored?
2. True or False. The fuel tank should always be drained when storing small engines.
3. Moisture will cause ________ of metal parts not protected during the storage of small engines.
4. During long periods of storage, the inside of a small engine should be protected by a ________ preventive.
5. In what position should the piston be in during proper storage of a small engine.?

ENGINE INFORMATION FORM

General Information ______________________________

____

Name of Equipment (on which engine is mounted) ______________________________

Name and Address of Equipment Manufacturer ______________________________

____

Name and Address of Engine Manufacturer ______________________________

____

Operating Position of Crankshaft ☐ Vertical ☐ Horizontal ☐ Multi-Position

Engine Cycle ☐ 2-Cycle ☐ 4-Cycle

Model Number or Name ______________________________

Serial Number ______________________________

Specification Number ______________________________

Type Number ______________________________

Horsepower ______________________________

TYPES OF ACCESSORIES AND MAJOR UNITS:

Carburetor Air Cleaner: ☐ Oil Bath ☐ Oiled Filter ☐ Dry Filter
Fuel Strainer: ☐ Combination screen and sediment bowl ☐ Screen inside the fuel tank
Crankcase Breather: ☐ Reed valve ☐ Floating disk valve ☐ Air Filter
Starter: ☐ Rope-wind ☐ Rope-rewind ☐ Wind up
☐ Electric, AC ☐ Electric, DC
Ignition System: ☐ Flywheel magneto ☐ External magnet ☐ Battery ☐ Solid State
Fuel Pump: ☐ Mechnically driven ☐ Differential pressure drive
Carburetor: ☐ Float ☐ Suction lift ☐ Diaphragm
Governor: ☐ Air vane ☐ Centrifugal

SERVICE AND MAINTENANCE SPECIFICATIONS

Fuel: Octane Number ______________
Mixture of oil and gasoline - 2-cycle (amount of oil per gallon of gasoline) ☐ 0.12 liters (¼ pint)
☐ 0.24 liters (½ pint) ☐ Other ______________
Oil: SAE Grade ______________ Service Classification ______________
2-cycle ☐ Yes ☐ No
Spark Plug: ______________
Gap setting: ☐ 0.5 mm (.020 in) ☐ 0.6 mm (.025 in) ☐ Other: ______________
Ignition Breaker-Point Gap: 0.3 mm (.012 in) ☐ 0.4 mm (.015 in) ☐ Other ______________

NOTES

Part 2

Maintenance and Repair of Small Gasoline Engines

Preface

This section, *Maintenance and Repair of Small Gasoline Engines*, is designed to bring the reader to a level of competency at the service entrance level of the maintenance and repair of small gasoline engines. Step-by-step procedures are given and illustrated for all the procedures.

From your study of this volume, you will be able to maintain and repair the following:

- Starters.
- Ignition systems.
- Fuel systems.
- Governors.
- Valves.
- Piston-and-rod assemblies.
- Lubricating mechanisms.
- Camshaft assemblies.
- Crankshaft assemblies.

NOTE: SI (metric) equivalents are given in approximate values. SI values are listed first and U.S. customary values are in parentheses. (Tool sizes are not given in metric).

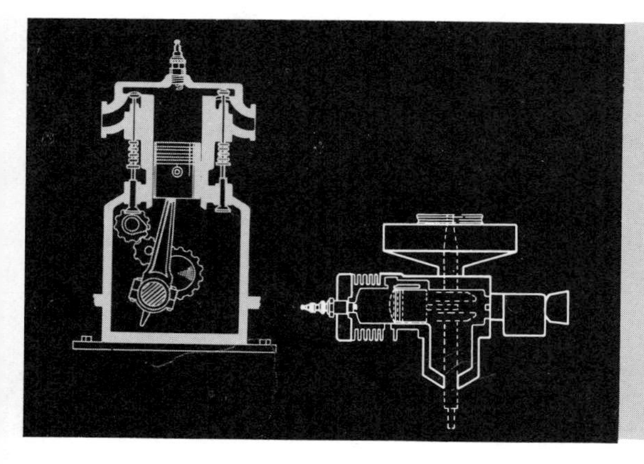

Introduction

The information in this section will help you to expand your knowledge beyond servicing and operating an engine into *major maintenance and repair.* It explains the operating principles of different units such as starters, ignition systems, valves and lubrication systems. It also explains how to tell when such units are not working properly and how to repair them. The principles and procedures are given for each unit in such a sequence that it is not necessary for you to disassemble completely the engine before you start to work on it.

Maintenance & Repair of Small Gasoline Engines

There is much personal satisfaction in understanding the more detailed operating principles of small engines, in becoming familiar with the common structural parts and their functions, and in knowing how to do the more difficult maintenance and repair jobs. Small engines are similar in principle and in operation to large internal combustion engines. Consequently, you gain a knowledge and an appreciation of how to maintain and repair large engines such as those used in trucks, tractors and automobiles.

You **save both time and money** by doing certain maintenance and repair jobs, or by knowing when they should be done. After you get some experience, you will be able to determine whether or not it is more economical to replace a complete assembly or to purchase parts and repair it. For example, an accessory such as a starter or a carburetor can usually be repaired at a reasonable cost. If you have trouble, however, with such parts as a piston, piston rod or crankshaft, it may be more economical for you to buy a "short block" assembly (Figure 1).

783

FIGURE 1. A "short block" assembly which may be purchased as a unit.

A short block assembly is a factory-built cylinder-block assembly which includes the crankcase, piston, piston rings, connecting rod, valves, cylinder and camshaft. It comes ready for you to add the remaining parts, such as the magneto, cylinder head, carburetor and starter. The cost is approximately one half that of a new engine—completely assembled, and it is approximately twice the cost of a new crankshaft.

There are other advantages to maintaining your small engine. You can **make your engine start more easily and run more smoothly, and develop its maximum horsepower.** In addition, there are maintenance jobs which, when done in time, will prevent serious damage to the engine and will make it last much longer.

PRINCIPLES OF GOOD WORKMANSHIP

A systematic approach is best for doing any job, but it is of particular importance when working on engines, A part left out, or incorrectly assembled, may cause more damage than existed originally.

Before working on your engine, *learn as much as you can about it.* Get the operator's manual, repair or service manual and the parts manual. You need these manuals for finding the specifications for your particular engine. Read and follow the instructions. Study the types of accessories on the engine and the functions of the various parts. These are also discussed in this text. In addition, the step-by-step procedures given here are a helpful guide for disassembling, reassembling and making adjustments on all engines. Reasons for doing certain jobs and using certain methods are explained.

Prepare a clean place to work. This helps prevent parts from collecting dirt. If you put dirty parts in the engine, it will not last long because dirt is an abrasive and causes wear.

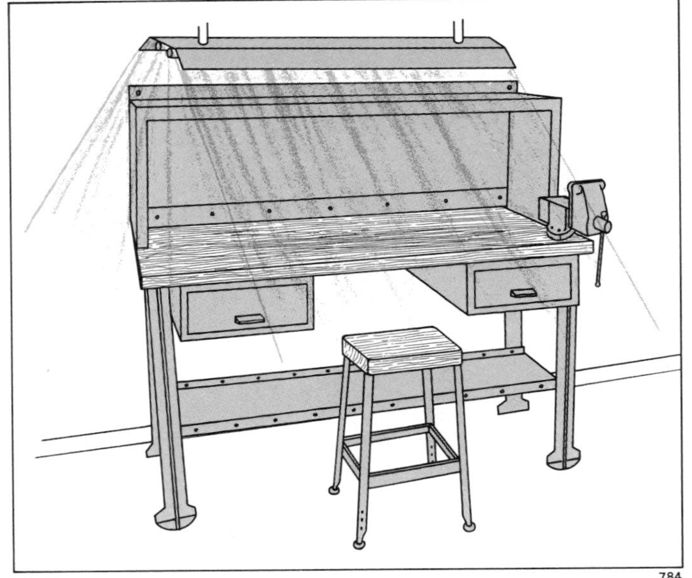

784

FIGURE 2. Provide a clean, convenient, well-lighted work area.

Use a workbench if possible (Figure 2). It provides a place to lay out parts and tools, and helps keep them clean. Also, working on a workbench is much more comfortable than working on the floor.

Provide plenty of light. Two 150-watt reflector-type bulbs or two 120 cm (48-in), 40-watt fluorescent tubes mounted 120 cm (48 in) above the work bench will give enough light spread—without shadows.

Get an engine stand, if possible (Figure 3). A stand holds the engine securely while you remove most of the parts.

Assemble the proper tools, materials and equipment before starting a job. The basic tools and materials needed for each job are listed in the text. You may need some special tools, however, in addition to this list.

Clean the engine before attempting to repair it. You can remove most of the dirt and grime from an engine before removing many parts. This makes the repair job much cleaner and reduces the possibility of getting harmful dirt particles inside the engine. For procedures, refer to "Cleaning Small Engines," Part 1.

Lay the parts out in order of their disassembly, and let them remain in this order throughout the cleaning and inspection process (Figure 4). This will assure correct reassembly. Another good practice is to **draw a sketch of the assembly be-**

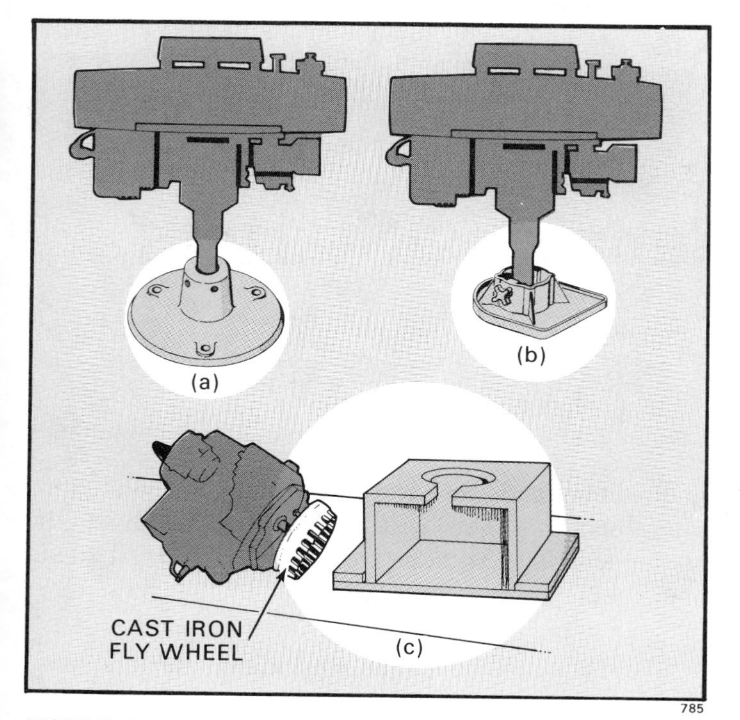

FIGURE 3. (a and b) Typical commercial engine stands used for holding the engine by the crankshaft while working on it. (c) A stand made from wood. Also a cast-iron flywheel bored for bench-testing vertical-shaft engines.

fore, or while, you take it apart. Then if the parts become mixed, you will have a record of how to put them together again.

Order parts as soon as you are aware of the need for them. When ordering parts for your engine, be sure to give your dealer sufficient information for him to determine the correct parts. Manufacturers of small engines make hundreds of variations in their models according to equipment manufac turer's needs. Therefore it is necessary for the dealer to have the model number and serial number to determine which part fits your engine. A type number is also given on some engines. You will usually find these numbers on the engine nameplate (Figure 5) or stamped on the cooling shroud. If the number is difficult to read, try rubbing it with a piece of chalk. Sometimes this brings out the raised or indented numbers.

If you can, take the old part for comparison when purchasing new parts. If you have access to the part number, give it to the dealer also.

While working on your engine, look for conditions that may cause future trouble. Check unusual wear or damage. You may be able to prevent a future breakdown by making an adjustment or repairing a part before it has failed.

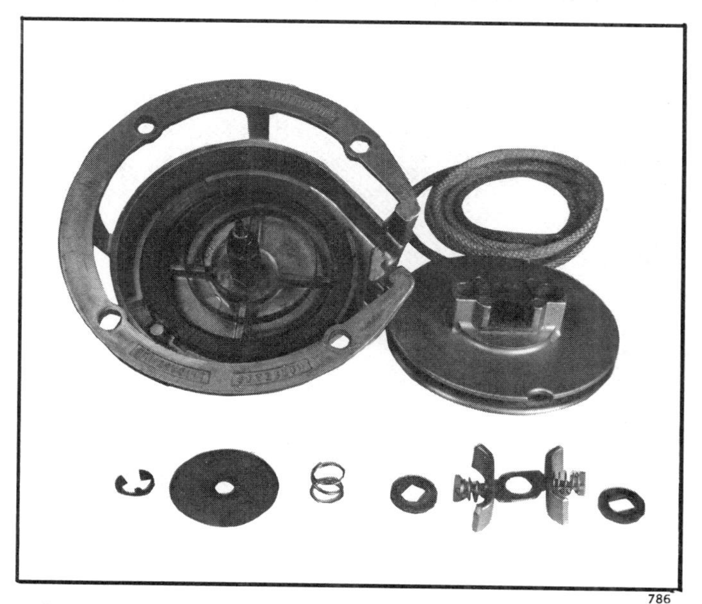

FIGURE 4. Always lay parts out in order of their disassembly.

The information presented in this section is included under the following headings:

I. Repairing starters.
II. Maintaining and repairing ignition systems.
III. Repairing fuel systems.
IV. Repairing governors.
V. Repairing valves.
VI. Repairing cylinders and piston-and-rod assemblies.
VII. Repairing lubricating machanisms in 4-cycle engines.
VIII. Repairing camshaft assemblies in 4-cycle engines.
IX. Repairing crankshaft assemblies.

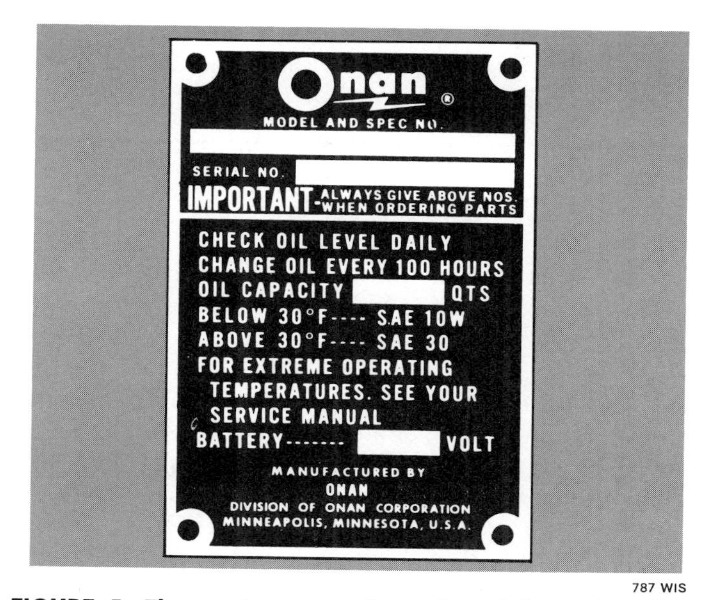

FIGURE 5. The engine nameplate gives information for ordering new parts.

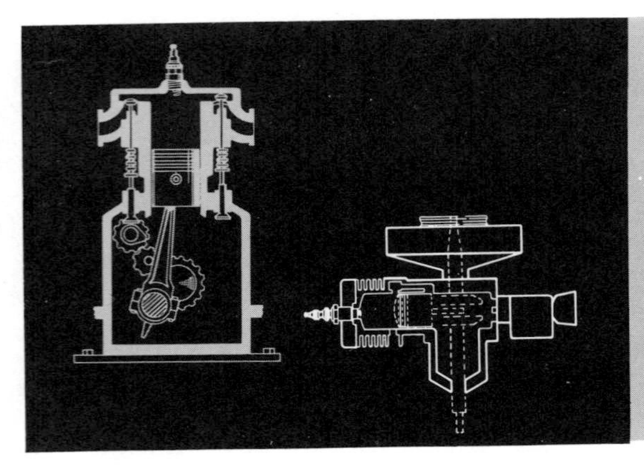

I. Repairing Starters

Starters take much punishment. Most operators neglect to maintain their small engines properly. As a result, they are difficult to start. The starter is overworked. For this reason, maintaining and repairing starters are discussed first in this section.

The information on starters applies to **both 4-cycle and 2-cycle engines.**

A good way to keep your starter maintenance at a minimum is to keep your engine in good running condition. This will not prevent, however, all starter troubles. Starters are subject to normal wear and occasional breakdown.

To do a good job when working on starters, you should first have an understanding of the *function of starters* and of the *different types* used on small engines.

FUNCTION OF THE STARTER

The purpose of the starter on an engine is to provide a means for turning the crankshaft until the combustion cycle is started. Once the cycle is started, it continues to repeat itself—the engine runs. The events that take place during each cycle are explained fully under "How Small Gasoline Engines Work," Part 1.

For starting, the crankshaft is *turned fast enough by the starter* for a combustible fuel-air mixture to be drawn into the combustion chamber, compressed and ignited. Once this takes place, the starter is released and the engine runs on its own.

A starter must have:

- A source of power—manual or electric.
- A drive mechansim for engaging the crankshaft during cranking and disengaging the crankshaft after the engine has started.
- A means for transmitting power from the source to the drive mechanism—usually a rope and pulley, V-belt and pulley, a gear drive, or a friction drive.

TYPES OF STARTERS

The starting functions are accomplished by two basic types of starters: **manual** and **electric.**

Most small engines are started by turning the flywheel with a *manual starter* (Figure 6a). *Electric Starters* (Figure 6b), however, are becoming more and more popular—especially on engines of 3.75 kW (5 hp) or more.

There are three common types of **manual starters:**

- Rope-wind (Figure 10).
- Rope-rewind or recoil (Figure 7a).
- Windup, or impulse (Figure 7b).

The **rope-wind starter** is the simplest of starters. It consists of a rope which has a knot on one end and a T-handle secured by another knot on the other end. For starting, the rope is wound around a flange on the flywheel and the engine is turned with a strong, hard tug of the rope.

The **rope-rewind starter** (Figure 7a) operates similarly to the rope-wind starter except it is a

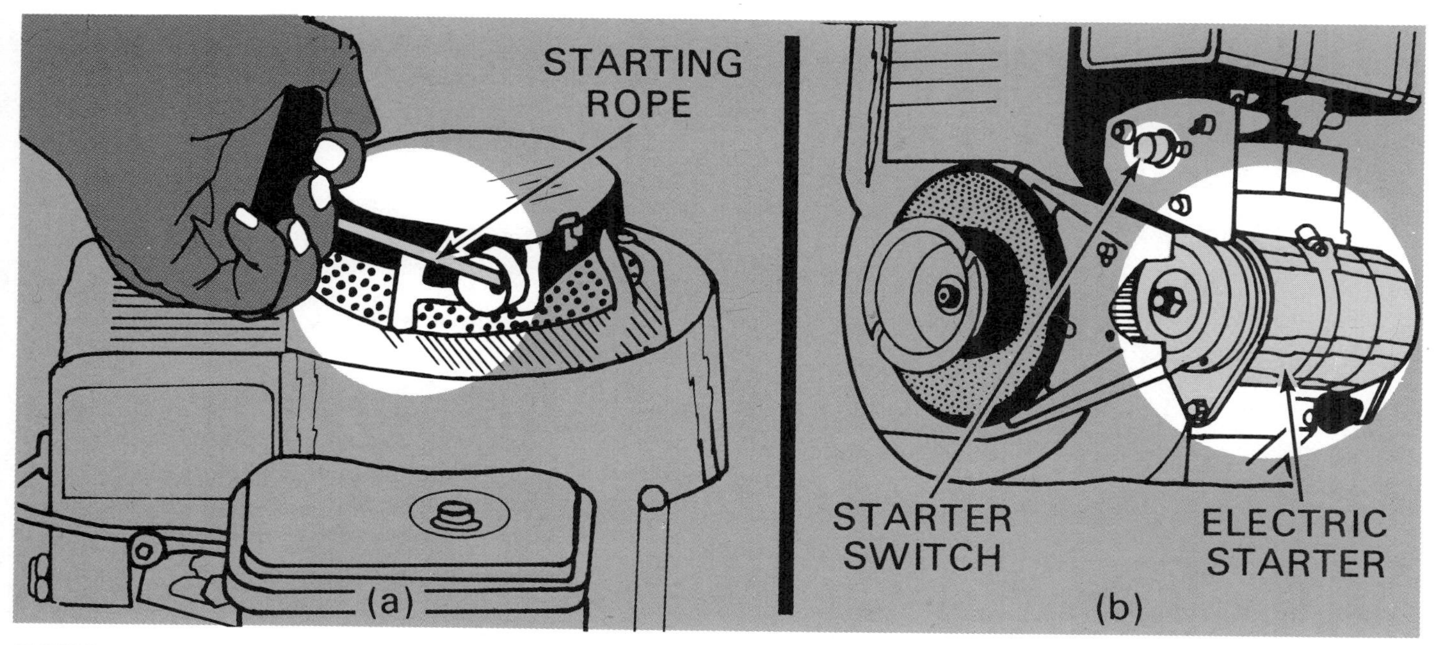

FIGURE 6. (a) One type of manual starter—rope-rewind. (b) An electric starter—combination starter-generator.

little more automatic. After each pull, the rope is automatically rewound by a recoil spring. Since the rope is permanently attached, there is a means of engaging and releasing the starter drive. A newer concept in rope-rewind starters is the **vertical-pull, rope-rewind, gear-driven starter** (Figure 7b).

The **windup starter** (Figure 7c) has no rope. It is an impulse-type mechanism which consists of a *heavy recoil spring, crank-holding rachet* and a *release mechanism.* When starting the engine, you wind up the spring with the crank; then you release the spring which spins the crankshaft.

The advantages and disadvantages of all three manual-type starters are listed in Table 1, page 80.

The use of **electric starters** is becoming more and more popular on small engines—especially on engines of 3.75 kW (5 hp) and above, such as those used on small tractors. Electric starters practically elimi-

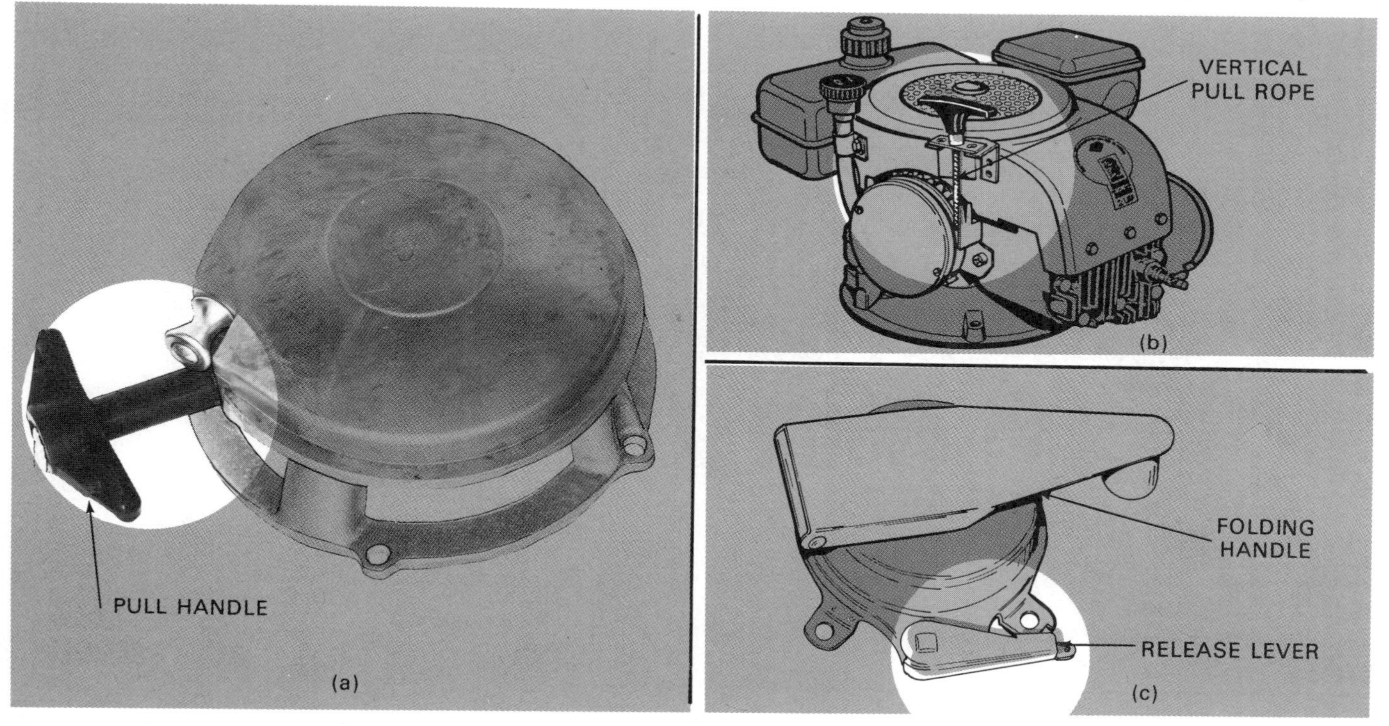

FIGURE 7. (a) Horizontal rope-rewind starter. (b) Vertical-pull, rope-rewind starter. (c) Windup starter.

TABLE I. ADVANTAGES AND DISADVANTAGES OF DIFFERENT TYPES OF MANUAL STARTERS

Type of Manual Starter	Advantages	Disadvantages
Rope-wind	Simple and economical Only maintenance required is to reknot and/or replace rope	Takes time to wind rope and requires, much physical effort to use Loose rope is easy to misplace and/or lose Danger of pulling engine over
Rope-rewind Ratchet Gear	Rewinds itself; saves time Plastic gear gives less trouble.	Requires occasional maintenance Engine cannot be started unless starter is operative Danger of pulling engine over
Windup	Requires very little physical effort	Some types are nonreparable Engine cannot be started when starter is out of order

nate manual labor for starting and, when maintained properly, provide a fast and convenient means for starting your small engines. An electric starter will save you much time if you are doing a job that requires frequent stopping and starting such as operating an elevator or a garden tractor (Figure 8).

Some electric starters are designed to operate on alternating current and some on direct current.

Alternating-current starter motors operate on 120-volt current from the home wiring system. An extension cord is used to connect the starter to a convenience outlet (Figure 9a).

Many **direct-current starters** (Figure 9b) are similar to the ones you have on your tractor, truck or automobile. Most early electric starters on small engines, however, are designed to start the engine and then to act as a generator when the engine is running. These types of starter-generators are discussed on page 42. Starter-generators are not as popular now as flywheel alternators and separate gear-drive starters.

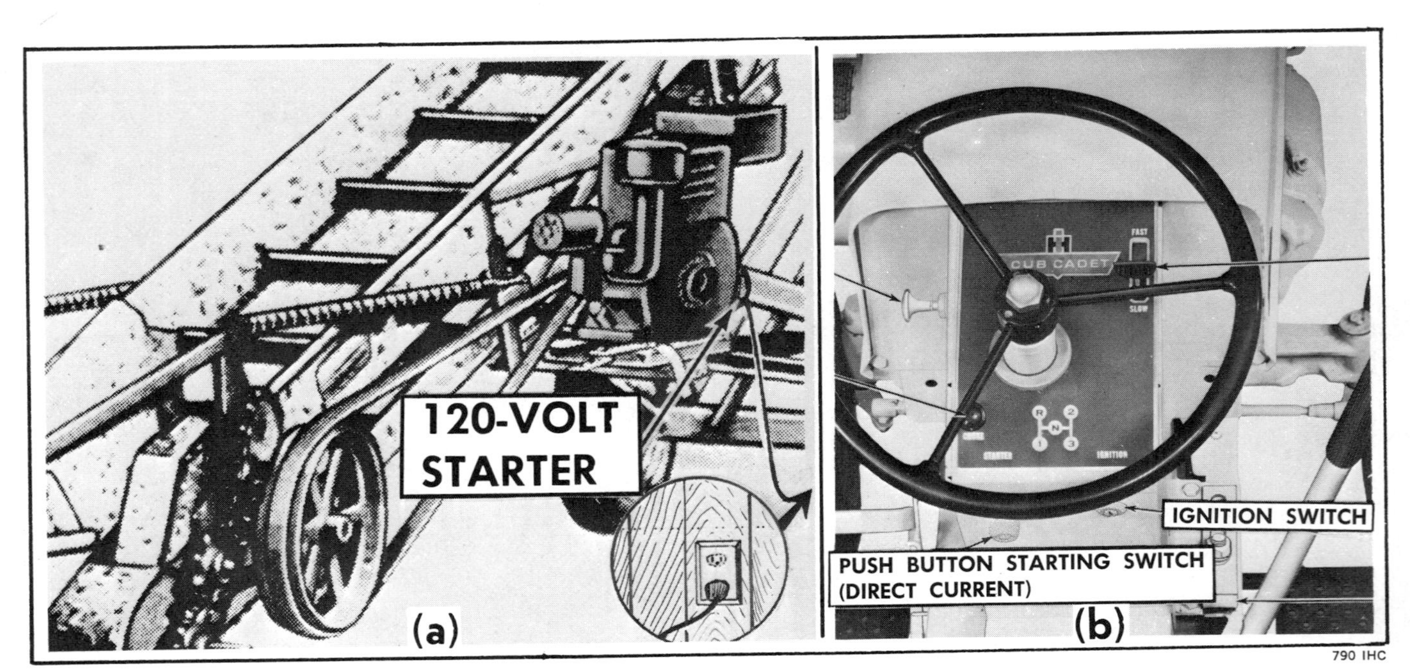

FIGURE 8. Applications of an electric starter. (a) A 120-volt alternating current electric starter on a small engine operating an elevator. (b) A 12-volt battery-operated direct-current starter on a garden tractor.

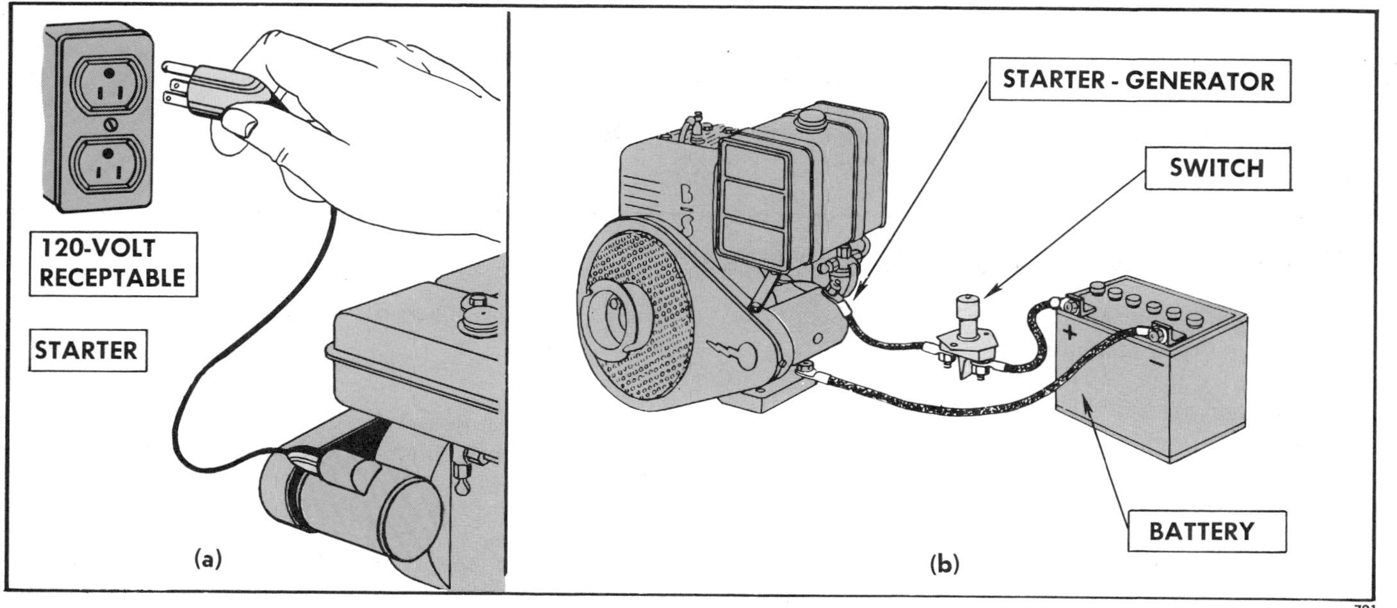

FIGURE 9. Sources of power for electric starters are (a) 120-volt alternating current from the home wiring system, and (b) 6- or 12-volt direct current from a battery-and generator system.

How starters work and how to maintain and repair them are discussed under the following headings:

A. Repairing rope-wind starters.

B. Repairing rope-rewind starters.

C. Repairing windup starters.

D. Repairing 120-volt alternating-current starters.

E. Repairing direct-current starting and generating systems.

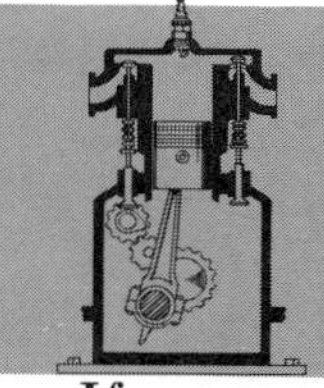

A. REPAIRING ROPE-WIND STARTERS

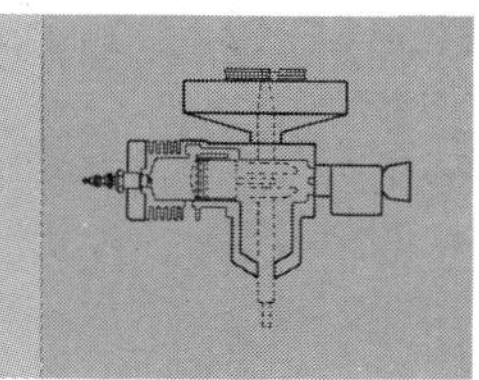

If your manual starter is the rope-wind type (Figure 6a), you will have little or no maintenance problem except with the rope. The information you need is given under the following headings:

1. Principles of operation.
2. Tools and materials needed.
3. Checking for proper operation.
4. Reknotting or replacing the rope.

PRINCIPLES OF OPERATION

Provisions for turning the crankshaft consist of a rope which is wound onto a pulley. The pulley is attached to the end of the crankshaft or on the flywheel. The rope is long enough for three to five turns on the pulley. This allows the crankshaft to be turned three or four complete revolutions during cranking—which should be sufficient for starting the engine.

One end of the rope is knotted and is engaged in a slot in the pulley flange for cranking. The slot is slanted away from the direction of rotation (Figure 10). When the engine starts, the rope is released.

A T-handle is attached to the opposite end of the rope for pulling by hand.

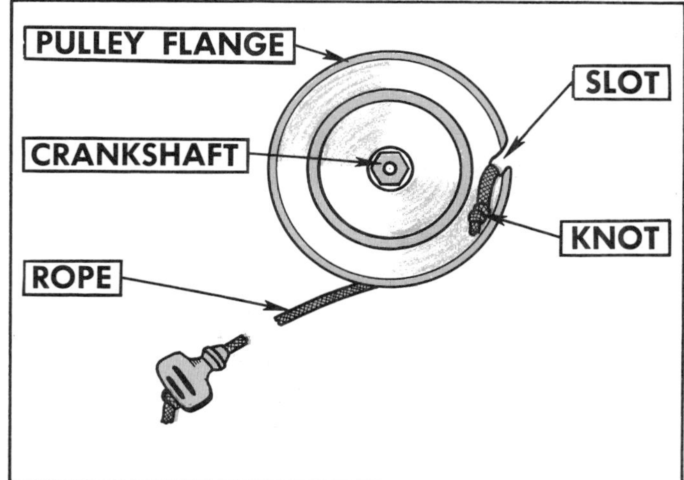

FIGURE 10. Starting rope and pulley. Slanted slot in the pulley flange is for engaging the rope during cranking and for disengaging it after the engine starts.

TOOLS AND MATERIALS NEEDED

1. Pocket knife or diagonal cutting pliers
2. Replacement rope
3. Matches

CHECKING FOR PROPER OPERATION

1. *Disconnect the spark-plug wire to prevent the engine from starting.*

2. *Engage the knot in the slot which is in the pulley flange.*

3. *Wind the rope onto the pulley.*

4. *Pull the rope slowly—turning the crankshaft —until the piston begins compression stroke.*

   You can tell by the increased resistance.

5. *Rewind the rope all the way.*

6. *Brace one hand against the engine to prevent pulling it over, if on light-weight equipment.*

7. *Pull the rope all the way with a sharp, firm pull.*

   The knot should hold firmly until the rope is completely unwound and released from the pulley.

   If for any reason the rope does not release and the engine starts, the rope will rewind on the pulley in the opposite direction. It is important that the rope does not bind.

   If the knot does not hold, remove and replace the knot; or if the rope breaks, replace it. See procedures that follow.

REKNOTTING OR REPLACING THE ROPE

1. *Select the proper size and length of rope.*

   Use a 0.5 cm ($\frac{3}{16}$ in) diameter nylon-braided rope. Usually starting ropes are 120 cm to 150 cm (14 ft to 5 ft) long. The length should be *long enough to wind around the flywheel flange pulley 3 to 5 times, plus about 15 cm (6 in) for lead*

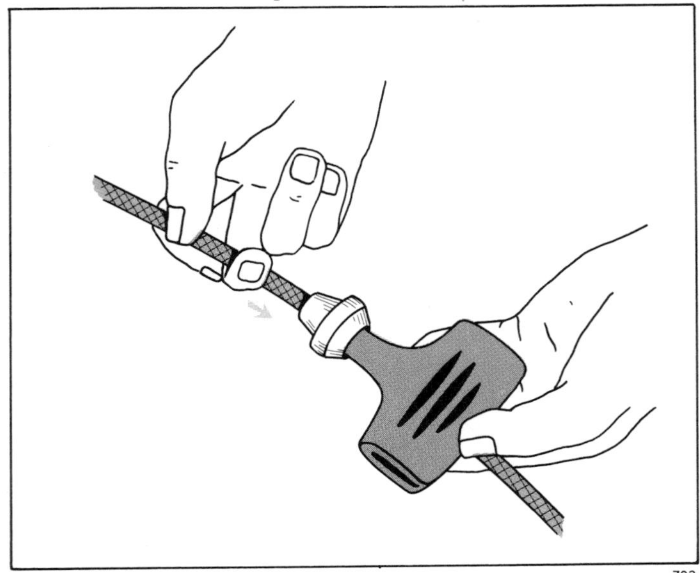

FIGURE 11. Insert the rope into the T-handle before knotting.

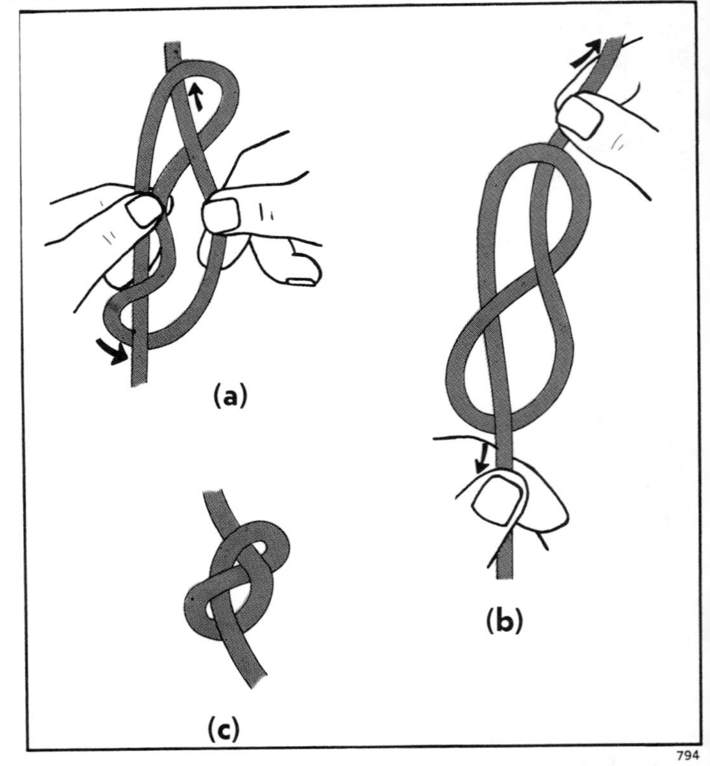

FIGURE 12. Procedures for making a "figure-of-eight" knot. (a) Form loop, wrap loose end around the rope and then feed the loose end through the loop. (b) Tighten knot. (c) Completed knot.

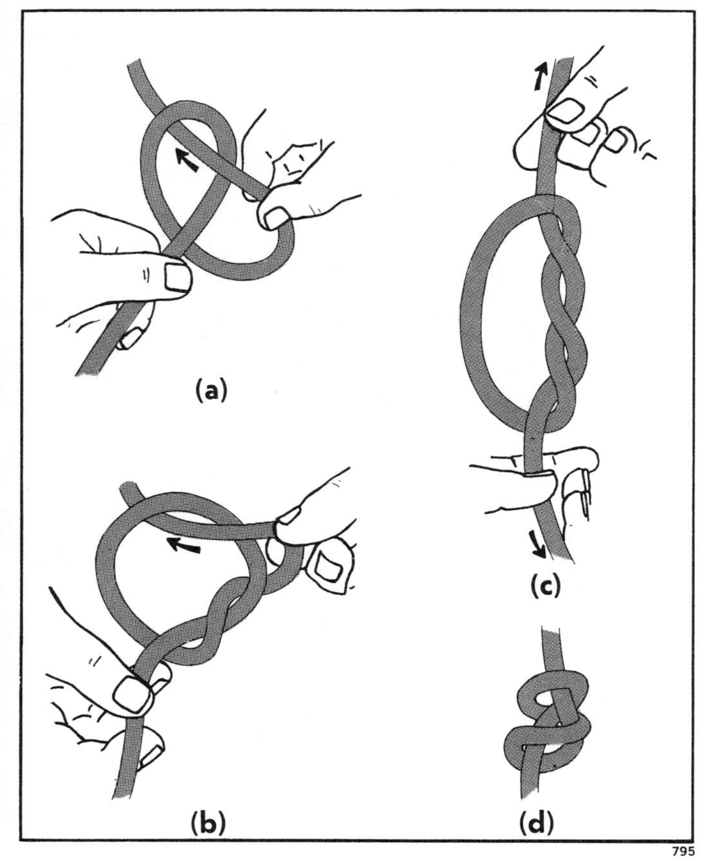

FIGURE 13. Procedures for making a "double-overhand" knot. (a) Feed end through loop. (b) Feed end through a second time. (c) Tighten knot and pull tight. (d) Completed knot.

between the pulley and T-handle, and *with an additional 15 cm (6 in) for knots.*

2. *Seal frayed ends of nylon rope by singeing.*

   Hold a lighted match under the end of the rope until it is fused. This prevents fraying.

CAUTION! Do not light a match near an open container of gasoline or spilled fuel.

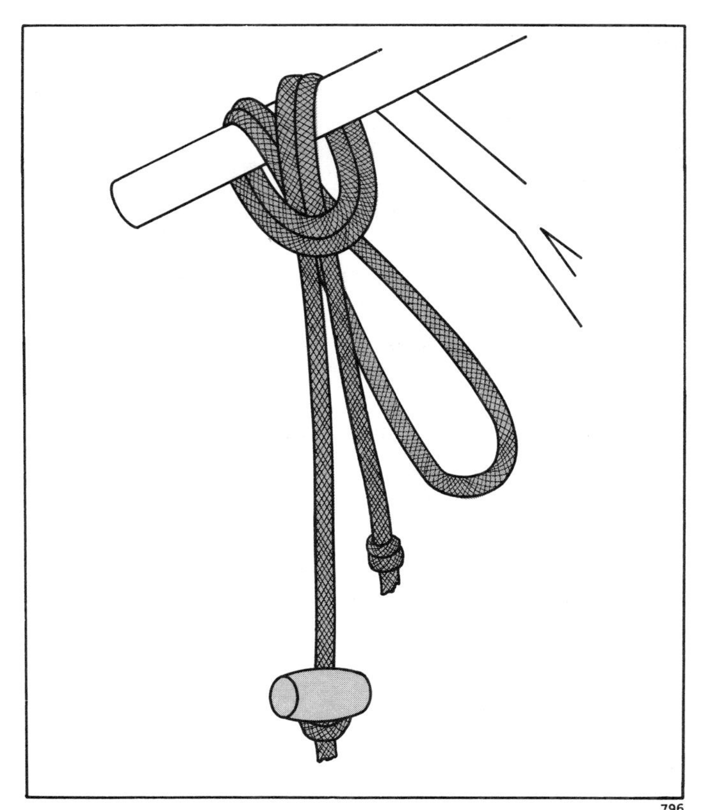

FIGURE 14. A method for stowing the crankrope.

3. *Insert one end of rope into the T-handle (Figure 11).*

4. *Knot both ends of the rope.*

   Use either a "figure-of-eight" knot (Figure 12), or a "double-overhand" knot (Figure 13).

5. *Stow rope on equipment (Figure 14).* Double rope twice. Make a "girth-hitch" knot and loop it over the handle bar or some safe part of the equipment (Figure 14).

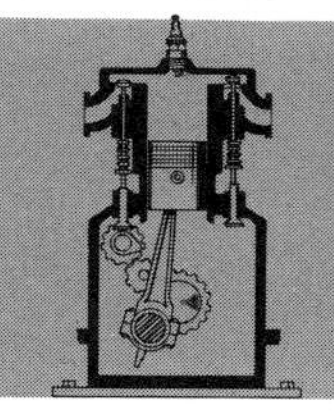

B. REPAIRING ROPE-REWIND STARTERS

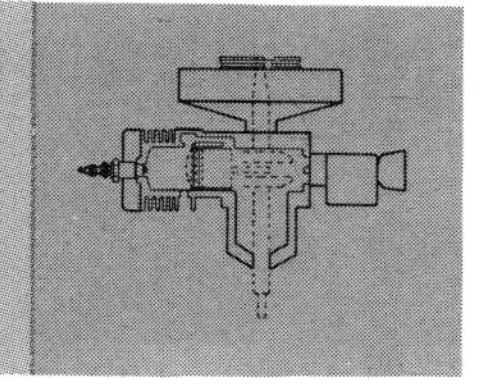

Dirt and grease, if allowed to accumulate in the starter, will cause wear and eventual failure. The springs will not flex properly and the parts will wear faster. One good cleaning each season will pay off in longer trouble-free service.

When you are repairing your starter, follow the instructions given by the manufacturer; but if they are not available, the discussions and procedures under the following headings will be helpful:

1. Principles of operation.
2. Tools and materials needed.
3. Checking for proper operation.
4. Replacing the rope.
5. Repairing the drive mechanism.
6. Replacing the recoil spring.

PRINCIPLES OF OPERATION

The rope-rewind starter saves you the trouble of rewinding the rope on the flywheel pulley each time you crank the engine. But you still have to pull the rope. Here is how it works (Figure 15):

When you pull the rope, the **starter pulley** is turned. The **pawls** fly out by centrifugal force and engage the crankshaft adapter so the engine is cranked as the rope is pulled.

At the same time, the **recoil spring** is wound because the inside end of the spring is attached to the pulley and the outside end to the starter housing.

When you release the rope, whether the engine has started or not, the **pulley** is turned in the opposite direction by the tension built up in the recoil spring. The rope is rewound on the pulley.

The **pawls** retract and a small spring holds each of them in the retracted position.

Procedures for repairing the drive mechanism are given under "Repairing the Drive Mechanism," page 90.

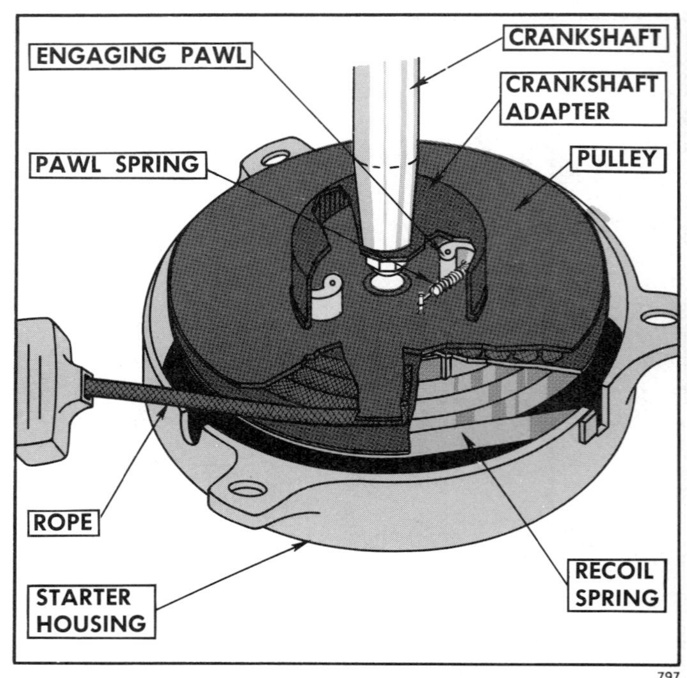

FIGURE 15. Basic parts of a rope-rewind starter are rope, pulley, recoil spring and drive mechanism.

TOOLS AND MATERIALS NEEDED

1. Vise-grip pliers or adjustable end-wrench (depending on your type starter)
2. Wire paper clip
3. Slot-head screwdriver — 8″
4. Nut driver — 7/16″
5. Long-nose pliers — 7″
6. Phillips-head screwdriver — 6″
7. Vise — 4″
8. 8-inch C-clamps (2)
9. 3/4″ x 3″ x 5″ woodstock
10. Small amount of multi-purpose grease
11. Matches
12. Rope
13. Gloves
14. Petroleum solvent (mineral spirits, kerosene, or diesel fuel)

CHECKING FOR PROPER OPERATION

1. *Disconnect the spark-plug wire to prevent the engine from starting.*
2. *Grasp the T-handle with one hand and brace yourself against the engine with the other.*
3. *Pull the rope out slowly (Figure 16).*

   The rope should unwind freely. **Binding** or **hitching** indicates the rope or *recoil spring* is *jammed.*

   If the engine crankshaft does not turn as the rope is pulled out, *the starter drive mechanism* is not engaging. It will need to be repaired or replaced. See procedures under "Repairing the Drive Mechanism," page 20.
4. *Inspect the rope for wear.*

   If the rope breaks, or is worn, replace it. See procedures under "Replacing the Rope," next heading.
5. *Allow rope to rewind.*

   Do not turn T-handle loose and allow it to fly back freely. There is a danger of break-

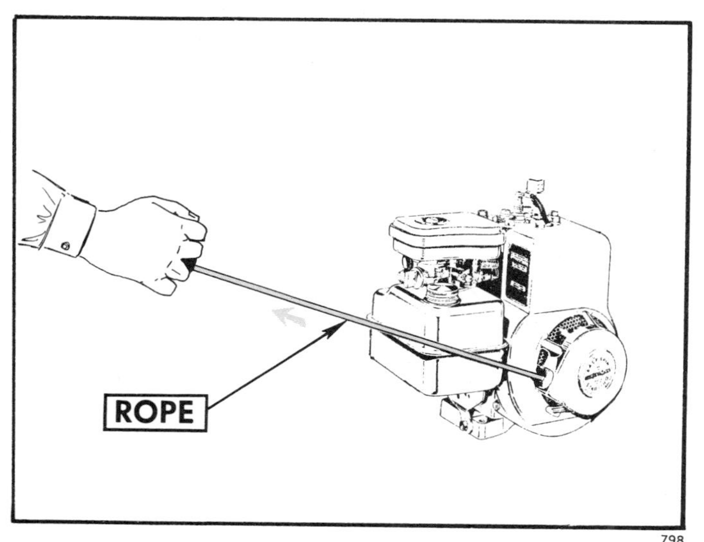

FIGURE 16. Pull the starter rope out slowly for inspection.

ing the recoil spring. This action also causes undue wear on the rope.

If the rope does not rewind, there may be several reasons:

- The rope and/or pulley may be binding.
- The spring may be bent, broken or disengaged.
- There may be insufficient spring tension.
- The starter may be assembled improperly.

Follow procedures under "Replacing the Rope," page 85 and, if necessary, "Replacing the Recoil Spring," page 93.

REPLACING THE ROPE

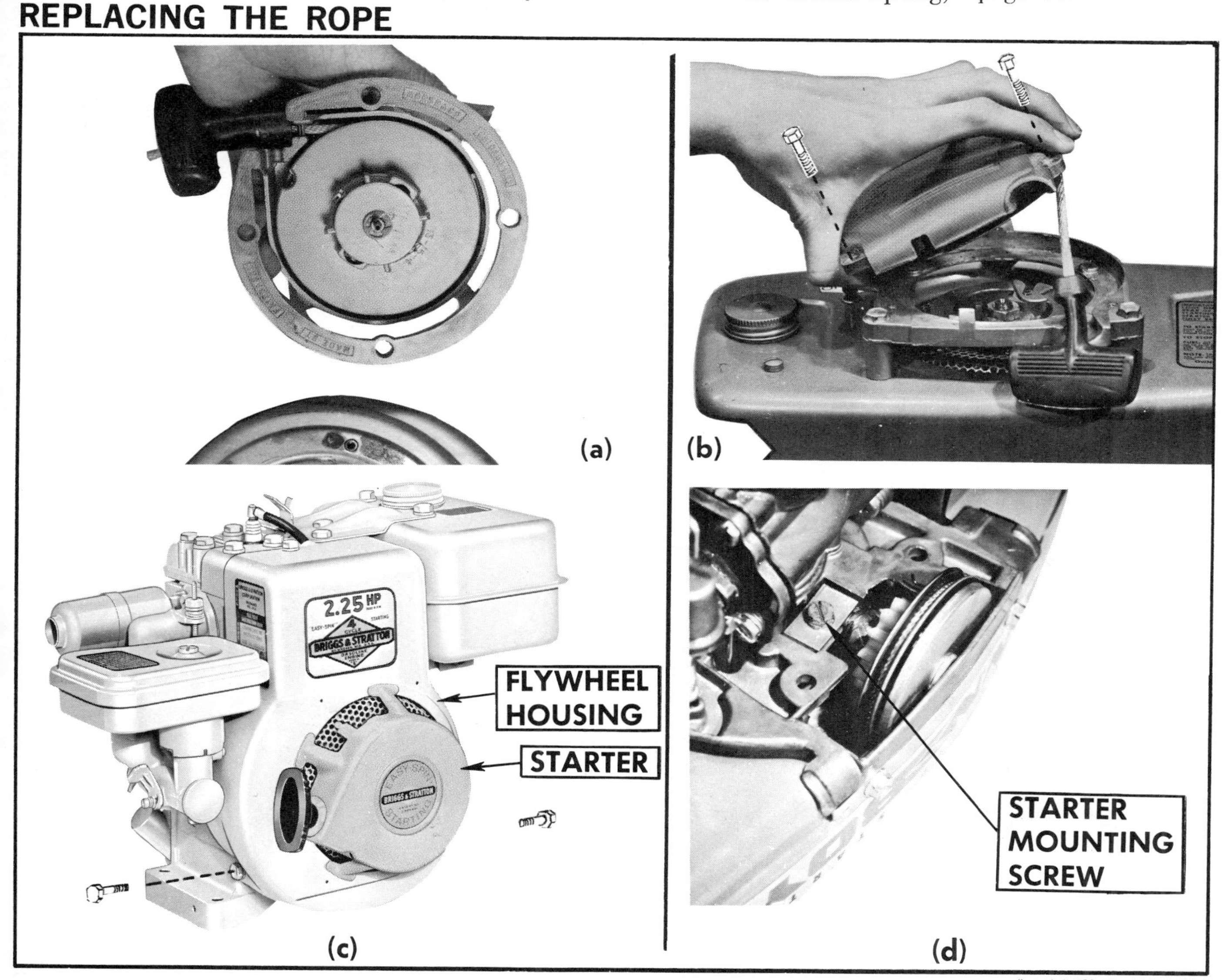

FIGURE 17. Common locations and methods for mounting rewind starters. (a) and (b) Starter surface-mounted and held by cap screws. (c) Starter surface-mounted but as an integral part of the flywheel housing. The flywheel housing must be removed before removing starter. (d) Starter partially enclosed by engine block casting and held by clamp and capscrew.

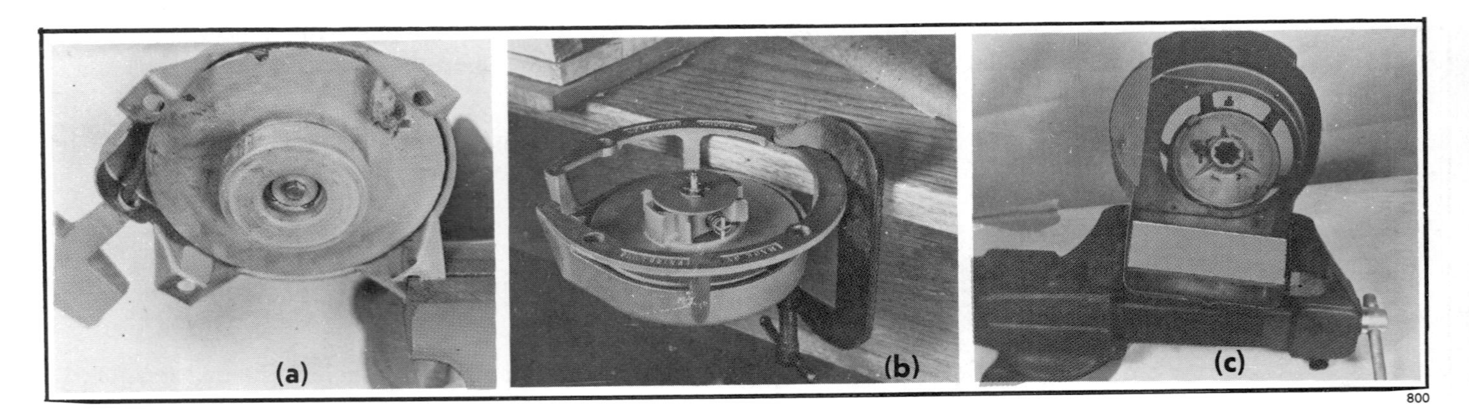

FIGURE 18. Secure starter housing before attempting to work on the starter. (a) Starter secured in a vise. (b) Starter secured by a C-clamp. (c) Starter and shroud held by a specially cut wooden block, and a vise.

1. *Determine how the starter is mounted.*

   Most starters are attached to the flywheel housing by three or four screws (Figure 17a and b). Some are enclosed in the flywheel housing (Figure 17c). Figure 17c illustrates a type of rewind starter that is secured by a mounting screw and clamp.

2. *Gain access to the starter if necessary.*

   It may be necessary to remove some part of the equipment or a part of the engine cowling.

3. *Remove the starter.*

   The starter is removed as a self-contained unit. The spring will not unwind.

4. *Clamp starter housing in a vise (Figure 18a), or use C-clamps (Figure 18b).*

   Use vise-jaw protectors to prevent damage in the starter housing.

FIGURE 19. A type of starter on which the rope can be replaced without disassembly. Note knot on the outside of the pulley.

5. *Remove and replace the rope.*

   There are two basic methods for replacing the rope on rewind starters. The method you use will depend on the starter design.

   If the knot is visible (Figure 19), you can most likely replace the rope without disassembling the starter. Proceed with steps under "a."

   If the knot is not visible, it will be necessary to disassemble the starter. Proceed with steps under "b."

a. If your starter is designed so you can **replace**

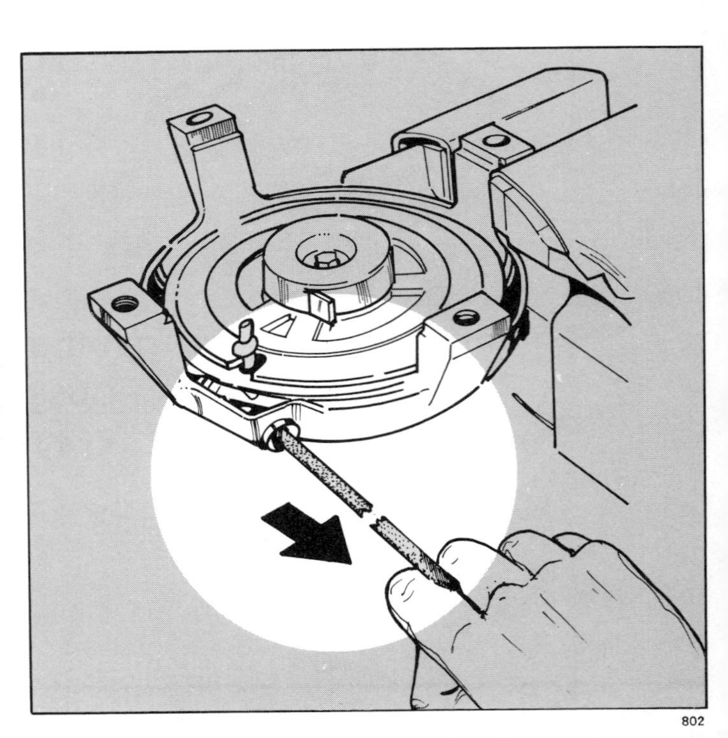

FIGURE 20. The first step in removing the starting rope: Pull the rope all the way out.

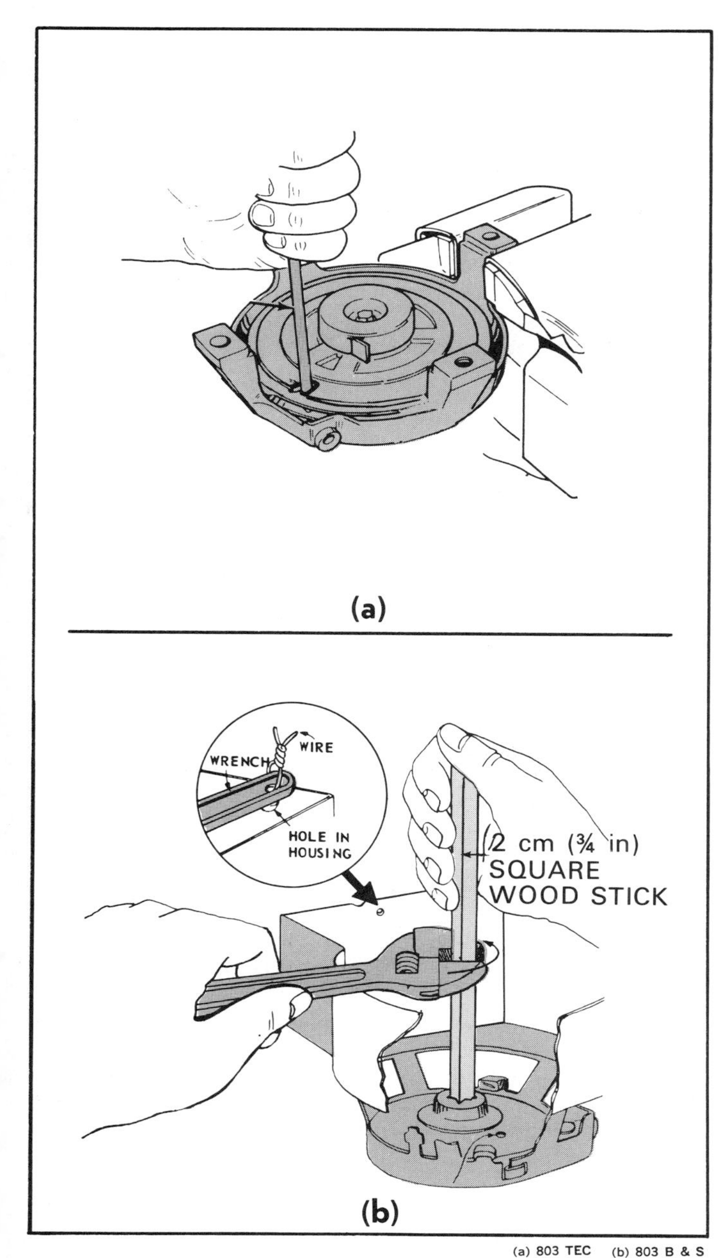

Figure 21. Rewinding the recoil spring. (a) Some starters may be rewound by a screwdriver. (b) Some require special tools.

the starter rope without disassembling the starter, proceed as follows:

(1) *Pull the rope all the way out (Figure 20) if not broken.*

(2) *Clamp the pulley with vise-grip pliers (Figure 22) or a C-clamp, or tie the wrench handle (Figure 21b, inset).* This keeps the spring from unwinding and holds the pulley in position for installing the rope.

(3) *Cut the knot and slide the rope out (Figure 23).*

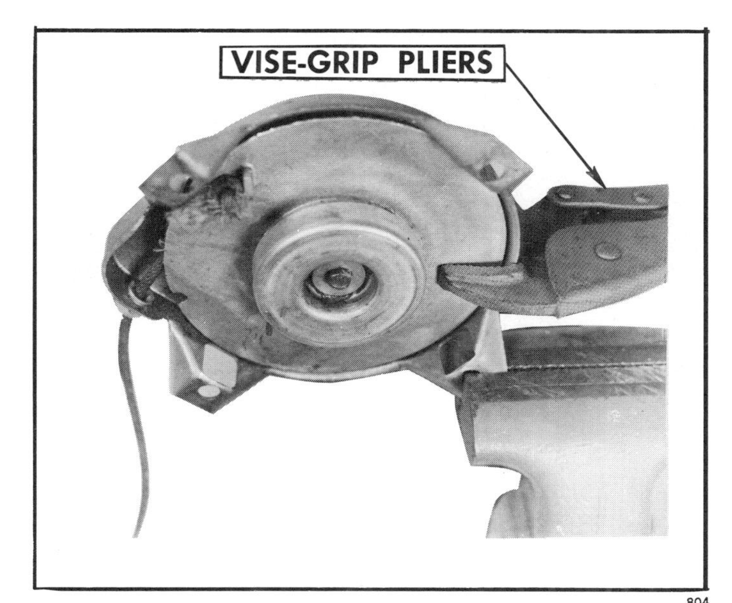

FIGURE 22. Use of vise-grip pliers to hold pulley.

If the old **rope is broken,** it will be necessary to wind the recoil spring before installing the new rope (Figure 21).

Follow instructions in your service manual if possible. If one is not available, wind the starter spring completely — as far as it will wind — then release one complete turn. This protects the spring from being overwound when the rope is pulled. Otherwise, the spring may break when you operate the starter.

FIGURE 23. Cut the knot and slide the rope out.

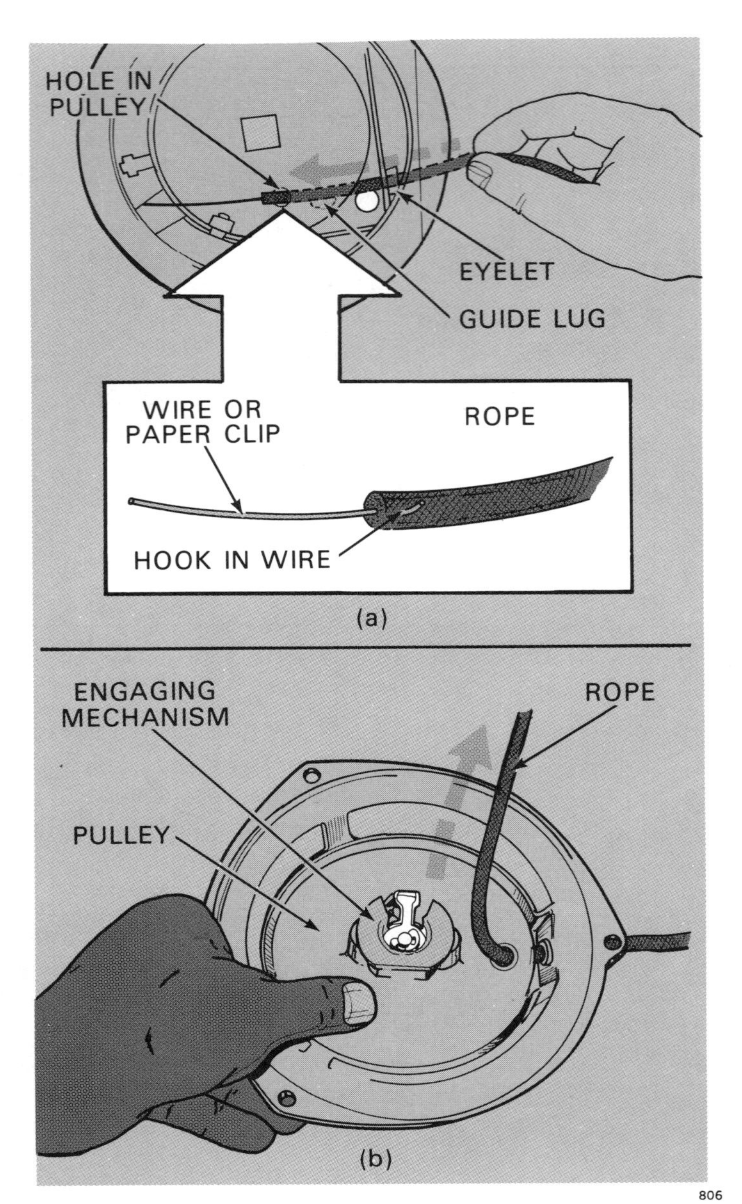

FIGURE 24. Threading the rope. (a) A wire or paper clip makes a convenient guide for threading the rope. (b) Sometimes no guide is needed if the holes are aligned.

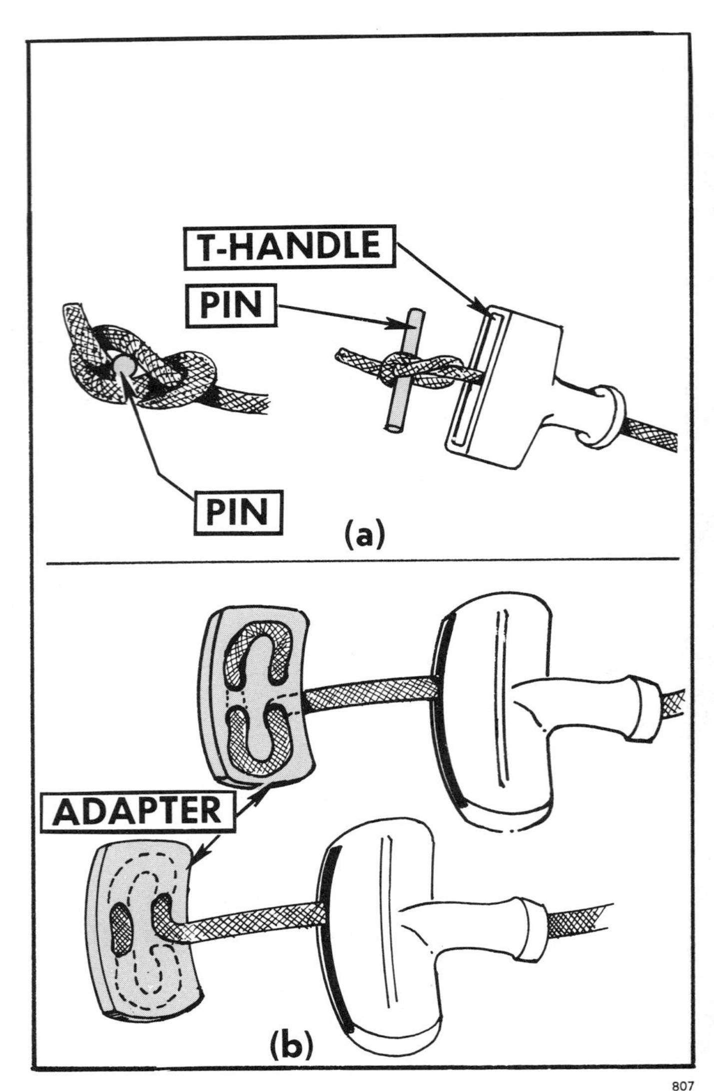

FIGURE 25. Methods for securing the starter rope T-handle. (a) A knot and pin. (b) A special insert adapter. Rope threads into insert and is held by clamping action.

(4) *Select new rope.*

Rope should be the same length and diameter as that which was used originally. Rope lengths for this type of starter vary from 90 cm to 120 cm (3 ft to 4 ft). If you do not know which length is correct, start with at least 150 cm (5 ft) and adjust the length as directed in steps 10 and 11.

Their diameters vary from .3 cm to 1.0 cm (⅛ in to ⅜ in). The most common diameter rope used is 0.4 cm ($^5/_{32}$ in). The rope should fill the pulley groove without binding.

(5) *Singe the end of nylon rope with a match flame.*

This prevents fraying.

(6) *Thread rope through housing eye and through pulley eye (Figure 24).*

It is helpful to use a paper clip or a small piece of wire, hooked into the end of the rope to thread the rope through the eye (Figure 24a).

(7) *Tie knot in rope and pull knot against hole in pulley.*

(8) *Allow rope to rewind on pulley until pulley groove is full.*

(9) *Attach T-handle (Figure 25).*

Pull out (unwind) enough rope to attach T-handle.

b. If your starter is designed so that it is necessary to **disassemble it before replacing the rope,** proceed as follows:

(1) *Disconnect the rope from the T-handle end (Figure 25) and allow the spring to unwind slowly.*

This action removes the tension on the spring.

Hold the pulley with a gloved hand, or cloth, to prevent injury.

(2) *Remove the starter drive from pulley assembly.*

For procedures see "Repairing the Drive Mechanism," next heading.

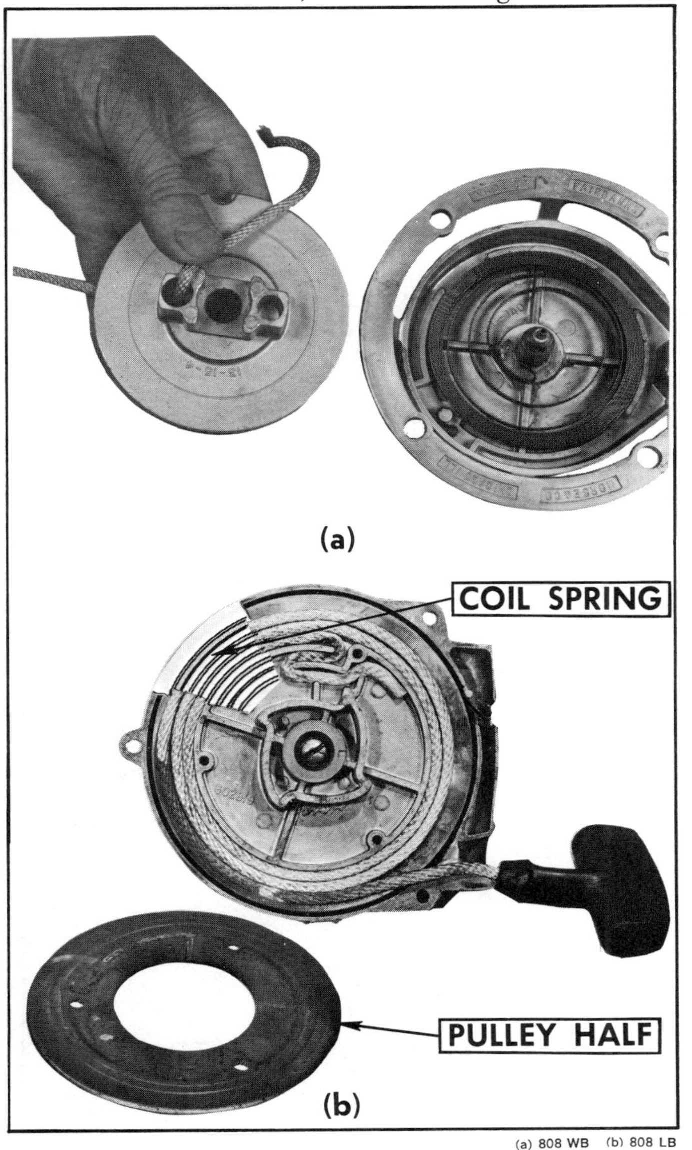

FIGURE 26. Gaining access to the rope. (a) A 1-piece pulley removed. (b) A 2-piece pulley with one half removed.

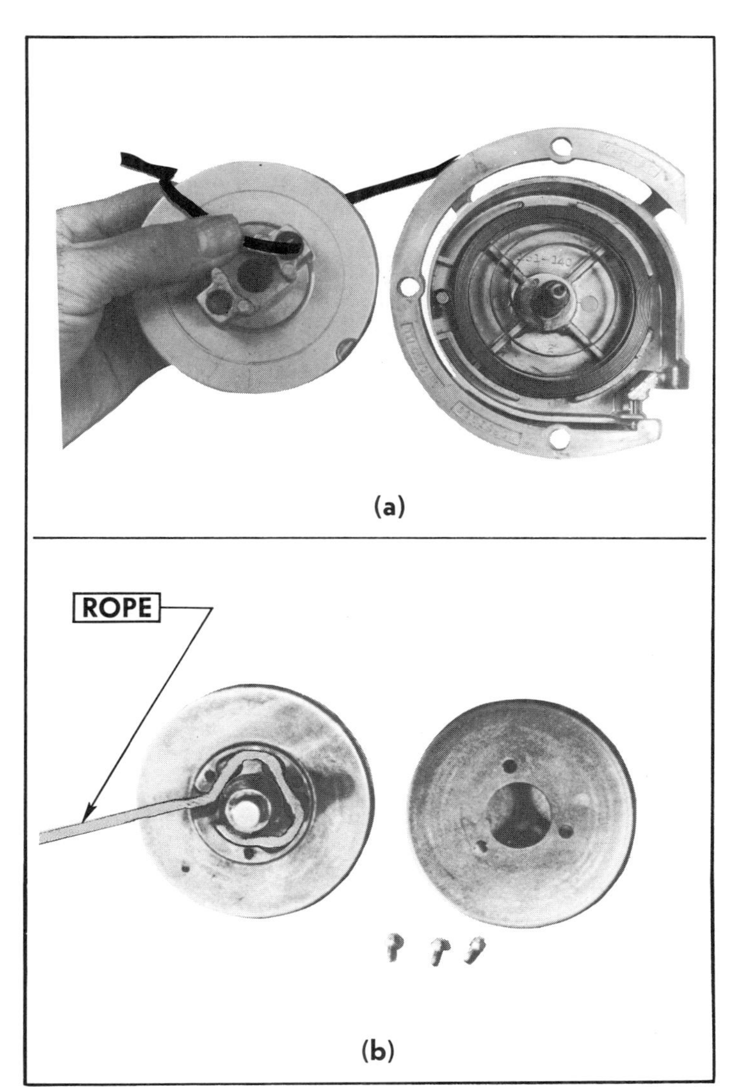

FIGURE 27. Installing the rope. (a) Threading the rope through the 1-piece pulley hub. (b) Anchoring the rope between pulley halves of the 2-piece type.

IMPORTANT! Always make a mental note of how an assembly is put together before taking it apart. Clean, inspect and lay the parts out in the order that they must go together for reassembly. If necessary, draw a sketch. Then you should have no difficulty reassembling them.

(3) *Clean and inspect the starter drive mechanisms for damage and wear.*

(4) *Remove or disassemble the pulley (Figure 26).*

There are two types of pulleys: (1) one-piece type (Figure 26a), and (2) two-piece type (Figure 26b).

If your starter has the **one-piece type pulley,** remove it to gain access to the pulley and leave the spring in the housing (Figure 26a).

If your starter has the **two-piece type pulley,** remove only one side of the pulley. The spring will remain intact under the other half of the pulley (Figure 26b).

(5) *Remove the rope.*

Cut knot from rope, or detach it from rope holder.

(6) *Select rope as described in step (4) under "a," preceding.*

(7) *Singe ends of rope to prevent fraying.*

(8) *Install new rope (Figure 27).*

There are various means for anchoring the rope to the pulley.

Examine the pulley to determine which method to use.

(9) *Wind rope on pulley, thread outer end through eyelet of the starter housing, and attach T-handle.*

6. *Reassemble and/or replace pulley.*

Connect spring to pulley if it was disconnected. Most service manuals illustrate the proper sequence for assembling starters and other mechanisms.

Apply light coat of graphite lubricant on the metal rubbing parts.

Do not lubricate nylon bearings or gears. Dirt sticks to the grease and causes wear.

7. *Install the drive mechanism.*

Refer to "Repairing the Drive Mechanism," under the next heading.

8. *Rewind recoil spring and secure it with vice-grip pliers or a C-clamp.*

For rewinding recoil spring, see "Replacing the Recoil Spring," page 93.

9. *Check starter for proper operation before installing the starter on the engine.*

Pull the rope out. It should pull easily with no binding.

Observe the drive mechanism. The mechanism that engages the flywheel cap should extend.

Allow the rope to retract. It should rewind all the way. The engaging mechanism should retract.

10. *Reinstall starter on engine.*

Be sure to align the drive mechanism with the crankshaft. Some are equipped with an alignment rod on the mechanism which is inserted into a hole in the center of the crankshaft, or in the keyway. If the starter drive is not properly aligned, it will not engage and disengage satisfactorily; neither will it last long.

11. *Recheck starter for proper operation.*

REPAIRING THE DRIVE MECHANISM

There are many different types of rope-rewind starter drives, but the basic principle of operation is the same for each. If you understand how they work, you will have no trouble repairing them.

All of them are some form of a ratchet drive. When the starter is operated, the ratchet engages a flywheel adapter which is attached to the crankshaft. When the engine starts and when the starter is not being operated, the ratchet disengages from the flywheel adapter.

The main difference in the drives is in the engaging mechanisms. Four types and how they work are described as follows:

Centrifugal pawls (Figure 28) — When

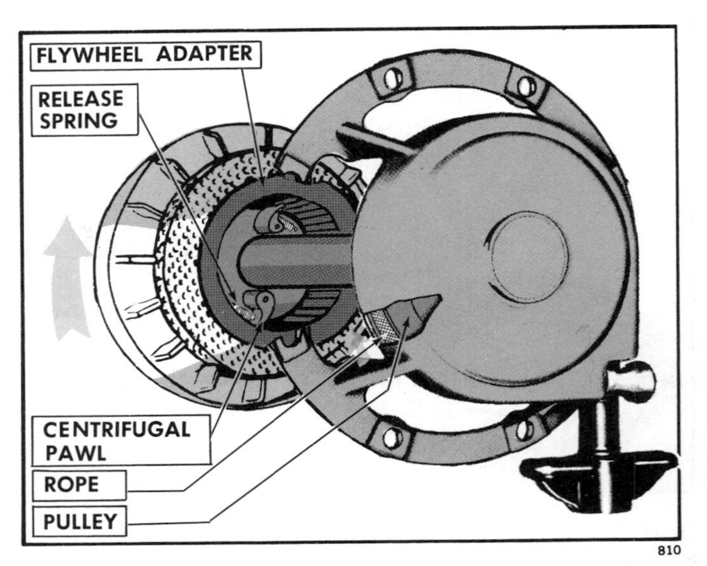

FIGURE 28. Centrifugal-pawl starter drive principle.

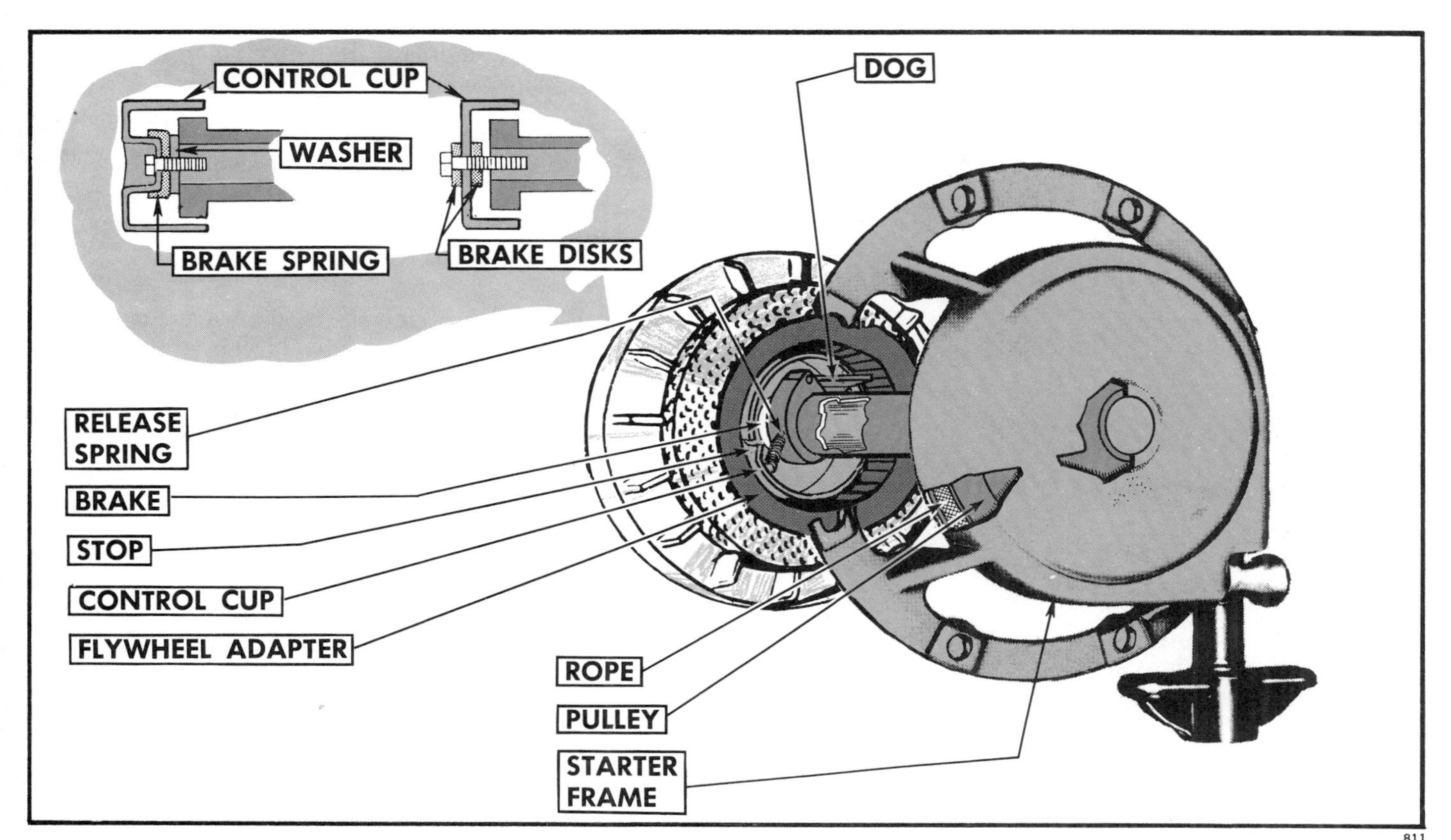

FIGURE 29. Cam-operated dog drive.

the starter is operated (rope is pulled), the pawls move out as a result of centrifugal force and engage the corrugated surface of the flywheel adapter, which is attached to the flywheel and the crankshaft. The crankshaft turns.

When the engine starts and the starter is released, the pawls are retracted by the release springs.

Cam-operated dog(s) (Figure 29) —

When the starter is operated the brake holds the control cup still until the dog is forced through the slot in the control cup far enough to engage the flywheel adapter.

The brake is attached to the starter frame at one end, and the brake shoe exerts pressure on the control cup. The friction is provided by brake disks or a spring clamp (Figure 29).

The flywheel adapter is connected to the crankshaft; so the crankshaft turns.

When the starter is released (rope is allowed to recoil) and the engine starts, the brake holds the control cup until the dog is retracted by the release spring. The stop prevents the dog from retracting all the way into the cup. If this should happen, it would not be aligned at the slot for the next operation. This type of starter drive may have one or more dogs.

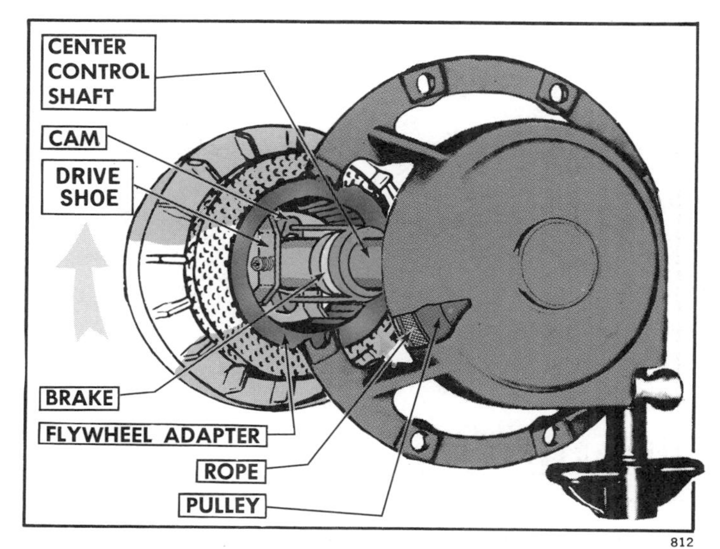

FIGURE 30. Cam-operated-shoe starter drive principle.

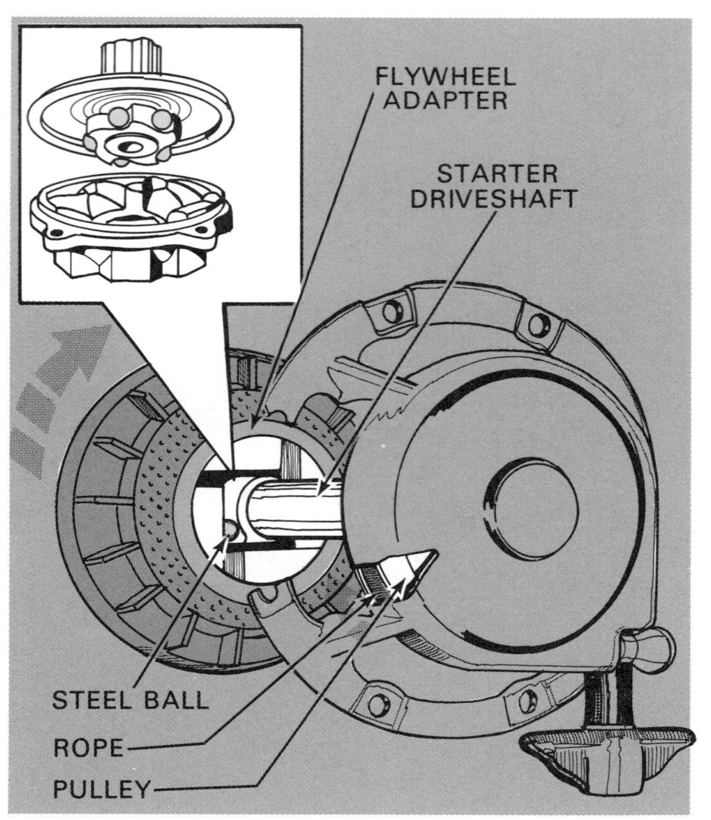

FIGURE 31. Wedging-steel-ball starter drive principle.

Cam-operated shoes (Figure 30)—When the starter is operated, the brake holds the spring-loaded cross member until the shoes are forced out against the flywheel cup by the cams. The crankshaft turns.

When the starter is released and the engine starts, the brake holds the spring-loaded cross member until the shoes are retracted. Springs on the ends of the cross member help to center it and to retract the shoes.

Wedging steel balls (Figure 31) — When the starter is operated, a steel ball is wedged between the cam on the starter driveshaft and the cam on the flywheel adapter. The crankshaft turns.

When the engine starts, the starter is released because the flywheel adapter runs at crankshaft speed. There is no longer any wedging action to hold the steel ball(s). They are forced into the recesses in the flywheel adapter by centrifugal force. They remain in this position as long as the engine is running.

Proceed as follows:

1. *Remove the starter and anchor it in a vise.* Follow steps 1 through 4, under "Replacing the Rope," page 85.
2. *Check the drive mechanism for proper operation.*

   Pull the rope and observe the action of the engaging mechanism. See that it extends when the rope is pulled and retracts when the rope is released.
3. *Remove the drive mechanism (Figure 32).*
4. *Disassemble the drive mechanism.*

   If you wish to identify the different parts, check your service manual or parts catalog. Most starters can be reversed for different directions of engine rotation. The engaging mechanisms for certain of these drives can

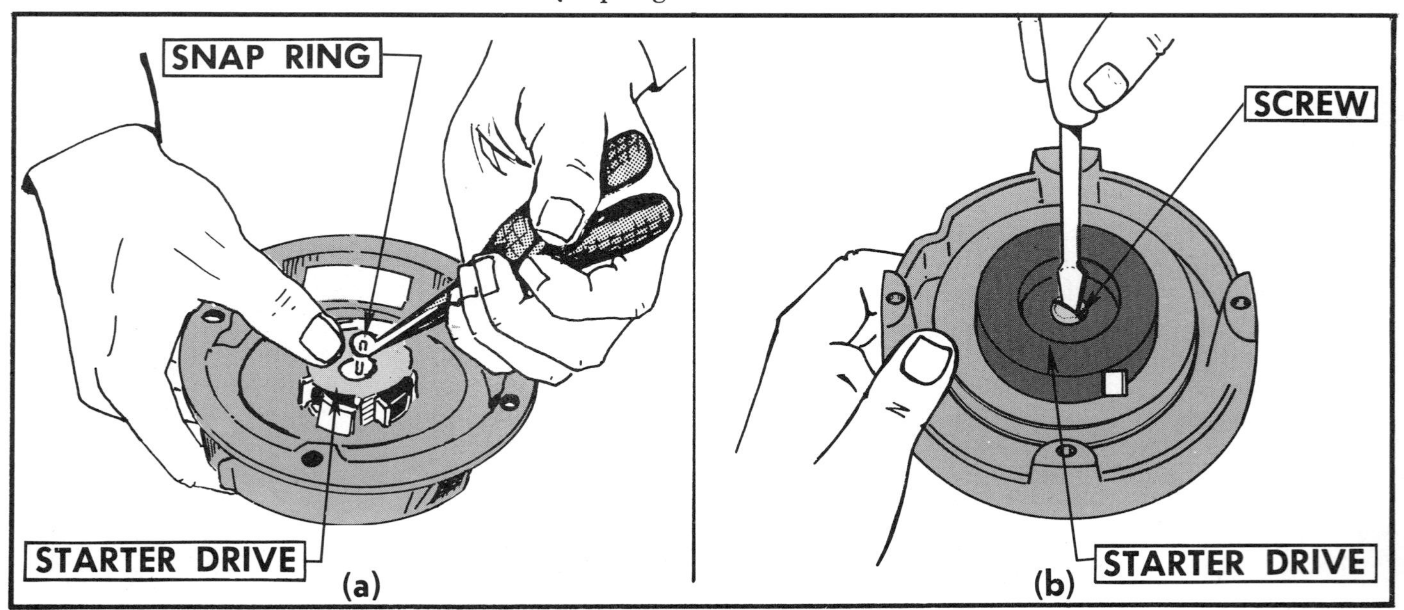

FIGURE 32. Removing the drive mechanism. (a) Some drives are held by a snap ring. (b) Some are held by a screw.

be easily reversed if not laid out in proper order during disassembly. If assembled in reverse, the engine-engaging mechanism will not engage the flywheel adapter when you pull the rope.

5. *Clean and inspect parts.*

   Look for worn or damaged parts, and bent or broken springs.

   Replace with new ones.

6. *Reassemble drive.*

   Be sure to install the starter spring and drive mechanisms properly for the direction of rotation of your engine. Most starter drives can be reversed to accommodate different directions of engine rotation. If you get the engaging mechanism in backwards, it will not work. Remember that the starter dog, shoe or pawl must point toward the direction of rotation. This is in order for it to engage the flywheel adapter while the starter is being operated.

   The friction shoes, such as the ones shown in Figure 30, must have the sharp edges toward the direction of rotation.

7. *Lubricate the drive lightly with graphite or multi-purpose grease—do not pack.*

   Do not put grease on the wedging steel balls as shown in Figure 31. Grease and dirt will cause them to stick. They may not engage the flywheel adapter. If they do, they may not release properly.

8. *Reassemble the starter drive.*

9. *Check for proper operation before installing the starter.*

10. *Install the starter.*

11. *Check for proper operation on the engine.*

REPLACING THE RECOIL SPRING

Two types of recoil springs are used in rewind starters: (1) a removable type, and (2) a packaged type.

One **removable type** of spring is designed to be unwound and stretched out in order to supply increased tension. You can recognize these by the end of the spring being accessible from the outside of the starter housing (Figure 33).

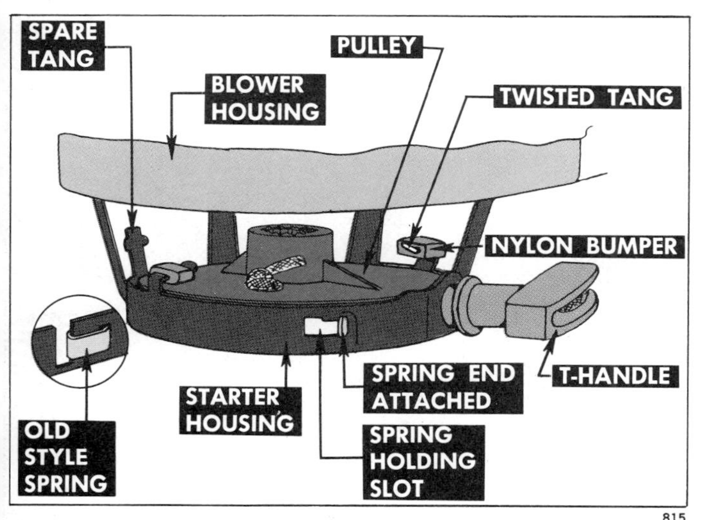

FIGURE 33. The removable type of recoil spring is visible on the outside of the starter housing (inverted view).

A **second removable type** spring is somewhat stiffer and comes precoiled, but not completely ready for installation. It must be compressed (partially wound) before installing in the starter housing. It is **not accessible from the outside of the starter housing.**

The **packaged type** spring is a stronger spring than the two previously described. It comes precoiled and compressed inside a retainer housing. Replacement springs for this type of starter comes precoiled. Some are incased in a permanent type retainer. Others are incased in a temporary retainer which is discarded upon installing the new spring. In either type starter, if the recoiled spring is damaged, broken or disconnected, it will be necessary to disassemble the starter to repair it.

a. If your starter recoil spring is the **removable type (accessible from the outside of the starter housing)**, proceed as follows:

   1. *Remove and secure the starter.*

      Follow steps 1 through 4 under "Replacing the Rope," page 85.

   2. *Release the spring tension.*

      Disconnect the rope from the pulley end if it is accessible (Figure 33). If it is not, disconnect it from the T-handle and allow the spring to unwind slowly. This is

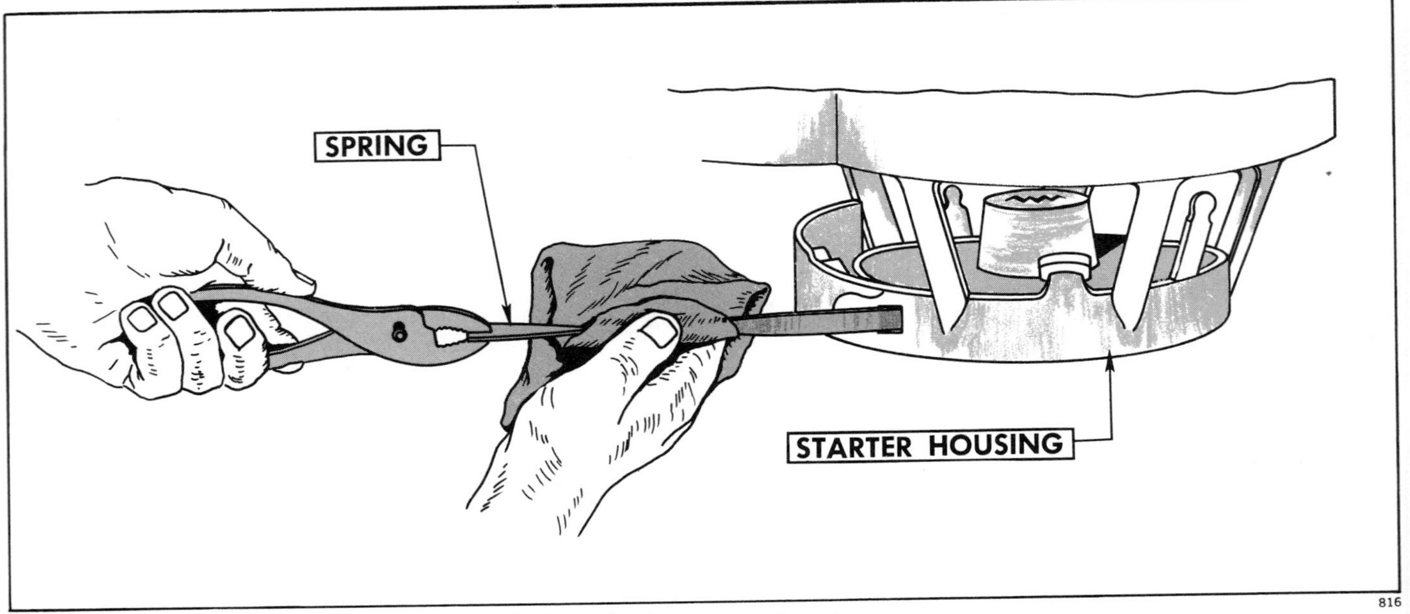

FIGURE 34. The recoil spring is unwound through the starter housing.

done by holding the pulley with gloved hands or with a cloth to protect your hands.

3. *Pull spring out as far as possible (Figure 34).*

   Grasp the end of the exposed spring with pliers, unhook it from the housing, and pull it out by hand. This unwinds the spring.

4. *Remove the pulley.*

   If the pulley is **held in place as the one in Figure 33,** raise one tang with a screwdriver and lift the pulley out.

   If the pulley is **held in place by the starter-drive mechanism,** remove it before attempting to remove the pulley.

5. *Disconnect the recoil spring from the pulley (Figure 35).*

   If the hook is broken, form a new hook with a grinder. Grind a notch on each side of the end of the spring.

6. *Straighten spring (36).*

   If you are going to use the old spring, clean and straighten it.

   Hold the spring coil in your left hand. With a cloth (or gloves) in your right hand grasp the spring between the thumb and fingers with the thumb on the outside curve of the spring. Pull the spring through, straightening it against your

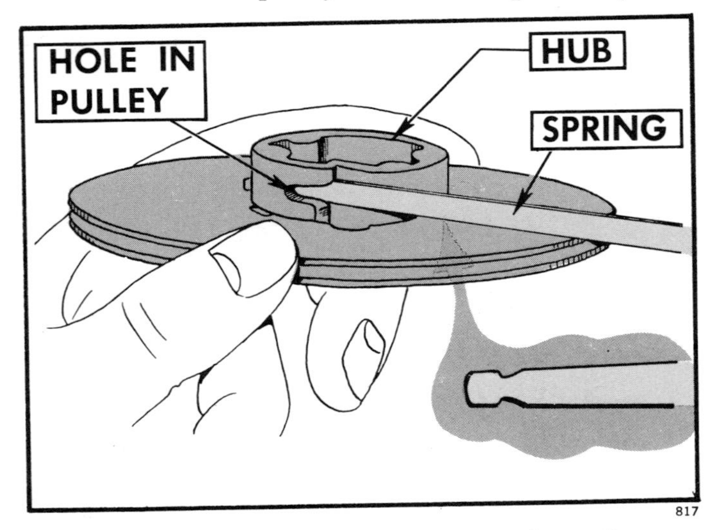

FIGURE 35. Unhook recoil spring from the pulley.

FIGURE 36. Straightening the recoil spring to provide more tension.

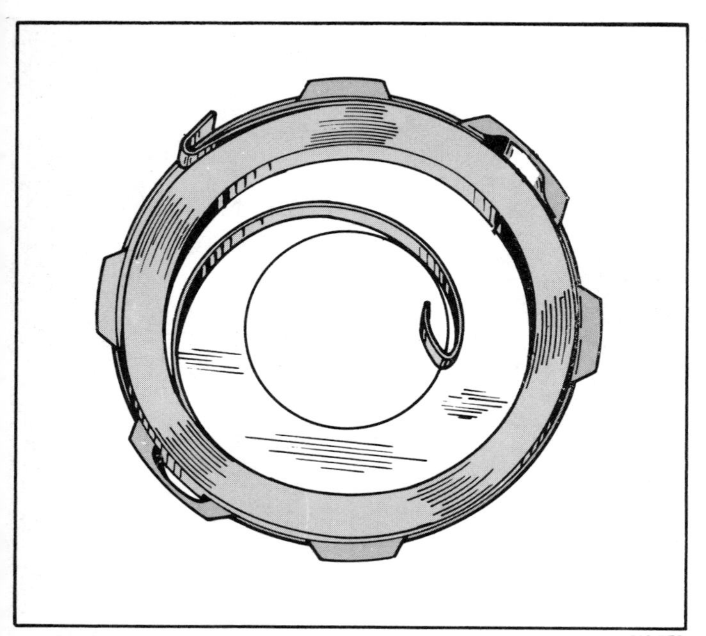

819 TEC

FIGURE 37. Most packaged springs are enclosed in a retainer housing.

thumb as you pull. This will strengthen the spring tension when it is recoiled.

7. *Clean and inspect other parts of the starter.*
8. *Attach new spring to the pulley (Figure 35).*

   Remember that you are tightening the spring as you crank the engine. Install it so that this will take place.
9. *Install pulley.*
10. *Wind the recoil spring (Figure 21).*

    A general rule is to wind the spring until it is tight; then back off one turn to prevent the spring from breaking when you operate the starter. Refer to step 5, under "Replacing the Rope," page 85, for additional information.

    Anchor the outer end of the spring during the rewinding process.
11. *Attach rope to pulley.*

    If rope is to be replaced, refer to "Replacing the Rope," page 85.
12. *Release pulley slowly.*

    Hold with a gloved hand or cloth to prevent injury.
13. *Check starter for operation.*
14. *Install starter on the engine.*
15. *Recheck starter for operation on the engine.*

c. If your starter recoil spring is of the **packaged type (Figure 37)**, proceed as follows:

1. *Follow steps 1 through 5 under "b" under the heading "Replacing the Rope," page 85.*
2. *Remove recoil spring (Figure 38).*

   Some springs are *packaged in a housing* (spring retainer) and should not be removed from the retainer housing. They are easy to replace (Figure 37).

   Other springs come in a *temporary retainer* for easy installation of the new spring, and the retainer is discarded.

   If the spring is *designed for removal* from the housing (pre-coiled removable type), be careful when removing it. When the tension is released, it will spring partially open.
3. *Clean and inspect all parts of the starter.*
4. *Lubricate the new spring and starter sparingly.*

   Use graphite or multi-purpose grease sparingly as recommended by your manufacturer. Do not lubricate nylon bearings or nylon gears. The lubricant is not needed and grease will pick up dirt.
5. *Install new spring.*

   If the spring is in a **permanent retainer,**

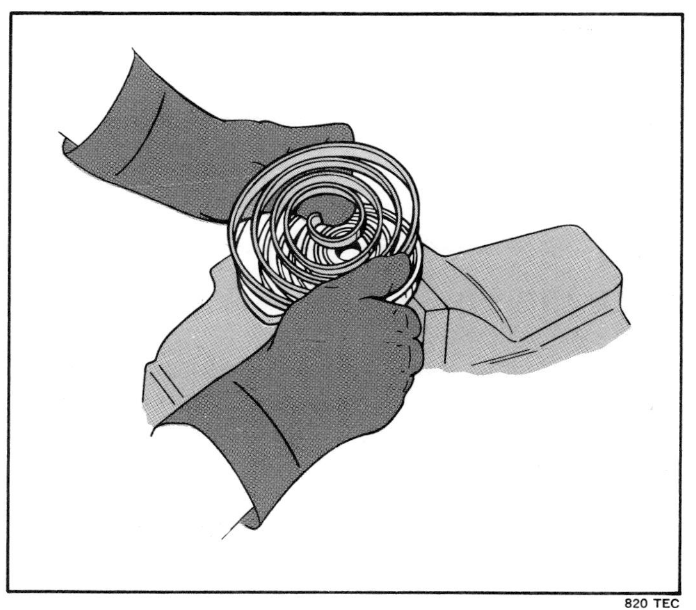

820 TEC

FIGURE 38. Some recoil springs are partially coiled but not packaged.

install retainer and spring (Figure 37).

If the spring is in a **temporary retainer,** press it into the housing and discard the retainer.

If it is necessary to **recoil the spring before installing it,** be sure to wear gloves to protect your hands (Figure 38).

Be sure to install the spring properly for the direction of your engine rotation. Wind it so it will tighten when wound in the direction of the engine rotation.

6. *Install pulley and rope if needed.*

   See "Replacing the Rope," page 85.

7. *Rewind spring (Figures 21a and 39).*

   If rope is replaced before assembling the pulley, the spring is wound by wrapping the rope around the pulley and pulling it (Figure 39). Repeat this procedure until the spring is wound tightly. Then release one turn and attach the T-handle.

8. *Secure the starter drive assembly.*

   Tighten nuts or screws.

9. *Check starter for proper operation.*

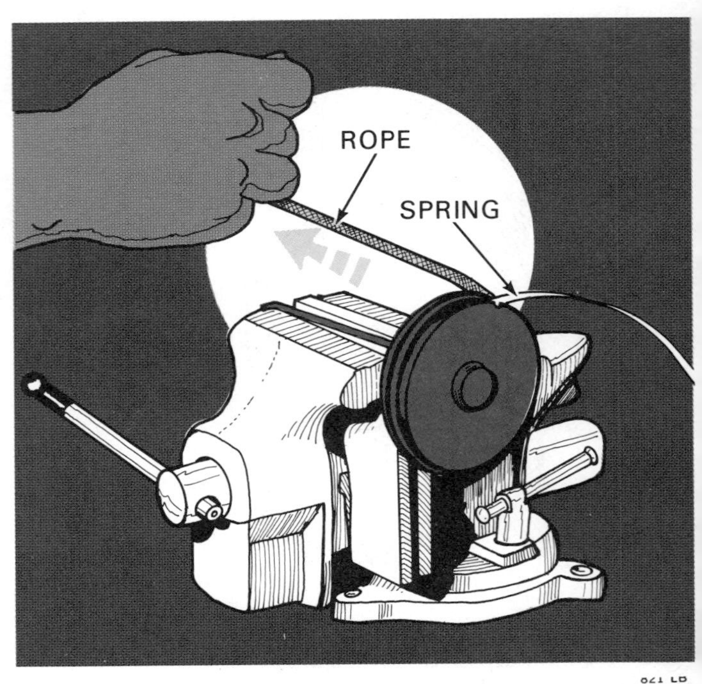

Figure 39. One method for rewinding a packaged-type spring.

10. *Install starter on engine.*

    Align drive properly.

11. *Recheck starter operation on the engine.*

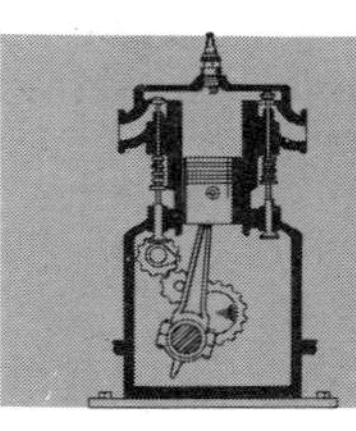

C. REPAIRING WINDUP STARTERS

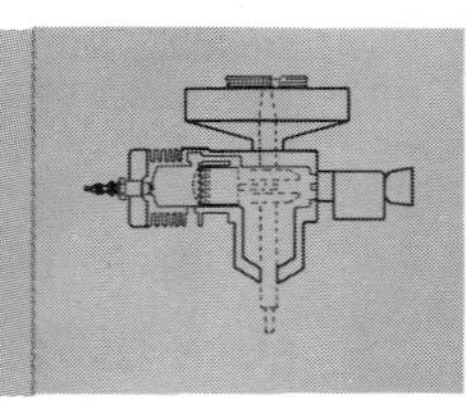

Windup starters are designed to reduce the amount of manual effort in starting your small engine. All you have to do is wind up a heavy recoil spring with a hand crank. When the spring is released, it turns the crankshaft for starting the engine.

Procedures for maintaining windup starters are discussed under the following headings:

1. Principles of operation.
2. Tools and materials needed.
3. Checking for proper operation.
4. Repairing the windup starter.

PRINCIPLES OF OPERATION

The principal parts of a simplified windup starter are shown in Figure 40. Although the parts of your starter may be different from those shown, they serve the same purposes.

The principal parts and their functions are as follows:

- A **crank handle** for winding the recoil spring — usually the handle folds for compactness.
- A **recoil spring,** which furnishes power for starting the engine.
- A **ratchet** (gear and spring) for holding the outside end of the recoil spring — it holds the crank handle in any position and thus prevents the handle from rebounding when it is released. It also provides for winding the spring with any length stroke.
- A **control mechanism** which holds the in-

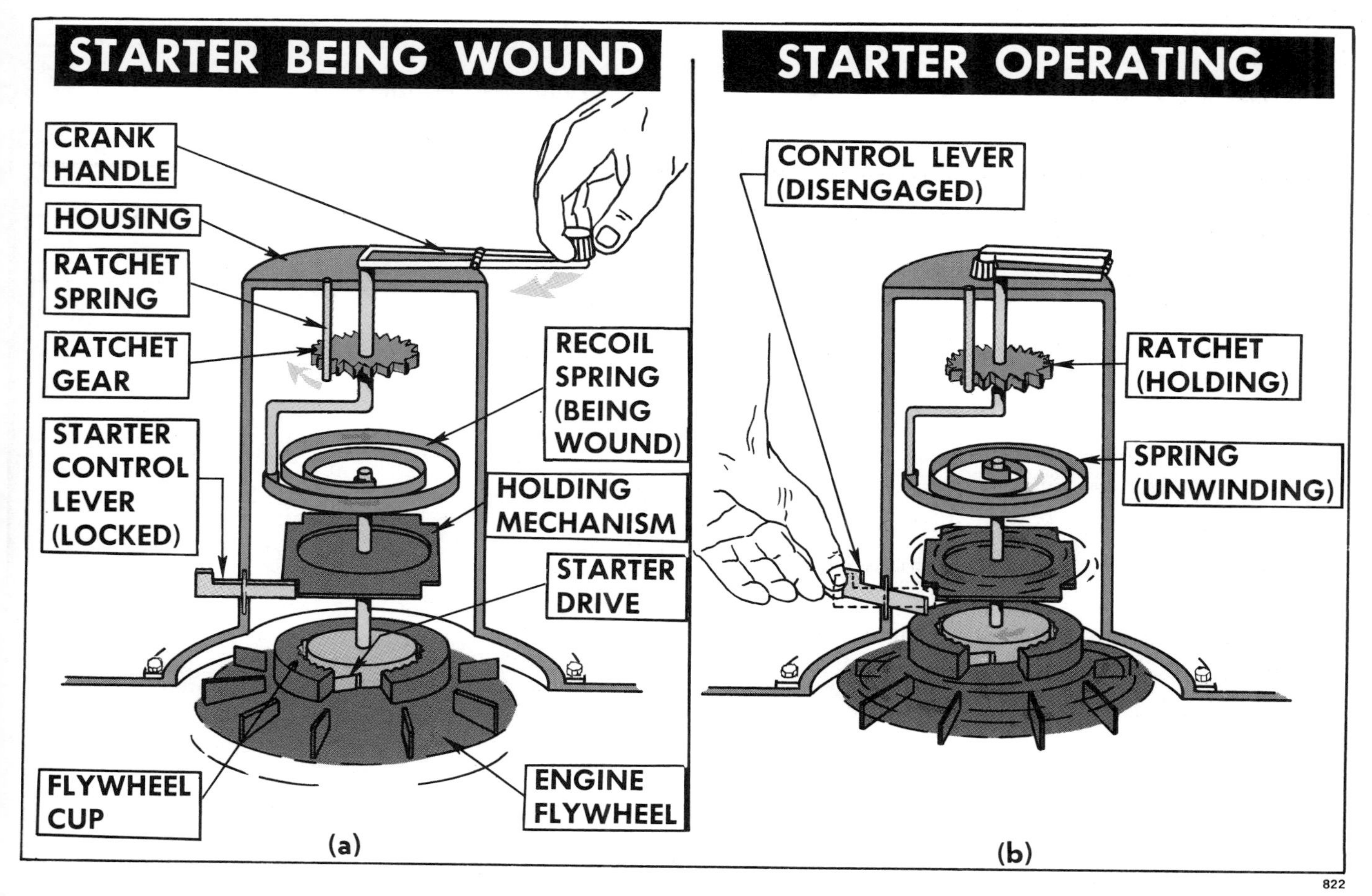

FIGURE 40. How a windup starter works. (a) The main spring is wound by a hand crank. As the recoil spring is wound, the starter drive is kept from turning by the starter-control lever. When the cranking action is completed, the recoil spring is held in position by the ratchet spring on the ratchet gear. (b) When the spring is wound and the starter control lever is disengaged, the spring unwinds and spins the starter-drive mechanism which in turn spins the flywheel.

side end of the spring and which also serves as a trip release for activating the starter.

- A **starter-drive** mechanism which transfers the rotary motion of the unwinding spring to the flywheel and crankshaft for starting.

Starter drives, on windup starters, are similar to those on rope-rewind starters except they are stronger. See Figures 28 and 29.

Some windup starters have provisions for winding the spring directly (Figure 40), while others are equipped with reduction gears (Figure 41). The purpose of the reduction gears is to make it easier to wind the recoil spring. This arrangement, however, requires more turns of the handle to wind the spring. For example, if the drive gear is one fourth the diameter of the driven gear it will require only one fourth the effort to turn the handle; but you will have to make four times as many turns to wind the spring as you would for a direct drive starter.

FIGURE 41. A reduction-gear windup starter.

TOOLS AND MATERIALS NEEDED

1. Vise—4″ jaws
2. 8-inch C-clamp (2)
3. Slot-head screwdriver—8″
4. Open-end wrenches—7/16″, 1/2″ and 9/16″
5. Needles-nose pliers—7″
6. Combination pliers—7″
7. Multi-purpose grease

CHECKING FOR PROPER OPERATION

1. *Lock the starter spring with the control lever (Figure 42a).*

   Some are locked automatically as the handle is raised.

2. *Remove any load from the engine.*
3. *Unfold the handle if it is of the folding type (Figure 42b).*
4. *Wind the spring (Figure 42c).*

   Most starters are wound in a clockwise direction.

   If the **spring does not wind,** it is probably broken.

5. *Fold the handle before starting.*
6. *Move release lever to starting position (Figure 42d).*

   A lever is the usual means of starting cranking action. It releases the holding mechanism in the center of the spring assembly. The engine should spin three or four revolutions.

   Some starters are actuated (set in motion) by folding the handle and pressing it down. If the starter **spins but does not crank the engine,** the drive mechanism is not engaging. Refer to "Repairing the Drive Mechanism," page 90.

   If the starter **engages the flywheel cup but does not crank the engine fast enough for starting,** the spring may be weak. Before replacing the spring, however, check the flywheel for freedom of rotation. See step 2.

   If there is **a slight hesitation on the part of the starter**—it starts to turn slowly and then spins the engine—it is because the starter is engaged at the time the piston is coming up on compression stroke. This is normal because of the high turning resistance of the engine during the compression stroke.

   If your engine **does not start after two or three tries** with the windup starter, check your starting procedures. Refer to "Starting the Engine," Part 1.

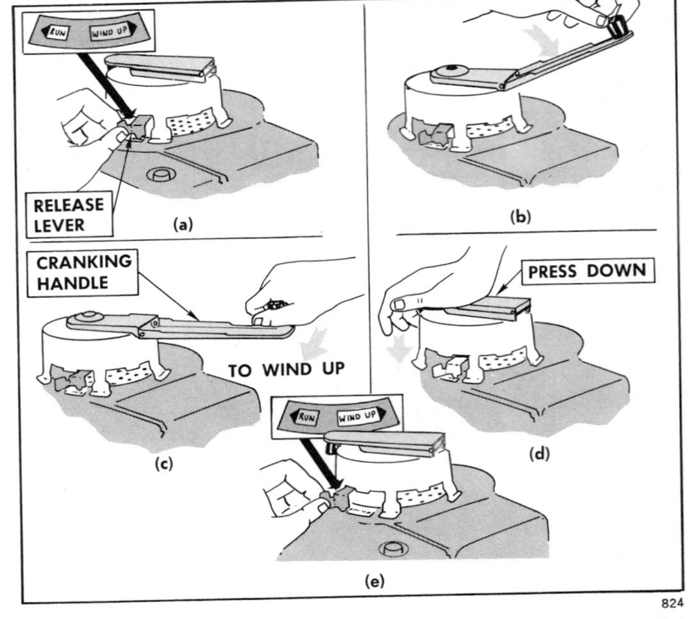

FIGURE 42. Steps in operating a windup starter. (a) Lock the spring by moving the control lever to "wind up." (b) Open the crank handle. (c) Wind the recoil spring. (d) Fold handle. (e) Move control lever to "run" to start the engine.

REPAIRING THE WINDUP STARTER

1. *Release the starter control (Figure 42d and e).*

   This relieves the tension on the spring.

   CAUTION! Never attempt to work on a windup starter without first deactivating (unwinding) the spring.

2. *Remove the starter assembly from the engine (Figure 43).*
3. *Disassemble the starter (Figure 44).*

   Some windup starters are disassembled from the drive end and others are disassembled from the handle end. If the handle is riveted or welded onto the shaft, you should remove the drive mechanism first. The drive mechanism and retainer screw are similar to those on rope-rewind starters—only heavier.

FIGURE 43. Removing the windup-starter assembly from the engine.

4. *Remove the drive assembly if necessary.*

   Refer to "Repairing the Drive Mechanism," page 90.

5. *Remove the main spring assembly (Figure 45).*

   Do not remove the spring from its retainer unless you have specific instructions from your manufacturer to do so.

   CAUTION! Very few main springs are intended to be removed from the retainer. If you are not certain yours is to be removed, do not do it. They are very strong and most of them will straighten out instantly (Figure 46).

   Included in this assembly will be a ratchet of some kind, perhaps a drive gear and washer. Watch for small springs and spacers. Clean, inspect and lay out the parts. Check the condition of the spring. Is it worn, bent or cracked?

6. *Clean and inspect the housing.*

   Look for breaks, cracks, worn gear teeth and a worn ratchet; also, for broken springs and worn bearing surfaces.

   Check control lever for condition and for proper operation.

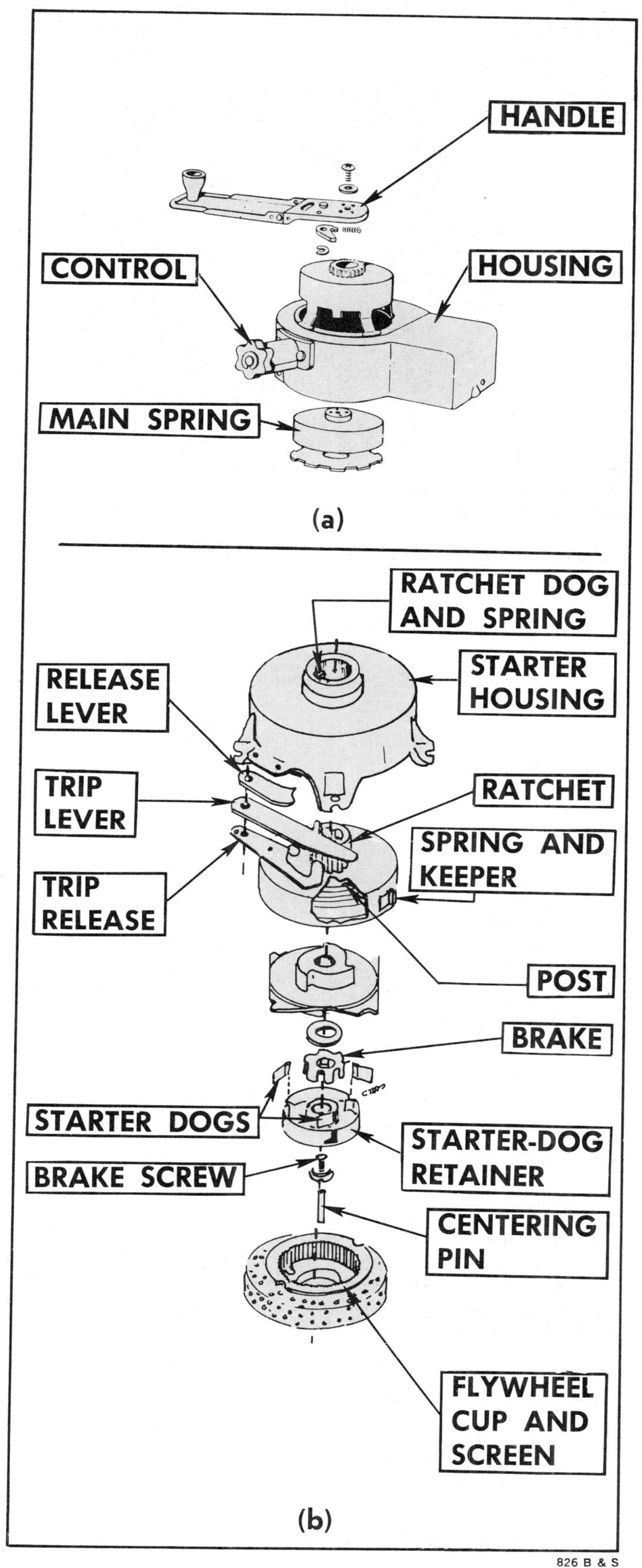

FIGURE 44. (a) A typical windup starter of the type that must be disassembled from the handle end. (b) One from which the drive mechanism is first removed.

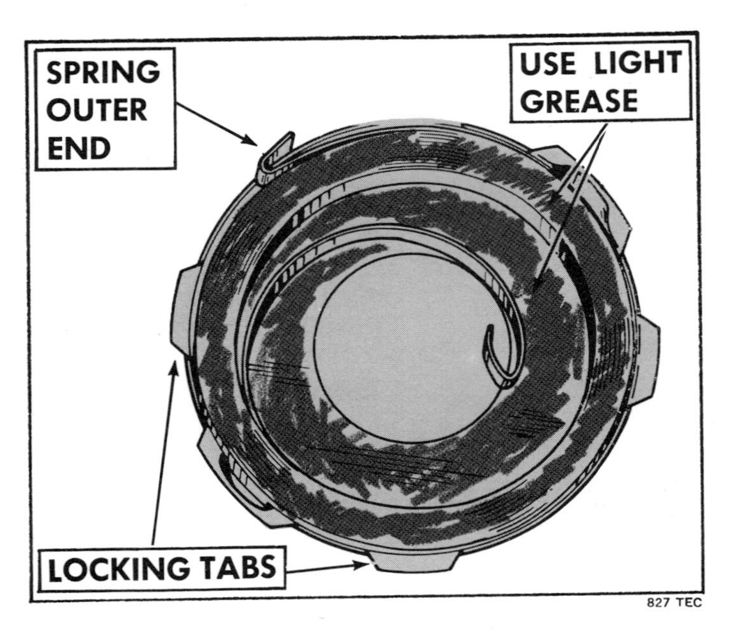

FIGURE 45. A main spring assembly.

FIGURE 46. Never remove a main spring from its retainer unless you are specifically instructed to do so by the manufacturer.

7. *Reassemble parts in reverse order.*

   Replace badly worn or broken parts.

   The main spring will most likely be in a retainer housing.

8. *Destroy the old main spring so that no one will be likely to tamper with it.*

   A good policy is to heat it with a torch. This will remove the tension.

9. *Reinstall the starter on the engine.*
10. *Recheck the starter for proper operation.*

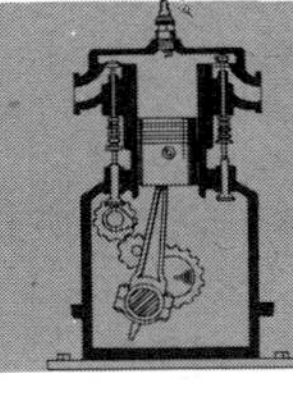

D. REPAIRING 120-VOLT ALTERNATING-CURRENT STARTERS

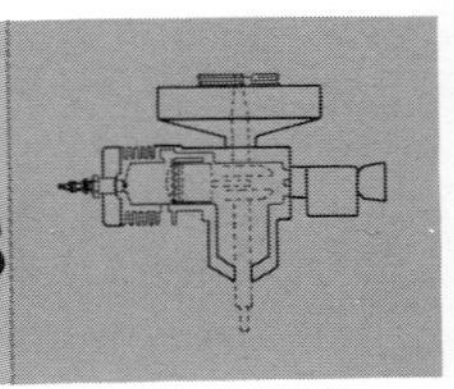

Alternating-current starters are operated on 120-volt current from your home wiring system. An extension cord is used to connect the starter to a convenience outlet.

Although most manufacturers offer alternating-current starters for small engines, these types of starters have not become very popular. The number in use is limited. For this reason, repair procedures are not given here. Only the two following topics are discussed.

1. Principles of operation.

2. Checking for proper operation.

PRINCIPLES OF OPERATION

The 120-volt alternating-current starting system consists primarily of the following:

- **An electric motor**—power to operate the starter.
- **A control switch.**
- **An extension cord** to reach the power source.
- **A starter-drive mechanism**—to engage the starter with the flywheel for cranking and for disengaging when the engine starts.

Basically, alternating-current starters can be classified according to the type of drive (engaging and disengaging) mechanism.

Three types are described here: (1) **cone-**

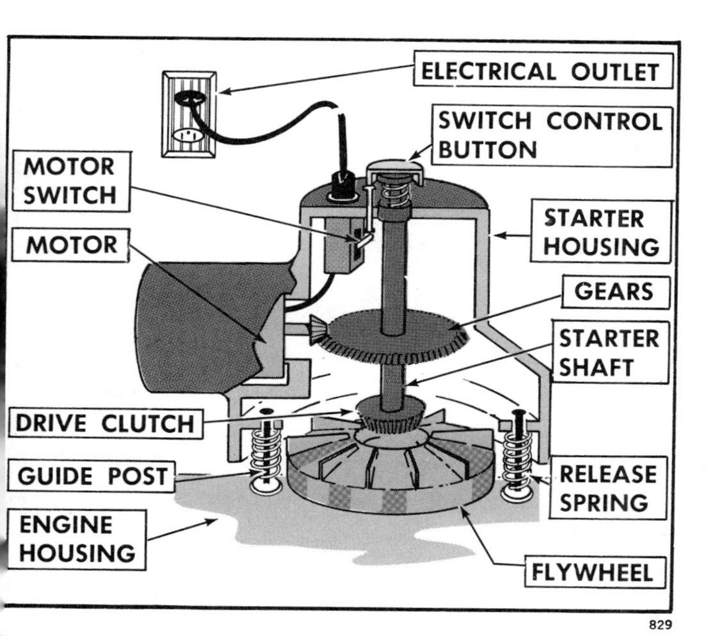

FIGURE 47. An alternating-current starter with a manually operated cone-shaped friction-clutch drive mechanism.

shaped friction-clutch, (2) split-pulley clutch and (3) bendix.

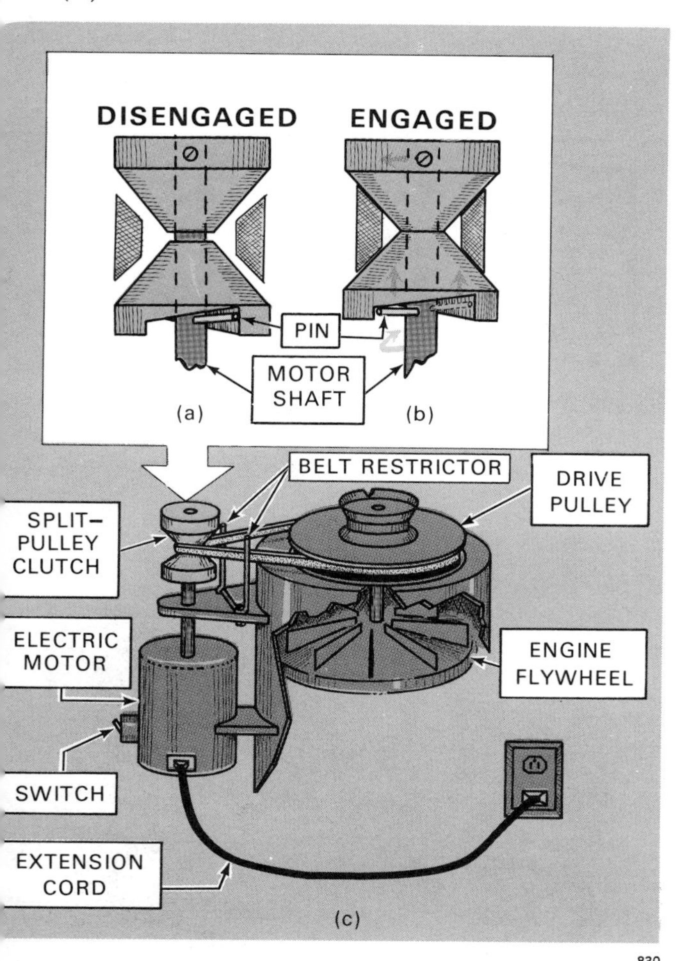

FIGURE 48. An alternating-current electric starter with split-pulley-clutch drive mechanism.

A starter with the **cone-shaped friction clutch** is shown in Figure 47. To operate this type of starter, press the switch-control button down until the electric motor starts. Hold it in this position until the electric motor gains speed; then push the entire starter housing down until the cone-shaped clutch is engaged. This engages the starter to the flywheel and cranks the engine.

When the engine starts, release the control knob. The starter housing is lifted by the starter-release spring, and the switch is turned off by the switch release spring.

The **split-pulley type clutch** (Figure 48) is also engaged by friction, but it is done automatically. The pulley halves are separated when the starter is not engaged (Figure 48 inset a). They close when the starter is engaged (Figure 48 inset b).

When you turn on the switch, the electric motor starts. The upper half of the pulley turns with the motor shaft. The motor shaft turns inside the lower half of the split pulley (Figure 48 inset a). The lower half does not turn momentarily because of inertia (resistance to changing position). As a result, the pin in the starter shaft pushes against the incline on the pulley and forces the lower pulley half upward. This closes the gap between the pulley halves (Figure 48 inset b). When this happens, the belt tension is increased; and friction is developed between the pulley flanges. The drive pulley rotates. The starter is engaged and cranks the engine.

When the engine starts and the starter switch is turned off, the electric motor stops. With the engine running, belt tension applied to the split pulley forces the movable (lower) half back down, thus releasing the belt tension and the starter. The belt rides loose in the open split pulley while the engine is running. A **belt restrictor** holds the belt in place when the starter is not operating.

The **bendix-type starter drive** (Figure 49) is a common type used on direct-current starters on all types of automobile and tractor engines. It is called "bendix" from the name of the inventor. Here is how it works. When you turn the switch on, the electric motor starts (Figure 49a). The pinion gear does not turn because of inertia (similar to the split-pulley type previously described). As a result, it moves endwise on the threaded shaft (Figure 49b) until it engages the flywheel gear. When this happens, the endwise motion stops and the pinion gear rotates

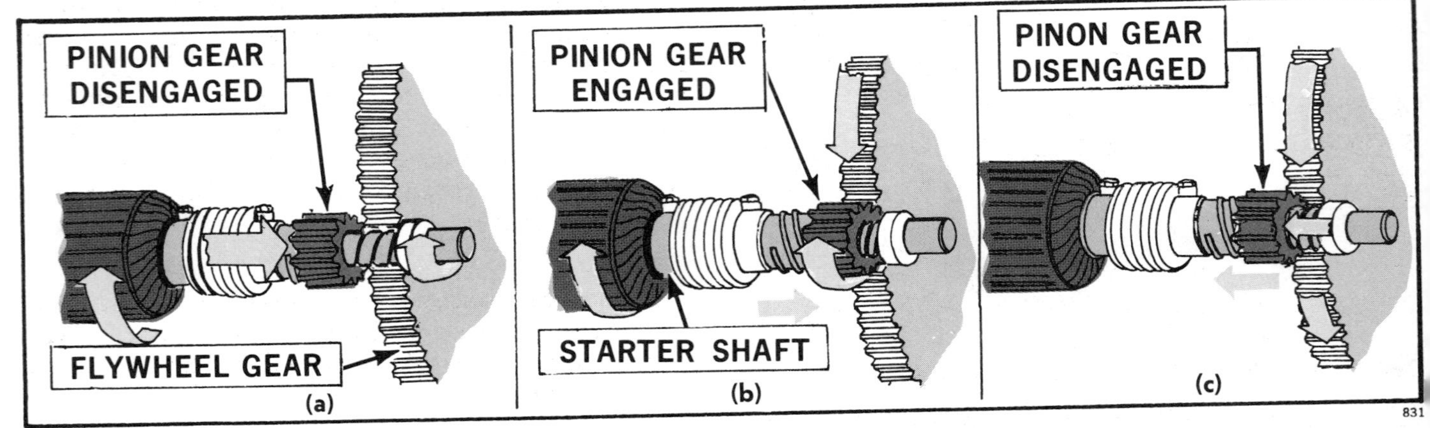

FIGURE 49. A bendix-type starter-drive mechanism. (a) Starter motor turned on, pinion gear moves endwise on threaded shaft. (b) Starter operating, pinion gear engaged in flywheel. (c) Starter turned off, pinion gear disengages.

with the shaft. The starter is engaged. The flywheel turns and cranks the engine.

When the engine starts, the starter switch is turned off. The pinion gear rotates faster than the starter shaft because it is now being driven by the flywheel. It spins back on the threaded shaft, away from the flywheel gear, thus disengaging itself (Figure 49c). The heavy spring on the starter shaft is used to help relieve the shock on the starter parts as the starter cranks the engine.

CHECKING FOR PROPER OPERATION

1. *Remove the load from the engine, if one is engaged. Disengage the clutch.*
2. *Disconnect the spark-plug wire to prevent the engine from starting.*
3. *Plug the extension cord into the starter motor.*

   CAUTION! Always plug the extension cord into the starter motor first. It is possible to get an external spark at the engine if you connect the extension cord to the convenience outlet first. This could start a fire.

   CAUTION! Use a three-wire extension if possible. The third wire is for grounding the equipment in case of a short circuit.

4. *Plug the extension cord into a 120-volt convenience outlet.*
5. *Turn on the starter switch.*

   If your starter is similar to the one in Figure 47, allow the starter motor to come up to speed; then press further down on the starter switch to engage the starter drive clutch. Other starter drives engage automatically.

   If the starter does not operate, the trouble may be in the source of electric power, in the electric motor and connections, or it may be in the drive mechanism. Proceed to step 7.

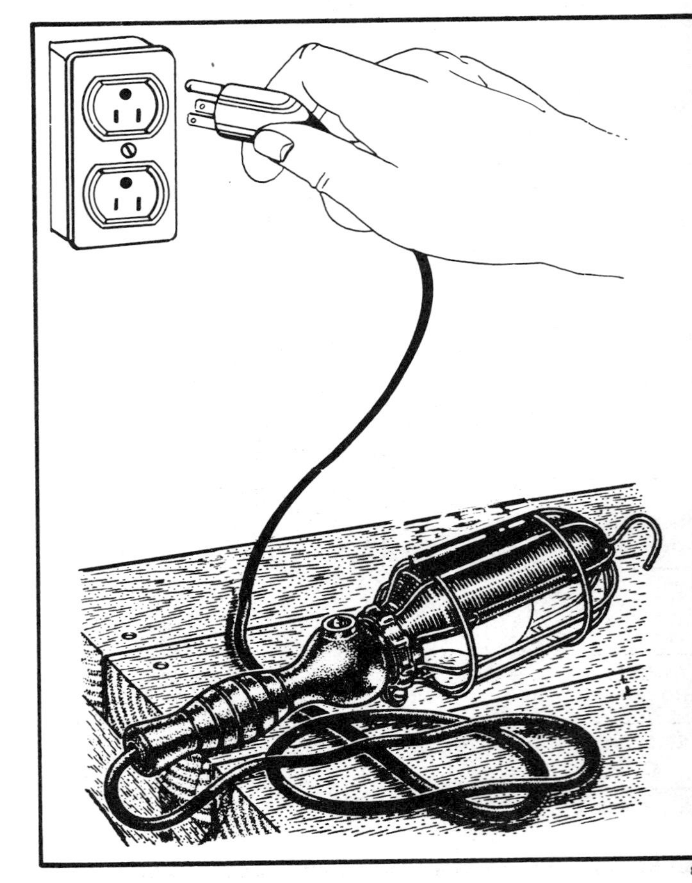

FIGURE 50. Checking the electric power source.

6. *Check starter operation.*

   Do not operate more than 10 seconds at a time.

   This is to prevent overheating the electric motor. Wait 30 seconds for the motor to cool before operating the starter again.

7. *Check the source of electric power (Figure 50).*

   Connect a portable lamp to the convenience outlet. If the *light does not burn*—and the bulb is good—you have no power available. The trouble is not in the starter.

   If the *light burns,* you know you have a source of power. Proceed to step 8.

8. *Check the electric motor and connections.*

   If the piston in the engine happens to be on compression stroke when the starter is engaged, the starter will be sluggish because it has trouble overcoming the resistance of compression at this point. Turn the flywheel by hand until the piston is off the compression stroke; then try the electric starter.

   If the electric motor still does not work and there is a hum in the motor, the trouble may be in worn brushes, worn bearings, or a short in the motor wiring.

   If there is no noise, the trouble may be a burned-out motor, worn brushes, or an open circuit—most likely in the extension cord.

   Procedures for replacing the brushes on direct-current starters are given under "Repairing Direct-Current Starting and Generating Systems," page 103.

   The same procedures may be used for the brushes in your alternating-current starter since they are of similar design. Or, refer to your service manual.

   If the starter motor **runs but does not engage,** proceed to step 9.

9. *Check the starter-drive mechanism.*

   Check the engaging mechanism for wear, broken parts and proper adjustment.

   See your service manual for procedures.

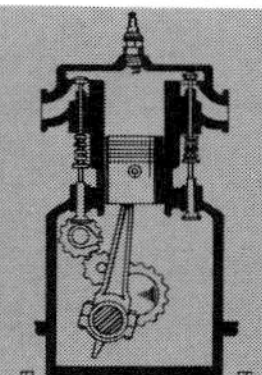

E. REPAIRING STARTING AND GENERATING SYSTEMS

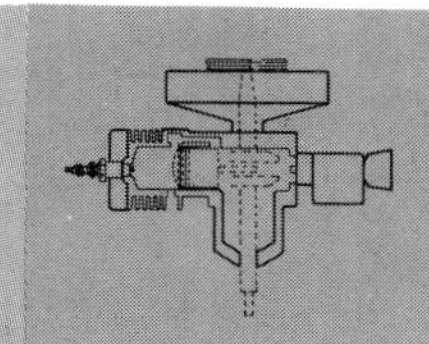

The direct-current starter on your small engine is similar to the one you have on your tractor, truck or automobile. It operates on power supplied by a storage battery (6 or 12 volts). The charge on the storage battery is maintained by an engine-driven generator. The **generator** may be (1) a *separate direct-current* unit, (2) a *separate alternating-current* unit—with a means for converting the alternating current to a direct current—or (3) it may be a *combination starter-generator* unit.

The combination unit is by far the most popular for small engines. It is more compact and self-contained, and does not require a disengaging mechanism.

Procedures for maintaining and repairing direct-current starting and generating systems are discussed under the following headings:

1. Principles of operation.
2. Importance of proper repair.
3. Tools and materials needed.
4. Checking for proper operation.
5. Checking and repairing the (12-volt direct-current) starting circuit.
6. Checking the generator control circuit (12-volt).
7. Testing the alternator system.
8. Repairing the direct-current starter and generator.
9. Repairing the flywheel alternator.

PRINCIPLES OF OPERATION

The direct-current starter-generator system consists of the following:

- A **direct-current starter** or a **combination starter-generator.**
- A **storage battery.**

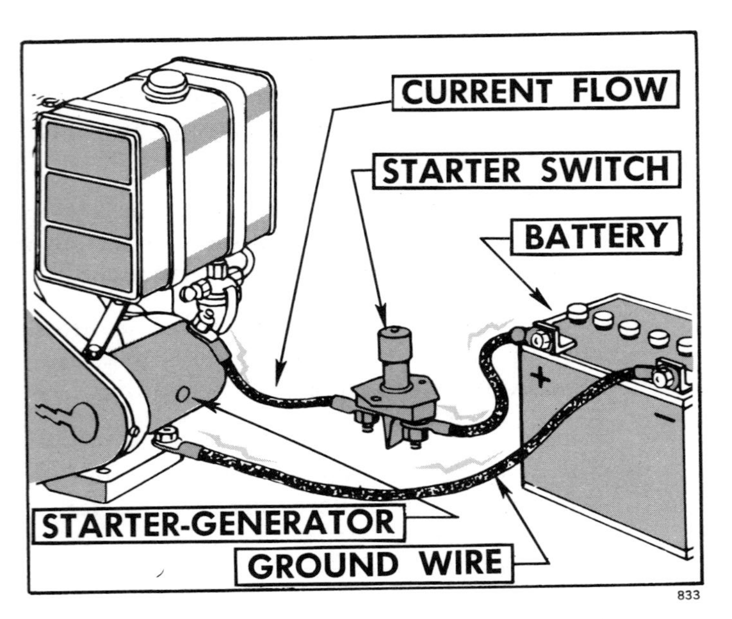

FIGURE 51. A 12-volt direct-current starter-generator system.

- A **direct-current generator** and **voltage regulator,** or an **alternating-current generator** and **rectifier** for changing the alternating current to direct current.
- A **starter switch.**
- An **ignition switch.**
- A generator **warning light, or an ammeter** (most systems).
- A **starter-drive mechanism**—a belt is used on the **starter-generator combination,** and it is never disengaged from the engine. **Direct-current starters**—without a built-in generator—normally have a bendix type of drive mechanism for engaging and disengaging the starter and the flywheel. It is described under "Principles of Operation," page 100.
- **Connecting wires.**

A **12-volt combination direct-current starter-generator** (Figure 51) is the most common type of electric starter used on small engines. When the starter switch is closed, the electric circuit is completed between the starter-generator and the battery. Current flows from the battery (negative to positive) through the ground wire to the starter-generator. The starter-generator acts as an electric motor. It turns the engine crankshaft through a belt drive.

When the engine starts, the starter switch is released. The starter-generator is *not* disengaged as with other starters. Instead, the engine drives the starter-generator. Then the starter-generator acts as a generator and produces electrical energy for the ignition system, for the lights—if installed—and for recharging the battery. The amount and the rate of charge to the battery is controlled by a current-voltage regulator (Figure 52).

Here is how it works.

A **voltage regulator** controls the amount of current going from the generator while the engine is operating. The regulator protects the generator from

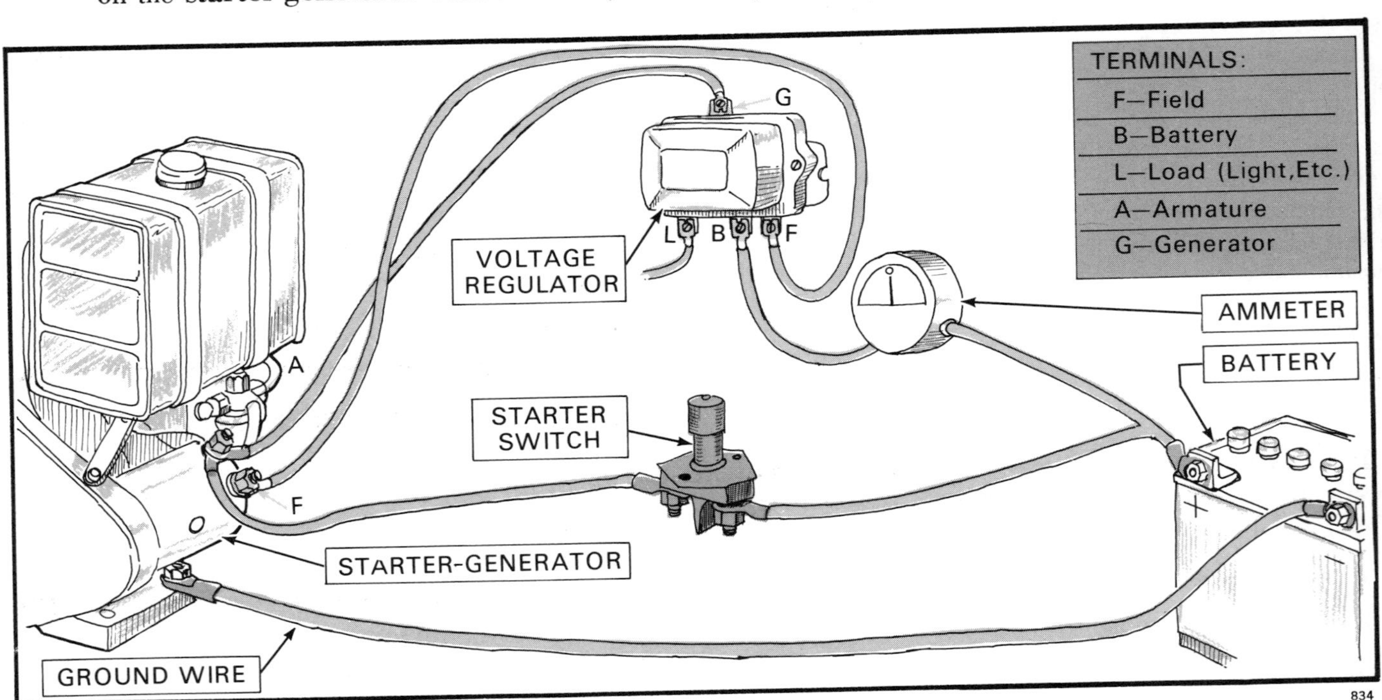

FIGURE 52. A 12-volt starter generator system showing the voltage regulator circuit.

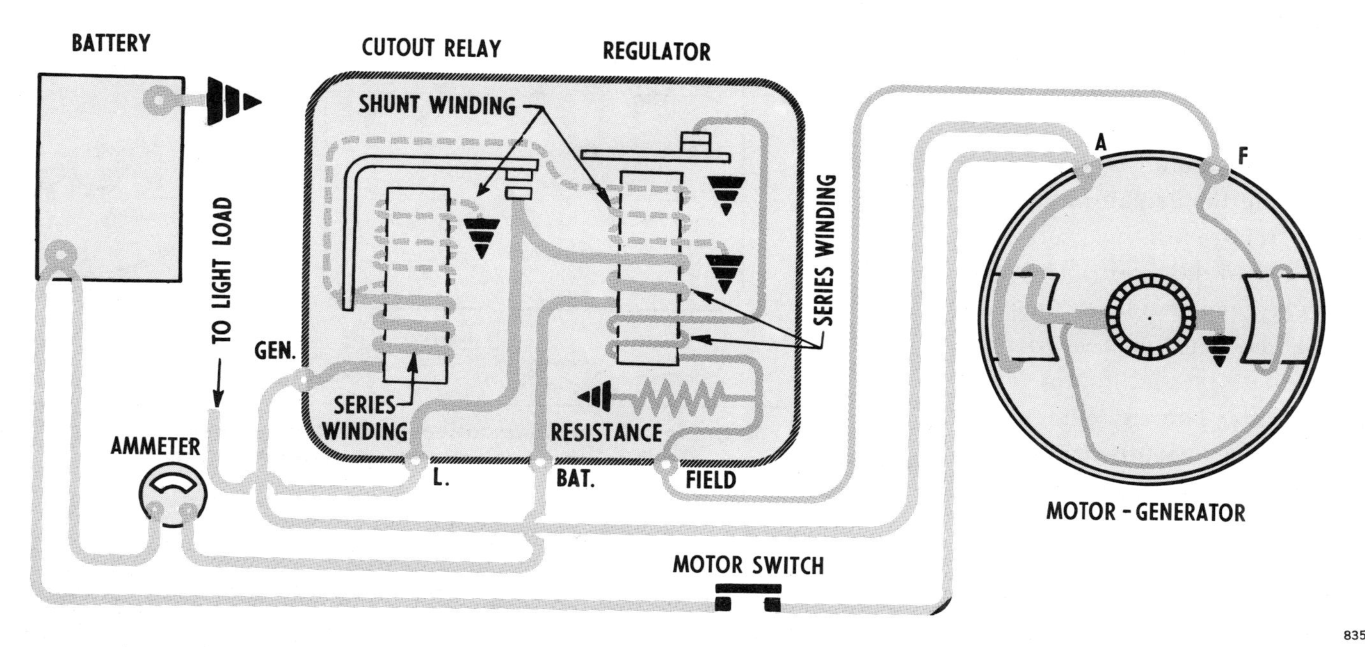

FIGURE 53. Diagram of a tractor-type, starter-generator system used on small engines, showing the current-voltage regulator circuit.

excessive output and possible failure. It also prevents the battery from becoming overcharged. Without some means of control, overcharging could cause the battery to get too hot, and the water in the electrolyte would boil away.

Wiring diagrams of voltage regulators used on direct-current generator systems are shown in Figures 53 and 54.

The first (Figure 53) is a **tractor-type, current-voltage regulator.** It is designed for light electrical loads and for normal operating conditions —moderate climate and infrequent starting. This type is most frequently used on small engines.

The second (Figure 54) is the type used on automobiles. It is designed for heavy electrical loads. The current regulator and voltage regulator are separate. This type of regulator allows the generator output to increase, or decrease, according to the demand (electrical load, or need for charging the battery).

The **tractor-type current-voltage regulator** operates as follows:

When the starter switch is closed, current flows from the battery to the starter, thus causing the starter to turn.

After the engine starts and comes up to 800-1000 r.p.m., with the starter switch off, current is generated. It flows from the generator armature through the shunt and series winding in the cutout relay of the regulator. This causes the cutout relay points to close and allows current to flow through the points to the battery, thus charging the battery. Increased current flow through the points and the series winding holds the points tightly closed.

As the generator voltage builds up, sufficient current goes through the series winding in the regulator to open the regulator points. The combination of currents through the series winding and the shunt-winding, act together to open the points. Therefore, high output of generator current and voltage are both controlled.

As soon as the regulator points open, the generator output (voltage) drops, thus allowing the regulator points to close again. The opening and the closing of the points continue at a rapid rate, thus regulating the generator output and voltage to match the requirements of the battery.

When the battery becomes fully charged, the regulator points remain open most of the time until the battery needs charging again. This prevents overcharging the battery.

The field current is directed through the resistor to ground. The resistor reduces arcing at the points and decreases the current flow from the generator.

The **automotive-type current-voltage regulator** operates as follows (Figure 54):

When the starter switch is closed, current flows from the battery to the starter, thus causing the starter to turn.

After the engine starts and comes up to 800 to 1,000 r.p.m., with the starter off, current is generated by the generator.

Current from the generator flows from the armature through the current-regulating coil and the cut-out-relay coil. The cut-out relay points close, thus allowing generator current to flow to the battery and to the electrical loads. The current regulator points do not open until the current flow is built up to a preset limit.

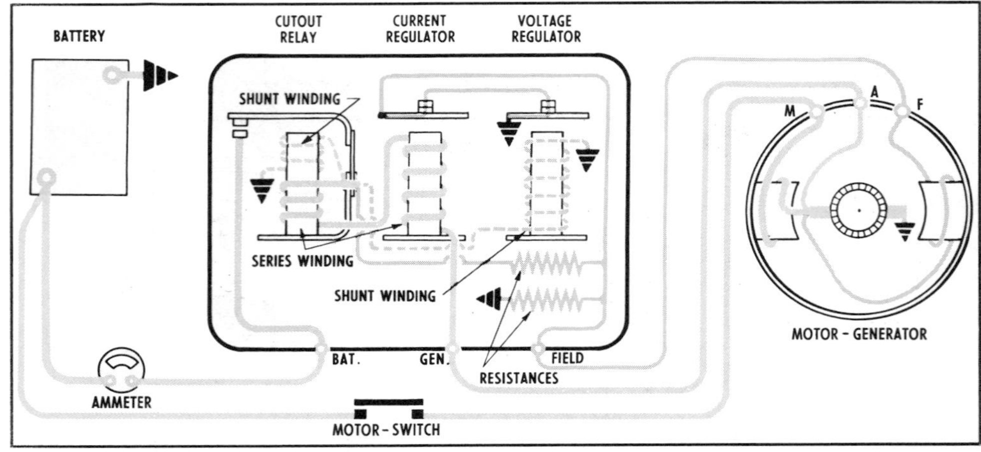

FIGURE 54. Diagram of an automotive-type, starter-generator system used an small engines showing the current voltage regulator circuit.

The generator voltage increases until it reaches the amount for which the two units of the regulator are set; then the points open.

If there is a **high current flow** to supply both the battery and various other uses, there is often not sufficient voltage to operate the voltage regulator coil. Then the current regulator limits generator output by making and breaking the field circuit.

If the electrical load is reduced, or if the battery comes up to charge, the system voltage will increase to the preset regulating voltage and open to voltage-regulator points. With the voltage regulator points open, the generator output is not sufficient to hold the current regulator points open, and they close. The voltage regulator points then open. The field current flows through the resistor to ground. The generator output decreases, thus allowing the regulator points to close again.

Some engines generate **alternating current** by means of a **magneto on the flywheel.** This supplies direct current to the starter motor and battery.

Alternating current differs, as the name implies, from direct current in that the direction of flow completely reverses itself at regular intervals. Alternating current cannot be used to recharge a storage battery. It is necessary to change it to direct current. This is done by rectifiers (or diodes) in this system (Figure 55). Figure 85 explains the principle of a diode rectifier.

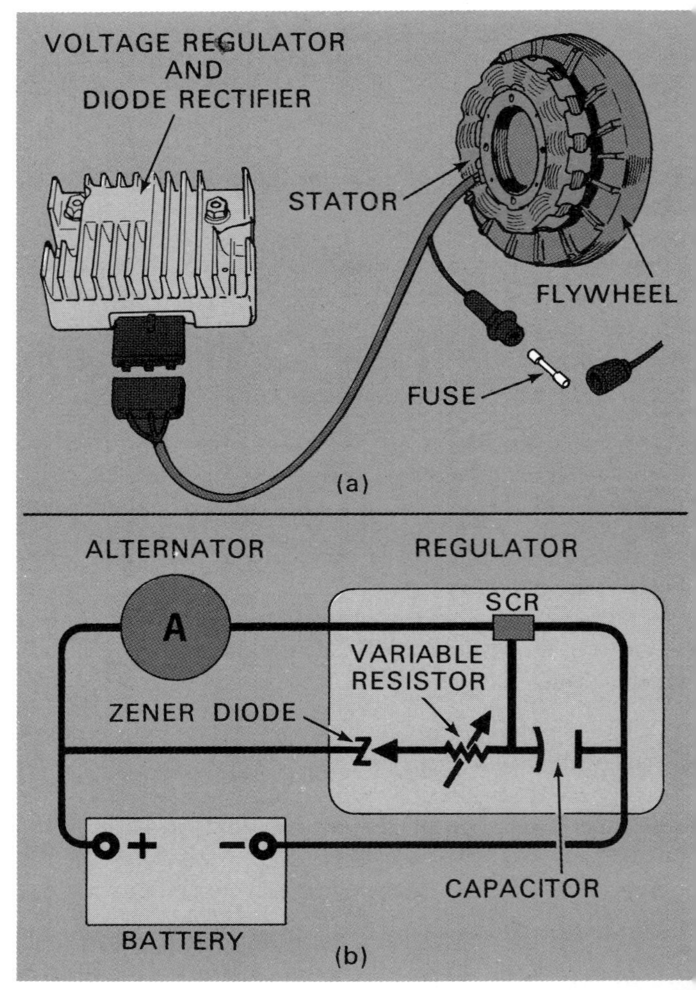

FIGURE 55. (a) An alternating current generating system consists of a stator, rotor (flywheel), and electronic voltage regulator and diode rectifier. The rectifier changes AC current to DC. (b) The voltage regulator contains a Zener diode, a variable resistor, silicone controlled rectifier (SCR) and a capacitor. When the battery needs charging, current flows directly from the alternator to the battery. When the battery current builds up to the preset limit, the Zener diode "breaks down." Current flows through the resistor and SCR. Alternator current is blocked off by the SCR until it is needed again.

This system is good for engines in intermittent service. But there is a danger of overcharging the battery when the engine operates too long at a time. It is sometimes necessary to remove one fuse to prevent overcharging of the battery. An overcharged battery gets too hot and the water in the electrolyte evaporates.

IMPORTANCE OF PROPER REPAIR

If you expect your starter to operate when you need it, there are a few maintenance and repair jobs you should do to the starter and generator system regularly. They are as follows:

- *Checking and servicing the battery*—for procedures refer to "Checking and Servicing Batteries," Part 1.
- *Lubricating the starter and generator bearings if needed*—some have factory sealed bearings and require no additional lubrication. If the bearings become worn, there is a danger of the starter or generator shorting out (grounding).
- *Checking and repairing the starter circuit*—much starter trouble is due to loose connections in the wiring circuit. The full charge of the battery never gets to the starter motor.
- *Replacing the brushes in the starter and generator*—occasionally brushes wear out in electric motors and generators. If you do not replace them in time, the starter or direct-current generator will be less effective. The segments in the commutator will become worn and pitted.

TOOLS AND MATERIALS NEEDED

1. 0-24 range DC voltmeter
2. 0-30 range ammeter
3. 0-50 range AC voltmeter
4. Slot-head screwdriver - 8″
5. Open-end wrenches - 7/16″, 1/2″, and 9/16″
6. Brush-seating stone, or sandpaper (No. 00) and a small wooden block with a ½″ square end
7. A clean cloth
8. Tachometer
9. Petroleum solvent (mineral spirits, kerosene or diesel fuel)

CHECKING FOR PROPER OPERATION

1. *Disconnect all driven equipment from the engine.*
2. *Turn on the ignition switch.*
3. *Check to see if the generator warning light comes on, or if the ammeter shows a discharge when the ignition switch is turned on.*

   If so, the instruments are working, and current is available from the battery.
4. *Shift power transmission into neutral if installed.*
5. Engage the starter switch.
6. *Turn the starter switch off as soon as the engine starts.*

   When the engine starts and picks up speed, look to see if the generator warning light goes out.

   If the warning **light does not go out,** the trouble is in the generator circuit. Follow procedures under the next heading.

   If you have an ammeter, check to see if the battery is charging (Figure 56).

   If the ammeter **does not indicate a charge,** the trouble is in the generator circuit. Follow procedures under the next heading.

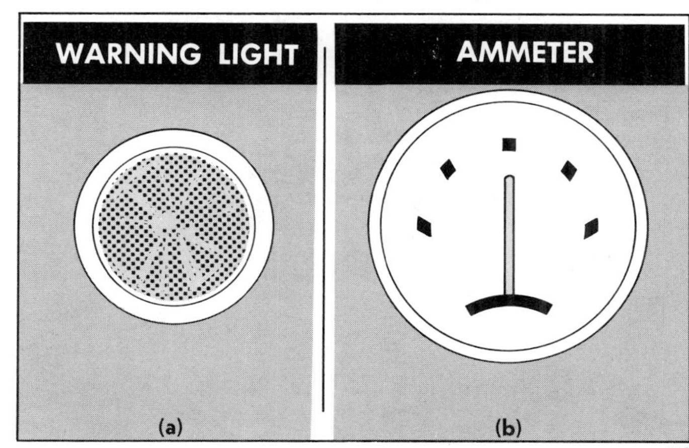

FIGURE 56. Generator warning signals. (a) A warning light indicates that the generator is not generating current. (b) An ammeter indicates the rate of charge going to the battery.

The **ammeter** indicates if the generator is working properly. The **warning light** comes on when the current is being supplied by the battery instead of the generator. This indicates the generator is not in good working order.

If the starter does not work or if it is sluggish, it is most likely due to low battery voltage or a faulty connection in the circuit. Check the battery first. Refer to "Checking and Servicing the Battery," Part 1.

If the battery is satisfactory, proceed with checking the starter circuit.

CHECKING AND REPAIRING THE 12-VOLT DIRECT-CURRENT STARTING CIRCUIT

Each step must be done in sequence and the condition corrected for this check to work. Otherwise, you will not be able to isolate the trouble.

1. *Check for proper ground connections at the starter-generator or at the battery (Figure 57).*

   NOTE: Be sure battery is good.

   (1) *Connect the negative lead of the voltmeter to the starter-generator mounting frame, and the positive lead to the negative battery post.*

   NOTE: This connection is for negatively grounded systems. If the starter circuit has the positive ground, the connections would be opposite.

   (2) *Close the starter switch and check the voltage.*

   The voltmeter should read approximately 10 volts, proceed with step 2.
   If the **voltage drops below 10 volts,** this indicates you have a poor ground connection in the system. Check for loose and/or corroded ground connections.

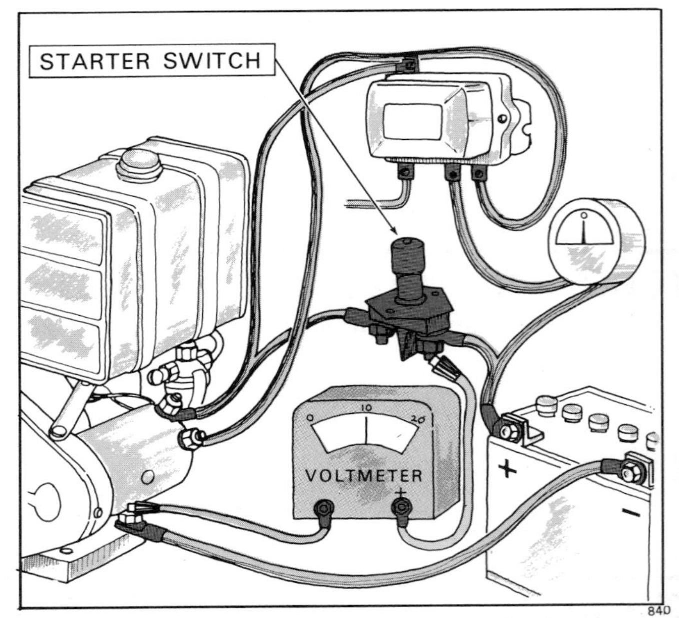

FIGURE 58. Checking the circuit between the battery and the starter switch with a voltmeter.

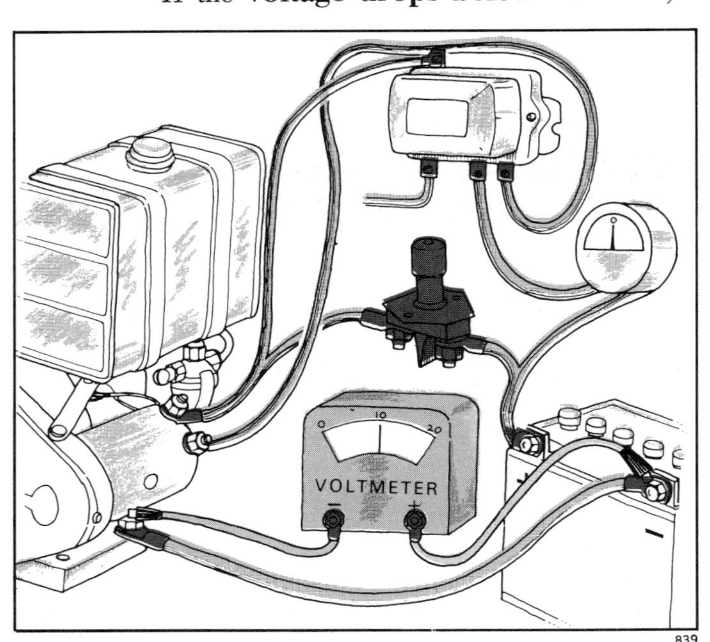

FIGURE 57. Checking the battery and ground connections with the voltmeter.

   (3) *Clean and tighten ground connections or replace ground cable if it is worn or corroded.*

   (4) *Close starter switch and recheck.*

   If the **voltage is still low,** proceed to step 2.

2. *Check the circuit between the battery and the starter switch (Figure 58).*

   (1) *Leave the negative lead grounded to the generator mounting frame and connect the positive lead of the voltmeter to the switch terminal nearest the battery.*

   (2) *Close the starter switch and check voltage.*

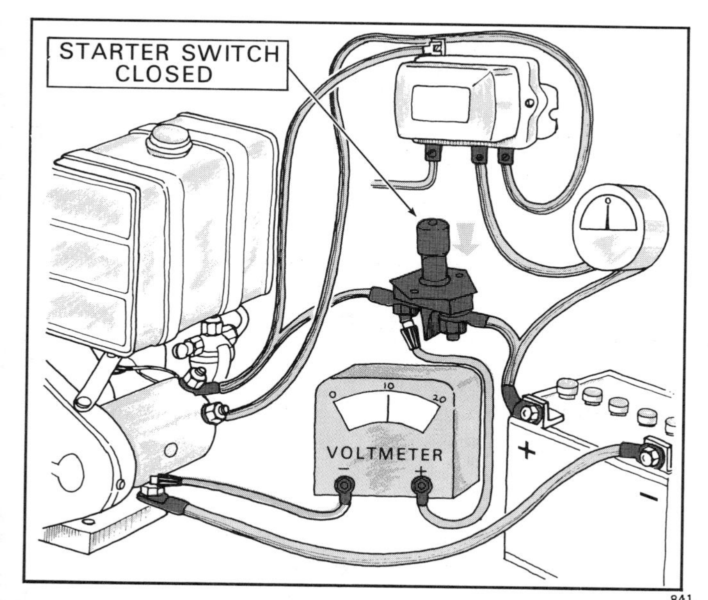

FIGURE 59. Checking the starter switch with a voltmeter.

Low (below 10 volts) voltage reading indicates a poor connection between the battery and the starter switch.

If the starter does not crank the engine, proceed to step 3.

(3) *Check for a loose connection or a worn battery cable.*

(4) *Clean and tighten the cable connections.*

(5) *Close starter switch and recheck the voltmeter.*

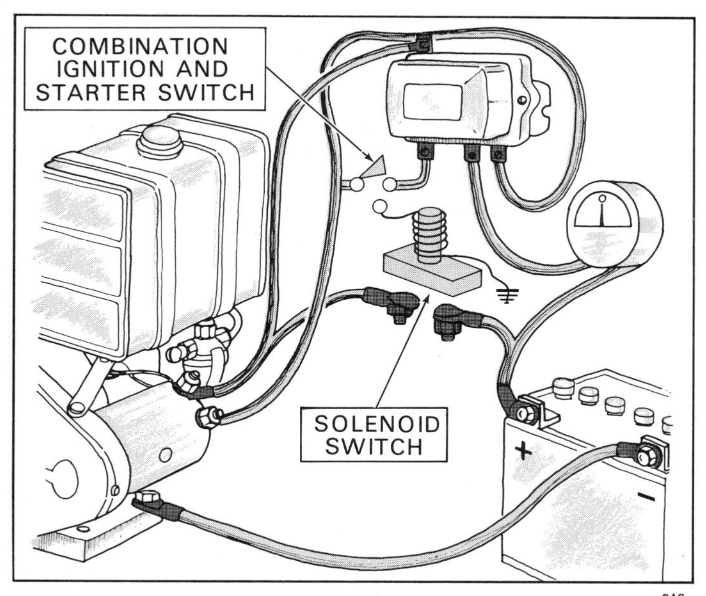

FIGURE 60. A solenoid starter switch. The solenoid is energized by a small amount of current from the battery when the starter switch is turned on. It becomes an electromagnet and closes the main starter switch.

3. *Check the operation of the starter switch (Figure 59).*

(1) *Leave the negative test lead connected to the starter-generator mounting frame; connect the positive lead to the switch terminal nearest the starter-generator.*

(2) *Close the starter switch.*

If you have **very little voltage or no voltage,** the starter switch is not closing the circuit properly. This would be true with either a plain switch as shown in Figure 59, or a solenoid switch (Figure 60).

(3) *Repair or replace the starter switch.*

(4) *Close the starter switch and recheck voltage.*

If the voltage is still low, proceed to step 4.

4. *Check the circuit between the starter switch and the starter (Figure 61).*

(1) *Connect the negative lead of the voltmeter to the generator frame (ground) and the positive lead to the "A" (armature) post on the starter-generator.*

(2) *Close the starter switch.*

The starter motor should turn satisfactorily and the meter reading should be nearly 11 volts.

If the starter motor **does not work and battery voltage is satisfactory,** this in-

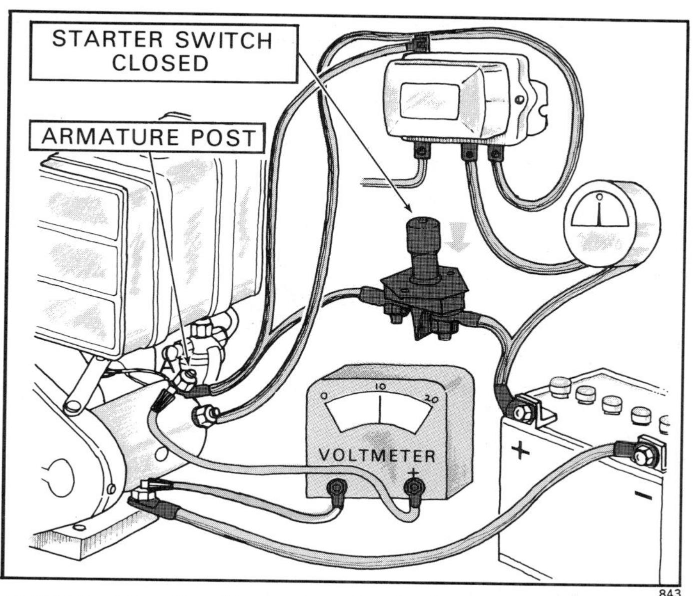

FIGURE 61. Checking the circuit between the starter switch and the starter with a voltmeter.

dicates the starter motor is at fault.

If the starter motor **does not work and there is little or no voltage,** this indicates that there is a loose or broken connection between the starter switch and the starter-generator connection.

(3) *Clean, tighten, and inspect or replace the wiring.*

(4) *Recheck with the voltmeter.*

CHECKING THE GENERATOR CONTROL CIRCUIT (12-VOLT)

If you have a **generator warning light or an ammeter,** you can tell at a glance if the generator system is working. But if the warning light comes on or the ammeter shows no charge while the engine is operating, you need to be able to determine if the trouble is in the generator or in the regulator.

Only the procedures for checking the tractor-type regulator (Figure 62) are given, since it is this type you are most likely to have on your small engine.

1. *Check the "A" and "B" leads to make certain they are not interchanged (Figure 62).*

   Sometimes this is done to bypass the regulator and get a quicker battery charge. The "A" lead should connect to the starter switch. The "B" lead should be connected to the positive terminal of the battery.

2. *Disconnect the "B" lead from the voltage regulator (Figure 62).*

3. *Connect one lead of the ammeter to the "B" lead on the voltage regulator and the other lead to the positive battery post.*

   Use an ammeter with a —30 to +30 amperage range.

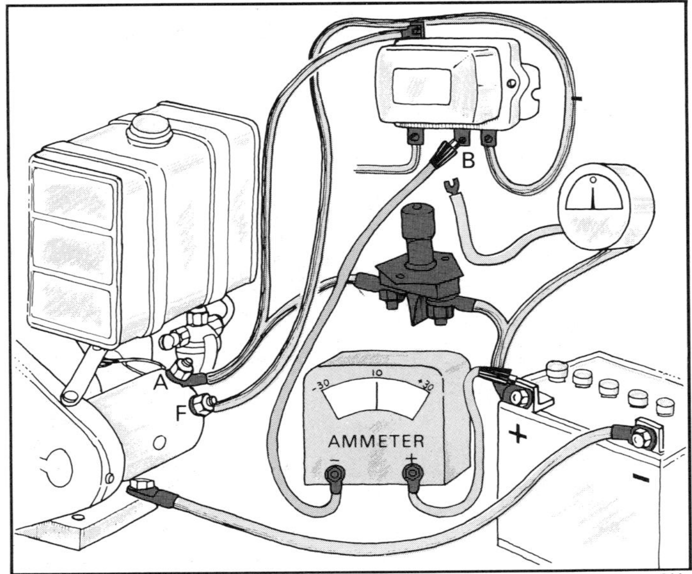

FIGURE 62. Checking the generator circuit with an ammeter on a tractor-type regulator.

4. *Start the engine and operate it at approximately 2,000 r.p.m. (fast idle).*

5. *Read the ammeter.*

   If the generator output is **too much** (10 amps *or more), proceed to step 6.*

   If the generator output is **too low** (5 amps or less—depending on the state of charge of the battery), the trouble is most likely in the regulator. A slipping drive belt or loose wiring connection might give the same indications. Check and correct these conditions first if they exist.

   Replace the regulator.

6. *Disconnect the "F" terminal on the regulator.*

   This opens the generator field circuit and should prevent generator output from building up.

   If the output **remains high,** the generator is defective. There is likely an internal short in the field winding. Refer to the heading, "Repairing the Direct-Current Starter and Generator."

   If the generator output is **zero,** the regulator is defective. The regulator points do not close or there is an open circuit. Further checks are given in steps 7 and 8.

7. *Short the "F" terminal to ground.*

   This completes the generator field circuit and the output should be excessive.

8. *Read the ammeter.*

   It should show 10 amperes, or more, because there is no control.

   If no charge is shown, you have a bad generator. Refer to the next heading, "Repairing the Direct-Current Starter and Generator."

   If you have **some charge but it is low,** the regulator needs adjusting.

NOTE: Do not tamper with a current-voltage regulator unless you know what you are doing. You can easily damage it beyond repair. Some service manuals give procedures for adjusting voltage regulators. But if you do not have procedures or if you are not thoroughly familiar with adjusting voltage regulators, it is advisable for you to see your service dealer.

TESTING THE ALTERNATOR SYSTEM

Alternator systems come in different sizes from 1.5 amps to 20 amps or more. Alternators that produce more than 4 amps are controlled by a solid state rectifier regulator. The procedures for testing them vary, so it is important that you follow procedures in your service manual when testing your system.

Procedures are given here for a **10-amp system that is commonly used.** There are two conditions which would cause you to test the system. They are (a) battery not charging or charging rate too low, and (b) battery charging rate too high.

a. If your battery is **not being charged enough (discharged),** proceed as follows:

1. *Check alternator-regulator output with DC voltmeter (Figure 62A).*

   Use a DC voltmeter (0 to 24 volts).

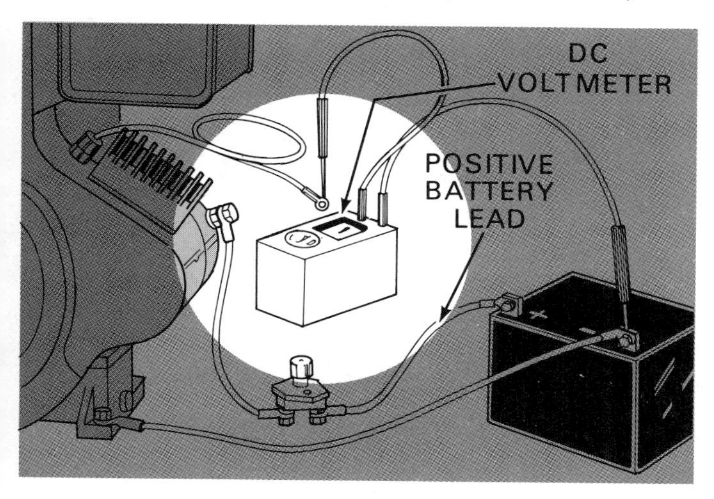

FIGURE 62A. Checking alternator-regulator output with DC voltmeter.

Disconnect the positive battery lead. Check voltage between battery lead and ground at full throttle.

If voltage is **above 14 volts,** this indicates the alternator is good. Check cables and leads for lose connections.

If voltage is **below 14 volts, but more than 0,** this indicates a defective regulator.

If voltage is 0 volts, this indicates a defective regulator or stator.

2. *Check alternator with an AC voltmeter (Figure 62B).*

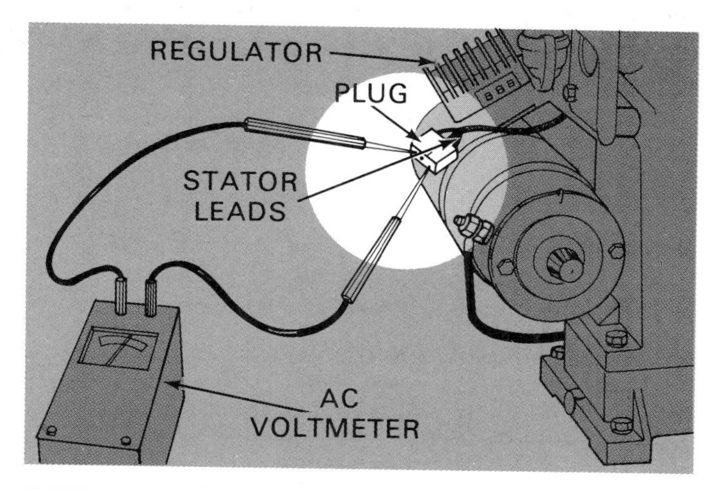

FIGURE 62B. Checking alternator output with AC voltmeter.

Use an AC voltmeter (0 to 50 volts). Disconnect plug from regulator. Connect AC voltmeter to AC terminals at the stator. Operate engine at full throttle.

If voltage is **less than 20 volts,** this indicates a defective stator.

If voltage is **more than 20 volts,** this indicates a defective regulator.

3. *Test regulator with a DC voltmeter (Figure 62C).*

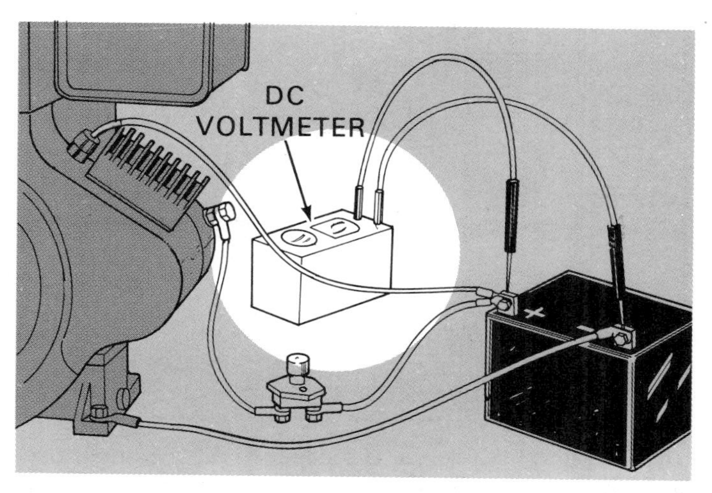

FIGURE 62C. Testing regulator.

With all leads connected, connect a voltmeter across battery terminals. Load by turning on lights or using a 12-volt headlamp.

Operate engine at **full** throttle.

If voltage is **above 13.8,** this indicates a system is good.

If voltage **does not increase,** this indicates a defective regulator.

b. If battery **charging rate is too high (you have to add water too often),** proceed as follows:

1. *Check voltage across battery terminals with a DC voltmeter (Figure 62C).*
2. *Connect voltmeter across battery terminals.*
3. *Operate engine at full throttle.*

   If voltage is **higher than 14.7,** this indicates a defective regulator.

REPAIRING THE DIRECT-CURRENT STARTER AND GENERATOR

Direct-current starters and generators are equipped with **commutators** and **brushes** that require occasional maintenance (Figure 63).

1. *Wipe dirt from starter or generator housing.*

   If it is covered with an oily film, use cloth dampened with petroleum solvent.

2. *Check for worn bearings.*

   Bearings are located at both ends of the armature shaft. You can check them for excessive play by moving the shaft up and down by hand. Usually worn bearings in a generator are noisy during operation.

   Another check for worn bearings is to look at the armature after it has been removed to see if it has been touching the field windings. Worn bearings will allow this to happen.

3. *Remove cover band (Figure 64a).*

   Some *generators* have no cover band. If yours is of this type, you will need to remove the through bolts and pull off the end frame on the commutator end (Figure 64b).

4. *Inspect for thrown solder (Figure 65).*

   If a starter or generator has been **overheated,** you can tell by the ring of solder that has been thrown against the band (Figure 65) or against the inside of the housing, if of the type without a band. If it appears that the starter or generator has been overheated, take it to a service shop. It may require expert repair.

5. *Check brushes for wear and binding action.*

   NOTE: When checking a brush for wear, do not pull on the wire that connects to the brush while it is being held under spring tension. This may loosen the wire connection to the brush. Remove the tension clip from the brush first. Avoid snapping the clips down on the brushes. This causes them to chip and crack.

   If you are checking a **generator,** look for either two or three brushes.

   If you have a **starter-generator combination,** look for two brushes. If the *old brushes are worn (Figure 66)* until the clips are pressing on the brush holders instead of the

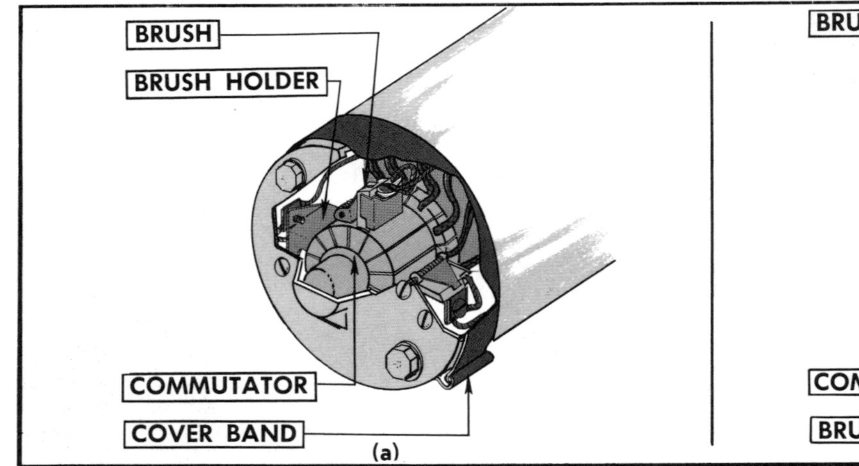

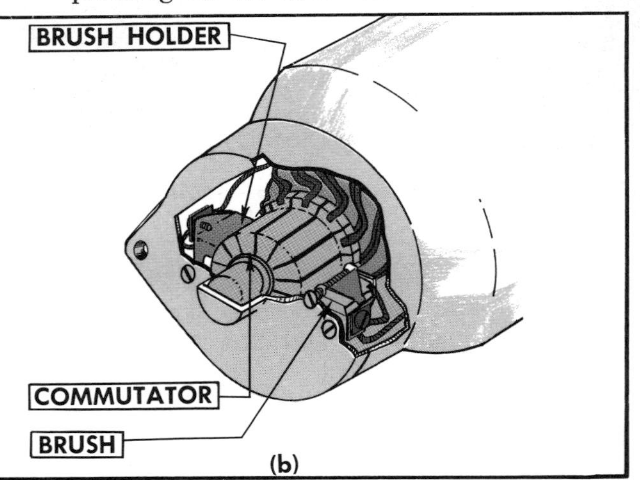

FIGURE 63. (a) The starter is equipped with a commutator, 4 brushes and 4 brush holders that require occasional maintenance. (b) The generator is equipped with a commutator and either 2 or 3 brushes and brush holders. Starter-generator combinations have 2 brushes.

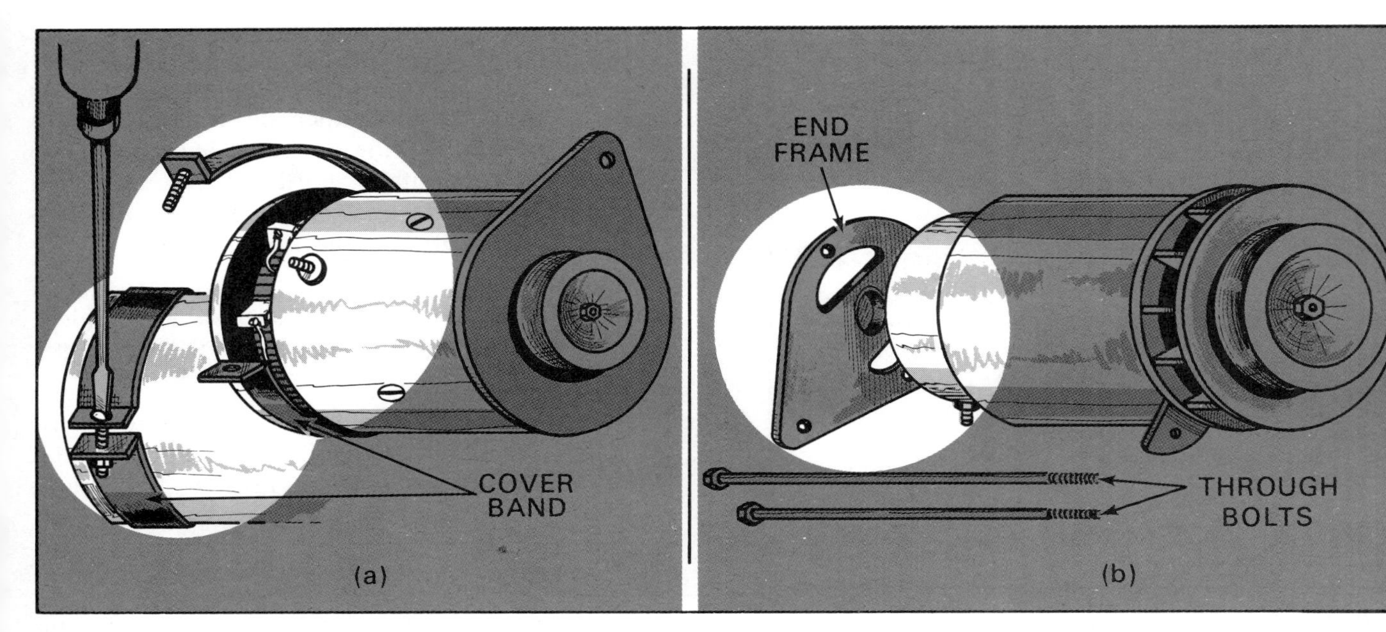

FIGURE 64. (a) Cover bands are generally held in position with a clamp screw. Unscrew and remove band. (b) Generators without cover bands must have through bolts removed and commutator end frame pulled from main frame.

brushes, or if they are worn to less than half their original length, replace them with new ones. Follow the procedure in step 6.

If you are checking a **generator** that has the end frame removed, the brushes are usually attached to it.

Disconnect the brush-wire lead wire where it fastens to the brush holder (Figure 67), lift brush tension clip, remove old brush and slip new one into place. Reconnect wire lead to brush holder.

6. *Replace worn brushes.*

   Install brushes with the beveled edge toward the direction of rotation. Otherwise, the brush edges will lock in the low segments of the commutator and break.

7. *Check brushes for binding action in holder.*

   If brush *tends to bind* in the brush holder, remove and wipe brush holder with a clean cloth. Do not use a petroleum solvent. It tends to soften the insulation on the wires.

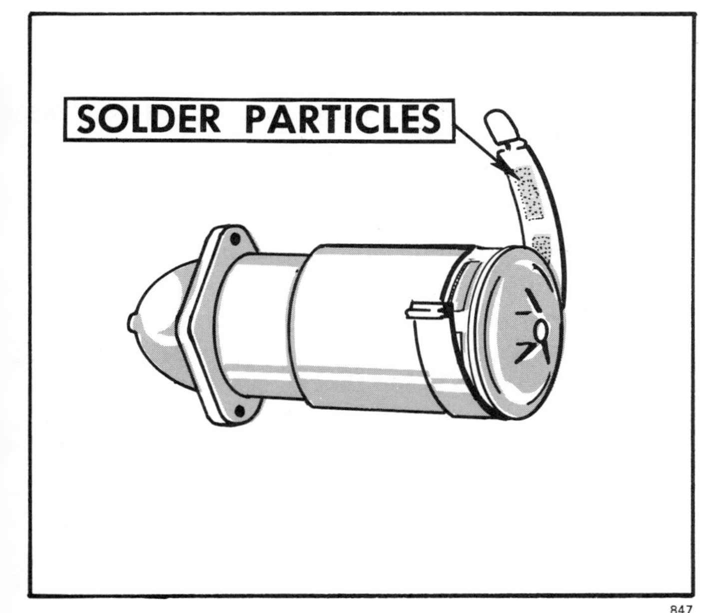

FIGURE 65. Thrown solder on cover band indicates starter or generator has been overheated.

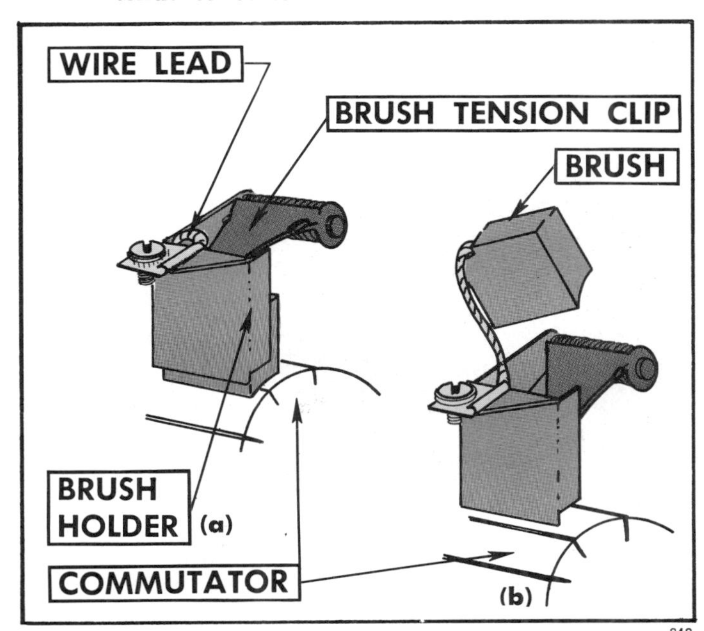

FIGURE 66. (a) Brush, brush holder and tension clip. (b) Brush removed from holder.

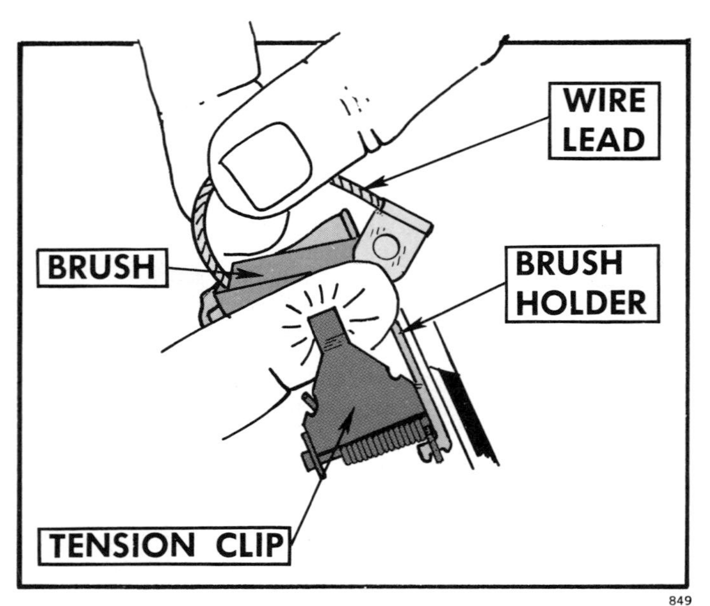

FIGURE 67. Removing a brush.

8. *Check electrical connections for tightness.*

9. *Inspect commutator for wear and roughness.*

   If the commutator is **rough or out-of-round**, it will have to be machined in a lathe. This must be done by an experienced service man.

   If the commutator **appears to have only dirt and a glaze** on the surface (Figure 68), proceed with step 10. (Replace end frame if it has been removed.)

10. *Remove dirt and glaze from commutator surface (Figure 68).*

    Clean the commutator on a **generator** while the engine is running slowly if possible. If for any reason you have disconnected a wire lead on your generator, reconnect it before starting engine. If left disconnected, the generator will burn out.

FIGURE 68. If commutator is dirty and has a glazed surface, it should be cleaned.

    Use No. 00 sandpaper on a stick with a square end, moving the stock back and forth on the commutator until all gum and dirt have been removed. You can also use a **brush-seating stone** for the same purpose.

    NOTE: Do not use emery cloth. Remaining particles will cause arcing, burning and rapid wear. Do not use a solvent. It damages the wire insulation.

    Clean the commutator on your **starter** while the starter is turning the engine. Disconnect the spark-plug wire to prevent the engine from starting.

11. *Seat new brushes on commutator.*

    In cleaning the commutator, there has probably been enough abrasive action on the brushes so that they are already seated and fitting squarely on the commutator surface. If not, continue to use the brush-seating stone on the commutator until the brushes are seated; or place a strip of sandpaper between the brushes and commutator with the abrasive side against the brush. Pull sandpaper strip in the direction that will cause the brush to fit the curvature of the commutator.

12. *Blow dust from commutator, brush holders and casing.*

    Removing the dust particles prevents further abrasive action on the brushes and commutator.

13. *Replace band if one is used.*

14. *Polarize the generator before starting the engine (Figure 69).*

    If any of the wire leads were disconnected while the starter or generator was being serviced, **polarize the generator.** On an **externally grounded generator,** reconnect the wires and touch a short jumper wire momentairly between the two posts on the regulator marked "B" and "G" (sometimes marked "A") (Figure 69a) You can recognize an externally-grounded generator by the external ground wire (Figure 69a).

    On an **internally grounded generator,** momentarily touch the ends of the "B" wire and the "F" wire together (Figure 69b). These procedures establish correct polarity of the generator.

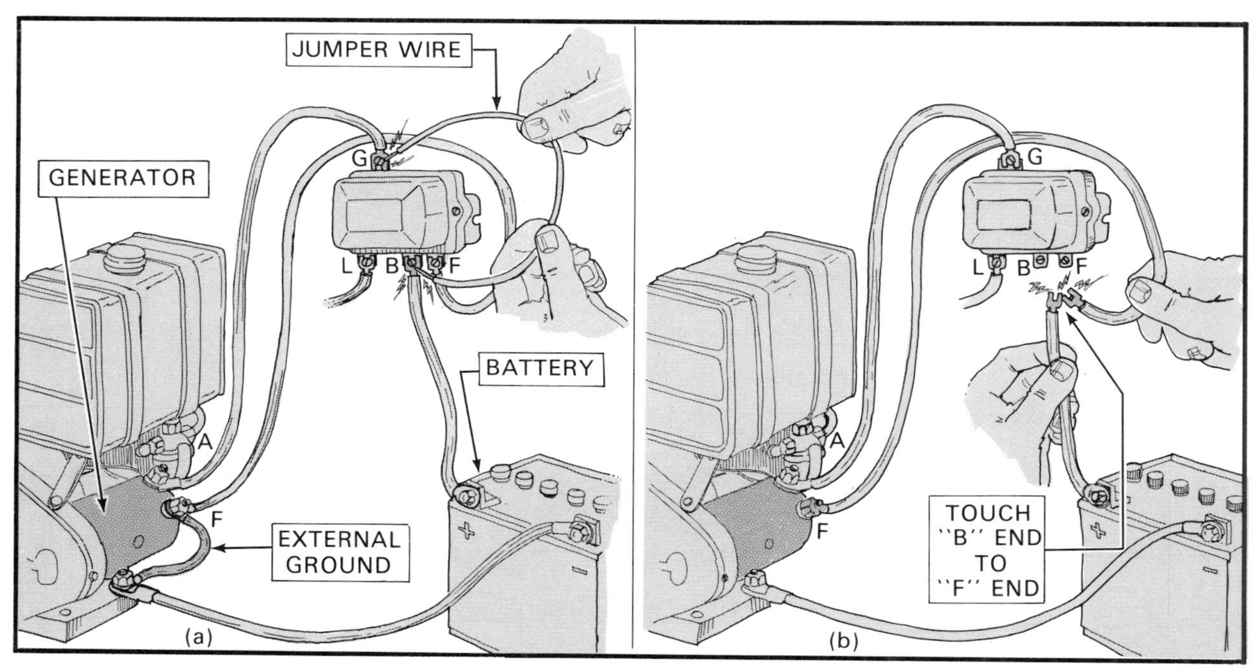

FIGURE 69. To polarize the (a) externally grounded generator, touch a short jumper wire to the "B" and "G" posts of the regulator momentarily. (b) On the internally grounded generator, touch the ends of the "B" wire and the "F" wire together momentarily.

If *you do not polarize the generator* and it has become reversed, you may run down the battery, burn out the generator or burn out the regulator points.

REPAIRING THE FLYWHEEL ALTERNATOR

To replace the alternator and/or rectifier (voltage regulator), follow procedures in your service manual. Procedures are given here for a type that is commonly used. Proceed as follows:

1. *Disconnect the leads.*

   Disconnect the leads to the rectifier (voltage regulator).

   If you have **solid state ignition,** disconnect the stator from the trigger coil.

2. *Remove the rectifier (70).*
3. *Remove the flywheel.*
4. *Remove the stator (71).*
5. *Replace stator, flywheel and rectifier.*
6. *Check for proper operation.*

Note: Alternator systems do not need to be polarized. Do not attempt to do so.

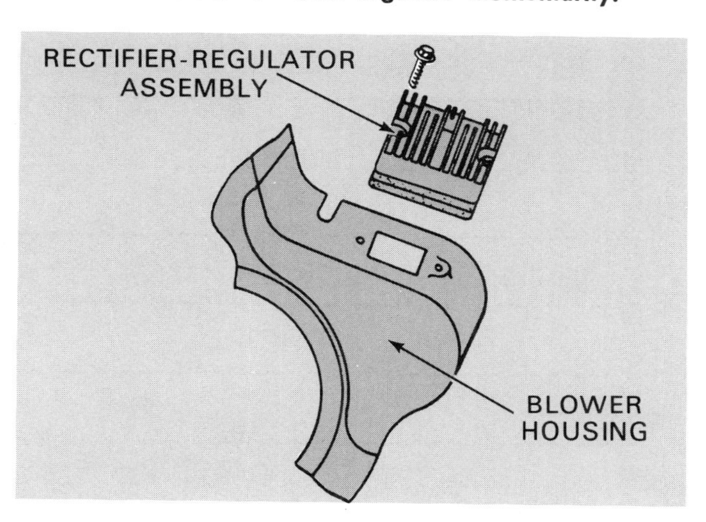

FIGURE 70. Rectifier removed.

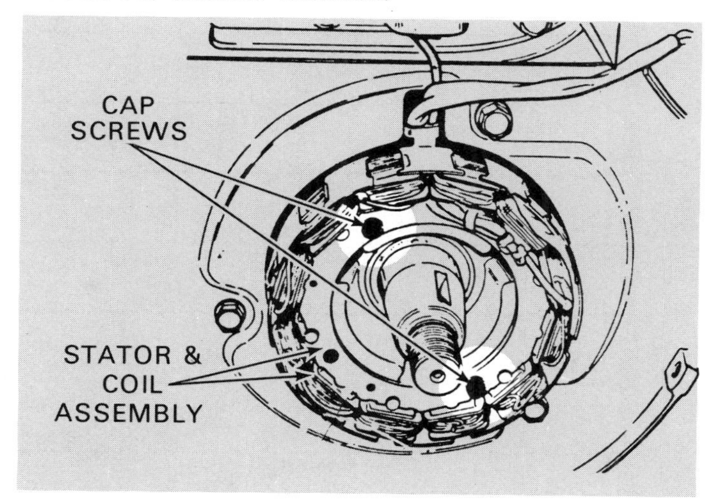

FIGURE 71. Usually cap screws hold the stator.

NOTES

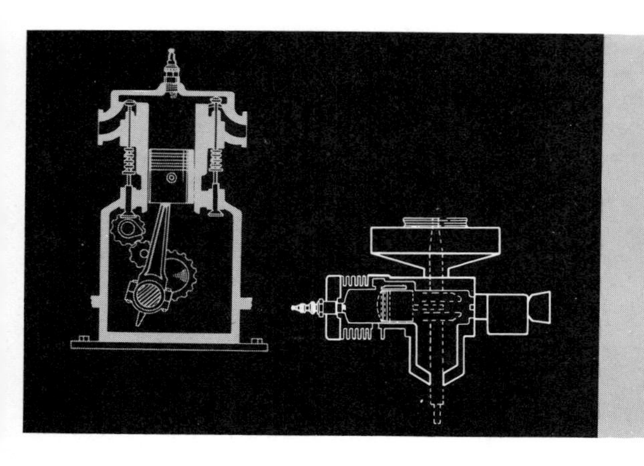

II. Maintaining & Repairing Ignition Systems

The purpose of the ignition system on your small engine — and on all spark-ignition engines — is to provide a strong spark in the combustion chamber at the proper time for igniting the fuel-air mixture.

The spark has two important requirements:

- **The spark must be of the proper strength.**

  It must be "hot" enough to ignite the fuel-air mixture in the combustion chamber.

  If the spark is too weak, the fuel will not ignite.

  If the spark is too strong, the spark-plug electrodes will burn.

- **The spark must take place at exactly the proper time.**

  The best time for the spark to occur on most engines is just before the piston reaches top-dead-center on the compression stroke. This gives the burning process a head start so the expanding gases (due to combustion) will be most effective in pushing the piston down. See Figure 5, Part 1.

TYPES OF IGNITION SYSTEMS

There are three types of ignition systems commonly used on small gasoline engines:

- **Magneto-ignition systems (Figure 72).**
- **Battery-ignition systems (Figure 73).**
- **Solid state ignition systems (Figure 74).**

855

FIGURE 72. Most magneto-ignition systems on small gasoline engines are located behind the flywheel shroud and are hard to reach.

Magneto-ignition systems are easily adaptable to small engines, especially those with no electrical load other than that of igniting the fuel in the combustion chamber.

The magneto produces its own electricity without the aid of a battery or generator. Magneto-ignition systems are simple and economical. They give very little trouble and are easy to maintain.

Battery-ignition systems are common on automobiles, trucks and tractors. They are also used on small engines where additional electrical loads are needed. Small garden-type tractors, with starter and lights, usually have a battery-ignition system. The presence of a battery and generator, however, does not necessarily mean you have a battery-ignition system. Some manufacturers continue to equip their engines with a magneto ignition but add a battery and a generator for starting and lights. You can recognize the battery-ignition system by the presence of a can-shaped **ignition coil,** in addition to the battery and the generator (Figure 73).

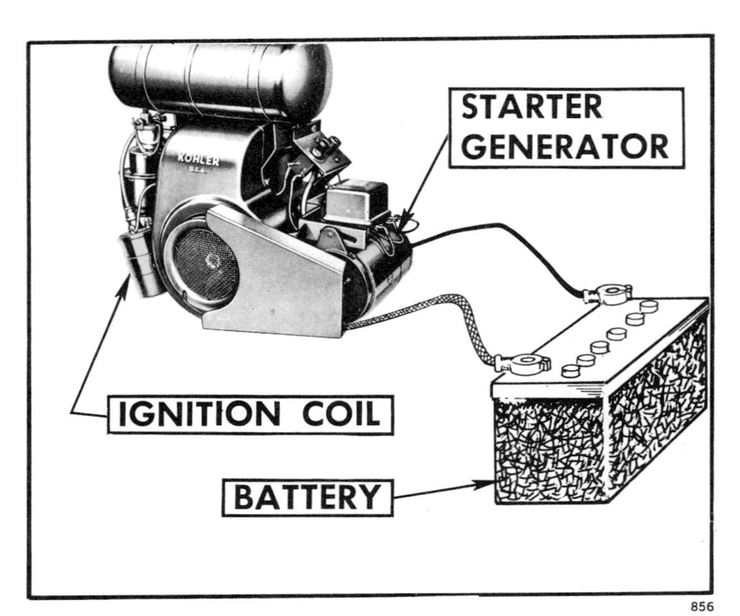

FIGURE 73. A battery-ignition system may be recognized by the presence of an ignition coil and a battery.

Solid state ignition systems have no mechanical means for breaking the circuit. The ones used in small engines receive electric power from a magneto-type alternator (Figure 74).

How it works is described on pages 52 and 53.

Maintaining and repairing ignition systems are discussed under the following headings:

A. Maintaining and repairing magneto and solid-state ignition systems.

B. Maintaining and repairing battery-ignition systems.

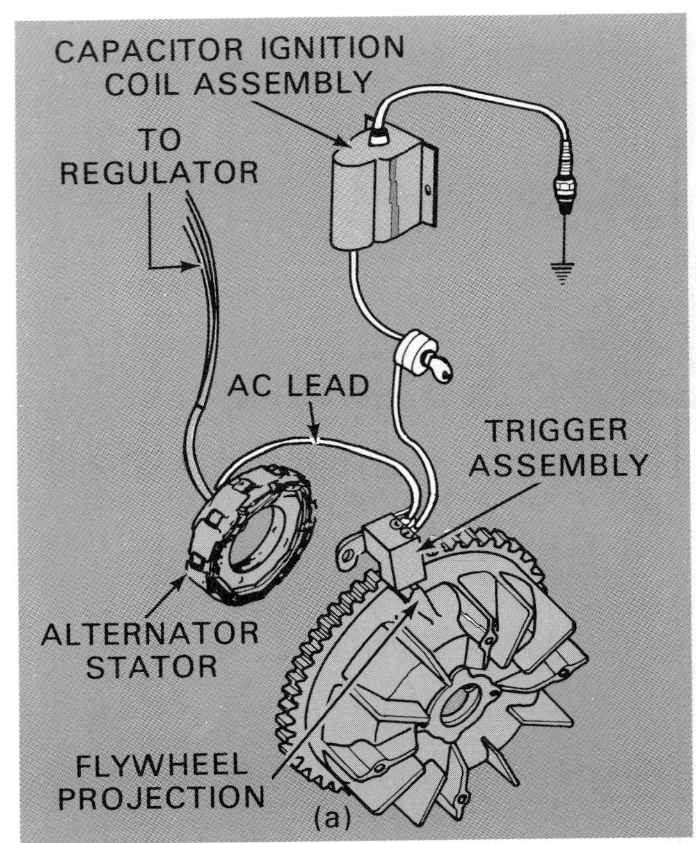

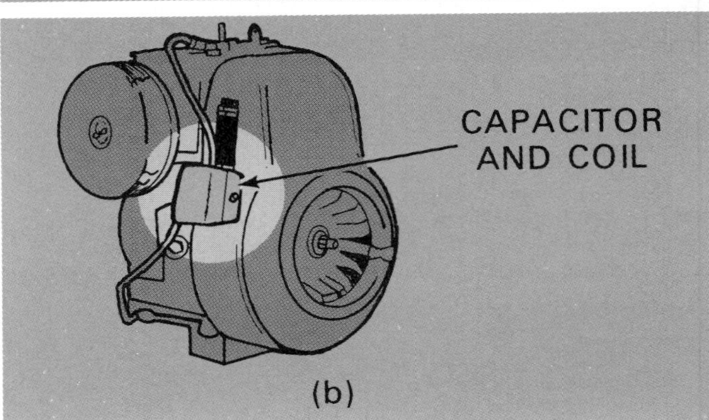

FIGURE 74. (a) Solid state ignition systems consist of an alternator, capacitor-coil assembly and trigger assembly. (b) What you see on the outside of the engine is the capacitor and coil.

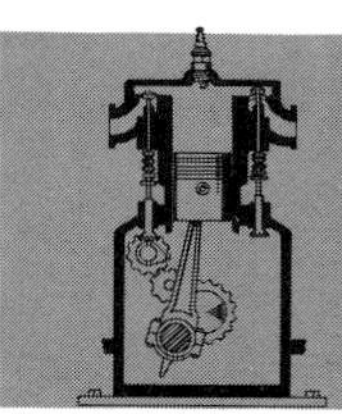

A. MAINTAINING AND REPAIRING MAGNETO AND SOLID STATE IGNITION SYSTEMS

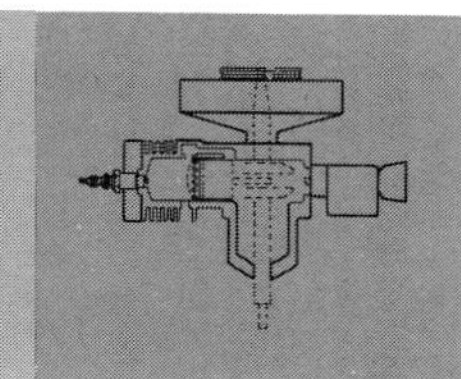

Magnetos are used on most small engines to supply the electrical power for the spark at the spark plug which ignites the fuel-air mixture in the combustion chamber. If the magneto on your engine is not functioning properly, your engine will be hard to start, and it will not develop maximum power.

Maintaining and repairing magneto-ignition systems are discussed under the following headings:

1. Types of magnetos.
2. Principles of operation.
3. Importance of proper maintenance.
4. Tools and materials needed.
5. Checking the magneto for proper operation.
6. Removing and checking the flywheel.
7. Checking and conditioning the breaker-point assembly.
8. Checking the condenser.
9. Adjusting the breaker-point gap.
10. Removing and replacing breaker points.
11. Timing the magneto to the engine.
12. Timing the breaker points to the magneto.
13. Checking and adjusting the stator-plate (armature) air gap.
14. Checking the magneto coil.
15. Removing and replacing the stator plate and coil.
16. Installing the flywheel.
17. Testing the solid state ignition system.

TYPES OF MAGNETOS

There are two types of magnetos used on small engines. They are (1) a **flywheel-type** magneto, which is not visible from the outside of the engine but is built in and around the flywheel (Figure 75); and (2) an **external type,** which is a self-contained unit mounted on the side of the engine (Figure 76).

SPARK PLUG
LAMINATED IRON CORE
FLYWHEEL MAGNETO
MAGNET
BREAKER POINTS
FLAT ON CRANKSHAFT
CONDENSER

857 B & S

FIGURE 75. A cut-away view of a flywheel-type magneto.

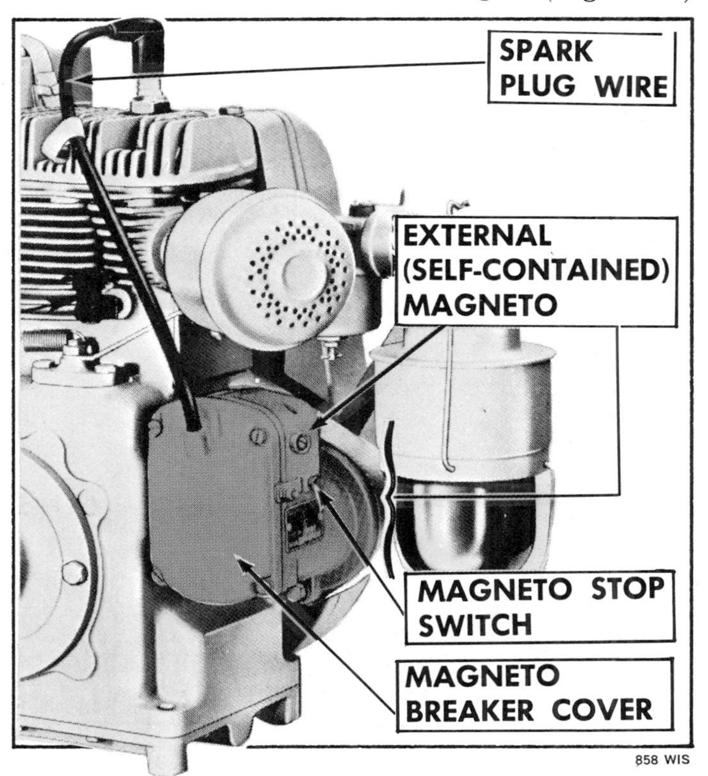

FIGURE 76. A typical external-type (self-contained) magneto.

PRINCIPLES OF OPERATION

Before you can understand the magneto-ignition system, you need to know how the magneto generates electricity.

To understand this, you need to know two basic principles of how electricity is generated and how voltage can be changed to meet different needs.

When magnetic lines of force are cut by a closed conductor, voltage (electrical pressure) is induced and current flows in the conductor. Magnetic lines of force are supplied by a permanent magnet (Figure 77). In Figure 78 the coil is stationary. The lines of force from the magnet are cut by the coil as the magnet rotates past it. This action induces voltage into the coil, thus causing current to flow. This is how current is generated. The mechanism is called a generator.

The amount of current flow is determined by three conditions:

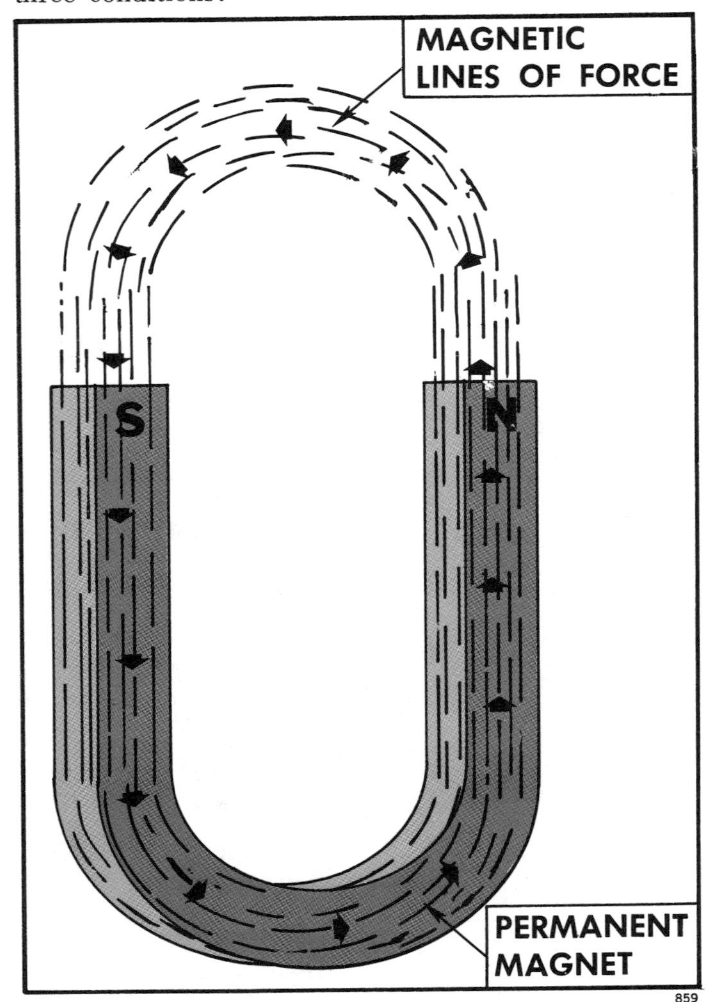

FIGURE 77. A permanent magnet has imaginary lines of force extending from the N (north) pole to the S (south) pole and through the length of the magnet.

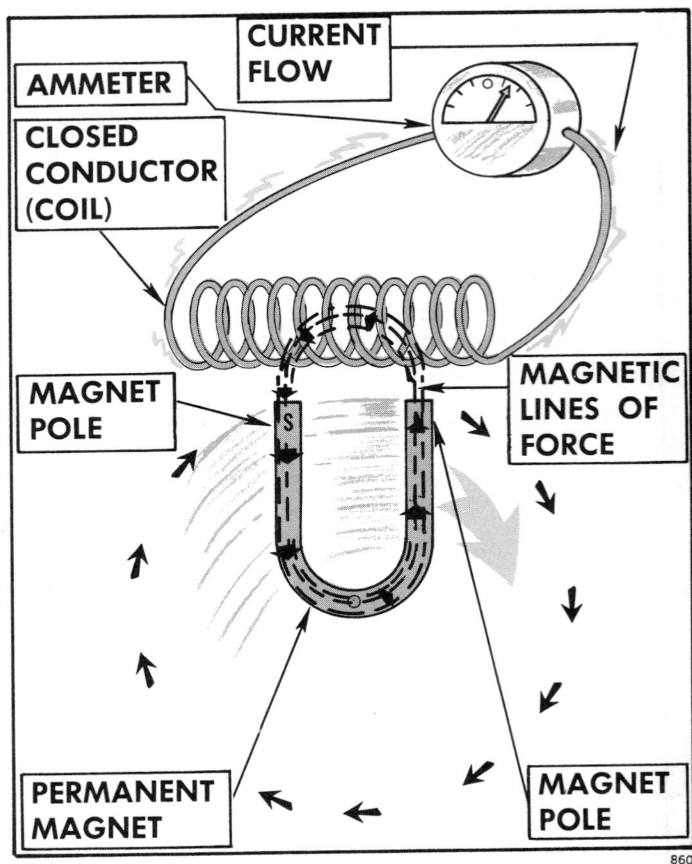

FIGURE 78. A simple generator. When these magnetic lines of force are cut by a closed conductor, current flows in the conductor.

- The *number of turns of wire in the coil.*
- The *intensity of the magnetic field.*
- The *speed at which the magnetic lines of force are cut.*

To develop the necessary voltage to jump the gap at the spark plug, a *second principle* is involved. It is that of a **transformer.** It consists of two coils wound together (one over the other) as shown in Figure 79. One coil is called a "primary" coil and the other, a "secondary" coil. The transformer primary coil is supplied with current by the generator (Figure 79). The current from the generator also flows through the primary coil. This causes a magnetic field to be developed around the coil. The iron core in the center becomes magnetized and establishes additional lines of force.

As the magnetic lines of force build up around the magnetized iron core, they are cut by the windings of the secondary coil. Voltage is induced in the secondary coil. Current flows momentarily. When the current flow in the primary coil is interrupted, the magnetic field collapses. Lines of force are cut by the windings of the secondary coil, and voltage is again induced in the secondary coil.

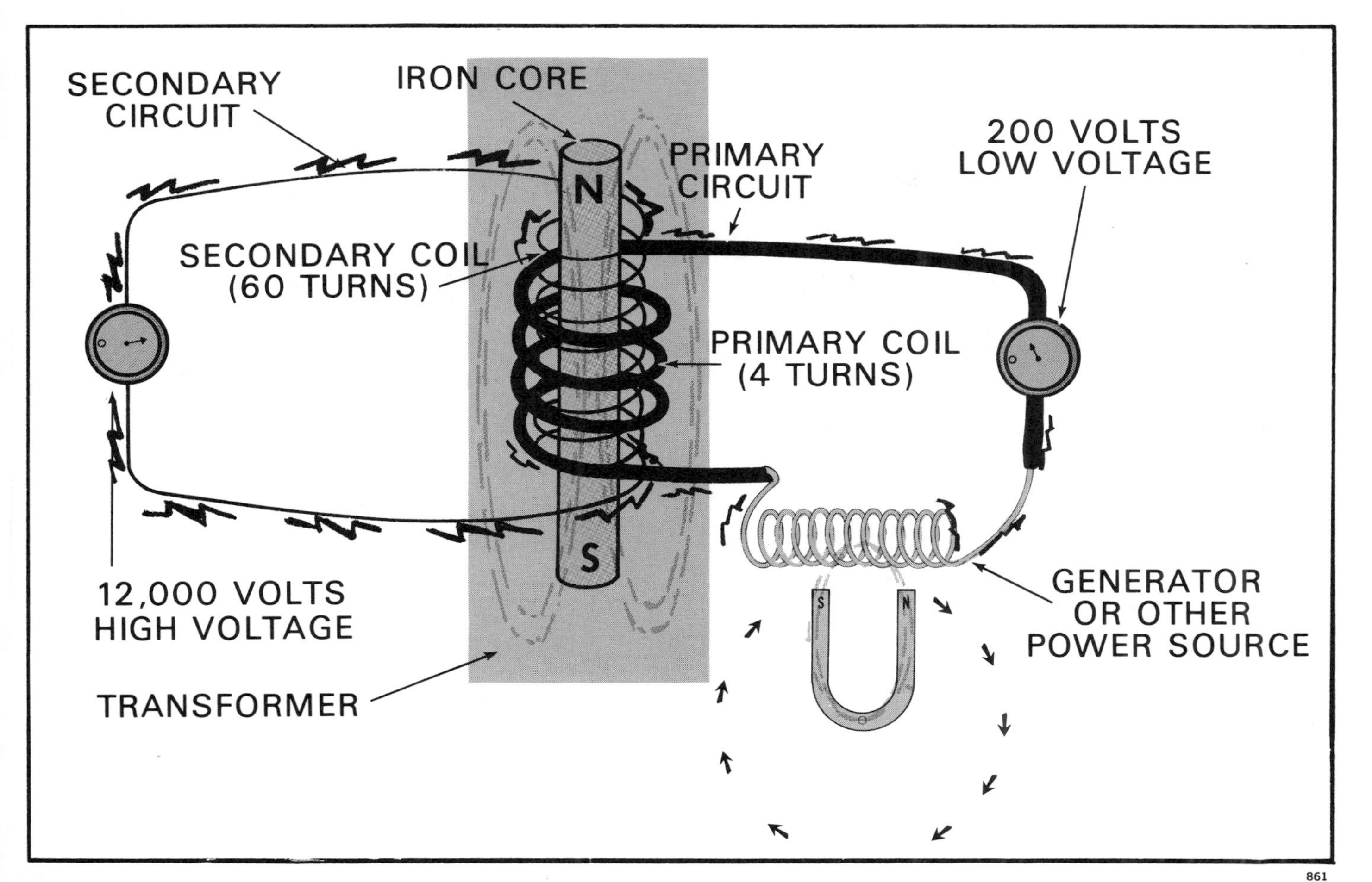

FIGURE 79. Principle of the transformer. When voltage is induced in the primary coil, it is also induced in the secondary coil. The amount is proportional to the ratio of the number of turns in the primary.

The amount of voltage induced in the secondary coil is directly related to the number of turns it contains in relation to the number of turns in the primary coil. If the secondary coil has twice the number of turns as the primary coil, the voltage induced will be twice that in the primary. For example,

Transformer coils used in ignition systems provide high voltage to assure a good spark at the spark plug. They are usually called "spark coils" or "ignition coils." The voltage for small engines is stepped up from approximately 200 volts to approximately 12,000 volts. This means there are 60 times more turns of wire in the secondary circuit than in the primary (12,000 volts $\div$ 200 volts $=$ 60 turns).

FLYWHEEL MAGNETO. The principles of the generator and the transformer apply to a flywheel magneto (Figure 80). A permanent magnet rotates past a coil. The coil consists of a stationary primary and secondary circuit. Both are wound around the center leg of a three-pronged soft-iron "core" called an armature or yoke. In a magneto the **primary serves as a primary coil and a generating coil.**

At the point where the 2 poles of the magnet start to align—one with the left-hand prong of the armature and the other with the center prong of the armature—magnetic lines of force flow through the center and left-hand prongs of the armature. Lines of force are cut by the spark coil (Figure 80a).

Voltage is induced in the primary coil and current flows in the primary circuit because the breaker points are closed and the circuit is complete. No current flows through the secondary circuit at this time because it is open at the spark-plug electrode, and there is not enough voltage generated to cause the current to jump the gap.

Current flow in the primary increases the strength of the magnetic field by adding more lines of force.

As the magnet continues to rotate to a position where the N pole of the magnet is a little to the right of center of the middle prong, the lines of force start to flow in the right prong and the center prong of the

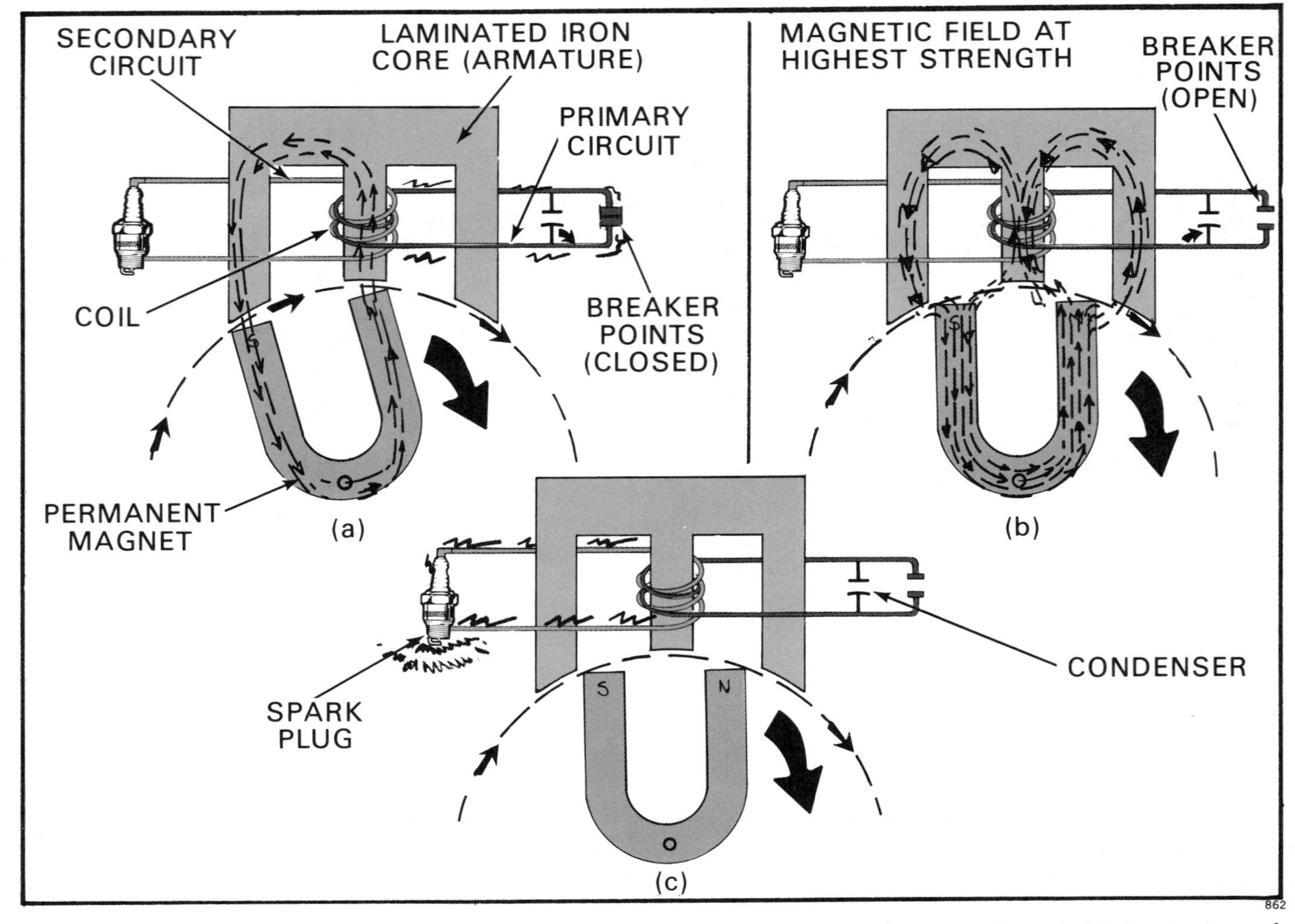

FIGURE 80. Principles of a magneto. (a) Magnetic lines of force build up when the poles of a permanent magnet, which is embedded in the flywheel, approach alignment with the left-hand prong and center prong of the armature. As the magnet continues to rotate, lines of force are cut and voltage is induced. Current flows in the primary circuit so long as the breaker points are closed. The magnetic field is strengthened. (b) A point is reached when the magnetic field is strongest. The breaker points open. Current flow stops in the primary. (c) The sudden collapse and reversal of the magnetic field induces enough voltage in the secondary coil to force current across the spark-plug electrode gap, thus causing a spark.

armature. When this happens, the lines of force change direction. *The magnetic field is strongest at this point. Therefore, the breaker points are timed to open at this instant (Figure 80b).* The flow of current stops in the primary circuit. The strong magnetic field collapses. This sudden collapse of the strong magnetic field induces enough voltage in the secondary coil to cause the current to jump the gap at the spark plug (Figure 80c).

One reason current does not jump the gap at the breaker points is that the voltage is not so great in the primary as it is in the secondary circuit. Another reason is that there is a **condenser** in the primary circuit (Figure 81).

Figure 82 shows one method by which the magneto parts may be arranged on a flywheel magneto.

The permanent magnet is embedded in the flywheel or in the flywheel rotor. It rotates past a coil which is wound around the center prong of a laminated iron core called the "armature" (Figure 82a).

To stop a magneto-equipped engine, the ignition switch is closed (grounded), or the secondary circuit is grounded at the spark-plug terminal. This bleeds off the electrical charge from the primary circuit so that no spark can develop. The engine stops.

EXTERNAL MAGNETO. **The operating principle of the external (self-contained) magneto** is the same as that of the flywheel magneto (Figure 83), except it looks different.

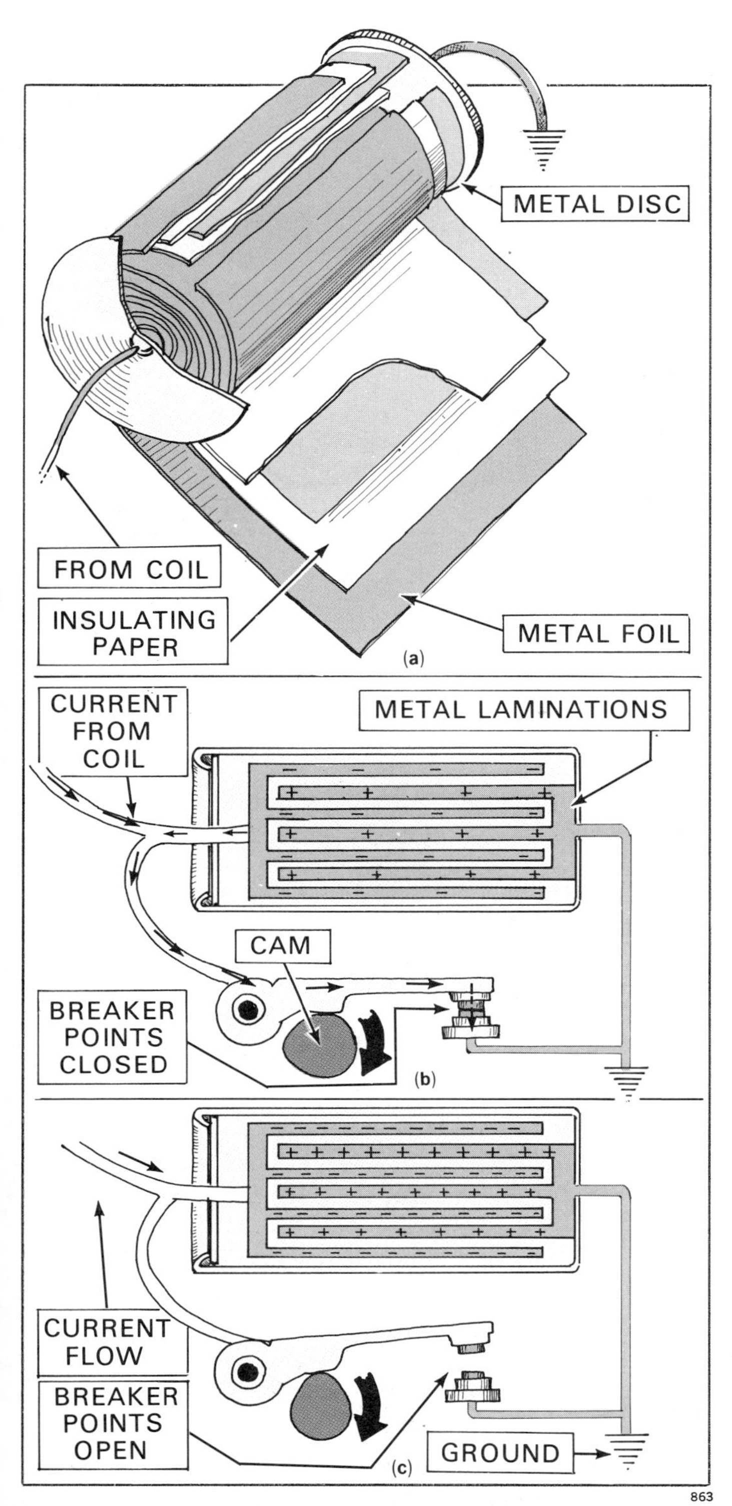

FIGURE 81. A condenser is a reservoir, or a "surge chamber," for electric current. (a) It consists of metal laminations (sheets) separated by insulating paper. (b) When the breaker points are closed, current by-passes the condenser and flows through the breaker points to the ground. (c) When the breaker points are opened, the condenser absorbs most of the excess current, thus preventing arcing at the breaker points. Some flows back into the primary circuit and helps in the reversal (breaking down) of the magnetic field in the iron core which makes the spark stronger. Without the condenser the breaker points would soon burn up.

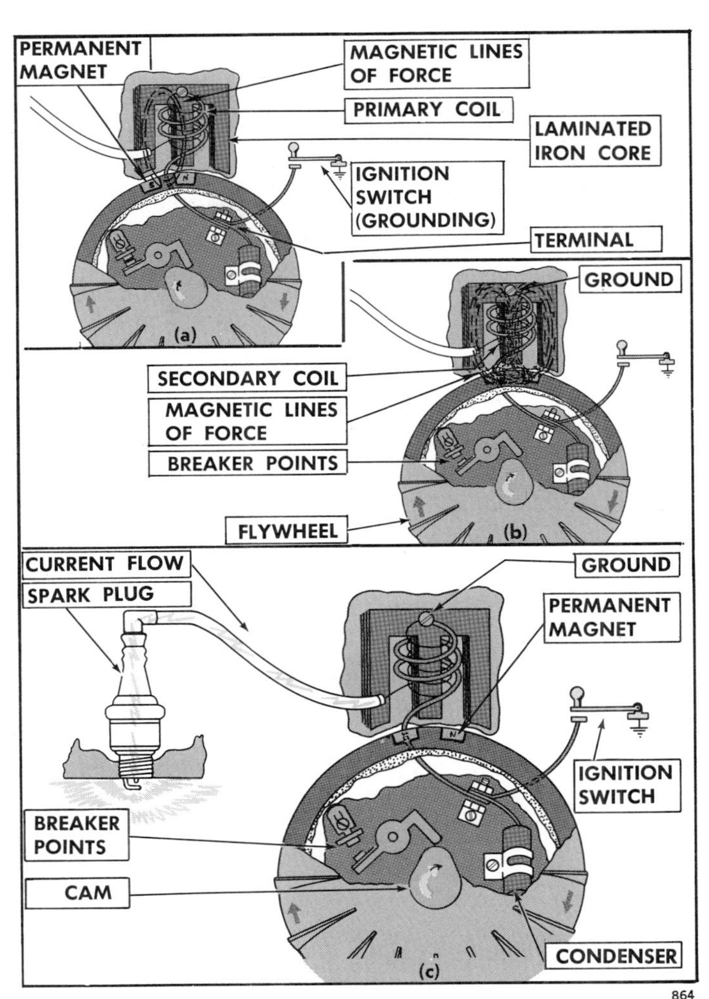

FIGURE 82. Principles of the flywheel magneto-ignition system. (a) Magnetic lines of force are built up around a coil by the movement of a permanent magnet past the armature. (b) As the permanent magnet passes the armature, the lines of force break down, thus causing a high voltage to be induced in the secondary circuit of the coil. (c) The high voltage causes a spark at the spark plug.

As the *magneto rotor rotates,* magnetic flux lines are established in the laminated iron frame (armature) (Figure 83a). The magnetic lines of force develop and collapse with each half turn of the rotor. The expanding and collapsing of the lines of force induce voltage into the primary winding, thus causing current to flow in the primary winding while the breaker points are closed.

When the current in the primary winding is greatest (Figure 83b), the breaker points open. The lines of force collapse and induce high voltage into the secondary coil. This is enough to cause a spark when it reaches the spark plug.

The **condenser** serves the same purpose here as in the flywheel magneto.

The **impulse coupling** on the rotor shaft is for starting. It has two functions: (1) It retards the

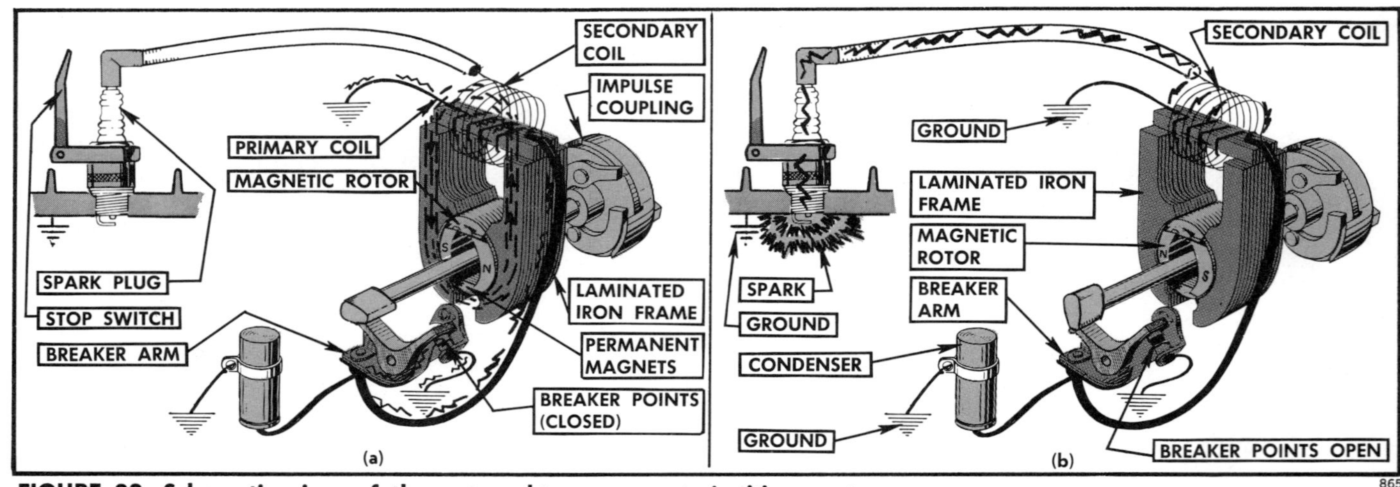

FIGURE 83. Schematic view of the external-type magneto-ignition system.

865

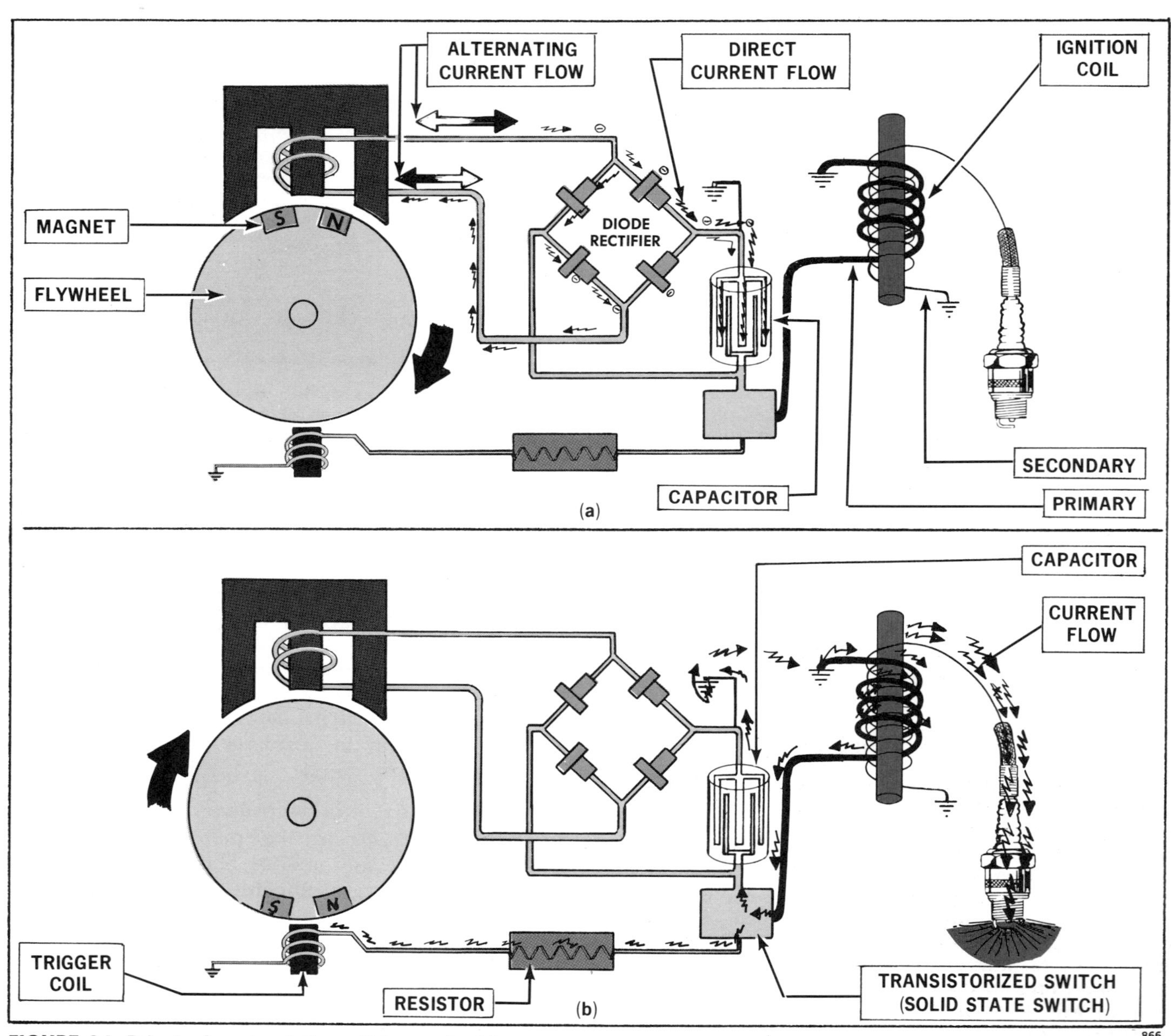

FIGURE 84. Principal parts of the solid-state ignition system.

866

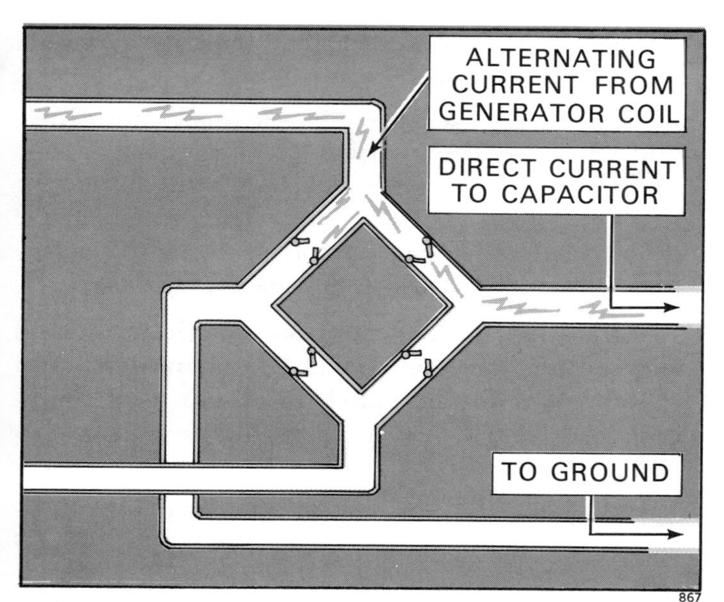

FIGURE 85. A diode rectifier is an electrical device that will allow current to pass in one direction only.

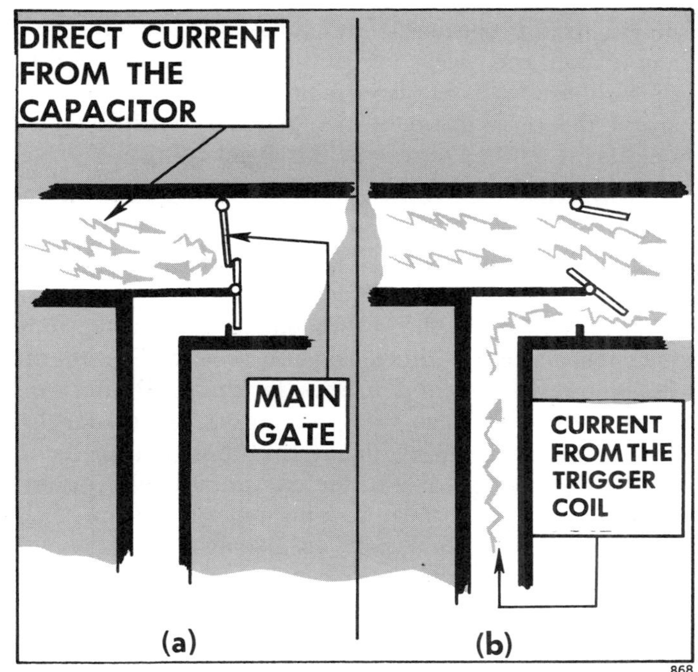

FIGURE 86. The trigger coil and the transistorized rectifier (solid-state switch) serve the same purpose as the breaker points on the conventional system, but there is no mechanical movement. (a) A high resistance (gate) in the transistor prevents current flow until the resistance is reduced (opened) by (b) a small current flow from the trigger coil.

spark for cranking, and (2) it flips the magnetic rotor at the proper time to produce a stronger spark through delayed spring action. After the engine starts, the impulse coupling automatically disengages.

SOLID-STATE IGNITION. The solid-state ignition system (Figure 84) used on small engines is the "capacitor-discharge type." It is similar in one respect to the flywheel magneto. That is, the initial current is generated in a coil by a magnet on the flywheel (Figure 82). But how the current gets to the spark plug at the proper time is different.

Current from the generator (input) coil is directed through a **diode rectifier** where it is changed from alternating current to direct current (Figure 85).

A transistor is an electrical device which can be used to control the flow of current in a circuit. It is a resistor whose resistance is changed by a small amount of current supplied to a third part of the transistor.

When explained electrically, a transistor is often difficult to understand. To help you develop an image of what happens, study Figure 86. If a transistor were a mechanical device, it would block the current from the capacitor with a main gate until it is needed. When needed, a second gate would be opened by a small current flow—enough to unlatch the main gate so the capacitor charge would be released. Once the charge is spent, the gates return to the original positions.

Direct current flows from the diode rectifier to a **capacitor** where it is stored momentarily (Figure 84a). A capacitor works on the same principle as a condenser (Figure 81). The term "capacitor" is just another name for a condenser.

The current remains stored in the capacitor until the flywheel rotates one half turn (Figure 84b). At this point the piston is in the proper position for combustion to take place—just before top-dead-center, compression stroke. As the **magnet** on the flywheel passes the **trigger coil,** a small amount of current is generated in the trigger coil. This current flows through the **resistor** and the **transistorized switch.** This small current flow through the transistorized switch completes the circuit which allows the charge in the capacitor to escape (Figure 84b and 86). It flows through the ignition coil where it is stepped up enough to jump the gap at the spark plug (Figure 84b). Thus, it is called the "capacitor-discharge type" of solid state ignition.

There are other types of solid state ignition systems but since they are not used on small engines, they are not discussed here.

IMPORTANCE OF PROPER MAINTENANCE

It is important that the ignition system on your engine be electrically sound and in good mechanical condition for your engine to function properly.

The breaker points must be clean and in good condition. Dirty, pitted or corroded points will slow the original build-up of current in the primary circuit because they make a poor electrical connection and resist current flow. When the points open, there will not be a sharp, clean-cut break to speed the breakdown of the primary current which results in a weak spark at the spark plug.

The breaker points must be adjusted properly. *Points that are too close together* will retard (slow) the timing because the cam must move farther to open the points. Retarded timing causes the spark to occur late, and the piston passes the point where the most effective ignition should occur. The engine loses power. *Points that are too far apart* (timing over advanced) will cause the spark to occur too soon before the piston reaches the optimum position for maximum compression. Again the engine will lose power. It will require a faster cranking speed and may cause the engine to "kick" — rotate backwards — when cranked. If the engine starts, it will run at a high temperature because the burned gases will remain in the combustion chamber longer. It will also "knock."

Either of the above conditions will cause the breaker points to open before or after the time when the magnetic field is at its highest intensity. A weaker spark will occur at the spark plug.

The condenser must be in good condition and of the proper capacity. If it is not, arcing will occur at the points; and they will burn. The magnetic field will not collapse fast enough in the primary to induce enough voltage in the secondary for a strong, hot spark at the spark plug.

Other difficulties, such as a partially shorted spark-plug wire, a partially grounded stop switch, or a fouled spark plug, can contribute to a weak spark. A weak spark results in poor combustion.

Continued operation with the ignition system out of order and improper combustion will eventually cause pre-ignition and knock, which results in overheating, valve burning, valve sticking and other complications.

If you know how to look for ignition trouble and how to recognize it when you find it, correcting the trouble is relatively simple.

TOOLS AND MATERIALS NEEDED

1. Socket-wrench set—1/4″ through 13/16″, 3/8″ drive
2. Flywheel puller(s)
3. Open-end wrenches—1/4″ through 1/2″
4. Slot-head screwdriver—6″
5. Phillips-head screwdriver—6″
6. Needle-nose pliers—7″
7. Tag-card stock or postal card for measuring air gap
8. Ignition tools
9. Feeler gage
10. Permate, for sealing around spark-plug wire at the coil
11. Ohmmeter
12. Flywheel holder
13. Coil tester
14. Continuity test light
15. Neon timing light
16. Clean rags
17. Cleaning solvent (denatured alcohol, mineral spirits, kerosene, or diesel fuel)

CHECKING THE MAGNETO FOR PROPER OPERATION

If your engine will not start, follow procedures given under "Starting Engines," Part 1.

If your engine runs but does not run properly, see "Operating Engines," Part 1.

If, after making these checks, you suspect the trouble is in the ignition system, proceed as follows:

1. *Check the spark plug.*

   Follow procedures under "Checking the Spark Plug for Proper Operation," Part 1.

2. *Check all visible wiring for looseness and the possibility of shorting.*

   Loose connections retard or disrupt current flow.

   Shorted circuits, caused from worn insulation and exposed wire touching some part of the engine, allow the current to flow directly to ground instead of going to the spark plug. Replace frayed, cracked or oil-soaked insulation, which may cause a short circuit in the spark-plug wire.

If the **ignition switch is grounded,** current from the primary winding goes through it to ground, and it is never interrupted when the points open. Therefore, no spark is developed (Figure 82).

3. *Disconnect the spark-plug wire from the spark plug but leave it loosely attached or touching.*

4. *Start the engine.*

5. *Remove the spark-plug wire and hold it .16 cm ($^1/_{16}$ in) away from the spark-plug terminal.*

6. *Operate the engine at various speeds: 1,000, 2,000 and 3,000 r.p.m.*

Do not prolong the test or hold the connector more than .16 cm ($^1/_{16}$ in) away from the plug as it is possible to damage the magneto coil. Current may have a tendency to jump the windings inside the coil and short the coil wires. The spark will occur at each revolution of the crankshaft on all 2-cycle engines and on some 4-cycle engines. If the breaker points are opened by a **cam on the crankshaft,** the spark will occur at each revolution of the crankshaft. This is always true on 2-cycle engines which fire each revolution. Single cylinder 4-cycle engines, however, use the spark only at every other revolution.

The breaker points on many 4-cycle engines are opened by a *cam on the camshaft.* They open and close every other revolution because the camshaft turns one half the speed of the crankshaft.

If a good spark occurs regularly, it is likely that the magneto is good.

If the engine runs better with the spark-plug wire loosely connected, the spark plug may be fouled. The .16 cm ($^1/_{16}$ in) air gap which you have provided in the test acts as a resistor-type plug and increases the spark intensity in the plug. This might cause a bad plug to fire. Here is the reason. The extra resistance causes a slight delay in the jump, and more voltage builds up. Then there is less time for the voltage to leak away before jumping the gap. Refer to "Servicing Spark Plugs," Part 1.

If the engine backfires, or kicks, when starting, the breaker-point gap may be too wide. Too wide a gap also causes a spark knock while the engine is operating.

If there is no spark or the spark is weak, proceed to step 7.

7. *Check the breaker points for condition and adjustment.*

   It may be necessary to remove the flywheel on some flywheel magnetos to get to the points. Procedures are given under the next heading for removing the flywheel.

   Checking the breaker points is discussed under "Checking and Conditioning the Breaker-Point Assembly," page 130.

8. *Check for a partially sheared flywheel key.*

   See step 3 under "Removing and Checking the Flywheel," next heading.

9. *Check the condenser.*

   See "Checking the Condenser," page 133.

10. *Check the ignition timing.*

    See "Timing the Magneto to the Engine," page 137.

11. *Check the stator-plate air gap for proper space.*

    See "Checking and Adjusting the Stator-Plate (Armature) Air Gap," page 141.

12. *Check the magneto edge gap (some flywheel magnetos).*

    See "Timing the Magneto to the Engine," page 137.

13. *Check the permanent magnet for strength.*

    See step 7, next heading.

14. *Check the magneto coil for continuity and strength.*

    See "Checking the Magneto Coil," page 143.

    A flywheel-magneto tester is available (Figure 86A).

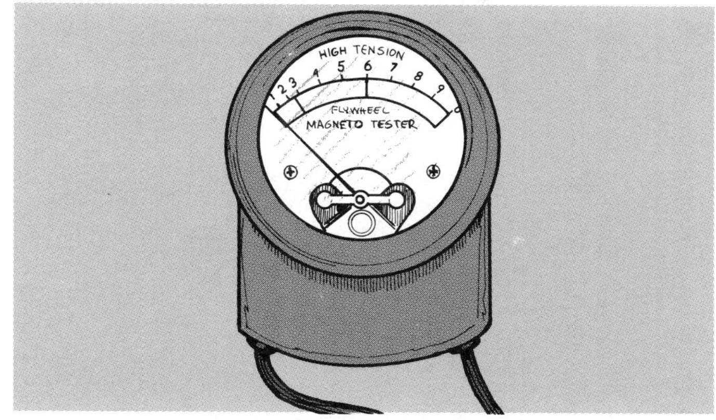

FIGURE 86A. A flywheel-magneto tester is easier to read than observing a spark.

REMOVING AND CHECKING THE FLYWHEEL

Engine parts can be damaged if you do not remove the flywheel properly. Follow procedures in your service manual if they are available. General procedures are given here.

If your breaker points are located underneath the flywheel, proceed as follows:

1. *Disconnect the spark-plug wire.*
2. *Remove the starter if one is installed.*

   If your engine has a **rope rewind** or a **windup starter,** it will be necessary to remove it first. See "Repairing Starters," page 78.

   If it has an **electric starter-generator,** it will have a stub-shaft, which must be removed (Figure 87).

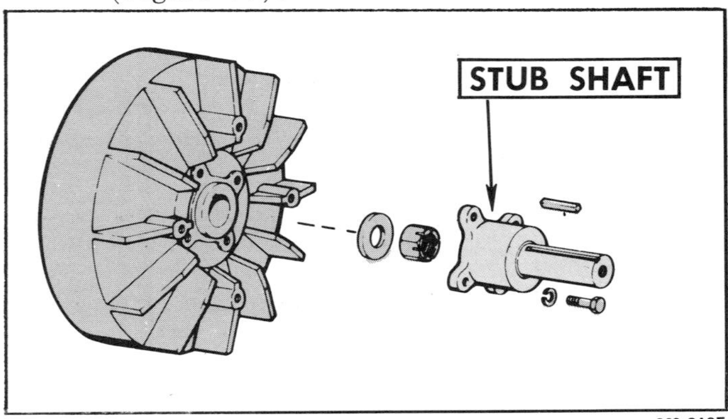

FIGURE 87. Stub-shaft for starter-generator drive pulley.

3. *Remove the flywheel shroud (Figure 88).*

FIGURE 88. Removing the flywheel shroud.

4. *Remove the flywheel nut (Figure 89).*

   A sharp blow on the wrench handle with a mallet will help break the flywheel nut loose (Figure 89a). Use a special flywheel holding tool (Figure 89b) to protect the crankshaft bearing while you are loosening the nut.

   Remember flywheel nuts have right-hand threads for clockwise rotating engines (viewing the engine from the flywheel side). They have left-hand threads for counter-clockwise rotating engines.

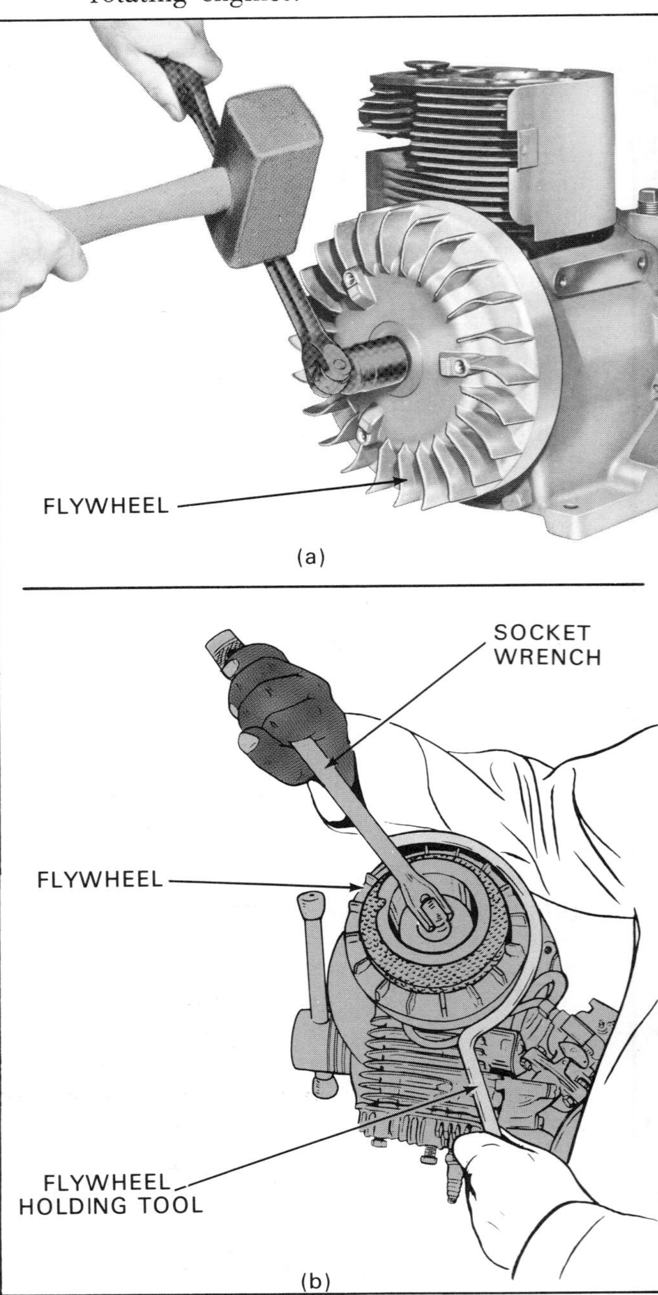

FIGURE 89. Removing the flywheel nut. (a) A sharp blow may break the nut loose. (b) A special flywheel holding tool should be used, if available.

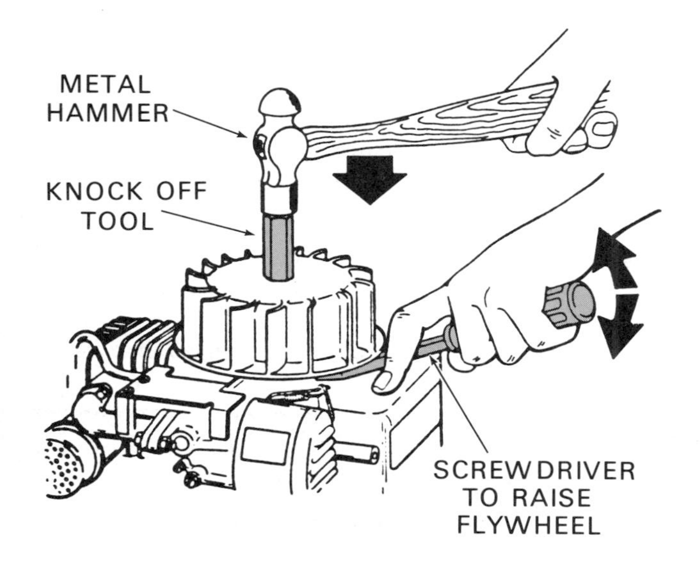

FIGURE 90. Removing the flywheel from a tapered crankshaft.

5. *Remove the flywheel.*

There are two methods for removing the flywheel. The method used depends on whether it is mounted on a tapered or a standard shaft.

(a) **If the flywheel is mounted on a tapered shaft,** remove it as follows:

(1) *Place special knock-off tool on the end of the crankshaft to prevent damage to the threads (Figure 90).*

(2) *Hold the flywheel by hand.*

(3) *Strike the end of the shaft with a sharp blow.*

Use a lead or plastic hammer. If the flywheel does not break loose after two or three tries, use a puller (Figure 91). Too much hammering will jar the magnetism out of the flywheel magneto and may damage the crankshaft bearings.

(b) **If your flywheel is not on a tapered shaft,** it will be necessary to use a puller. All of them are equipped for a special puller. Proceed as follows:

(1) *Attach special puller to the flywheel (Figure 91).*

Be careful not to damage the crankshaft threads and do not drop the flywheel. Dropping the flywheel

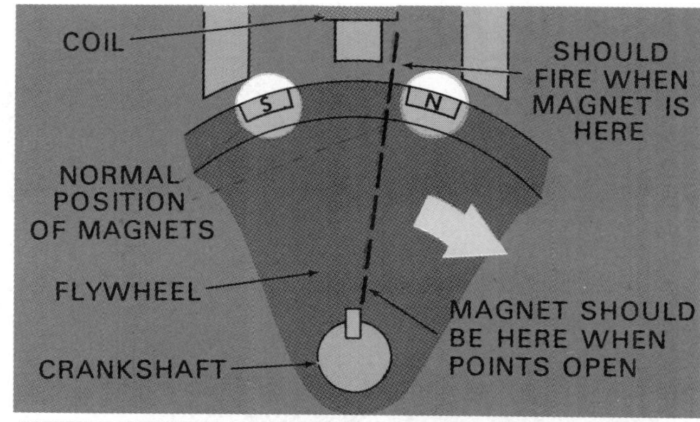

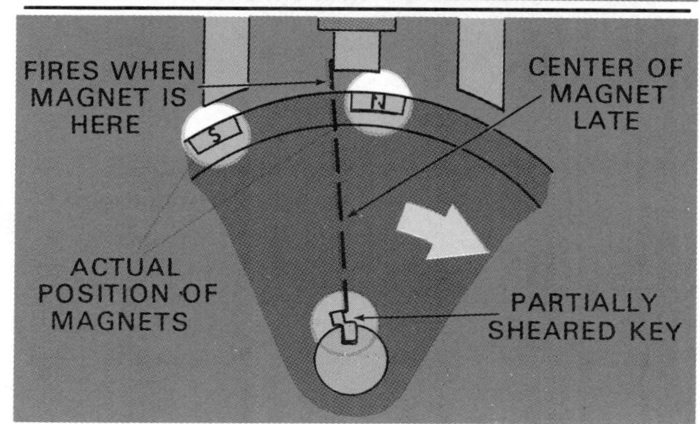

FIGURE 92. A partially sheared flywheel key can cause the ignition spark to be weak. The points open before the magnet reaches the point where the maximum magnetic-field intensity is developed.

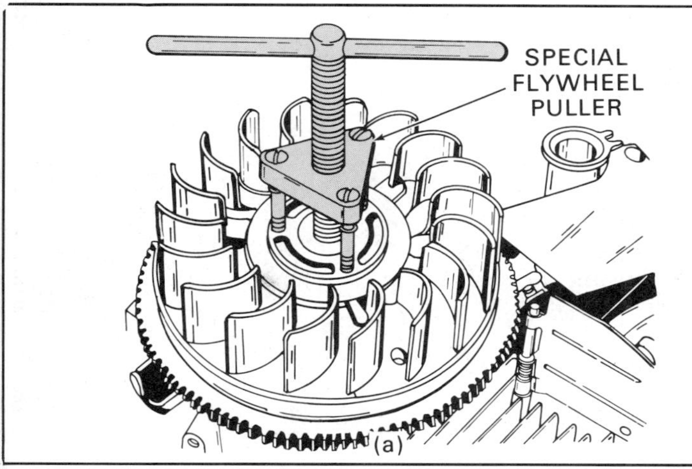

FIGURE 91. Special flywheel pullers. (a) A T-handle type. (b) A puller to be turned with a wrench.

will jar the magnetism from the magnets. Do not store it near iron or steel.

(2) *Turn puller screw with a wrench until flywheel is loosened.*

6. *Check for a partially sheared flywheel key (Figure 92).*

7. *Check strength of the magnets.*

   Hold a steel tool near the magneto. There should be a strong magnetic pull. There is no accurate measurement given by manufacturers for measuring the intensity of the magnetic field. One method suggested is to dangle a 15 cm (6 in) screwdriver with the blade 2.5 cm (1 in) from the magnet. The magnet should pull the blade against itself.

8. *Proceed with checking the breaker-point assembly.*

CHECKING AND CONDITIONING THE BREAKER-POINT ASSEMBLY

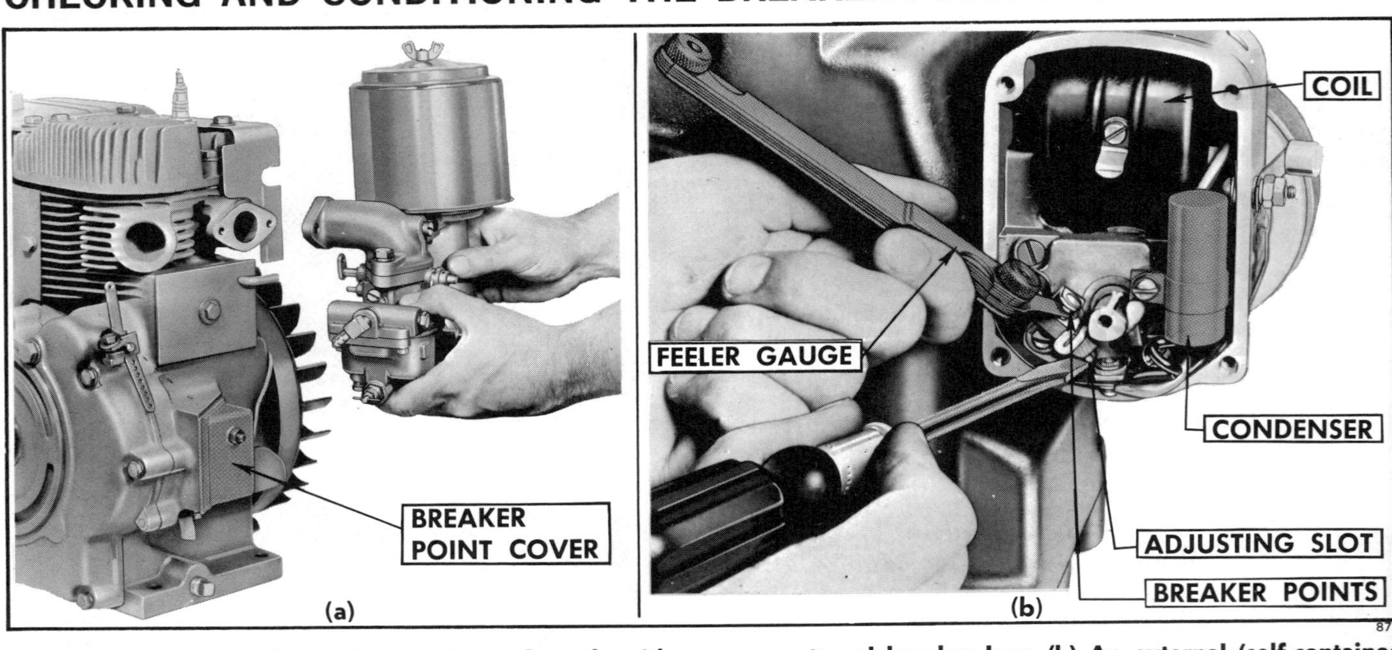

FIGURE 93. Some breaker points are located on the side of the engine. (a) Engine with a flywheel magneto but an external breaker box. (b) An external (self-contained) magneto showing breaker points.

1. *Locate the breaker-point assembly (Figures 93 and 94).*

   Some breaker points are easily accessible. They are **located on the side of the engine** and may be reached by removing a protective cover (Figure 93).

   Other breaker points (on flywheel magnetos) are **located underneath the flywheel,** and and it is necessary to remove the flywheel to gain access to them (Figure 94). Most of those located underneath the flywheel are also protected by a dust cover, which must be removed.

2. *Gain access to the breaker-point assembly.*

   If your points are located on the **outside of the engine,** remove the protective cover. Before removal, wipe the dust from the cover with a clean rag to prevent dirt from getting into the breaker box.

FIGURE 94. A cut-a-way view of a flywheel magneto showing the breaker points.

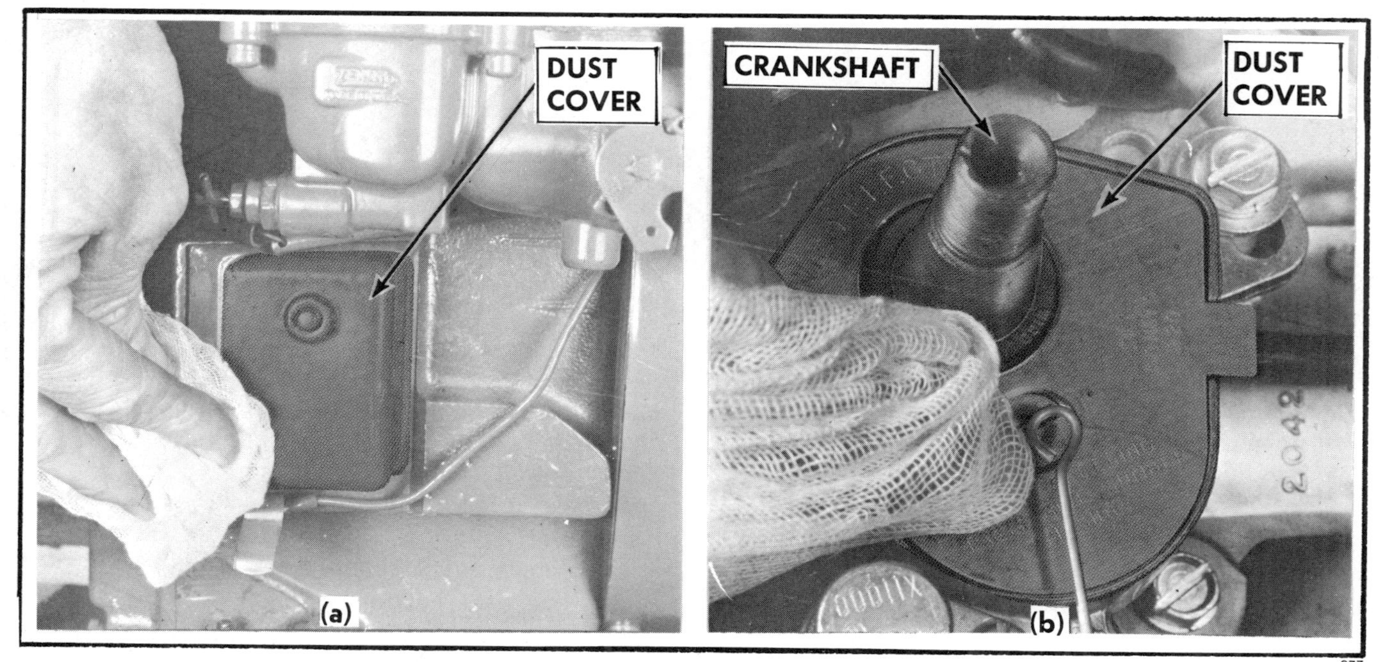

FIGURE 95. Wipe away dirt, dust and grime before removing the breaker-box cover. (a) Outside cover. (b) Cover under the flywheel.

Use a clean cloth with a small amount of petroleum solvent to remove grease. Do not use water. It may cause short circuiting (Figure 95).

If the points are located **underneath the flywheel,** remove it.

3. *Check and remove, if necessary, the centrifugal spark-advance mechanism (Figure 96) if installed.*
4. *Check and recondition or replace the breaker points.*

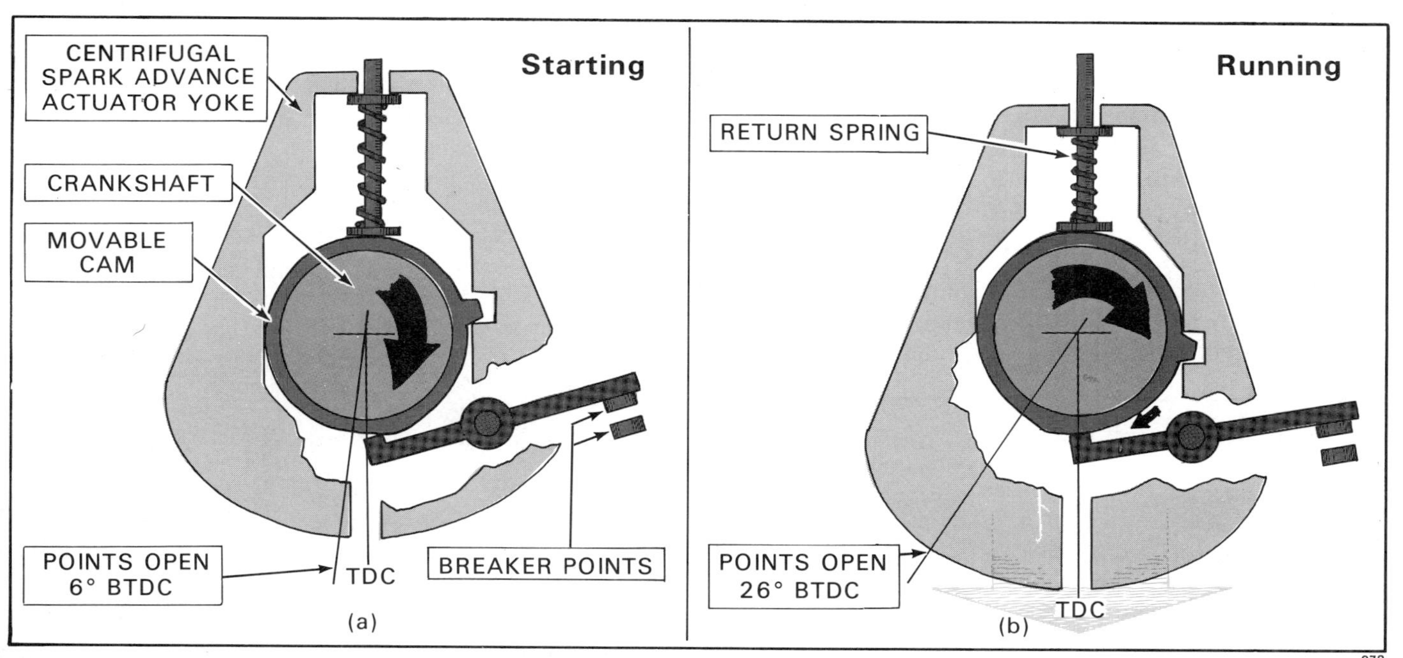

FIGURE 96. One type of centrifugal spark-advance mechanism moves the cam, which opens the points earlier when the engine is running above 1,000, to 1,500 r.p.m. (a) For starting, the movable cam remains stationary and opens the points when the piston is near top dead center.

(b) When the engine speed increases, the yoke moves because of centrifugal force. When it moves, it rotates the cam in the direction of the crankshaft rotation. Then the points are opened earlier.

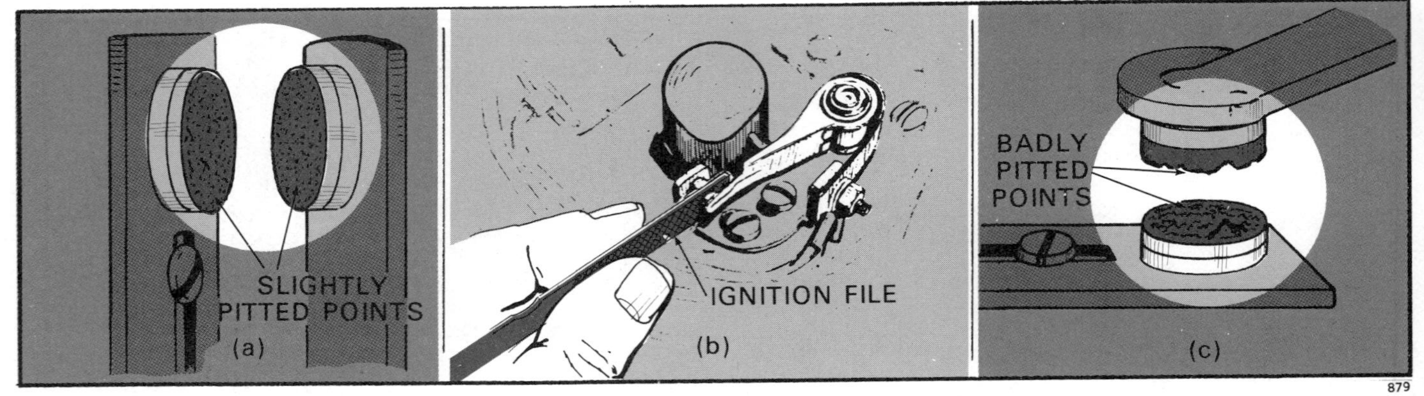

FIGURE 97. (a) Unplated contact points that are slightly burned or pitted may be (b) smoothed with an ignition file. (c) Badly pitted points should be replaced with new ones.

If the contact points are rough but show only a **slight pitting and metal deposits** (Figure 97a), they can be smoothed with an ignition file (Figure 97b). It is not necessary to have the points completely smooth with all pits removed. Just remove the high spots. When the engine is operating properly, there is a small amount of pitting and metal deposits. Make certain a large area of contact is still being maintained.

Blow out the dust after you have completed the filing.

Do not use sandpaper or emery cloth. The particles may embed in the contact surface and cause the points to burn.

Some points are plated with a high-melting-point metal, like tungsten. They are so hard there is little you can do to them if they are worn or damaged. They must be replaced. If points are **badly pitted and/ or worn,** (Figure 97c), replace them with a new set. Follow procedures under "Removing and Replacing Breaker Points," page 136.

Badly burned points may be due to (1) oil or foreign material on the contact surfaces, (2) a defective condenser, (3) contact-point gap out of adjustment, or (4) points out of alignment.

5. *Check the condenser.*
   See next heading.

   If the **condenser is not working properly,** or if it is not of the proper capacity for your system, the points will not last long.

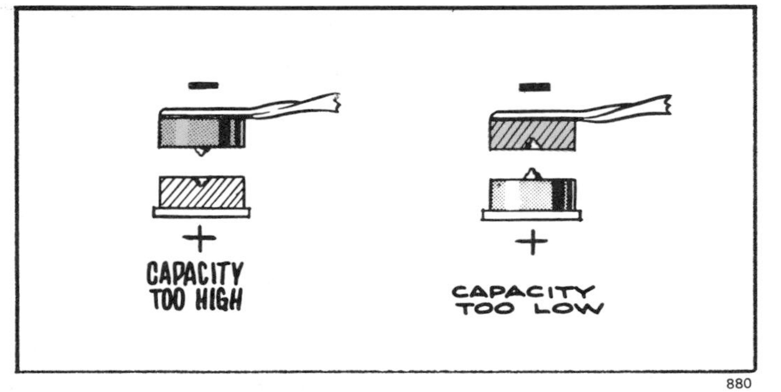

FIGURE 98. Effect on contact points when condenser is of improper capacity.

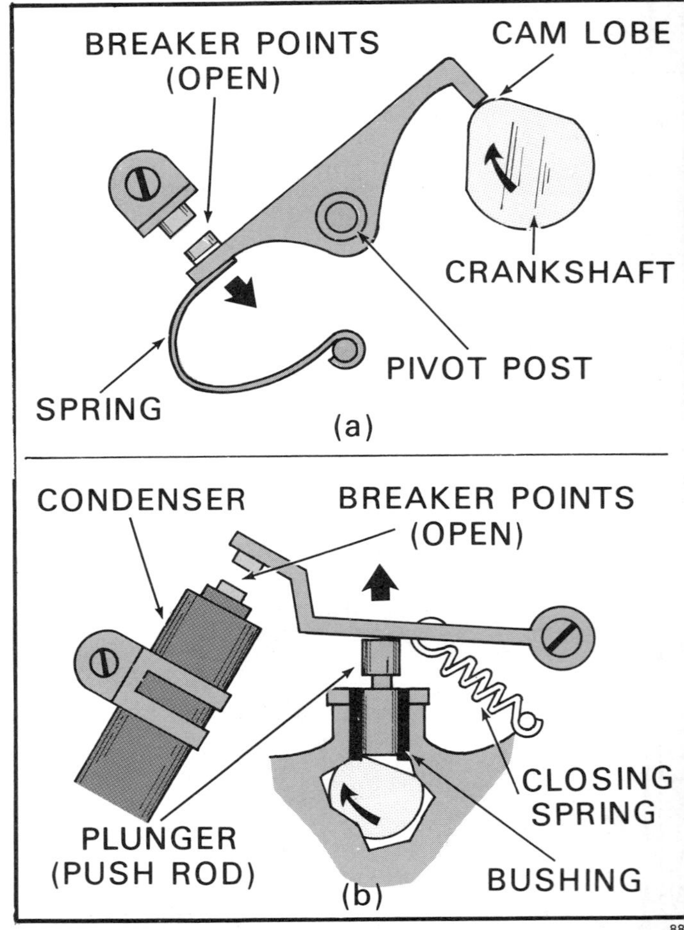

FIGURE 99. (a) Breaker points opened by a cam on the crankshaft and closed by a spring. (b) Breaker points closed by a spring and a flat spot on the crankshaft and then opened at the proper time by the crankshaft and a plunger.

6. *Check the breaker cam for wear.*

   Some breaker points are opened by a cam on the crankshaft (Figure 99a). The cam may be machined on the crankshaft, or it may be a separate collar locked by a key on the crankshaft. With the latter type, it is possible to install it up-side-down. If you do. the points will open 180° ahead of time.

   Other breaker points are opened by push rods which are operated by the crankshaft (Figure 99b). A flat space on the crankshaft closes the points in time for the induction coil to build up a charge and then opens them for the spark discharge.

   Check push rod for wear and for freedom of movement.

   Some are operated from a cam on the camshaft which trips a lever arm on the breaker arm shaft (Figure 100).

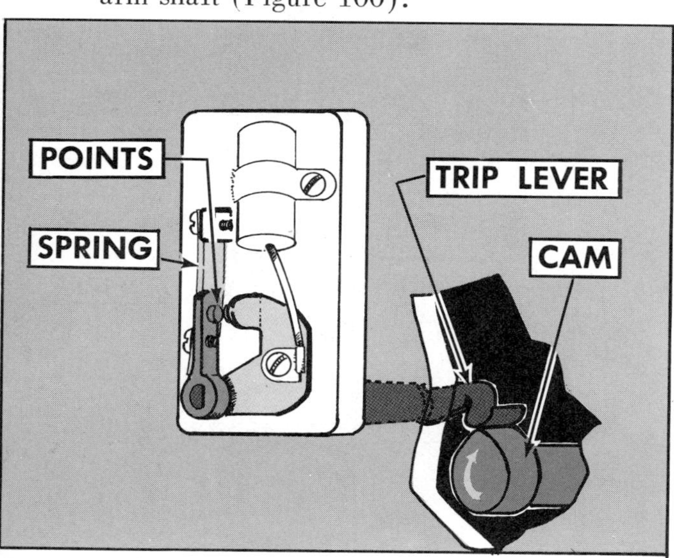

FIGURE 100. Some breaker points are opened by a trip lever which is tripped by a cam on the camshaft.

7. *Check rubbing block for wear.*

   When wear takes place on the rubbing block, this changes both the point gap and the timing. This is one of the main reasons for an engine getting out of time.

8. *Check for a leaking crankshaft seal.*

   This may allow oil to get on the points of flywheel type magnetos and ground them out.

9. *Check condition of wiring and tightness of connectors.*

   Look for frayed insulation and loose connections.

10. *Check point assembly for tightness.*

11. *Check and adjust the breaker-point gap.*

12. *Check for a worn crankshaft bearing.*

    The major problem with the internal flywheel magneto is caused by a worn bearing on the flywheel end of the crankshaft. This causes the point gap to vary.

    Check by feel. Any side movement indicates a worn bearing.

13. *Check plunger and plunger hole for wear, if installed (Figure 99b).*

    A leaking plunger will cause oil to get on the points.

    If the **points are opened by a plunger,** check plunger hole for wear with a special gage.

    If the **hole is worn,** ream hole, and install a bushing and a new plunger (Figure 100A).

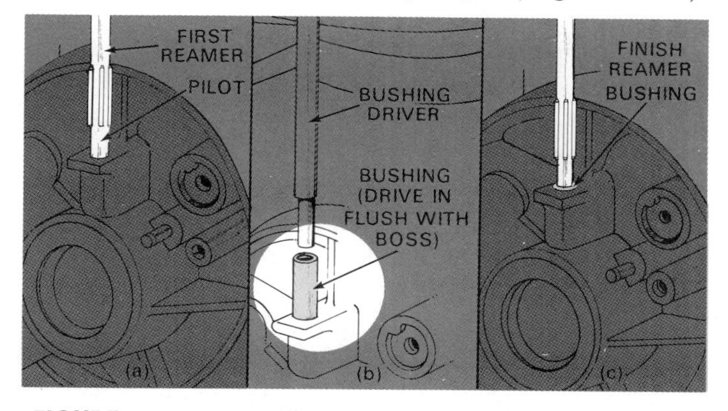

FIGURE 100A. Installing plunger bushing.

CHECKING THE CONDENSER

Some manufacturers recommend changing the condenser every time you change points. They say the price of a condenser is not worth the risk of burning up a new set of points.

Other manufacturers say the odds of a condenser going bad are 1,000 to 1, and you are better off to use the old condenser. Some points, however, are attached to the condenser, and you are forced to replace condenser when you replace points (Figure 101).

Badly pitted breaker points indicate the condenser is bad or is of the improper capacity.

1. *Remove and check condenser for capacitance (.10 to .30) microfarads, depending on your type of magneto)*

If you are not equipped to check the condenser, have it checked at your dealer's or replace it with a new one.

2. *Replace the condenser if it is bad.*
3. *Check the tightness of the clamp that holds the condenser.*

The clamp is also a ground connection for the condenser. It must be tight to provide a good electrical connection. This helps prevent arcing at the points.

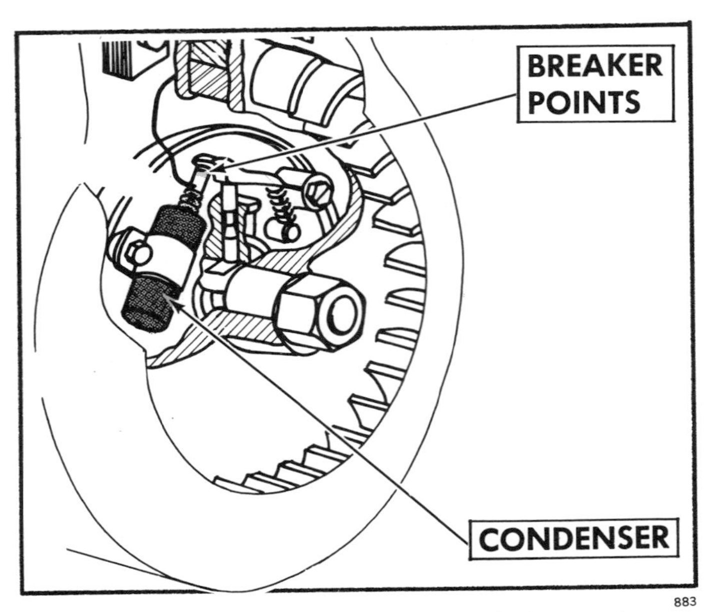

FIGURE 101. Some breaker points and the condenser are assembled as one unit. It is necessary to replace one when you replace the other.

Follow procedures given with the particular instrument you use and the manufacturer's specifications.

Check the old condenser first. If it is not good, replace it — but also check the new one.

Condensers should be checked for capacity, shortage (or leakage) and resistance (Figure 102). Heat the condenser to approximately 38°C (100°F) before testing. This is more nearly operating temperature. A condenser that checks out good when cold may not when heated. For example, if it has a leak, the short will show up better at higher temperatures. Do not overheat. The expansion may crack some of the insulation. Hold it in the palm of your hand for a couple of minutes, or put it in an oven with a thermometer control.

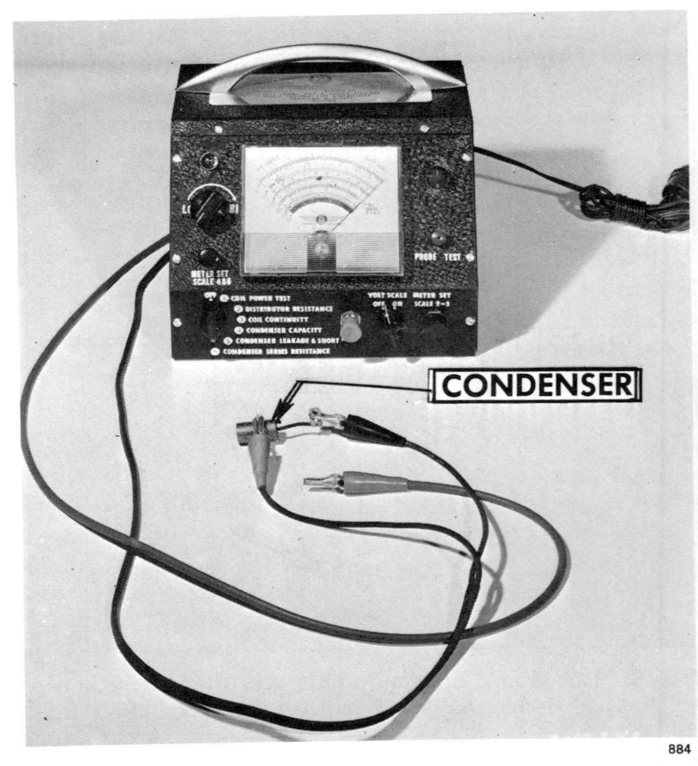

FIGURE 102. A typical instrument for testing coils and condensers.

ADJUSTING THE BREAKER-POINT GAP

1. *Turn the crankshaft in direction of normal rotation until cam opens breaker points to the widest position (Figures 99 and 100).*
2. *Check points for proper gap spacing (Figure 103).*

   Use a clean feeler gage of the thickenss recommended for your engine. This may vary from .038 cm to .050 cm (.015 in to .020 in). Usually the point gap can be found on the engine nameplate and/or on the breaker-point dust cover. If not it will be given in your operator's manual. The proper width gap i[s] provided when there is a slight drag on th[e] thickness gage as you pull it between th[e] contact points. If adjustments are needed, pr[o]ceed to step 3.

3. *Loosen locking screw on bracket that provides for the point adjustment (Figure 106).*
4. *Tighten snugly but not completely.*

   Tighten just snug enough so that the adjusting plate and breaker cam will remain i[n] place when the stationary point is moved

5. *Adjust points for proper spacing and alignment (Figure 103).*

Four different means are provided for adjusting breaker point gap. The one used depends on the design of the point assembly: (1) a **slot** for moving the back plate, or stationary points (Figure 103) with a screwdriver, (2) a **screw** with a cam for moving the back plate with a screwdriver, (3) a **slot in the back plate with no special provision for moving the back plate,** and (4) a **clamp** on the condenser if the points are part of the condenser Figure 103c).

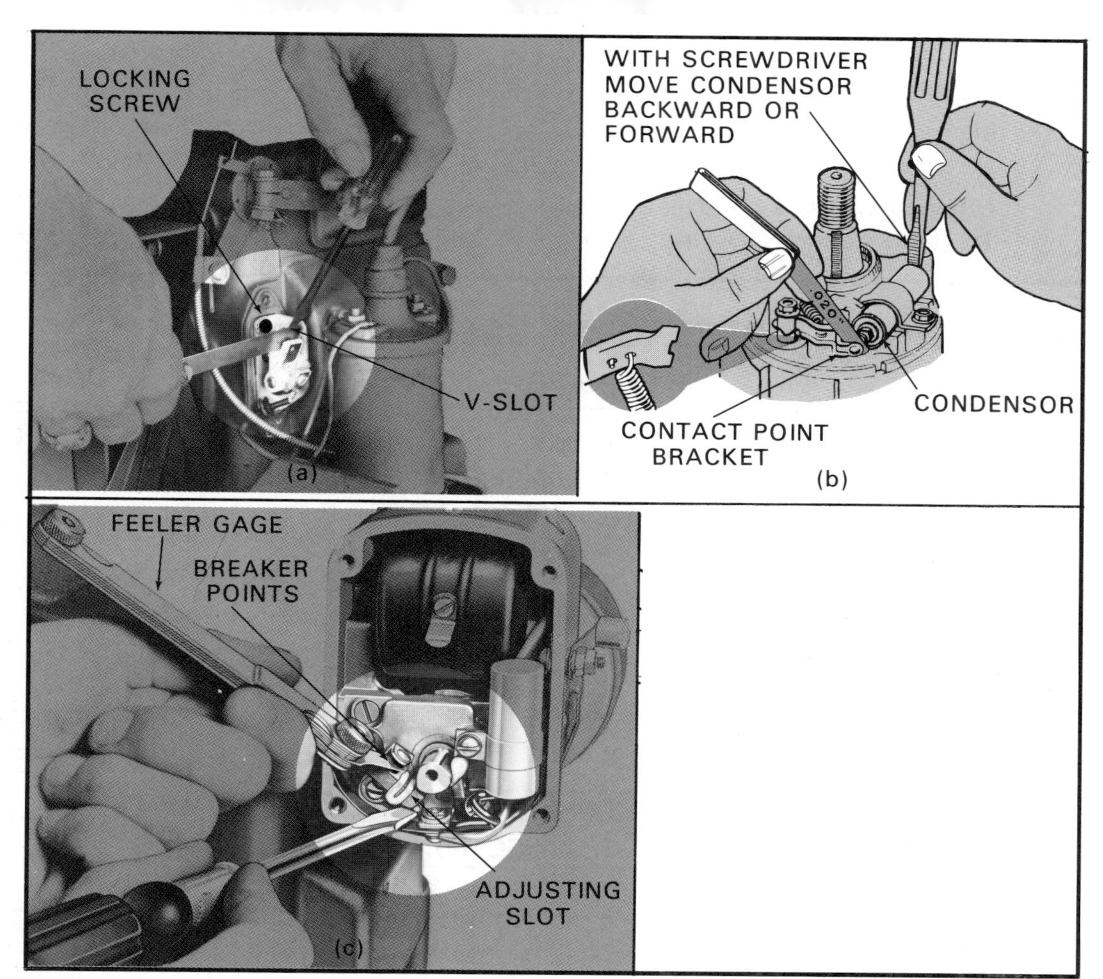

FIGURE 103. Adjusting breaker points. (a) Points adjusted by loosening the stationary point and shifting it with a screwdriver at the adjusting slot. (b) Moving the condenser on which the stationary point is attached. (c) Similar adjustment on external magnet on a flywheel magneto.

If the gap width recommended for your engine provides for a width range such as .038 cm to .045 cm (.015 in to .018 in) use the wider recommendation. As the cam and rubbing block wear, the points will gradually decrease in gap width.

Be sure the breaker point faces fit together squarely (Figure 140a). If they fit as shown in Figure 104b and c, the points will burn, pit and wear unevenly. It is possible to bend some metal-arm breaker points into position to get the points to fit squarely. This action, however, is not generally recommended.

6. *Lock breaker points into position with locking screw (Figure 103).*

7. *Recheck gap between points and wipe points clean.*

Rechecking the gap insures that no change was made in the gap setting when you tightened the lock screw.

Running a strip of paper between the contacts will remove dirt particles. Do not use cloth because it may leave enough lint to provide insulating "fuzz."

If points are oily, clean them with a small amount of petroleum solvent.

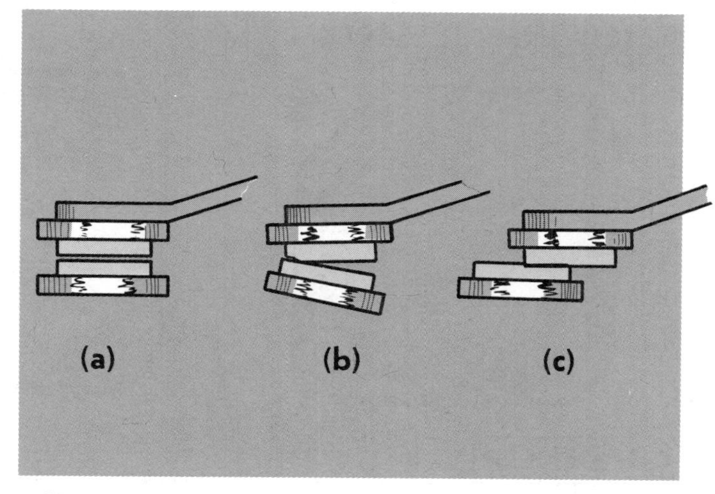

FIGURE 104. (a) Breaker points should fit together squarely to keep down water. (b) and (c) Only partial contact causes arcing and uneven wear.

8. *Clean and lubricate cam with a grease of high melting point.*

   If the breaker-point assembly has a **cam lubricating wick,** replace it with a new one. If the assembly has **no wick,** use *only* enough grease to provide a thin film on the cam lobe. This avoids getting it on the breaker points, thus causing them to burn.

   A very small amount greatly reduces wear on the rubbing block and cam.

REMOVING AND REPLACING BREAKER POINTS

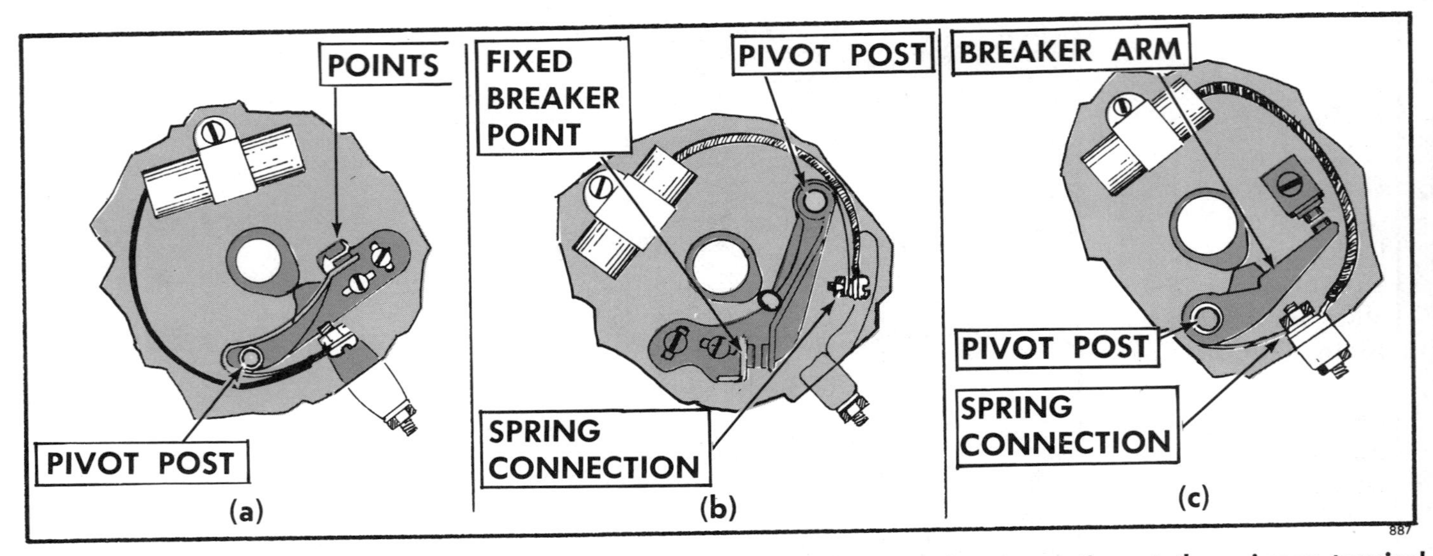

FIGURE 105. Typical mounting arrangements for removable breaker arms on different magnetos. In all three types — (a), (b) and (c) — the breaker arm is removed by loosening a spring connection at the primary terminal and lifting the arm from the pivot post.

1. *Open the breaker points.*

   Turn the flywheel by the starter. This prevents scratching the new points when installing them.

2. *Remove breaker arm and spring.*

   The exact procedure you use depends on the type of installation. Figure 105 shows three common assemblies. Study the installation before taking the assembly apart.

3. *Remove stationary breaker point and bracket (Figure 106).*

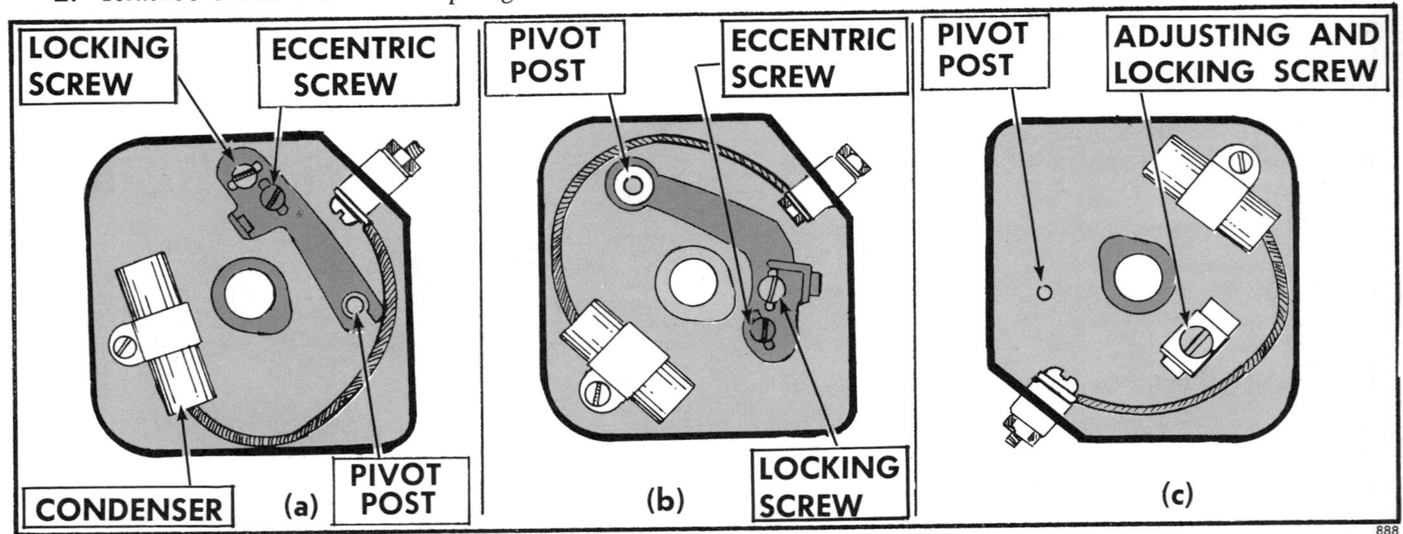

FIGURE 106. (a) and (b) Stationary breaker point and bracket are removed by loosening the locking screw and lifting out the bracket. The adjusting or eccentric screw remains in position. (c) Loosen the adjusting-and-locking screw to remove bracket.

On most installations, one end of the breaker-point bracket is held by the same pivot pin that holds the breaker arm. The other end is held by a locking screw (Figure 106a and b). The second screw is for adjusting the position of the points and is called an "eccentric screw," or "adjusting screw." On others the bracket and adjustable point are held by one screw (Figure 106c). With this type, the hole in the bracket is slotted so that the breaker point can be adjusted to the proper position.

There are some installations (Figure 101) that have one point attached to the condenser. Loosen the retaining clamp and remove the condenser.

4. *Install new points in reverse order.*

   Be sure electrical connections are tight.

   It is important on most installations that the breaker-arm spring be connected next to the retaining nut to give proper spring tension.

5. *Adjust point clearance.*

   Refer to "Adjusting the Breaker-Point Gap," page 134.

TIMING THE MAGNETO TO THE ENGINE

Very seldom will your magneto get out of time. But *if your engine is running rough* and the spark plug and breaker points are in good condition, check the timing. Or, if you replace the coil or have moved the stator plate, it will be necessary to adjust the timing.

Timing the magneto to the engine consists of adjusting the position of the stator plate (coil and laminated yoke, or armature). The center line of the rotating magnet should be just past the center line of the stator plate, when the piston is at the desired position for ignition to take place — on or before top-dead-center on the compression stroke. This position is set by the manufacturer.

Many small engines have no provision for adjusting the position of the armature.

The distance between the center lines is known as the **"edge distance"** because it can only be measured at the edges of the armature and magnet (Figure 107). It is at this point that the spark is greatest. Only a few manuals give this measurement because the edge distance is preset at the factory; and when the engine is in proper condition, the edge distance will be right.

There are several reasons why **the edge distance may be off.** Some of them are a *worn flywheel key and/or key way, partially sheared flywheel key, or twisted crankshaft.* But the most probable one is the magneto being out-of-time.

For **checking and adjusting the magneto timing** on your engine, proceed as follows:

1. *Determine what provisions are made for adjustment.*

   On **flywheel-type magnetos** the stator plate (laminated iron yoke and coil, or armature) may be (1) *stationary* with no provisions for adjustment (Figure 108a); (2) it may have *straight adjusting slots* for adjusting the armature air gap (Figure 108b) — which does not affect the timing; or (3) it may have *curved slots,* or slots on the mounting flange, for adjusting the timing (Figure 108c).

   External-type magnetos (self-contained types) may or may not have adjusting slots in the mounting flange.

   If your magneto stator plate has **curved slots for rotational adjustment,** proceed to step 2.

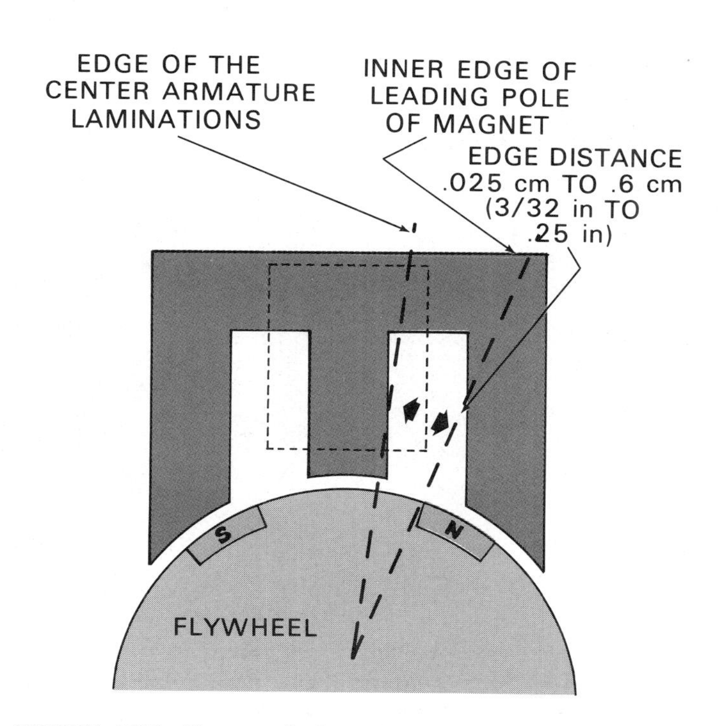

FIGURE 107. The spark is greatest if the points open when the center of the magnet is just past the center line of the coil. This distance is known as "edge distance."

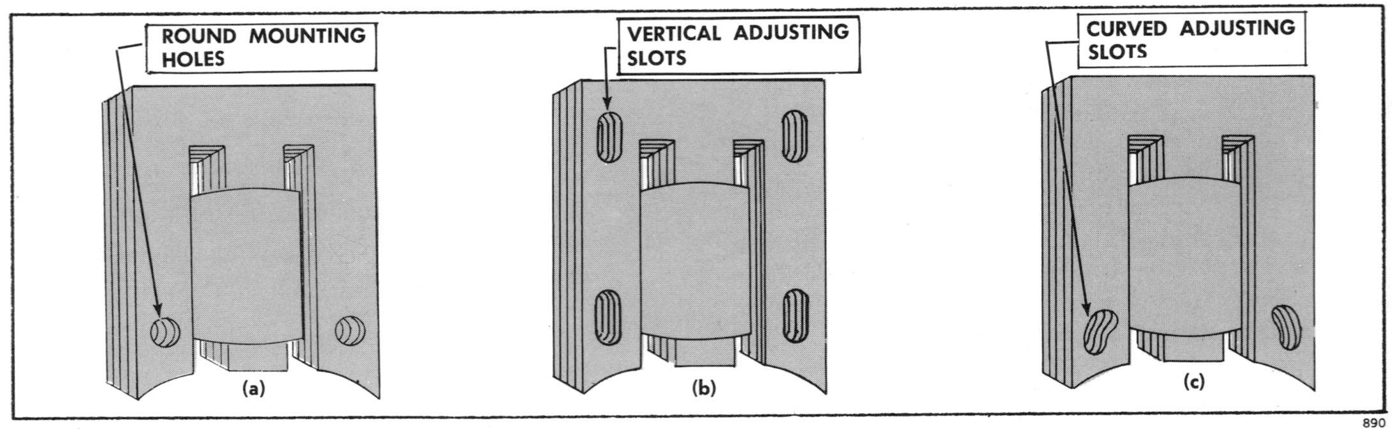

FIGURE 108. The stator plate (laminated iron yoke and coil or armature) on your flywheel magneto may be (a) stationary with no means for adjustment, (b) designed for vertical adjustment, which does not affect timing, or (c) slotted for rotational timing adjustment.

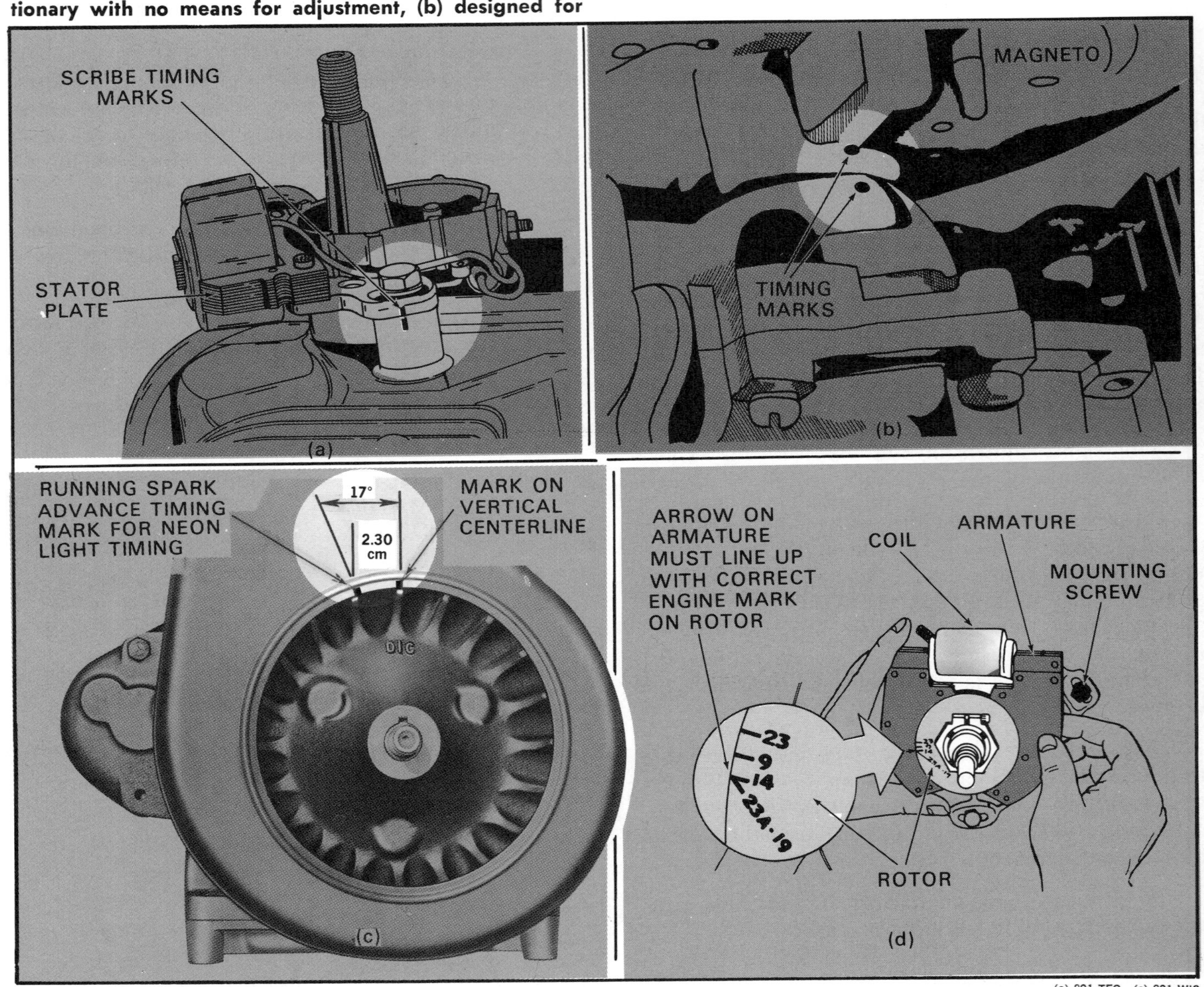

FIGURE 109. Some engines have timing marks which tell you the proper relative positions of the rotating magnet and the stator plate. (a) The timing marks may be scribed marks matching up the stator plate with a point on the engine block. (b) Timing marks may be drilled holes. (c) Timing marks may be scribed marks on the blower housing and on the flywheel. (d) Some magnetos are used on different engine models, and the different timing for each model is indicated.

2. *Locate the timing marks (Figure 109).*

   Timing marks on flywheel magnetos are on the stator plate and the engine block. They may be in the form of scribe marks (Figure 109a) or drilled holes (Figure 109b). Some have no timing marks.

   Timing marks are also located on the flywheel and flywheel housing (Figure 109c). More than one timing mark may be present. One for static timing and one for timing with the engine running. In some cases multiple timing marks are provided for magnetos which are used on different engine models (Figure 109d).

3. *Time the magneto.*

   If your engine **has timing marks,** loosen the capscrews which hold the stator plate and rotate it until the timing marks are aligned. If there are **no timing marks,** and your stator plate is adjustable, proceed as follows:

   (1) *Remove the spark plug.*

   (2) *Adjust the breaker-point gap.*

   See "Adjusting the Breaker-Point Gap," page 134.

   (3) *Set the piston at the recommended position for firing.*

   Check your service manual for the distance in inches or degrees from top-dead-center on the compression stroke. Measure with a ruler or dial indicator if given in inches. Measure with a disk indicator if given in degrees of crankshaft travel.

   (4) *Adjust the magneto stator plate so the points start to open at this point.*

   Check with a continuity timing light. See next heading for procedures.

TIMING THE BREAKER POINTS TO THE MAGNETO

Timing the breaker points is adjusting them so hey will open at the exact time the magnet is ligned with the coil for the greatest spark, when he piston is at the desired position.

There are two types of lights used for timing the reaker points: (1) **a continuity light** and (2) **a eon timing light.**

A continuity light (Figure 110) is a light similar o a flash light. It is wired so that when the circuit is roken, the light will go out. When the light is conected to the breaker points, you can tell when they irst open. It must be used when the engine is stopped.

A neon timing light (Figure 111) is a specially esigned neon light which flashes when the points pen. With the flash directed to the timing marks you an tell if the points are opening at the proper time hen the engine is running.

a. If you adjust the breaker-point timing with a **ontinuity light** (Figure 110), proceed as follows:

1. *Rotate the crankshaft by hand until timing marks are aligned (Figures 109 and 110a, inset).*

   Piston is at or near top-dead-center, compression stroke.

2. *Disconnect primary-coil lead wire at the terminal stud (Figure 110a).*

3. *Connect one alligator clip of the continuity light to ground—any convenient place on the crankcase.*

4. *Connect the other lead to the breaker-point terminal stud (Figure 110a).*

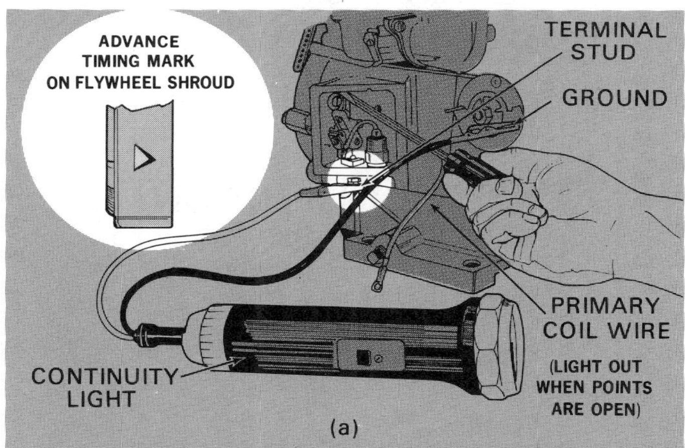

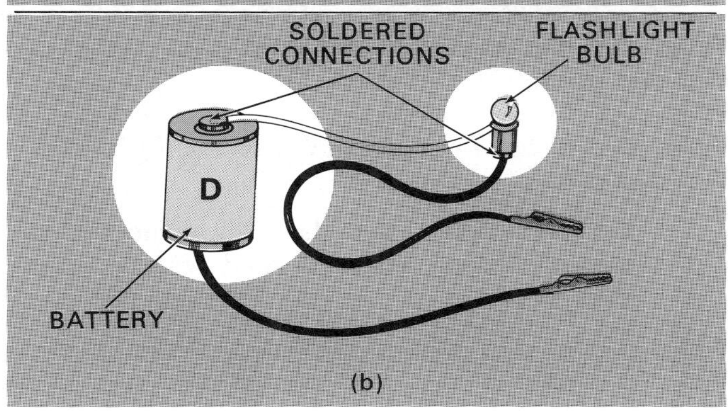

FIGURE 110. Timing the magneto-ignition system with a timing light. (a) Static timing with a continuity light. (b) You can make a continuity light from a flash light battery and bulb.

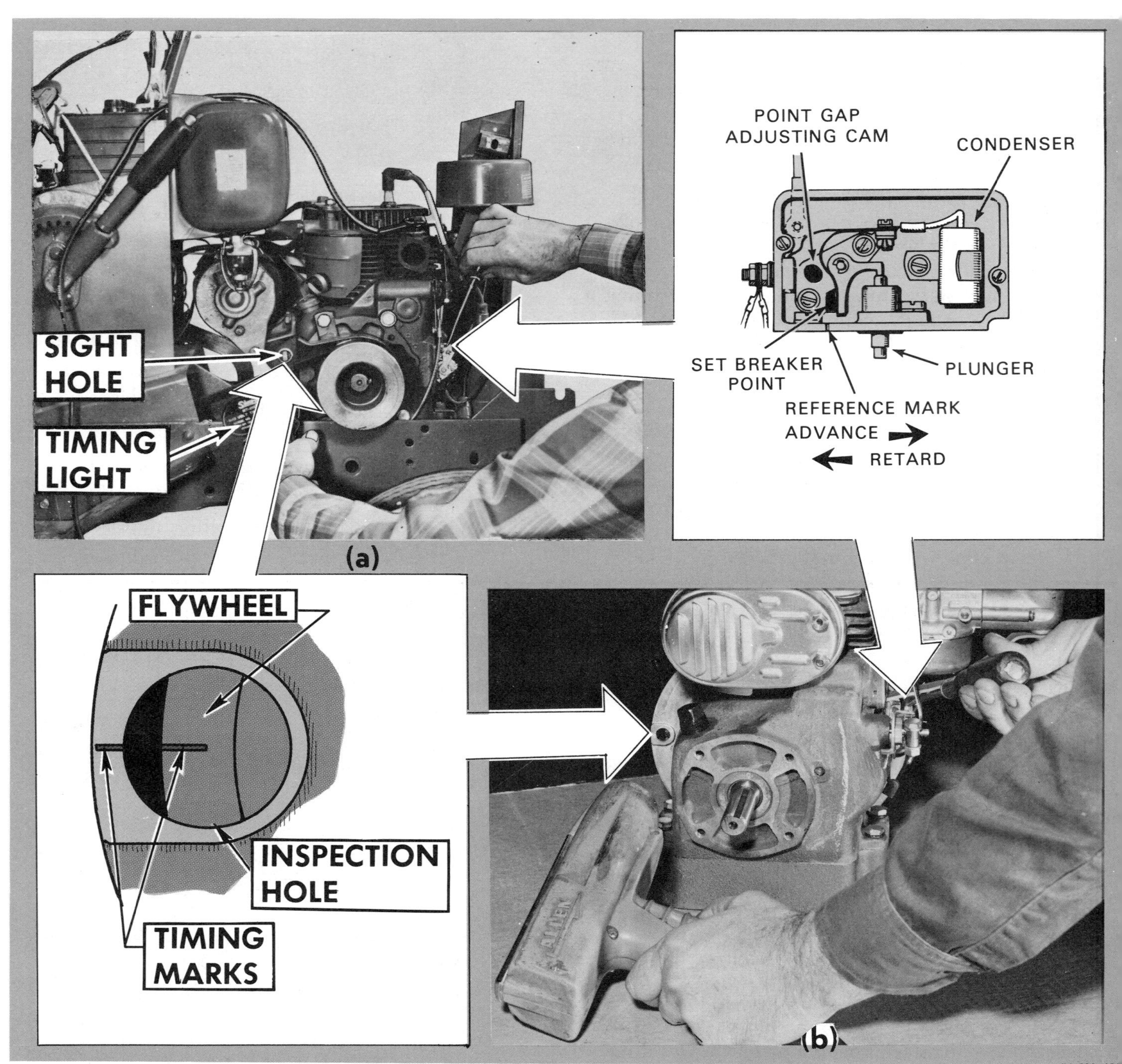

FIGURE 111. Checking the ignition timing with a neon timing light. (a) Engine installed on a tractor. (b) Engine mounted on a test stand.

5. *Loosen breaker-point lock screw and retighten snugly.*

   Tighten enough so the plate can be moved for adjustment but will still hold without slipping.

6. *Adjust points so that the light will just go out.*

7. *Tighten points.*

8. *Check adjustment.*

   This is done by rotating the flywheel backwards—opposite from normal direction of rotation—until the light comes on again. Then rotate the flywheel forward until the light goes out. The timing marks should be aligned. If they are not, readjust the points and recheck.

9. *Remove continuity light and reconnect primary lead.*

b. If you adjust the breaker-point timing with a **neon timing light** (Figure 111), proceed as follows:

1. *Adjust points to recommended gap at full open position.*

   Refer to "Adjusting the Breaker-Point Gap," page 134.

2. *Remove inspection plug or the flywheel shroud, if necessary, to see the timing mark on the flywheel (Figure 109c).*

   Remember this check will be made while the engine is operating. Most engines have a spark-advance mechanism. If so, there will be two timing marks: one for static timing—when the engine is not running—and one for timing while the engine is running. This is called the advance timing mark.

3. *Accent the advance timing mark with chalk so that it may be seen easily.*

   Refer to your service manual.

4. *Connect the neon timing light.*

   Connect one lead to ground and the other lead to the spark-plug terminal. Do not disconnect the spark plug.

5. *Start engine and operate it above 1,000 r.p.m.*

6. *Direct the timing light on the timing mark of the flywheel.*

7. *Loosen and adjust the points until the timing mark on the flywheel and the one on the engine frame appear to be exactly in line when the light flashes.*

8. *Stop the engine.*

9. *Disconnect timing light.*

CHECKING AND ADJUSTING THE STATOR-PLATE (ARMATURE) AIR GAP

The stator-plate air gap is a space between the laminated-iron arms (yoke) of the coil and the rotating magnet (Figure 112). Since the yoke is known as "armature," "yoke heels" and "pole shoes," the air gap is sometimes referred to as "armature heel space," or "pole-shoe clearance." The closer the spacing (without the two members touching), the stronger the spark — more magnetic lines of force are cut.

Some engines have no provision for adjustment of the gap spacing (Figures 108a and 108c), while others do (Figure 108b).

There are three different types of armatures that can be adjusted: (1) the type which is located **outside the flywheel** (Figure 112a), (2) the type which is located **under the flywheel,** and (3) the type which is located **under the flywheel, but with a magnet in a rotor** which turns inside the stator plate.

Only stator plates with straight elongated mounting slots provide for air-gap adjustment. Those with curved (timing) slots provide for timing adjustment only.

When those without adjustment provisions get out of adjustment, it is usually caused by worn crankshaft bearings or a bent crankshaft.

a. If your **stator plate is located outside the flywheel (Figure 112a),** proceed as follows:

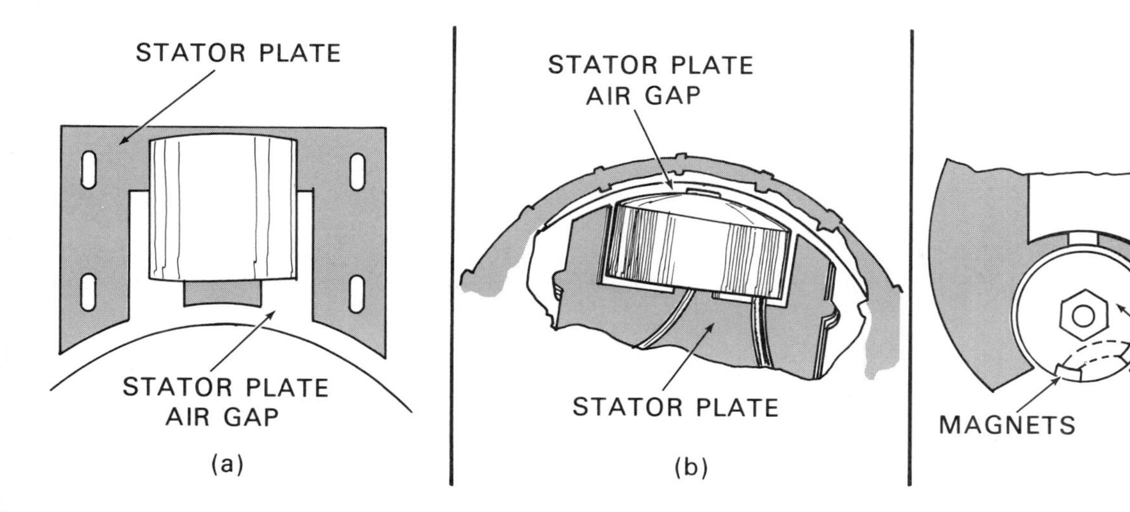

FIGURE 112. The stator-plate (armature) air gap is the clearance between the laminated iron arms and the rotating magnet. (a) Stator plate outside the flywheel. (b) Stator plate under the flywheel. (c) Stator plate under the flywheel with the magnets in a rotor which turns inside the stator plate.

1. *Check the clearance.*

   Use a feeler gage or postal card. The thickness of a postal card is within the range of .18 mm to .30 mm (.007 to .012 in) which is the recommended gap width.

   Rotate the flywheel as you check the clearance to make sure it is clear on all sides.

   If you use a **metal feeler gage,** check clearance at points away from the magnet. Otherwise, the magnet will attract the gage and give you a sense of binding when it is not.

   If the **air gap is correct,** proceed to checking the coil described under the next heading.

   If the clearance is **too little or too much and the stator plate is adjustable,** proceed to step 2.

   If the **air-gap clearance is not adjustable,** the problem may be a warped armature, warped flywheel, or worn bearings.

2. *Adjust the clearance if possible.*

   (1) *Put a postal card, or a tag card, between the stator plate and the flywheel.*

   (2) *Turn flywheel by hand until magnet aligns with the armature.*

   (3) *Loosen the stator plate.*

   The magnetism will pull the stator plate against the magnets, thus closing the gap to the thickness of the card.

   (4) *Tighten the stator plate (Figure 113).*

3. *Remove the card.*

4. *Recheck the clearance.*

   The clearance should be accurate.

b. If your **stator plate is under the flywheel** (Figure 112b), proceed as follows:

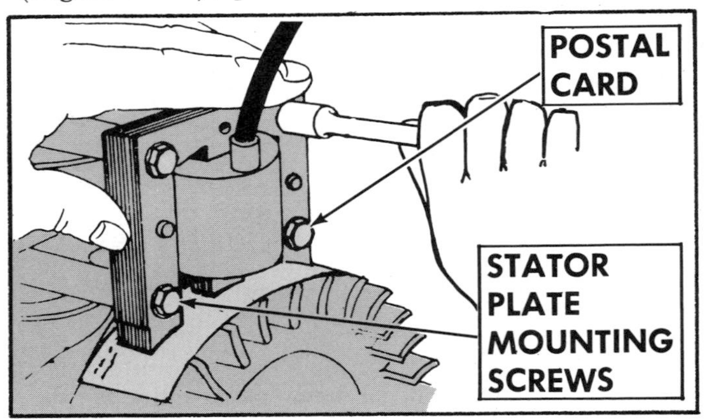

FIGURE 113. Adjusting the stator-plate air gap.

1. *Remove the flywheel if it is not already removed.*

   Refer to "Removing and Checking the Flywheel," page 127.

2. *Place a piece of electrician's tape on the inner rim of the flywheel.*

3. *Reinstall the flywheel.*

4. *Rotate the flywheel 10 or 12 times by hand to check for scuffing on the tape.*

5. *Remove the flywheel.*

6. *Check to see if the tape is scuffed.*

   If tape is scuffed, the clearance is too small. Adjust if possible.

   If tape is not scuffed, proceed to step 7.

7. *Add a second piece of tape to the rim of the flywheel.*

8. *Replace the flywheel.*

9. *Turn the flywheel 10 or 12 times by hand.*

10. *Remove the flywheel.*

11. *Check to see if the tape is scuffed.*

    If it is not scuffed, the clearance is too great. Adjust if possible as described in step 12. If not possible, check the crankshaft bearings.

12. *Adjust the stator plate if possible.*

    Loosen the mounting screw and shift the stator plate up or down as needed.

13. *Recheck the clearance.*

c. If your **stator plate is located under the flywheel and the magnet is on a separate rotor** (Figure 112c), proceed as follows:

1. *Remove the flywheel if not already removed.*

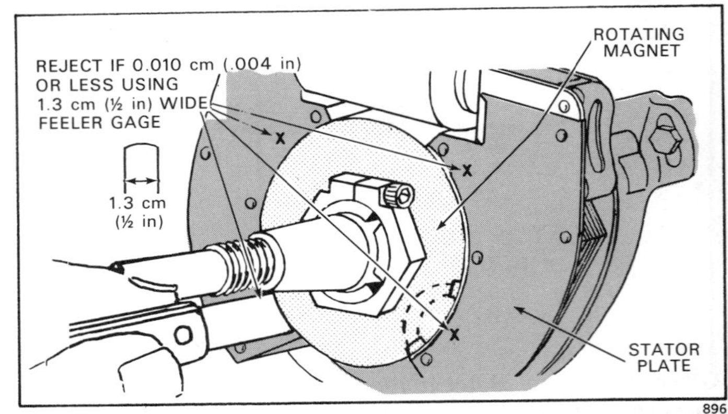

FIGURE 114. Check the stator-plate air gap in four places.

2. *Check the clearance (Figure 114).*

Check in four different places as shown in Figure 114. Move the magnets away from the gage when checking. This is done by turning the crankshaft by hand. Removing the spark plug will make the crankshaft easier to turn because it relieves the compression.

If the clearance is **too small** or **too large,** check the crankshaft and bearings. There is no adjustment on this type of stator plate.

CHECKING THE MAGNETO COIL

Magneto coils seldom give trouble. Consequently, you should make a thorough check of the remainder of the system before you decide to replace one. A weak spark, or no spark, at the spark plug may be due to a damaged or defective coil. The coil can usually be checked and tested without removing it.

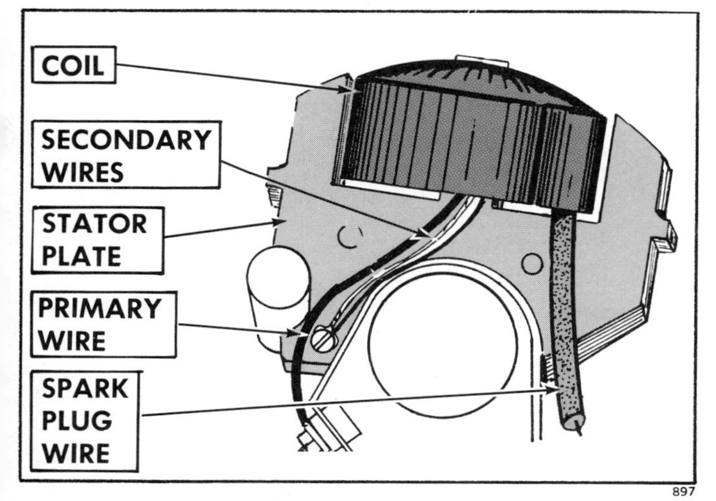

FIGURE 115. Make a thorough check of the coil, coil wires and connections before deciding to replace it.

1. *Inspect coil assembly for damage (Figure 115).*

   Look for cracks or gouges in the insulation, evidence of overheating or other damage. Make sure the electrical leads are intact, not shorted, and are tightly connected.

2. *Check coil spark.*

   a. If you have an **approved coil tester,** follow instructions given by the tester manufacturer and the specifications given for your coil. Examples of approved testers are shown in Figure 116. Check with the coil assembly installed on the yoke. If the coil fails to check out properly, it should be replaced.

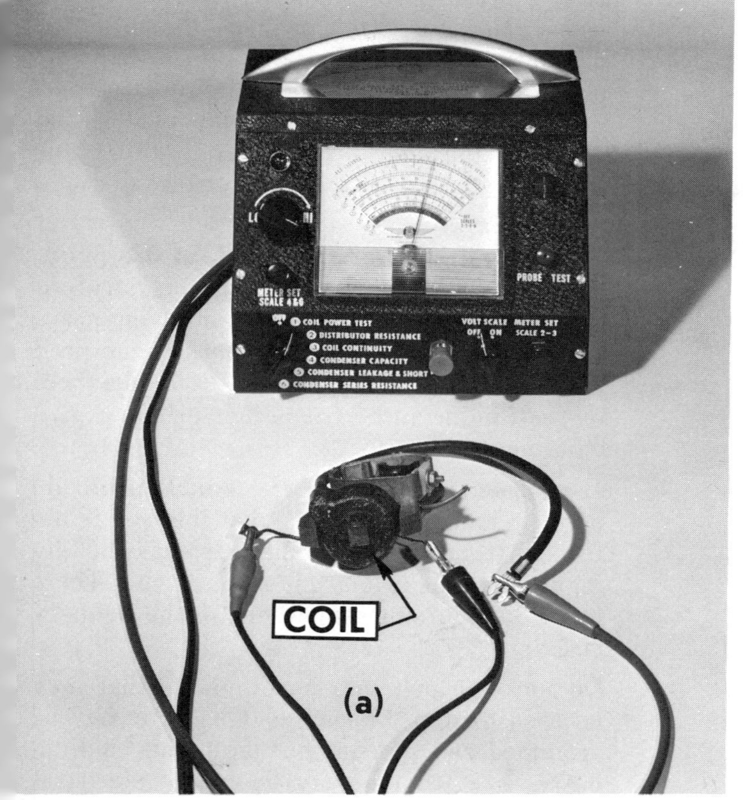

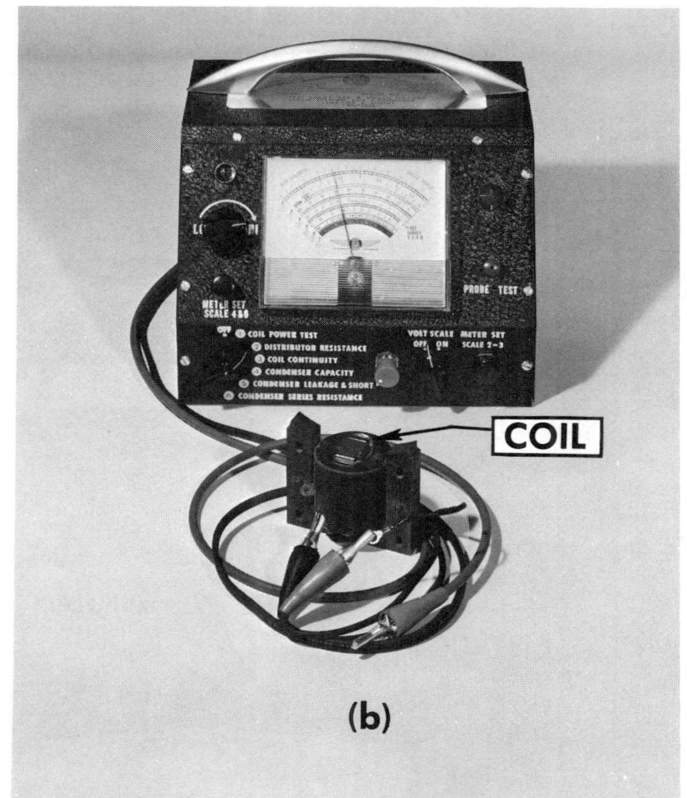

FIGURE 116. Testing coils. Hook up for testing a coil that is is designed for mounting (a) underneath the fiywheel, and (b) outside the flywheel.

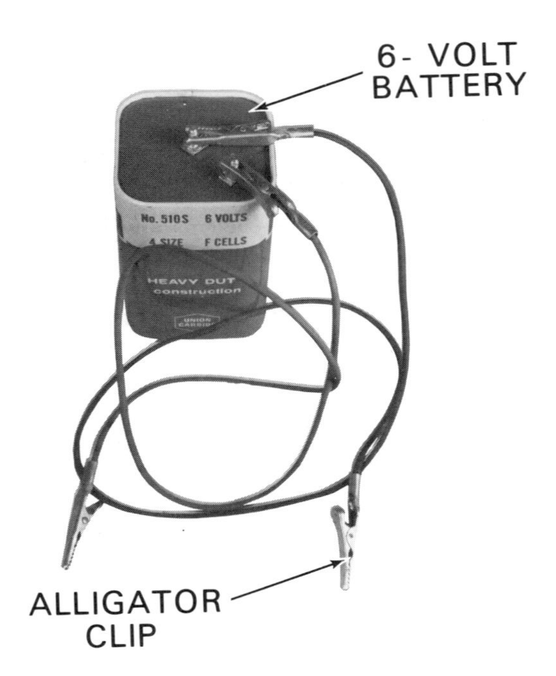

FIGURE 117. A 6-volt lantern battery rigged for checking coils.

CAUTION! Do not use a coil tester on a metal workbench. There is a danger of your getting shocked by the high voltage from the secondary coil carried by the metal in the bench.

b. If you **do not have a tester available,** the coil and ignition circuit can be tested with a **6-volt lantern battery** as shown in Figure 117. It is convenient to have an alligator clip on the end of one battery

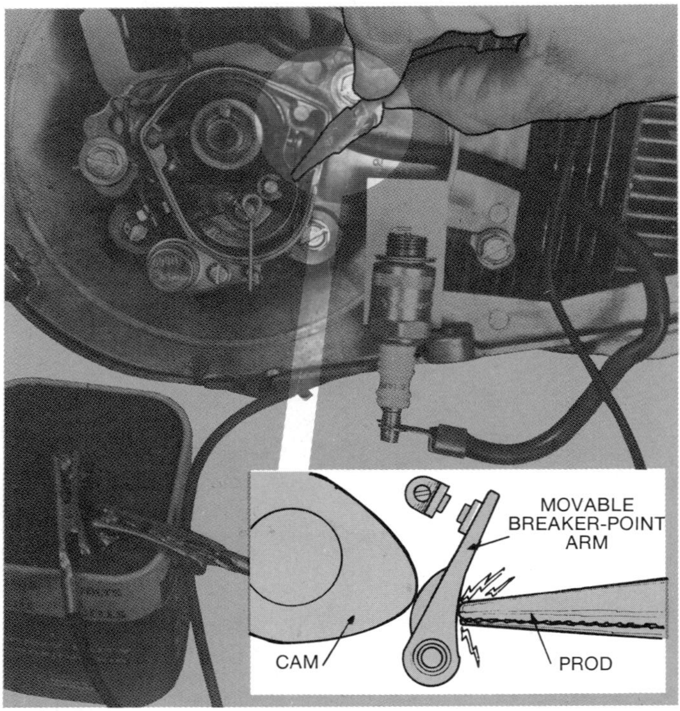

FIGURE 118. Checking the coil with a 6-volt lantern battery tester.

wire and a test prod on the end of the other wire. Continue with step 3.

3. *Remove the spark plug and ground it on the engine as shown in Figure 118.*
4. *Connect the spark-plug wire to the spark plug.*
5. *Clip one battery wire to the engine frame.*
6. *Turn the engine until the points are fully open.*
7. *Strike the movable breaker-point arm with the test prod (Figure 118).*

You should get a strong spark at the prod. The spark is most intense when you touch the prod to the insulated point for a moment and then break the contact quickly. This is the same action as that when the engine is running.

What you are doing is energizing the primary circuit and building up a magnetic field in the coil. When you break this circuit by suddenly removing the prod, the primary field collapses. This induces a high voltage in the secondary and provides a spark at the spark plug. It also gives a momentary surge of current back through the primary circuit, thus causing a bright blue spark at the prod.

If you **get a good spark at the spark plug** when you scratch the prod on the insulated breaker point, with the points open, both the primary and secondary coil circuits are functioning properly. If you do not get a good spark at the spark plug, then examine the spark at the prod in an effort to pinpoint the trouble.

If you get **a bright blue spark at the prod,** this generally indicates that the primary circuit is all right, whether or not you get a spark at the plug. A bright spark at the prod means you are getting perhaps 200 volts there by self-inductance from the primary coil winding.

If you get **no spark at the spark plug,** do not conclude immediately that the coil is defective. It merely means that current is not going through the primary ignition circuit. There is an open circuit somewhere in the primary circuit.

Do not overlook the shorting wire that may be used to stop the engine. This wire may be grounded and shorting out the points continuously. The trouble may be in the electrical leads to the coil, such as in the **primary wire** leading from the points to the coil, in the

coil grounding wire, or in the **secondary wire** leading from the coil to the spark plug. Look for a poor connection or a break in the wire leading from the point assembly to the coil. A loose connection at the points where the coil winding is grounded will also prevent the current from flowing in the primary.

If *after rechecking the connections* you still get **no spark,** you may assume that the coil is bad. Have it checked at your dealer's before replacing it. Proceed with the steps under the next heading.

If you get a **weak spark at the spark plug,** most likely the insulated breaker-point assembly is grounded somewhere. This may be due to the lug on the wire, attached to the insulated breaker point, touching the frame, or to a break in the insulation on the wire leading from the point to the coil.

If you get a **weak spark at the prod**—one comparable to what you get by scratching the two battery leads together—it generally means that the insulated breaker point is grounded or there is a high-resistance ground somewhere in the primary circuit.

REMOVING AND REPLACING THE STATOR PLATE AND COIL

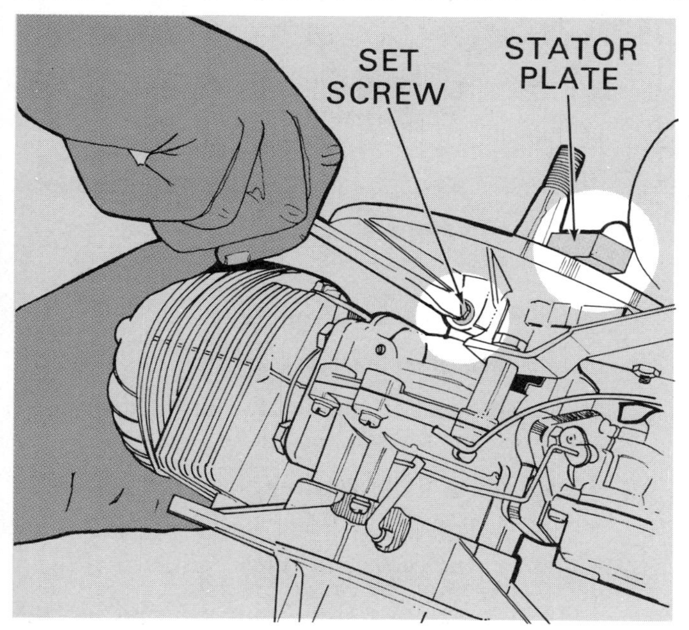

FIGURE 119. Some stator plates are held by a set screw.

1. *Mark the position of the stator plate on the crankcase (Figure 109) if it is not already marked.*

   Use a sharp chisel and mark both the stator plate and engine block. This is very important. It will save you from having to locate its proper timing position when replacing it.

2. *Disconnect the spark-plug wire at the spark plug.*

3. *Remove the spark-plug wire frame clamp if attached.*

   This is a clamp which holds the wire to the engine frame.

4. *Disconnect the primary-ground wire if attached.*

   This wire goes to the ignition switch.

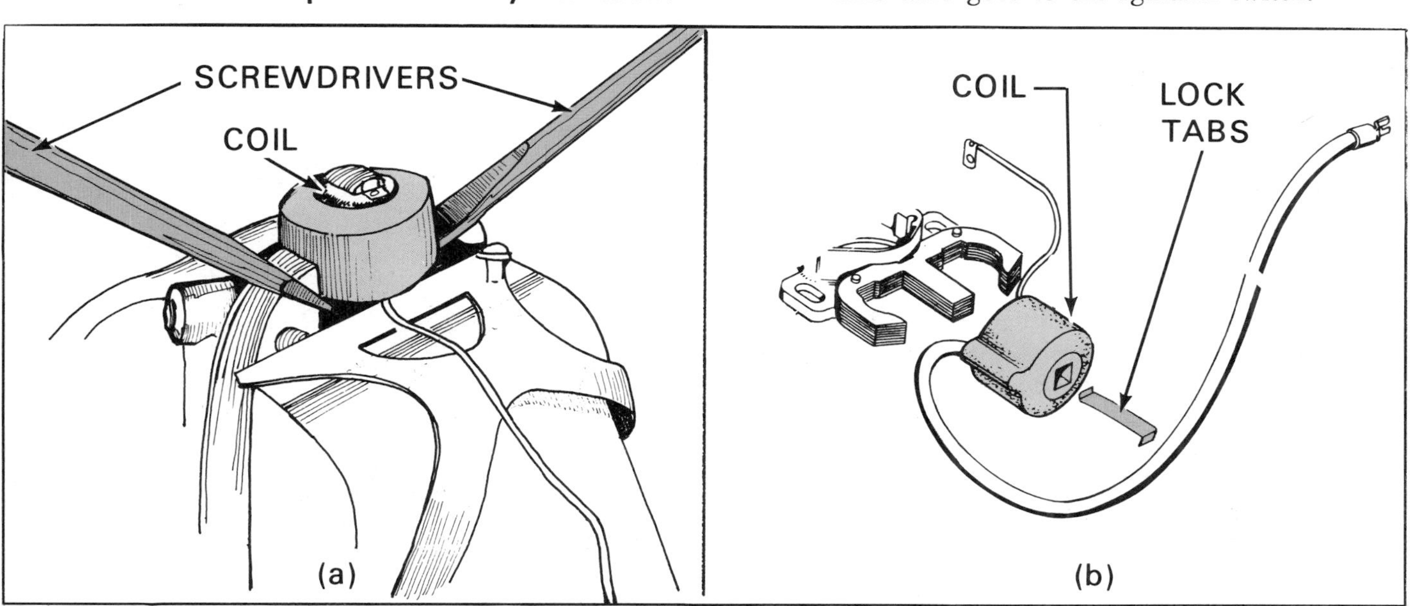

FIGURE 120. Most coils are removable from the iron core. (a) Removing a coil. (b) A coil removed.

5. *Loosen or remove the screws holding stator plate (Figures 108 and 119).*
6. *Remove magneto coil.*

   Most coils are removable from the iron core (Figure 120).
7. *Replace coil and stator plate, using reverse procedures.*

INSTALLING THE FLYWHEEL

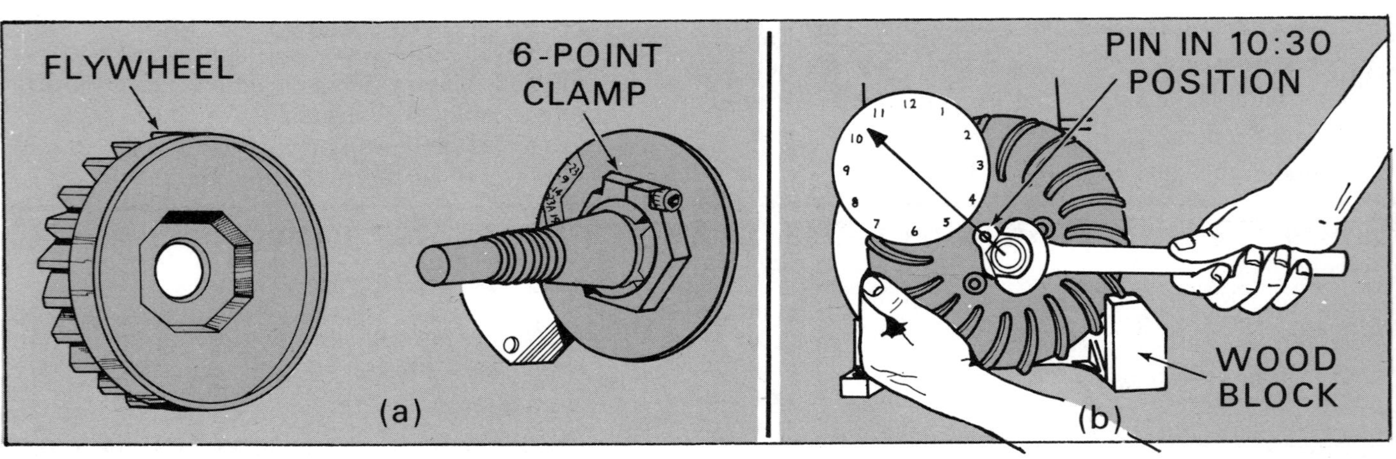

FIGURE 121. A flywheel that is not keyed directly to the shaft. (a) Six-point clamp acts as a flywheel locking device, and (b) starter pin is aligned to the 10:30 o'clock position.

1. *Check all electrical connections.*

   Make sure no wires are pinched and that connections are tight.
2. *Recheck to see that all nuts, screws and bolts under the flywheel are tight.*
3. *Remove all dust, grease and foreign particles from inside the magneto and breaker points.*

   Use denatured alcohol to clean the points. It will clean oil from the contacts.
4. *Install washer if applicable.*

   Some have a spacer washer.

   Some have a concave washer. If the nut is not tight, the key will shear.
5. *Install flywheel.*

   Most flywheels are keyed onto the crankshaft, and there is only one way to put them on.

   If your engine has a **magnetic rotor** under the flywheel (Figure 121a), there may be several positions in which the flywheel will fit. If your engine is of this design (Figure 121a), check the service manual before installing the flywheel. Some specify aligning the starter pin to the 10:30 o'clock position (Figure 121b).
7. *Install starter pulley if applicable.*

   Refer to "Repairing Starters," page 78.
8. *Install lock washer.*
9. *Install retainer bolt and tighten.*

   See your service manual for the proper torque. Do not tighten too much.

   If you use a **lock washer,** turn the nut until the lock washer flattens.
10. *Check crankshaft end play (Figure 122).*

    Correct end play varies from .05 mm to .48 mm (.002 in to 0.019 in). You can get some idea

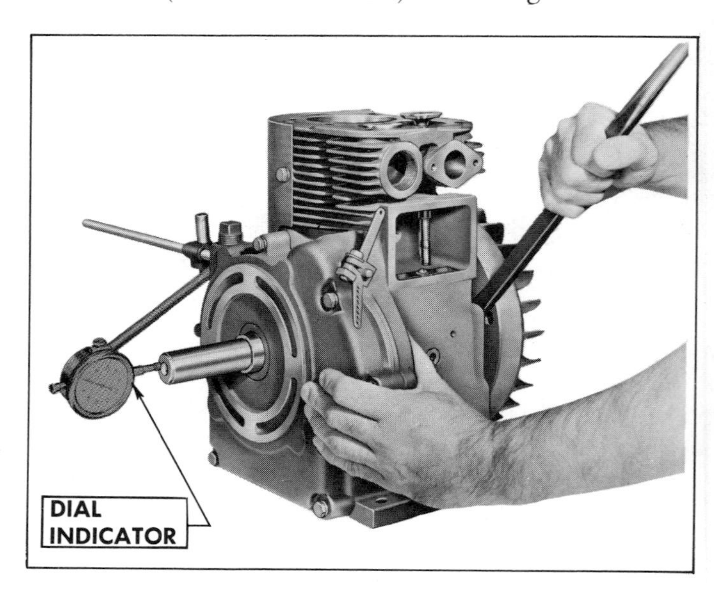

FIGURE 122. Checking the crankshaft end play with a dial indicator.

if it is too little or too much by feel, but the best way to check it is by the use of a dial indicator as shown in Figure 122.

See your service manual for directions on adjusting end play. Some use shims or gaskets under the bearing plate. Some are not adjustable. Adjustments are described under "Repairing Crankshaft Assemblies," page 248.

TESTING THE SOLID STATE IGNITION SYSTEM

1. *Check ignition switch for a short circuit.*

   Use a continuity light. Turn the switch through the various positions. If the light burns when the switch should be off, replace the switch. This could cause the battery to burn out the solid state ignition unit also.

2. *Check leads for grounding or loose connections.*

3. *Check for spark at spark plug.*

   Use a spark tester, or remove spark plug and ground it to the engine.

   If no spark, try a new plug.

   If still no spark, proceed to step 4.

4. *Check air gap (Figure 122A).*

   Loosen screws holding ignition unit. Insert a .13 mm to .25 mm (.005 in to .010 in) non-metal card. Press unit against card and tighten.

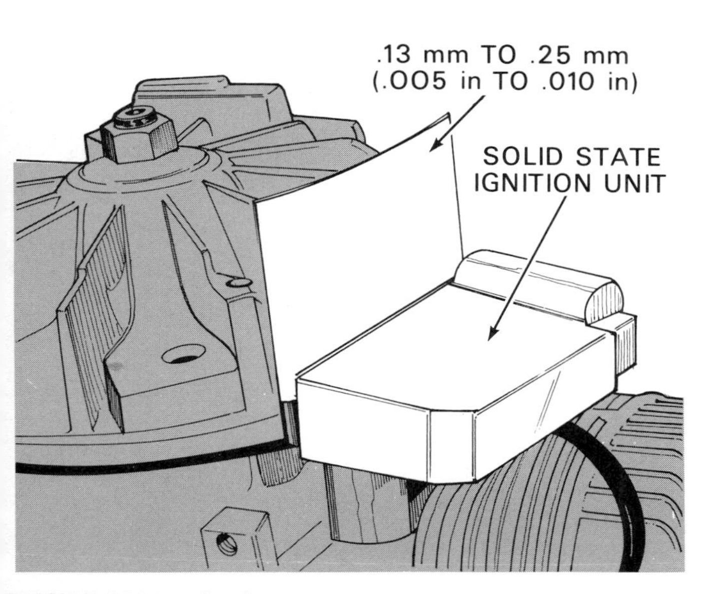

FIGURE 122A. Checking air gap.

   Some units have adjustable trigger pins in the flywheel. They can also be adjusted, if needed.

5. *Check coil if accessible (Figure 122B).*

   On some engines, the resistance of the coil is given in the service manual. If so, a check of the resistance will tell you something about the condition of the system.

   If the **resistance is less than the rated resistance of the coil,** this indicates the coil is shorted. Replace the coil and stator.

   If the **resistance is greater than the rated resistance of the coil,** this indicates the trouble is in the ignition unit. Replace the unit.

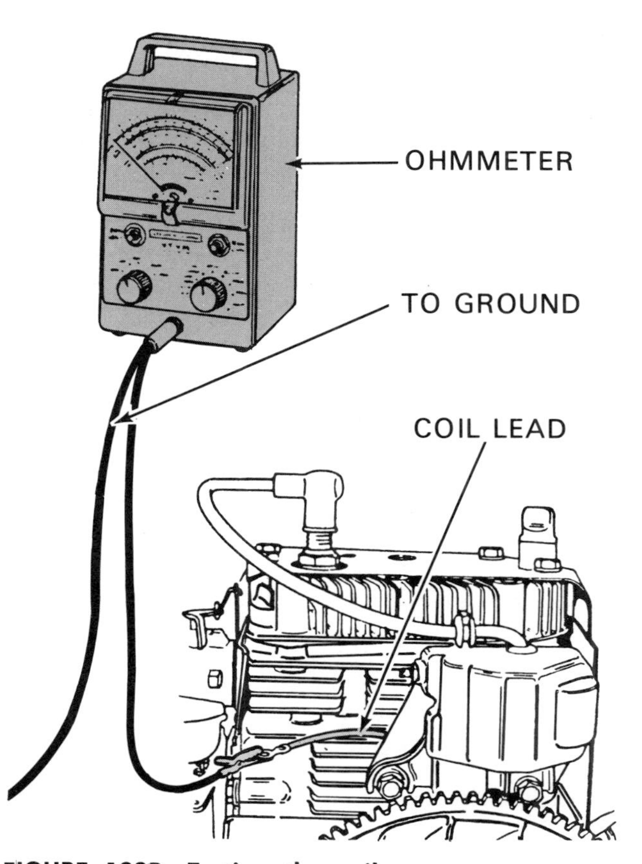

FIGURE 122B. Testing the coil.

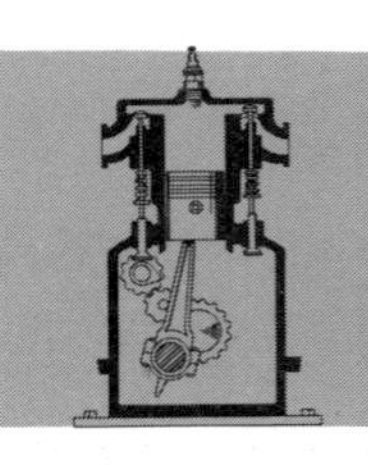

B. MAINTAINING AND REPAIRING BATTERY-IGNITION SYSTEMS

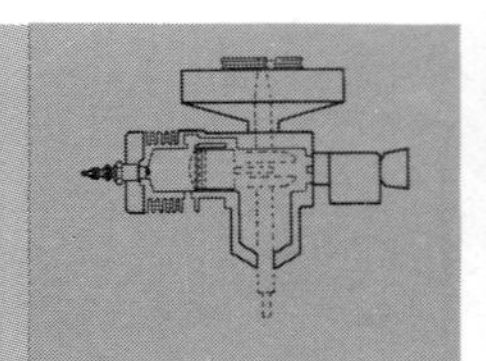

Many small engines come equipped with a battery ignition instead of a magneto ignition. Since a battery and generator are necessary for self-contained electric starting, it is a simple matter to equip the engine with an ignition powered by the battery and generator. This is accomplished by the installation of a few additional parts (Figure 123) to the battery-generator system.

PRINCIPLES OF OPERATION

The battery-ignition system differs from the magneto-ignition system in two ways. They are as follows: (1) **current is supplied** to a spark coil *from the battery and/or the generator rather than a magneto,* and (2) the **ignition switch must be closed** in the battery system for current to flow through the coil. (The switch must be open in the magneto system.) The function of the ignition system is the same — to produce a hot spark at the plug, and at the right time for igniting the fuel. See "Maintaining and Repairing Magneto and Solid State Ignition Systems," page 119.

Here is how the battery-ignition system works (Figure 123).

- With the ignition switch closed (Figure 123a), *current flows from the battery or generator to the primary winding* of the coil, and through the closed breaker points to ground.
- At the right moment, when the piston is just before top-dead-center (compression stroke), the breaker points are opened by a cam, or push rod.
- When the points are opened (Figure 123b), the primary *electric circuit is broken* and the magnetic field breaks down.
- *The magnetic lines of force cut across the conductor (coil) and a high voltage is induced in the secondary winding* because it has many more turns of wire than the primary.
- *A spark occurs at the spark plug* which is in the secondary circuit. The high voltage developed in the secondary coil causes the current to jump across the spark-plug electrode gap, thus making the spark.
- *The residual current in the primary winding is taken up by the condenser (Figure 82).* It eliminates arcing at the points and aids in producing a stronger spark at the spark plug.

Maintaining the battery-ignition system is very much the same as maintaining the magneto-ignition system. For procedures, refer to the appropriate heading under batteries, generators and magneto-ignition systems.

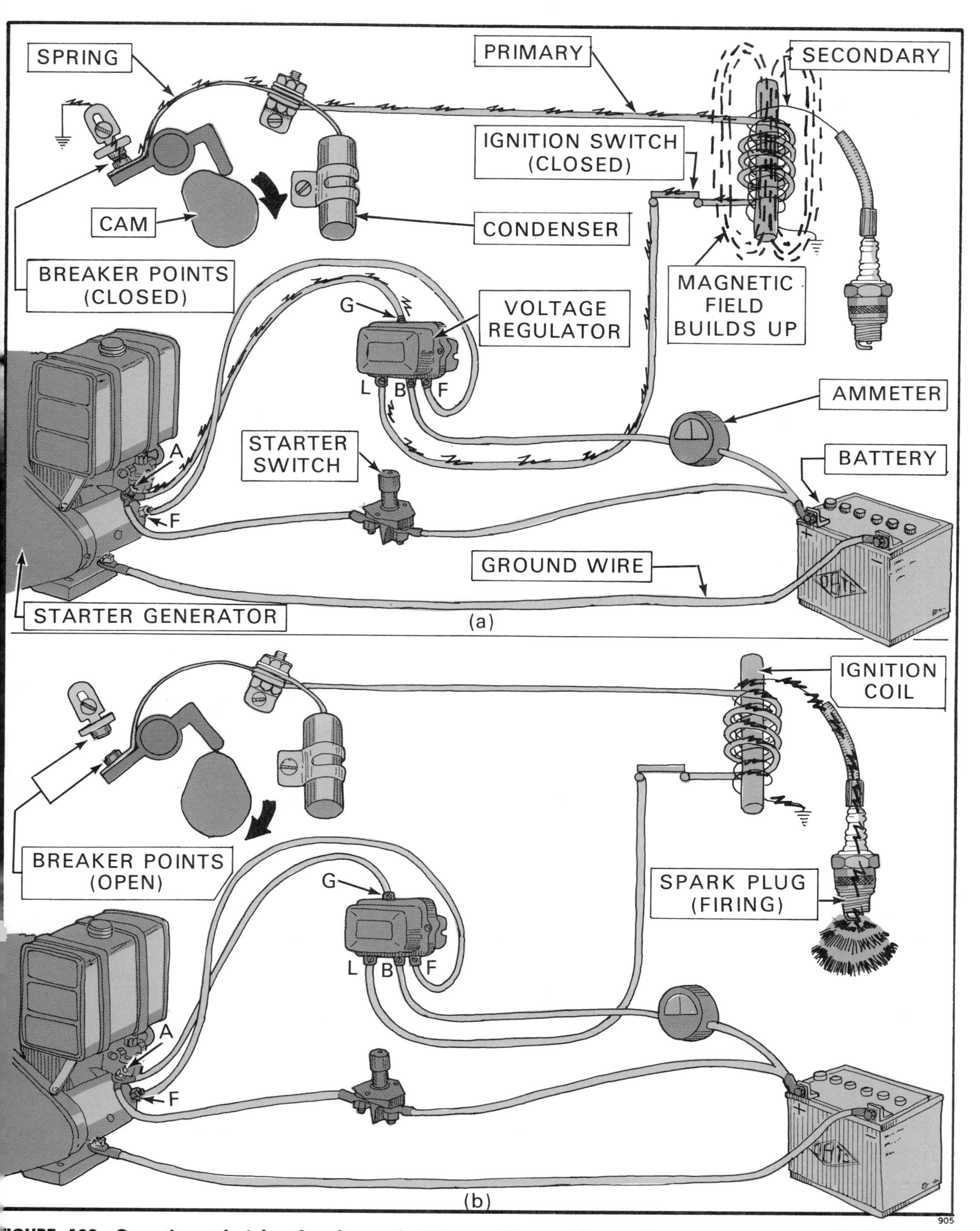

FIGURE 123. Operating principle of a battery-ignition system. (a) Breaker points closed. Current flows from battery through points to ground. Electromagnetic lines of force develop around the ignition coil. (b) Points open. Magnetic lines of force collapse. High-voltage current, induced in the secondary coil, jumps the gap at the plug, thus causing a spark.

NOTES

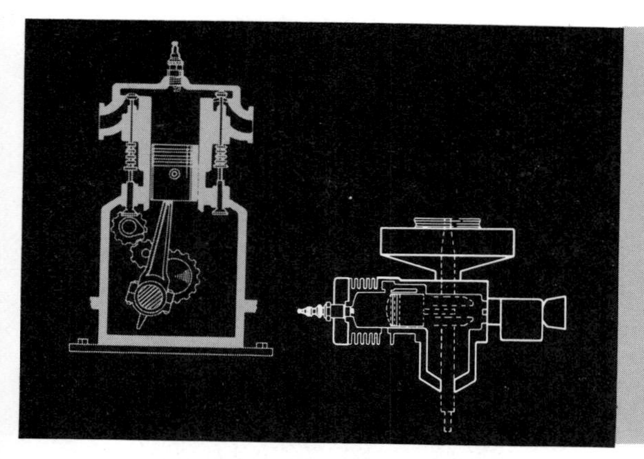

III. Repairing Fuel Systems

You will probably experience more trouble with the fuel system than with any other system on your small engine. But most of the trouble is minor and can be corrected easily. For example, a common problem is attempting to start the engine without fuel in the tank. Another one is failing to open the fuel shut-off valve in the supply line. Such troubles are the fault of the operator and are easily corrected.

Failure to perform some maintenance jobs, however, can result in more serious troubles. For example, **trash, flakes of rust or gum** in the fuel tank and supply lines restrict the flow of fuel. This causes intermittent operation, such as popping and skipping of your engine. Or, it may cause the engine to stop. **An air leak on the suction side of the fuel line may also cause it to stop.**

If the *carburetor is out of adjustment,* it may result in either **too rich or too lean a mixture of fuel.** *Too rich a mixture* (too much fuel for the amount of air) allows liquid gasoline to wash down the cylinder walls, thus resulting in poor lubrication. The fuel-air mixture also fails to burn completely and leaves carbon deposits in your engine. These carbon deposits develop "hot spots." These hot spots will cause preignition — fuel ignited before the piston reaches the desired position during the compression stroke (Figure 124a). This can cause knock, loss of power, scoring and scuffing of the cylinder

FIGURE 124. (a) Preignition is ignition taking place before time for the spark plug to fire. One cause is glowing carbon spots or high temperature buildup in the combustion chamber. (b) Piston damage caused by preignition knock.

and piston walls, and damage to the piston (Figure 124b).

Too lean a mixture (too little fuel for the amount of air) burns more slowly and at a higher temperature. This high temperature raises the temperature of the pistons, rings and valves. The oil may break down on the cylinder wall, and scuffing and scoring will occur (Figure 125). Valve stems stick, and valve faces and seats burn.

FIGURE 125. Piston scored by continued operation with too lean a fuel mixture.

FUNCTIONS OF THE FUEL SYSTEM

The purpose of the fuel system **is to provide a constant supply of clean fuel and air to the combustion chamber.** The fuel-air mixture must be of *sufficient quantity* and of the *proper proportion* to meet the demands of your particular engine under all conditions. The proportion of fuel-to-air varies with the load being put on the engine.

The fuel system must perform its functions, regardless of constantly changing atmospheric pressure, humidity (moisture in the air) and the outdoor temperature.

The *principal parts of the fuel system and their functions* are as follows:

- A **fuel tank assembly,** which serves as a fuel reservoir.
- A **carburetor,** which provides the proper mixture of fuel and air.
- A **fuel pump,** on some engines, which provides fuel under pressure (higher than atmospheric) to the carburetor. On most small engines fuel is provided to the carburetor by gravity feed.
- **Screens and strainers** in the fuel tank, between the tank and the carburetor, which help provide clean fuel.
- The **carburetor air cleaner,** which cleans the air going into the carburetor. See "Servicing Carburetor Air Cleaners," Part 1, for complete explanation.

TYPES OF FUEL SYSTEMS AND HOW THEY WORK

There are *three types of fuel systems* used on small engines: (1) **gravity-feed** (Figure 126), (2) **pressure-feed** (Figure 127), and (3) **suction-feed** (Figure 128).

The **gravity-feed fuel system** has the fuel tank mounted above the carburetor, and fuel flows by gravity to the carburetor (Figure 126).

The **pressure-feed fuel system** may have the fuel tank mounted either above or below the carburetor. A **fuel pump** is installed between the tank and the carburetor to supply fuel to the carburetor under pressure — above atmospheric pressure (Figure 127). A mechanically driven diaphragm-type pump is used. It is the same type that is used on automobiles and tractors. Figure 127 shows how it works.

A *differential-pressure-driven diaphragm-type fuel pump* is used on engines that may be operated at extreme angles, such as on chain saws. This type pump is usually incorporated in the carburetor. It is explained under "Types of Carburetors and How They Work, Part 1.

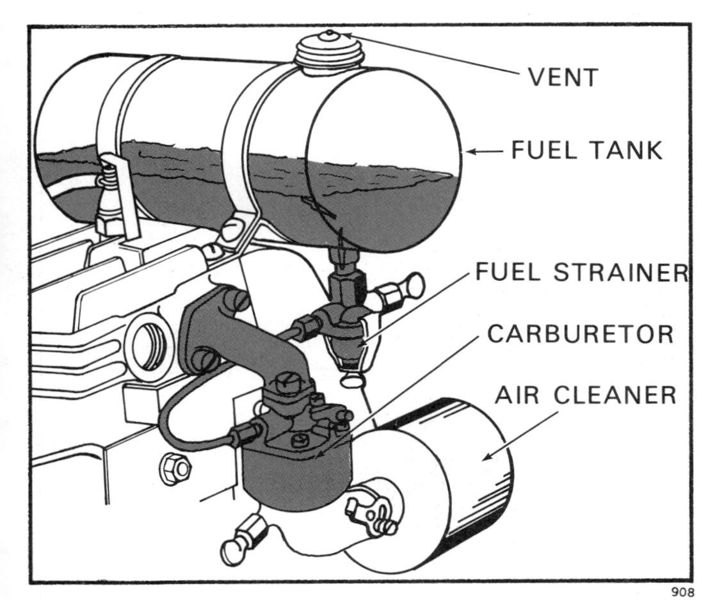

FIGURE 126. A gravity-feed fuel system. Fuel flows by gravity — its own weight — to the carburetor.

In the **suction-feed fuel system**, fuel is drawn directly from the fuel tank into the low-pressure area (suction) developed in the carburetor (Figures 128 and 128A). Air is drawn through the carburetor by the engine suction—the partial vacuum created by the downward movement of the piston. As the air passes the venturi section of the carburetor, a suction is created that draws the fuel from the fuel tank.

The *fuel tank* is mounted below the carburetor. Sometimes it is attached to the carburetor and takes the place of the carburetor storage bowl.

A *foot valve* in the bottom of the tube keeps the fuel from running back. This keeps the tube full of fuel at all times.

Maintaining and repairing fuel systems are discussed under the following headings:

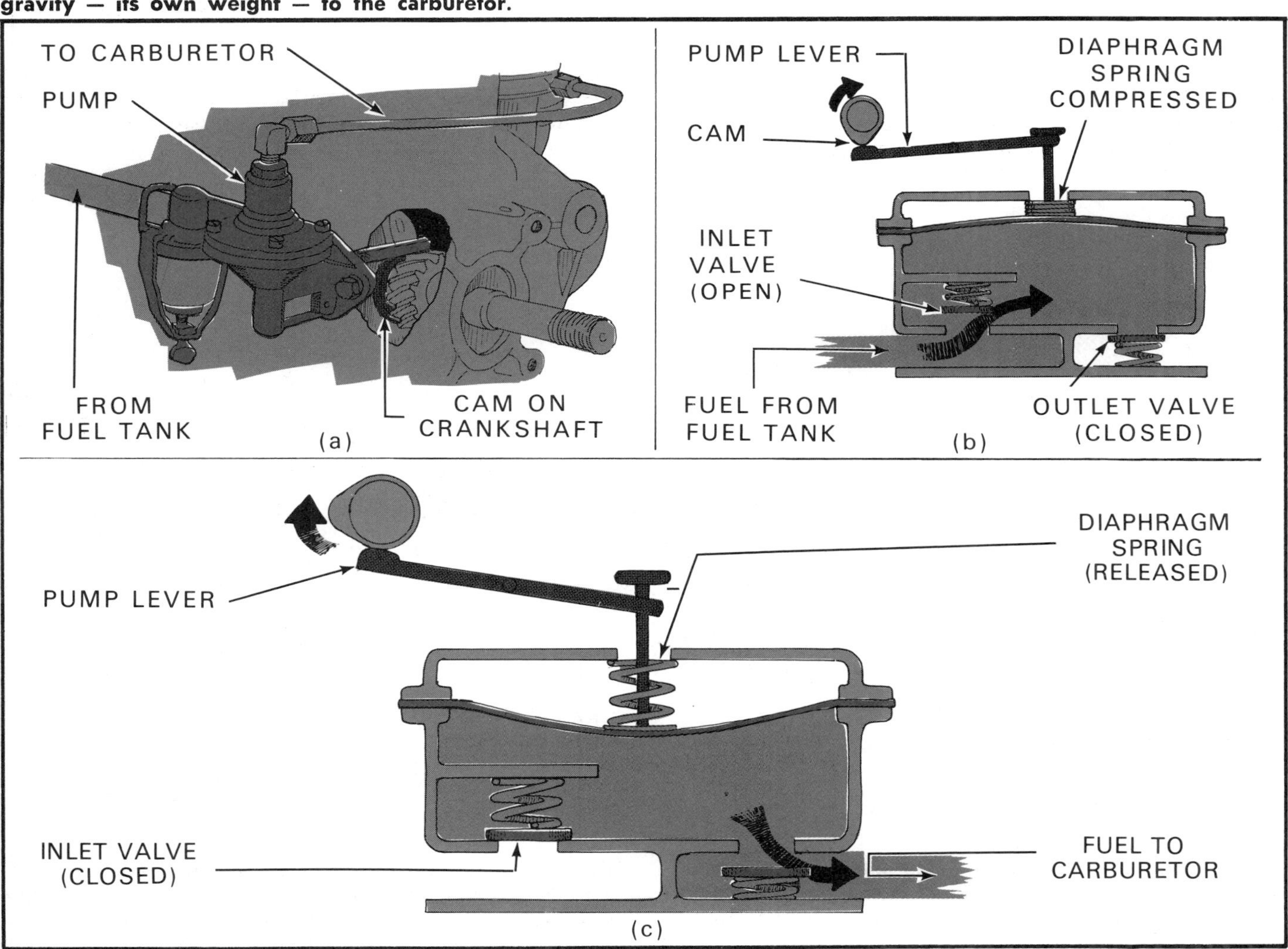

FIGURE 127. A pressure-feed fuel system. (a) A diaphragm-type fuel pump is operated from a cam on either the crankshaft or the camshaft. (b) and (c) Cut-a-way of a diaphragm fuel pump showing how it operates. (b) When the cam turns, it presses the pump lever. The lever raises the diaphragm. A partial vacuum is created in the pump chamber. The inlet valve opens and fuel flows into the pump chamber from the fuel tank. (c) When the cam releases the pump lever, spring tension forces the diaphragm in the opposite direction. The inlet valve closes and the outlet valve opens. Fuel is pumped to the carburetor.

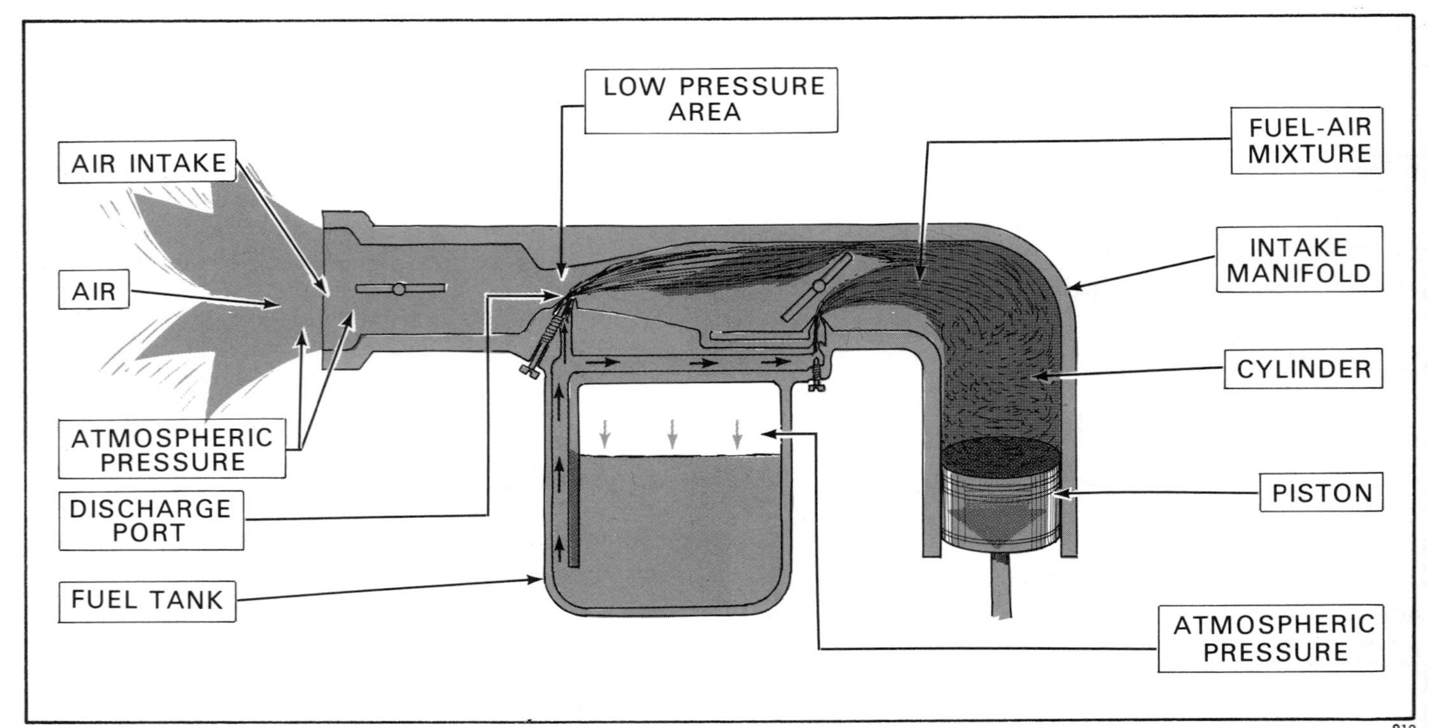

FIGURE 128. Principle (schematic) of a suction-feed fuel system. As the engine piston moves down, a vacuum tends to develop in the cylinder. Atmospheric pressure forces air through the air intake, past the fuel port and through the intake manifold to the engine cylinder (combustion chamber). As the air passes the fuel port, fuel is drawn in and mixed with the air.

A. Repairing fuel-tank assemblies.

B. Repairing fuel pumps.

C. Repairing carburetors.

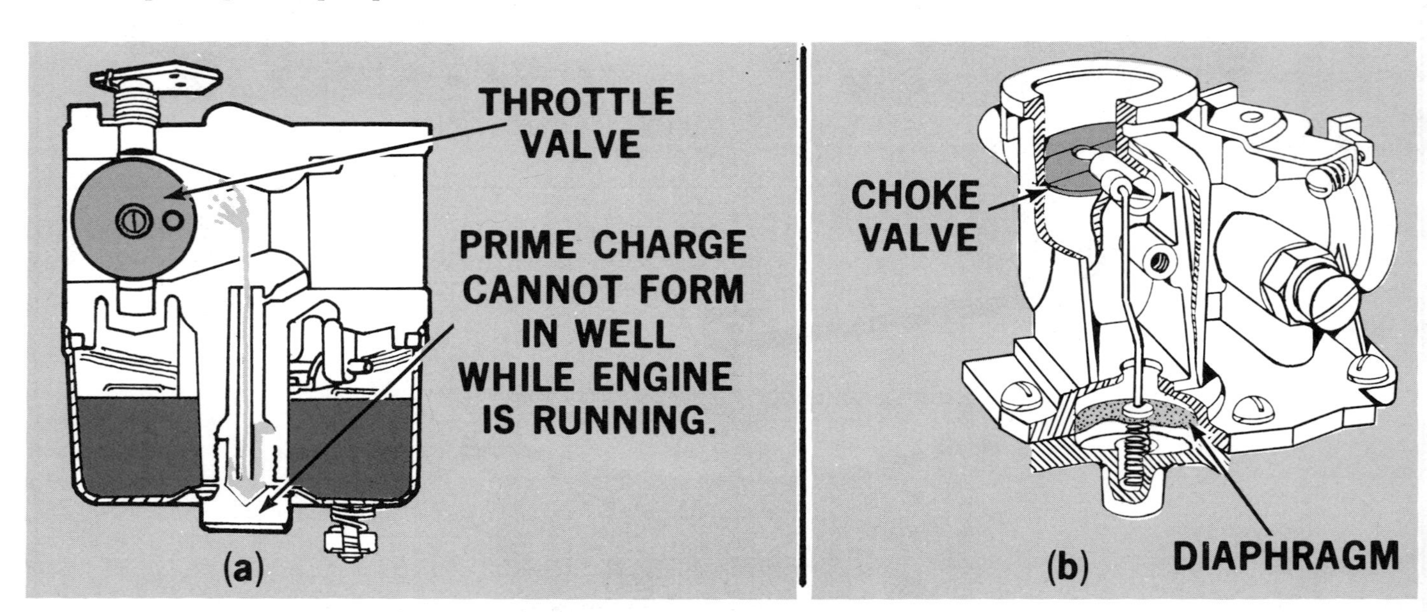

FIGURE 128A. (a) Suction lift carburetor with no choke valve. Suction, created by the intake stroke of the piston, draws the fuel into the carburetor air horn and on in to the combustion chamber. Since the air bleed can not supply air until the fuel charge in the well ahead of it is pushed into the main nozzle, the fuel-air mixture at this point is predominantly gasoline. This initial charge of fuel is the enriched mixture needed for starting. When the engine starts and runs, the fuel level stabilizes at the lower end of the main nozzle so that the customary power blending of the gasoline and air at the nozzle tip is accomplished. Note there is no choke valve. (b) Carburetor with automatic choke. Upon starting, vacuum created during the intake stroke is routed to the bottom of the diaphragm, through a calibrated passage, thereby opening the choke. A diaphragm under the carburetor is connected to the choke shaft by a link. A calibrated spring under the diaphragm holds the choke valve closed when the engine is not running.

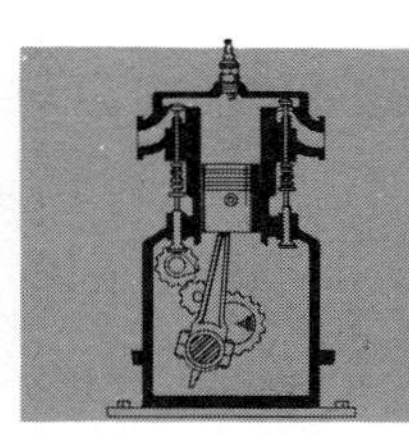

A. REPAIRING FUEL TANK ASSEMBLIES

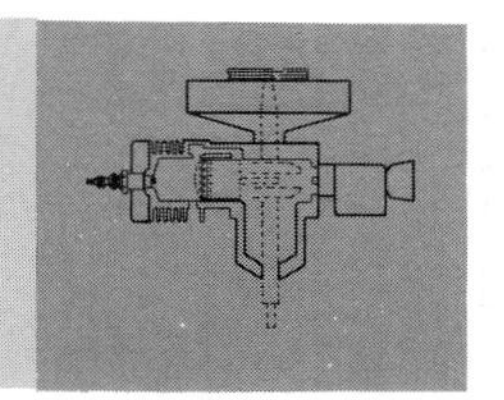

The fuel tank on your small engine serves other functions in addition to providing a readily available source of fuel (Figure 129).

Fuel tank assemblies provide for:

- A *fuel reservoir* — some are separately mounted, while others are connected directly to the carburetor (Figure 130).
- *Venting the fuel tank* to atmospheric pressure.
- *Fuel strainers* for filtering out trash and dirt.
- *A fuel shut-off valve* — on most assemblies.

The primary function of the **fuel tank** is to *provide a reservoir of gasoline on or near the engine.* The *size of the tank* is determined by the size and use of the engine. For example, the average lawn can be mowed with a quart of gasoline. Therefore, a large tank is not necessary on a small lawn mower, while a small tractor can use a tank of several quarts capacity to advantage.

The *shape of the tank* may be determined by the design of the equipment. It may be molded for compactness, appearance and convenience.

Tanks are made from untreated steel, galvanized steel, annodized steel, nylon, plastic and other materials. Rust is a problem in non-treated steel tanks.

All tanks have a **vent** — usually the vent is in the cap (Figure 129b). This is to provide for atmospheric pressure to enter the tank and push the fuel through the carburetor into the engine. Otherwise, a partial vacuum would form in the tank and fuel would not flow.

Here is how it works. As a vacuum is developed in the tank, the poppet valve is forced down, and air enters the tank as shown in Figure 129b. If the tank is tilted or turned upside down, gravity plus the pressure of the fuel against the poppet valve closes it. This prevents fuel from leaking through the vent.

Tanks on chain saws, and other applications of extreme angle operation, have a flexible fuel pick-up line (Figure 129f) with a weight on the end. It shifts to the low side of the tank, as the position of the engine is changed, and remains submerged in the fuel.

Some tank caps are equipped with a quantity gage in addition to the vent (Figure 129c).

Most gravity-feed and pressure-feed fuel tanks have a **strainer** at the bottom (Figure 129d and 129e). It collects dirt and trash. One tiny speck of dirt can clog a carburetor adjusting valve.

As part of the strainer assembly, most tanks have a **fuel shut-off valve** (Figure 129d). This provides for draining the fuel from the carburetor and supply line.

A combination fuel strainer, shut-off valve and sediment bowl is used on many small engines. This combination provides the surest method for preventing trash, dirt and water from entering the carburetor. Water, being heavier than gasoline, settles to the bottom of the sediment bowl along with dirt particles.

Maintaining and repairing fuel tanks is discussed under the following headings:

1. Tools and materials needed.
2. Checking and cleaning the fuel tank.
3. Repairing leaks in the fuel tank.

TOOLS AND MATERIALS NEEDED

1. Open-end wrenches — 7/16″ through 9/16″
2. Slot-head screwdriver — 6″ regular
3. Clean rags
4. Petroleum solvent (mineral spirits, kerosene, or diesel fuel)
5. One-foot length of wire — approximately 14 gage
6. Fuel container

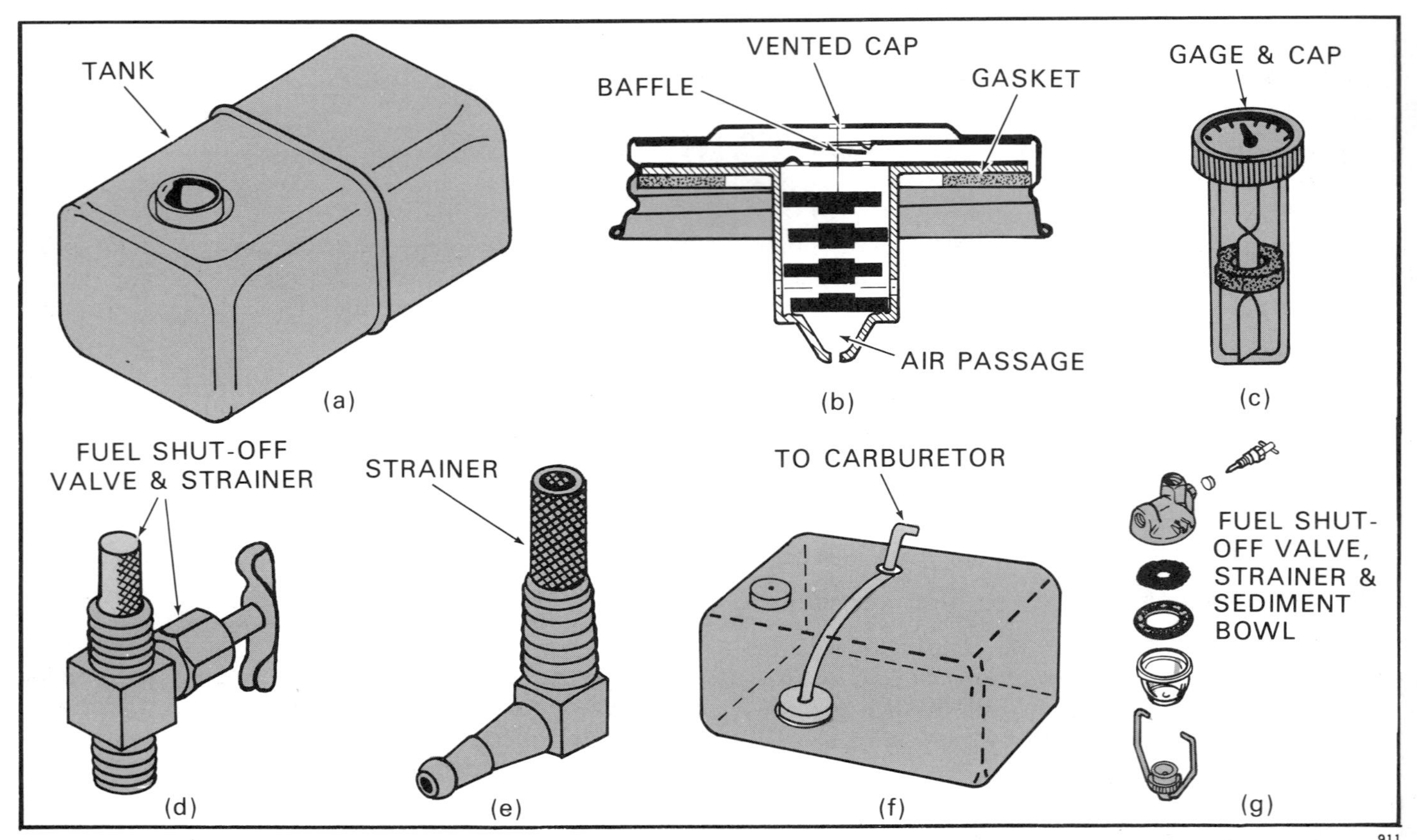

FIGURE 129. A typical fuel tank and accessories. (a) Tank. Supply line accessories may consist of (b) vented cap, (c) combination cap and gage, (d) combination shut-off valve and fuel strainer, (e) fuel strainer and fitting, (f) fuel strainer mounted on the end of a flexible hose, or (g) combination fuel shut-off valve, fuel strainer and glass sediment bowl.

CHECKING AND CLEANING THE FUEL TANK

1. *Determine if carburetor must be removed with tank.*

   If your carburetor is **mounted directly on the fuel tank,** as the one in Figure 130, you will have to remove the carburetor. See "Removing the Carburetor," page 160.

2. *Drain the fuel from the tank.*

   If the tank **has no fuel shut-off valve,** tilt the engine and pour the fuel out the filler hole.

   If the tank **has a fuel shut-off valve,** close it. Disconnect the fuel line from below the fuel shut-off valve, and drain the fuel from the tank. Large engines, and those mounted on heavy equipment, have fuel shut-off valves.

3. *Remove the fuel strainer.*

   If the strainer is **similar to the ones in Figure 129d and 129e,** unscrew the fitting from the tank.

   If the strainer is **on the end of a flexible pipe** fish it out with a wire hook.

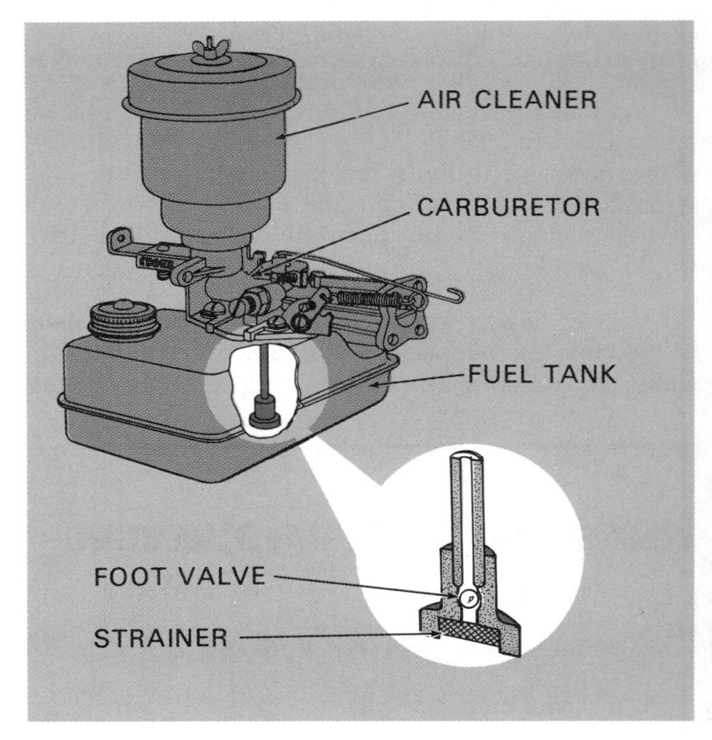

FIGURE 130. A fuel tank serving as a carburetor fuel reservoir.

If you have a **sediment bowl and strainer,** see "Servicing the Sediment Bowl Type of Fuel Strainer," Part 1.

Refer to "Servicing Fuel Strainers," Part 1, for more information on fuel strainers.

4. *Remove the tank.*

   Most tanks are held by capscrews.

5. *Clean the tank strainer and fuel lines in petroleum solvent.*

6. *Inspect tank for damage and wear.*

   If you have a **steel tank,** small leaks can be soldered. See next heading for procedures.

   But usually if the leak is due to rust, the tank is too corroded to repair. Replace it with a new one. There is little you can do to repair **aluminum** and **plastic** tanks.

7. *Reinstall the strainer.*

   If your tank is to be repaired for leaks, leave the strainer out until the leak has been repaired.

REPAIRING LEAKS IN THE FUEL TANK

If you have a **steel tank,** you can solder it.

1. *Remove the tank (if not already removed).*

2. *Soak the tank in water to remove liquid fuel and gas fumes.*

   Fill the tank with water and let it stand a few minutes.

   CAUTION! Never apply heat to a tank that has gas fumes inside it. It will explode.

3. *Clean and solder the hole.*

   Use a 40 - 60 lead solder.

4. *Fill tank with water and recheck for leaks.*

5. *Drain tank and dry as soon as possible to prevent rust.*

   Use forced air if available. Keep tank in a place where dirt and trash will not get inside it.

6. *Repaint outside of tank.*

7. *Reinstall strainer.*

8. *Reinstall tank on engine.*

9. *Connect fuel lines and replace cap.*

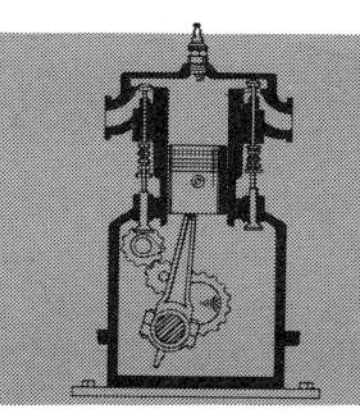

B. REPAIRING FUEL PUMPS

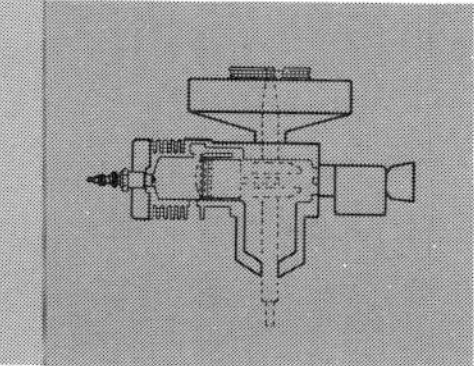

Fuel pumps are generally used on installations where the fuel tank is located beneath the carburetor. Their purpose is to supply fuel to the carburetor at a constant pressure. They are also used on multi-position equipment such as chain saws.

Most fuel pumps have a **flexible diaphragm** for a pumping element. This is the part that is likely to cause the most trouble. If the diaphragm leaks, the pump will not function properly, and fuel will not be supplied to the carburetor.

Maintaining and repairing fuel pumps are discussed under the following headings:

1. Types of fuel pumps and how they work.
2. Tools and materials needed.
3. Checking the fuel pump for proper operation.
4. Repairing the fuel pump.
5. Installing the fuel pump.

TYPES OF FUEL PUMPS AND HOW THEY WORK

Fuel pumps on small engines have a diaphragm pumping element. The diaphragm may be driven by either (1) **mechanical** means (Figure 127a and 131) or (2) **differential pressure** (Figure 132).

Either type of pump has two check valves: an inlet and an outlet valve. Some are spring-loaded steel balls. Some are floating discs and some are flapper-type valves. The *mechanically driven type* is explained in Figure 127.

The **differential-pressure type** works in much the same manner. The difference is in the way force is applied to the diaphragm. With this type, the differential (changing) pressure is developed from the alternating (partial) vacuum and pressure in the crankcase of 2-cycle engines (Figure 132). When the piston moves away from the crankshaft, a partial

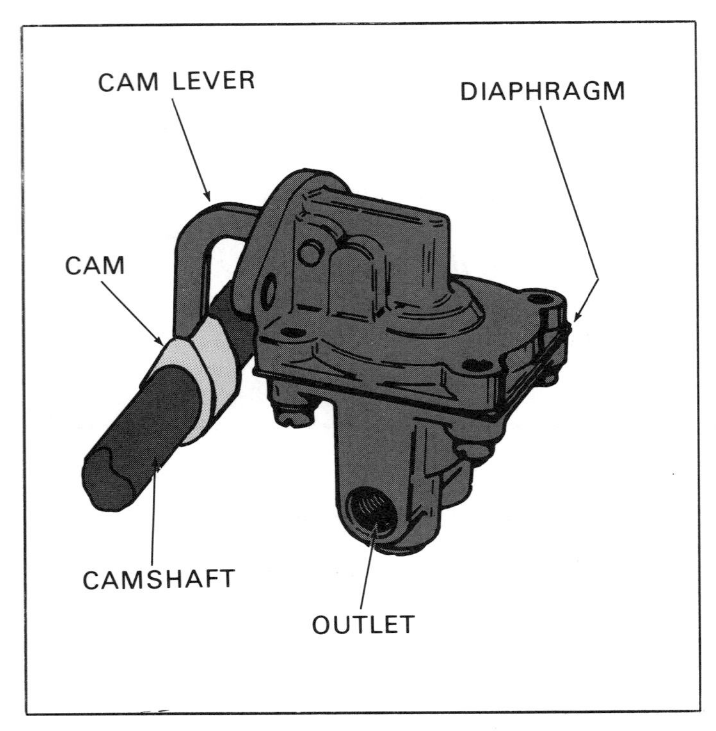

FIGURE 131. Typical mechanically driven fuel pump.

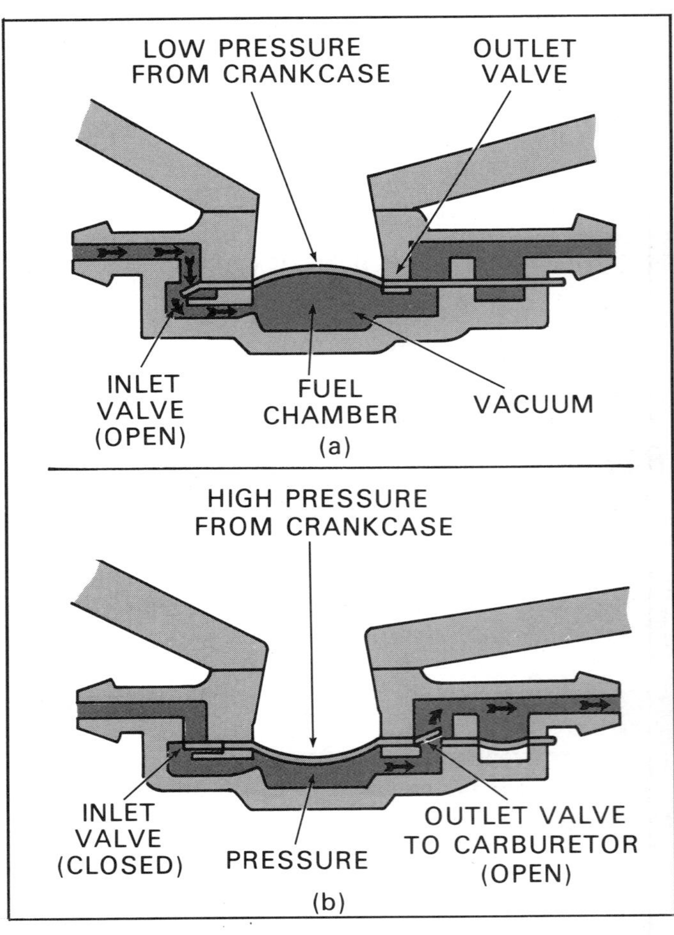

FIGURE 132. Cross section of typical differential-pressure driven fuel pumps used on 2-cycle engines. (a) Low pressure from the crankcase (2-cycle engines) allows the diaphragm to be forced upward. Fuel from the tank enters the pump chamber through the inlet valve. (b) High pressure from the crankcase (2-cycle engines) forces the diaphragm down. The inlet valve closes and the outlet valve opens. Fuel goes to the carburetor.

vacuum develops in the crankcase. As it moves toward the crankshaft, pressure develops.

When a partial vacuum develops on the diaphragm, the diaphragm moves in the direction of least pressure (Figure 132a). A partial vacuum develops in the fuel chamber. This causes the inlet valve to open and the outlet valve to close. Fuel rushes into the fuel chamber. Atmospheric pressure on the fuel in the fuel tank forces the fuel into the fuel chamber.

When the diaphragm is moved in the opposite direction by crankcase pressure, the outlet valve opens and the inlet valve closes. Fuel is forced out of the pump to the carburetor by the pressure exerted on the diaphragm.

TOOLS AND MATERIALS NEEDED

1. Open-end wrenches — 7/16″ to 9/16″
2. Slot-head screwdriver — 6″
3. Phillips-head screwdriver — 6″
4. Petroleum solvent (mineral spirits, kerosene or diesel fuel)
5. Clean rags
6. Wire brush

CHECKING THE FUEL PUMP FOR PROPER OPERATION

1. *Disconnect the spark-plug wire.*
2. *Disconnect the fuel line at the carburetor.*
3. *Rotate the engine crankshaft with the starter 15 to 20 revolutions.*
4. *Observe the fuel flow.*

Fuel from the pump should flow strongly and in regular squirts.

If **fuel does not flow, or if** the **flow is weak and/or erratic,** the trouble is most likely a leak in the fuel pump diaphragm. You can get repair kits for most pumps. There are some, however, for which no repair parts are available, and the entire pump must be replaced. See your dealer.

Procedures for repairing the two types of fuel pumps are given under the next heading.

REPAIRING THE FUEL PUMP

Procedures for rebuilding are approximately the same for all types. But if your pump is different, follow instructions given in your service manual.

1. *Close the fuel shut-off valve.*
2. *Disconnect fuel lines from each side of pump.*
3. *Remove screws holding the pump to the engine block.*
4. *Remove the pump.*
5. *Clean the fuel pump unit in petroleum solvent.*
6. *Mark the alignment of the two main sections of the pump housing with a file or a center punch.*

   Some mechanically driven pumps are built so that the fuel inlet-and-outlet section may be rotated 1/4 turn, in relation to the cam-lever section, to fit different installations. After disassembly you may not remember in which position yours was installed. The alignment marks will serve to remind you.
7. *Remove the screws that hold the pump together.*
8. *Separate the two sections.*

   Remember when disassembling a unit, such as the fuel pump, always lay the parts out in the order they came apart. This will help you remember how to put them back together again.
9. *Remove pin from pump lever on mechanical driven pumps.*

   Use a small pin punch.
10. *Remove the diaphragm.*

    If your pump is **similar to the one shown in Figure 131,** hold the connecting link steady with the thumb of the left hand. With the heel of your right hand, press down on the diaphragm. Turn it clockwise one quarter turn. This should unhook the diaphragm from the link so it can be removed.
11. *Clean and inspect parts.*

    Check parts for damage and excessive wear. Lay them out in a clean, well-protected place.
12. *Install new diaphragm, spring (if used) and gasket.*

    Diaphragm spring should fit into the cup under the diaphragm.
13. *Align the two housing sections by the alignment marks you made before disassembly.*

    Tighten screws loosely.
14. *Insert the pump lever on mechanically driven pumps (Figure 133a).*

    Hold diaphragm down while inserting the pump lever.

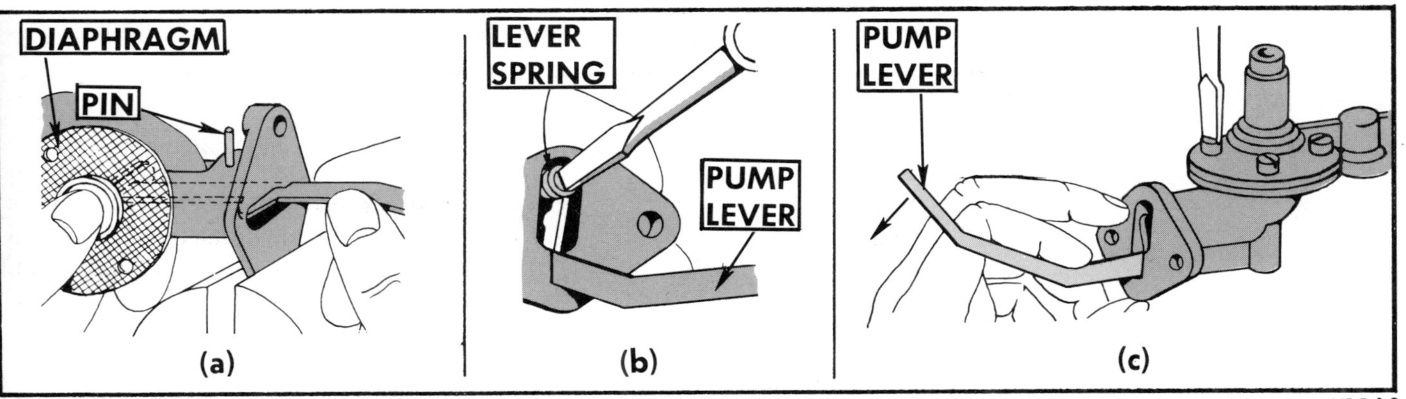

FIGURE 133. Reassembling one design of a mechanically driven fuel pump. (a) Installing the pump drive lever. (b) Installing the drive-lever spring. (c) Installing the pump housing.

15. *Install drive-lever pin on mechanically driven pumps (Figure 133a).*

16. *Install drive-lever spring (Figure 133b) on mechanically driven pumps.*

17. *Reassemble pump housing.*

18. *Press down on lever as far as possible on mechanically driven pumps (Figure 133c).*

    The purpose of this is to give the diaphragm flexibility.

INSTALLING THE FUEL PUMP

1. *Lubricate the part of the lever arm that contacts the cam on mechanically driven pumps.*

   Use multi-purpose grease.

2. *Attach the pump to the engine.*

   Use a new gasket.

   Keep mounting face of pump parallel to mounting face of engine. This is to make sure the drive arm seats properly on the drive cam.

3. *Crank engine to a position where there is pressure on the diaphragm.*

4. *Tighten the assembly screws securely.*

   This gives the diaphragm maximum flexibility.

5. *Turn crankshaft and check to see if drive arm is properly seated on the cam before tightening pump.*

   Put your thumb and finger over the intake and discharge ports and observe pressure and suction.

6. *Secure pump to engine.*

7. *Attach fuel lines coming from fuel tank.*

8. *Recheck pump for operation.*

9. *Attach fuel line to carburetor.*

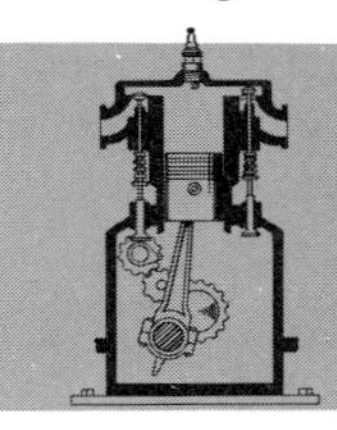

C. REPAIRING CARBURETORS

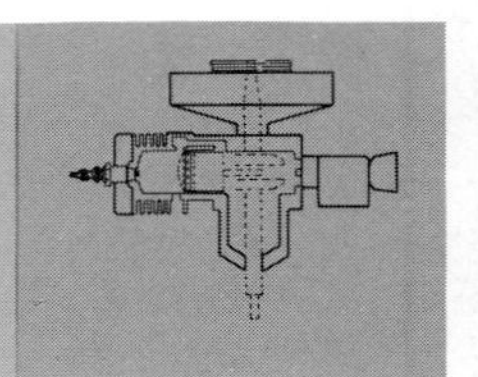

The different types of carburetors and their principles of operation are described in Part 1. If you have forgotten how they work, it would be well to review the operating principles before proceeding with repairing a carburetor.

Repairing carburetors is discussed under the following headings:

1. Removing the carburetor.
2. Repairing the float- and diaphragm-type carburetor.
3. Repairing the suction-lift carburetor.
4. Installing the carburetor.

TOOLS AND MATERIALS NEEDED

1. Slot-head screwdriver — 6″
2. Phillips-head screwdriver — 6″
3. Open-end wrenches — 1/4″ through 1/2″
4. Needle-nose pliers — 7″

REMOVING THE CARBURETOR

1. *Close fuel shut-off valve.*

2. *Remove carburetor air cleaner.*

3. *Disconnect governor linkage (Figure 134).*

   Remember how the governor is connected. Usually there are several holes in the governor linkage. You must reconnect it to the same hole. Sometimes it is easier to disconnect the linkage after disconnecting the carburetor.

4. *Disconnect the throttle linkage.*

   Sometimes the ignition grounding switch is also connected to the throttle-control linkage so that when the throttle is closed, the switch is off (Figure 135)—magneto is grounded. Disconnect linkage to it also.

5. *Remove cowling if necessary.*

6. *Disconnect fuel line.*

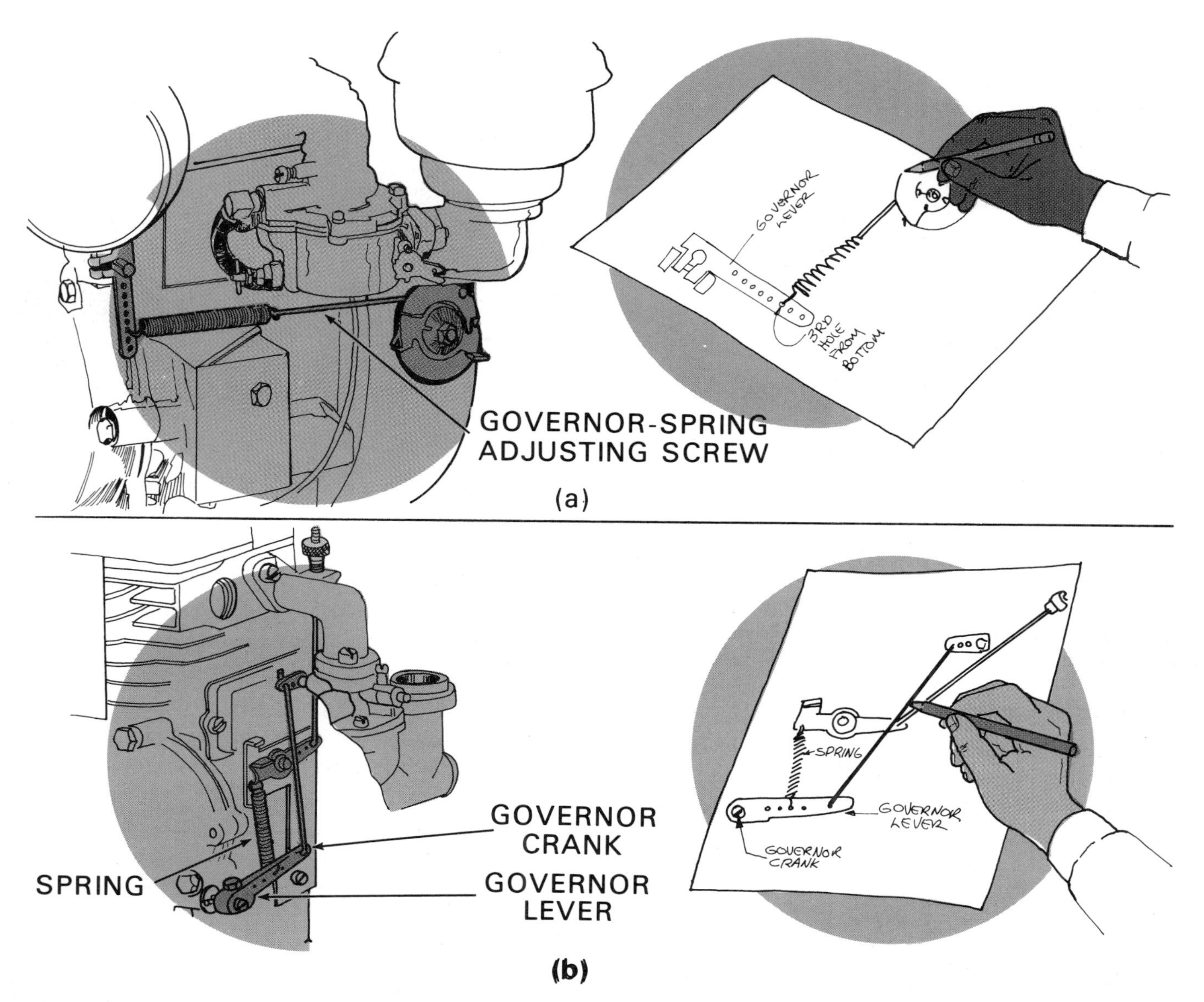

FIGURE 134. Make a note of how your governor and throttle are connected so that you can reconnect them properly. (a) Indicate at what point the spring attaches to the lever arm. (b) Indicate spring and linkage attachment points.

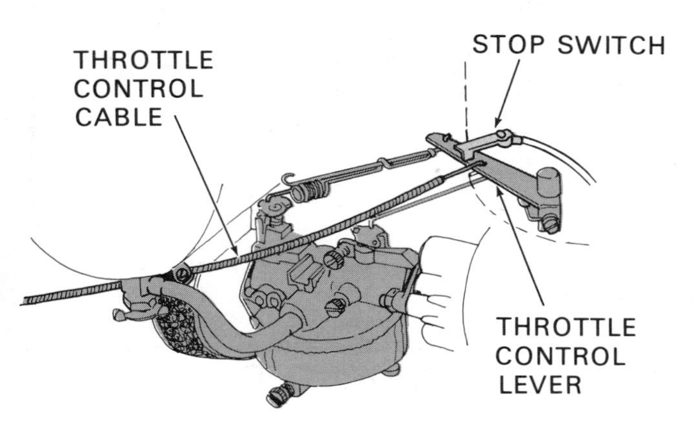

FIGURE 135. Throttle linkage combined with an ignition-grounding switch.

If the fuel lines are made of **neoprene,** they are usually pressed onto hose nipples. Pull hose from nipple.

Check for deterioration, cracks and breaks. Do not replace neoprene hose with rubber hose. Gasoline will soften rubber.

Metal tubing is attached with a flange nut. Use two open-end wrenches to prevent twisting the tubing. Hold the fitting with one while loosening the jam nut with the other.

7. *Remove breather return line, if installed (Figure 136).*

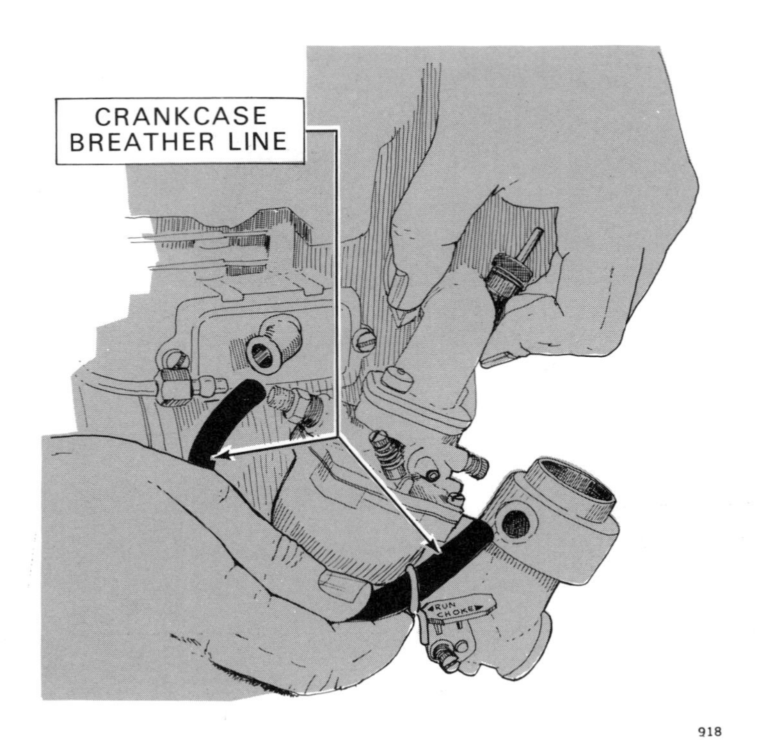

FIGURE 136. Breather return line on a two-piece float-type carburetor.

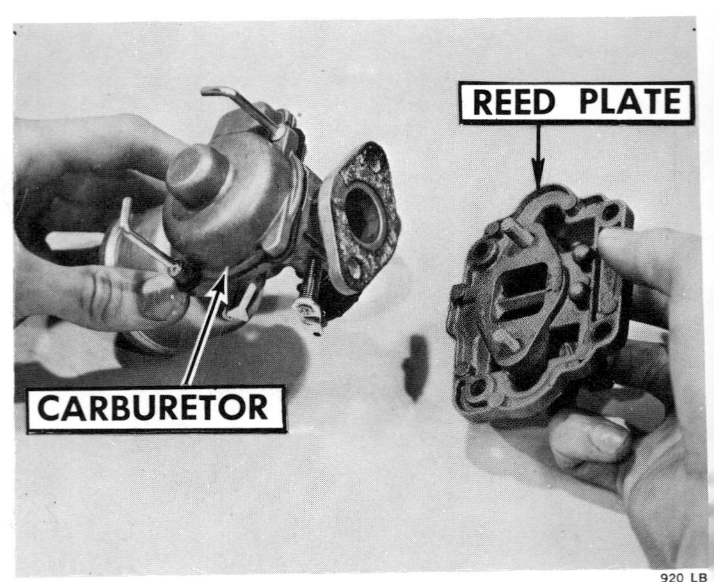

FIGURE 138. Carburetor separated from a reed plate (2-cycle engine).

8. *Remove carburetor.*

Disconnect any control linkage which was not disconnected before. The carburetor may be attached directly to the engine (Figure 137a), or to an intake manifold (Figure 137b). With the latter, it is usually easier to disconnect the manifold from the engine and separate it from the carburetor later.

On 2-cycle engines, it may be attached to the intake-valve plate on the crankcase, or to the reed plate. On some, the reed plate must be removed first with the carburetor attached; then separate the two (Figure 138).

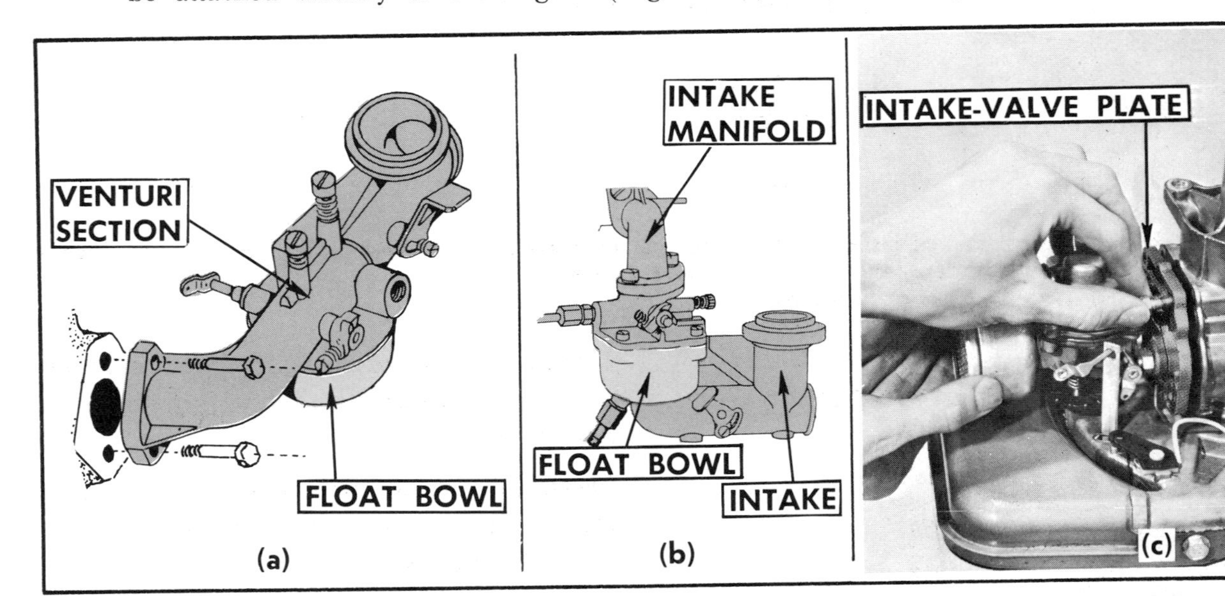

FIGURE 137. Most carburetors are attached by two cap screws. (a) Carburetor mounted directly to the engine. (b) Carburetor mounted to an intake manifold. (c) Carburetor mounted to the crankcase intake-valve plate (2-cycle engine).

REPAIRING THE FLOAT- AND DIAPHRAGM-TYPE CARBURETOR

The designs of carburetors vary so much that it is impossible to give specific steps and illustrations for all types and models. For this reason follow procedures given in your service manual for your particular carburetor, if you have a manual. The following steps, however, will give you a general idea

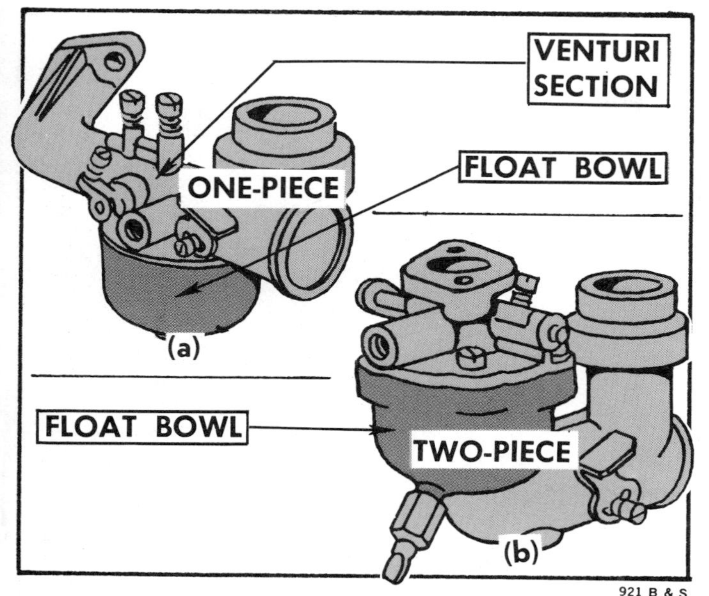

FIGURE 139. Typical float-type carburetors. (a) One-piece float type. Bowl can be removed from the bottom. (b) Two-piece float type. The venturi section is surrounded by the float. The housing can be separated into two parts.

of how to repair your carburetor even though the carburetor may not be exactly like either of the ones described here.

You may have a **1-piece float-type** carburetor (Figure 139a), or a **2-piece float-type** carburetor (Figure 139b), or a **diaphragm-type** (with or without a built-in fuel pump) carburetor (Figure 140).

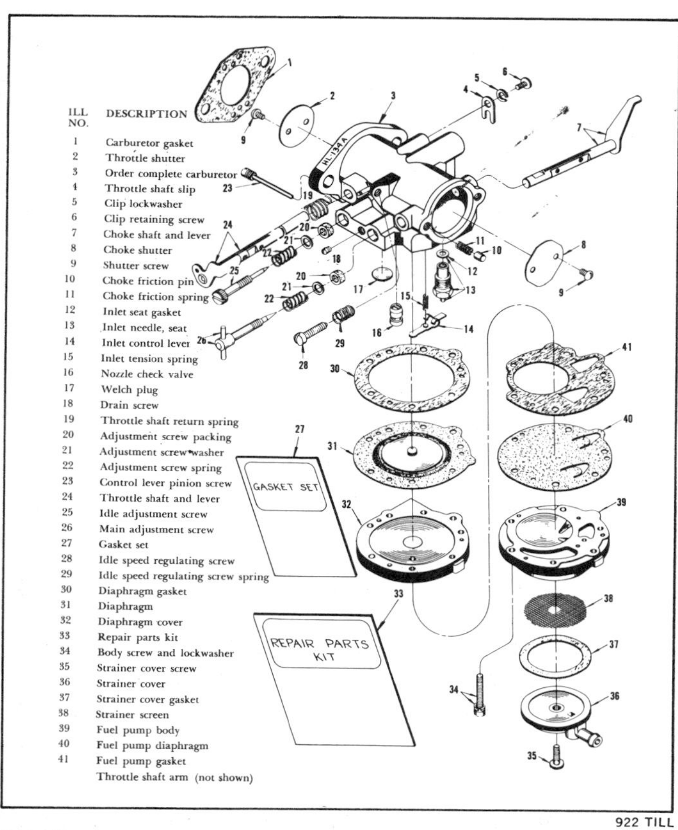

FIGURE 140. Parts of a diaphragm-type carburetor with a built-in fuel pump.

With few exceptions repair procedures are similar for either of the *float-types* (Figure 139) and for the *diaphragm type* (Figure 140) carburetors.

a. If your carburetor is a **float-type or diaphragm-type (without a built-in fuel pump)**, proceed as follows:

1. *Check throttle shaft for wear (Figure 141).*

   Place a metal block on the carburetor adjacent to the throttle shaft.

   Check play in **throttle shaft** with a feeler gage. If the wear is more than .25 mm (.010 in), replace the bushing. That is, if it has a bushing and the remainder of the carburetor is reparable. See step 11.

   If the carburetor **has no throttle-shaft bushing** and the hole is worn excessively, it will be necessary to replace the top housing of the carburetor. Before replacing the carburetor housing, check the rest of the carburetor to see if it is worth repairing. A worn shaft and/or bushing in either type will allow air to enter the carburetor.

   This will provide too lean a mixture and make the engine run roughly.

2. *Remove and check the condition of the needle valves (Figure 142).*

   Normally there are two needle valve adjustments—one for high speed and one for idle speed. You may or may not have an idle-mixture adjustment.

   Needles become worn thin from damage and wear. A worn rim on the needle will occur, and you cannot adjust the needle properly. If valves are **worn or damaged,** replace with new ones.

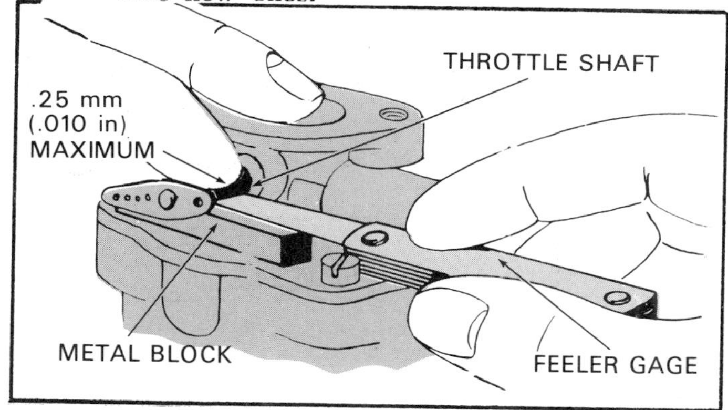

FIGURE 141. Checking the throttle shaft and/or bushing for wear.

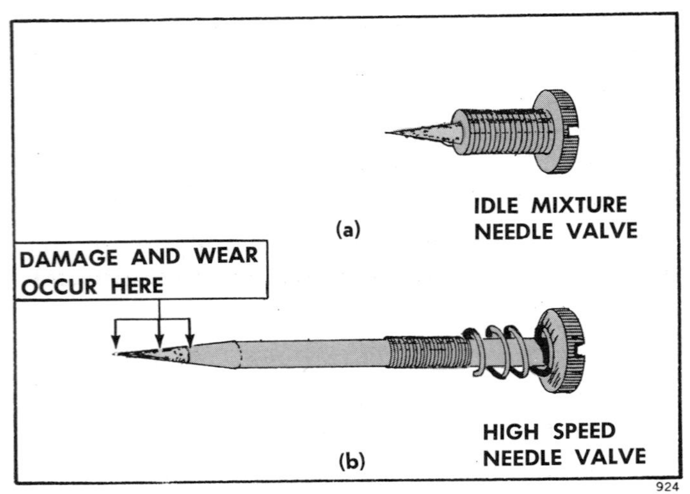

FIGURE 142. Where damage or wear occurs on needle valves.

If you have a **2-piece float-type carburetor and the main fuel nozzle is diagonally installed,** remove and check main fuel nozzle (Figure 143). Do this before separating the two halves to prevent damage to the diagonally inserted nozzle.

Be sure to use the proper size screwdriver. If you do not, threads and brass fittings will be damaged. Check for gum, dirt and mechanical damage. Make sure openings and passageways are clear.

3. *Remove fuel bowl (1-piece carburetor), or disassemble carburetor.*

If you have a **2-piece float-type carburetor,** separate the two pieces of the carburetor

FIGURE 143. Remove main fuel nozzle with care.

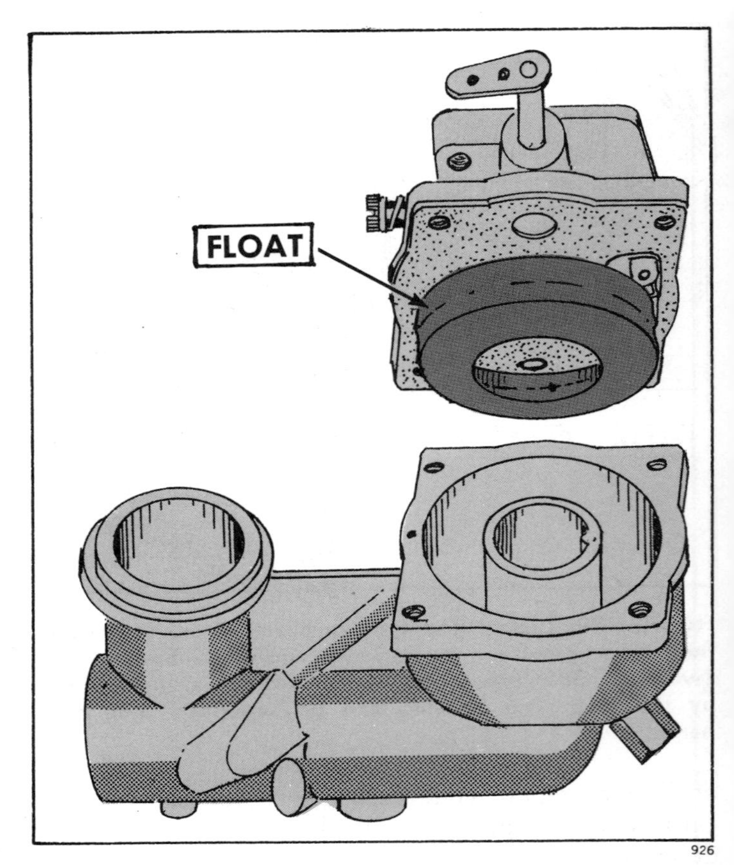

FIGURE 144. The two sections of a two-piece float-type carburetor.

body (Figure 144). Take care not to lose the needle valve. It may fall free.

4. *Remove the gasket(s).*

Be sure to notice how the gasket(s) fit(s) and how the holes in the gasket(s) are aligned with the holes in the carburetor. If you install the gasket wrong, you will restrict some vital passageways.

If the **float-chamber vent is clogged or restricted,** the mixture will be too lean.

If the **internal air bleed is restricted,** you will get a rich mixture.

5. *Remove float or diaphragm, and check condition (Figure 145).*

If you have a **float-type carburetor,** pull pin from float hinge. Check float valve and hinges for wear and check float for mechanical condition and leaks. Most floats are **airtight metal capsules** (Figure 145a). If this type float contains gasoline or if it is crushed, it must be replaced with a new one.

Some floats are **varnished cork** (Figure 145b). Check to see if the varnish coating is good. Do not puncture the protective coating

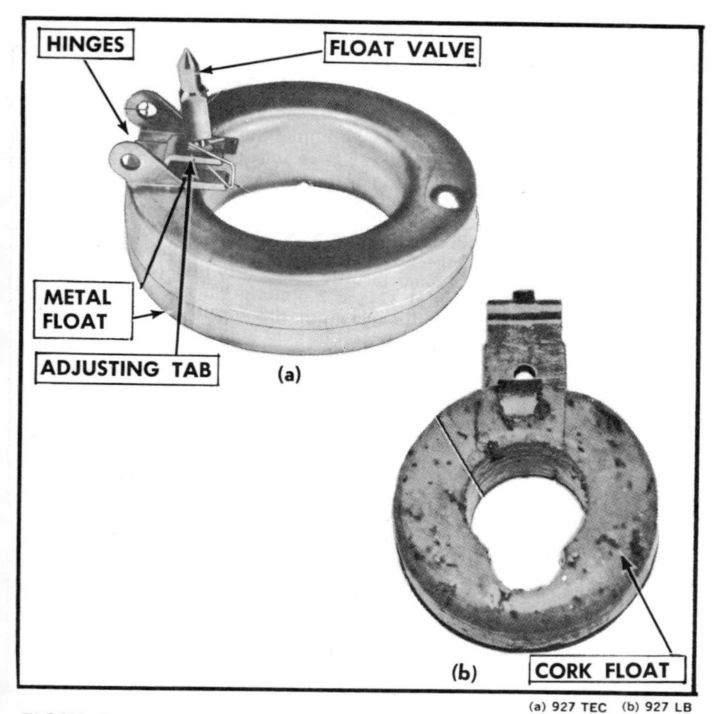

FIGURE 145. Types of floats. (a) Metal float with needle valve and adjusting tab. (b) Cork float.

or soak it in strong solvents. If you do, it will break the coating so that the float will absorb gasoline.

If you have a **diaphragm carburetor,** remove and check the diaphragm. Inspect the diaphragm for condition and leaks.

6. *Determine if float or diaphragm needs adjustment or replacement.*

   Figure 146 shows three methods for checking different *float adjustments.* Specifications vary for different carburetors. Each manufacturer supplies the measurements that are needed to determine when your carburetor float is in proper adjustment.

   If the float is **set too high,** the valve will not close completely and the carburetor will flood.

   If float is **set too low,** the valve will close too early and the carburetor will starve. (*High* and *low* refer to the assembled carburetor).

   If the float **valve sticks open,** the carburetor will flood.

7. *Adjust tab (tang) on float, or diaphragm control lever, if necessary.*

   Bend tab under float needle. You may have to remove the hinge pin to reach the tab.

   If you have a **diaphragm-type carburetor,** check the diaphragm inlet valve to see if it

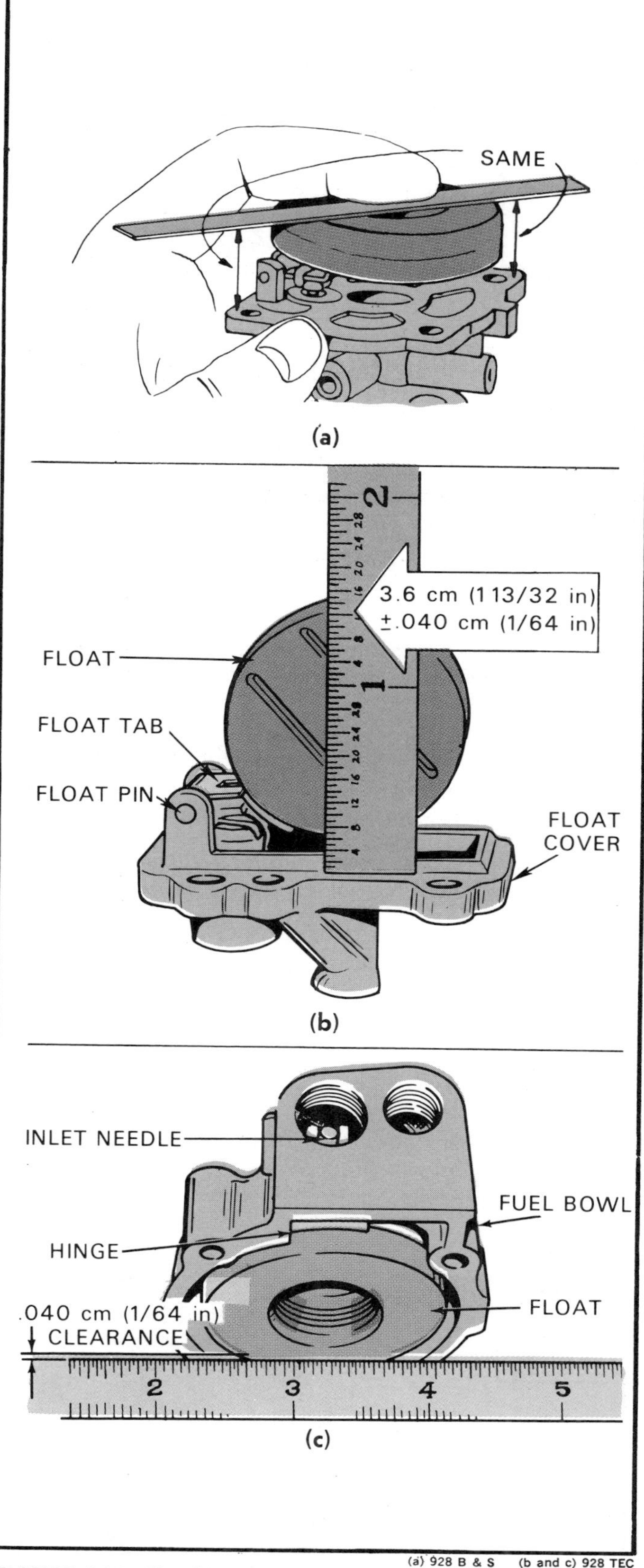

FIGURE 146. Checking floats for adjustment. (a) Checking distance from carburetor housing to top of horizontal float to determine if it is the same on both sides. (b) Checking distance on a vertical float from housing to top side. (c) Checking bottom of float for proper clearance in relation to float housing.

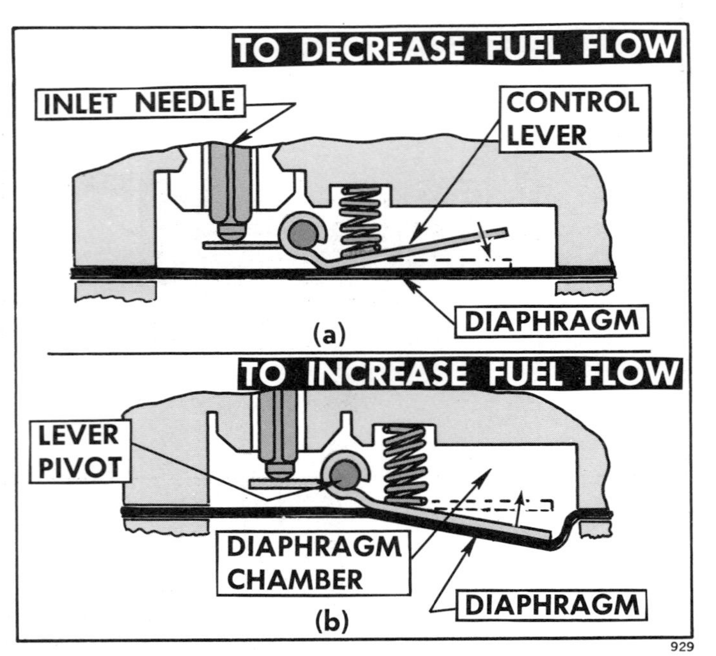

FIGURE 147. Adjustment of the control lever position on a diaphragm-type carburetor. Tab and control lever should be parallel.

opens and closes properly (Figure 147). If necessary, bend inlet control lever until the ends of lever are parallel.

8. *Remove and check main discharge nozzle and valve seats, if not already removed.*

   Be sure to use the proper size screwdriver. The brass fittings are easily damaged. Inspect for wear and damage. Replace with new ones if needed.

9. *Remove primer pump if installed (Figure 148).*

   Remove cotter pin from primer-pump shaft and remove shaft from the bottom.

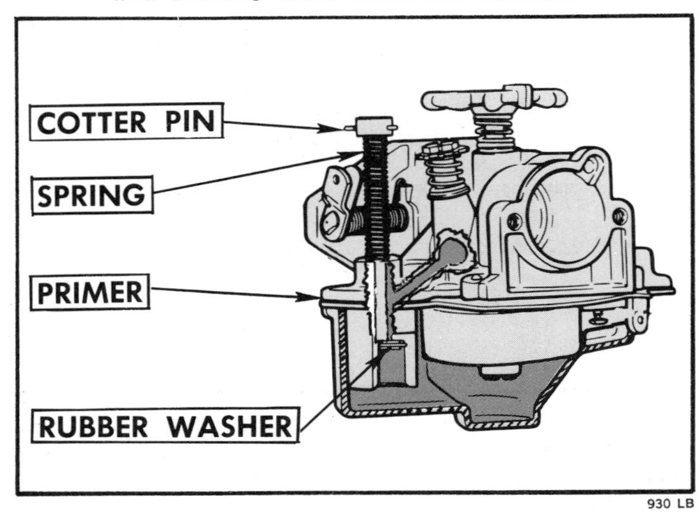

FIGURE 148. A one-piece float-type carburetor with primer pump.

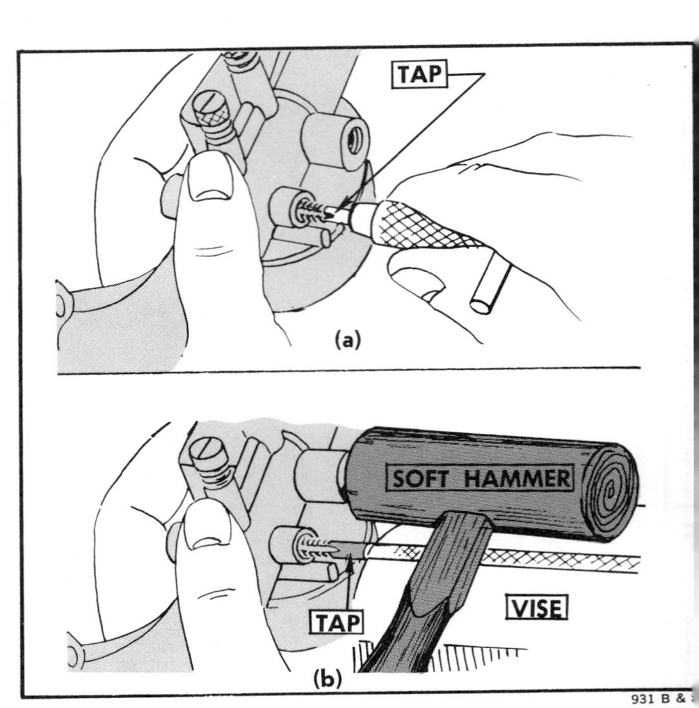

FIGURE 149. Removing the carburetor throttle-shaft bushing. (a) Inserting a tap in the bushing. (b) Driving the carburetor from the bushing.

10. *Remove and replace throttle-shaft bushing if necessary (Figure 149).*

    To remove the throttle-shaft bushing, thread a tap of the proper size into the worn bushing. Hold tap in vise and drive carburetor off with a soft hammer. Press new bushing into place. Ream to size if necessary. Be sure to use the proper size drill for reaming. Use a drill .025 mm (.001 in) larger than the new shaft.

11. *Remove choke valve and throttle valve if necessary.*

    If your carburetor has a **welch plug** on the elbow like the one shown in Figure 150a, push it out with a screwdriver to get to the choke valve.

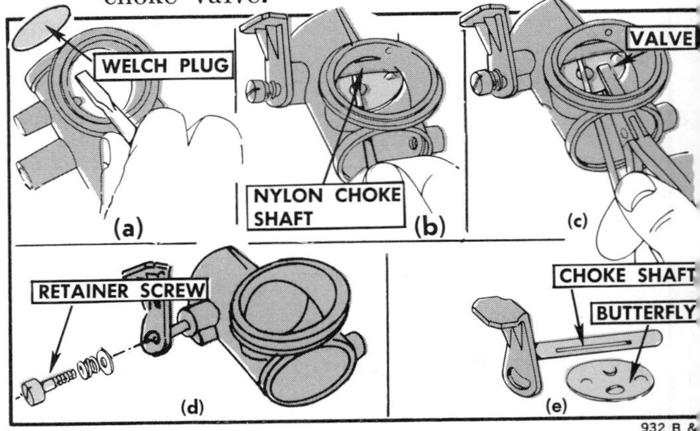

FIGURE 150. Removing choke valve from a nylon choke valve shaft. (a) Pry out welch plug. (b) Insert sharp tool under choke valve. (c) Pull out choke butterfly with long nose pliers. (d) Remove retainer screw, spring and washer. (e) Remove shaft.

Some choke valves have a **nylon shaft.** The valve is wedged in a slot in the shaft. Remove them in the manner shown in Figure 150.

12. *Soak carburetor parts.*

    Use Bendix Cleaner, Petisol or other recommended commercial carburetor cleaners.

    Nylon and other plastic parts may be washed in the same carburetor solvent, and soaked for *not more than 30 minutes.* If left for longer periods, the chemical may soften them.

13. *Rinse in petroleum solvent.*

    Clean with air hose. Use 700 kPa (100 psi) or more for efficiency.

14. *Reassemble carburetor in reverse order of disassembly.*

    Use new gaskets.

    Be sure to install the gaskets properly.

15. *Adjust carburetor for initial setting.*

    For procedures refer to "Adjusting Carburetors," Part 1.

16. *Install the carburetor.*

    See "Installing the Carburetor," page 96.

REPAIRING THE SUCTION-LIFT CARBURETOR

Some simple suction-lift types carburetors — those with only one fuel lift tube — have no means for adjusting either fuel mixture. Some have a high speed adjustment but no idle mixture adjustment. About all you can do for these types of carburetors is to clean them and inspect the high-speed, load needle valve and seat, if installed. They usually have the fuel tank mounted directly beneath the carburetor.

A suction-lift carburetor with a diaphragm pump and fuel chamber is a little more complicated (has more parts).

Repair procedures vary for different suction-lift carburetors (Figure 152). Check your service manual for those that apply to your particular carburetor.

1. *Remove the carburetor from fuel tank.*

   Lift the carburetor straight from the tank to avoid damaging the suction tube. Check fuel tank for flatness (Figure 152).

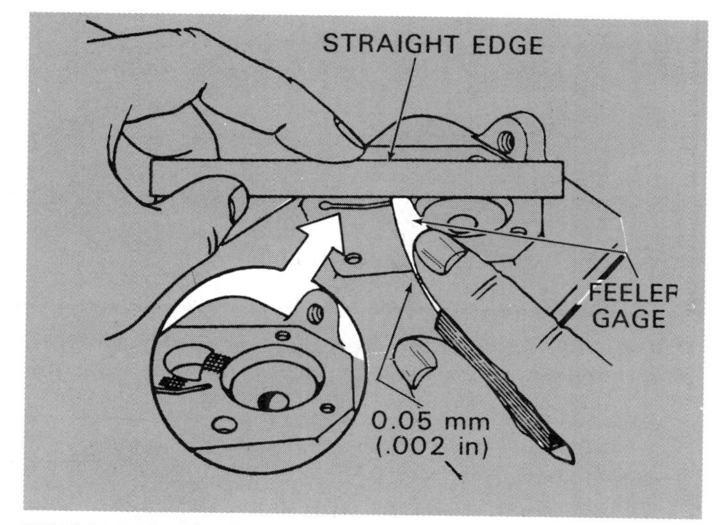

FIGURE 152. Checking fuel tank for flatness.

2. *Check throttle-valve shaft for wear (Figure 141).*

3. *Remove throttle valve (Figure 153).*

   Valves such as those in Figure 153a and 153b are removed by turning them until the flange

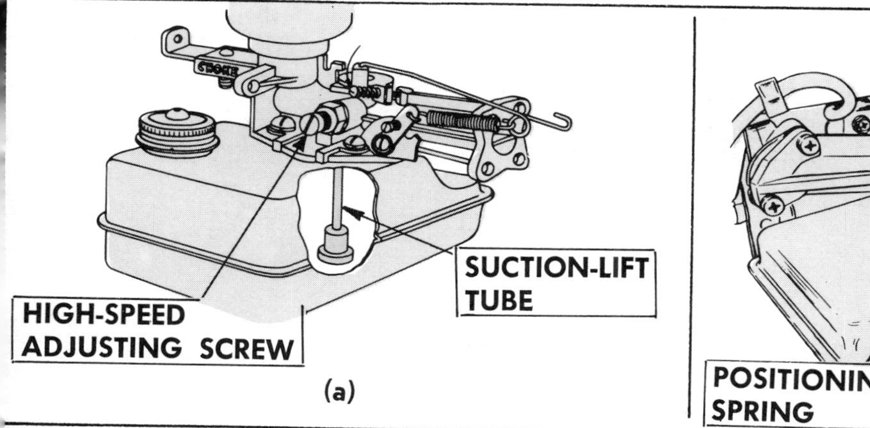

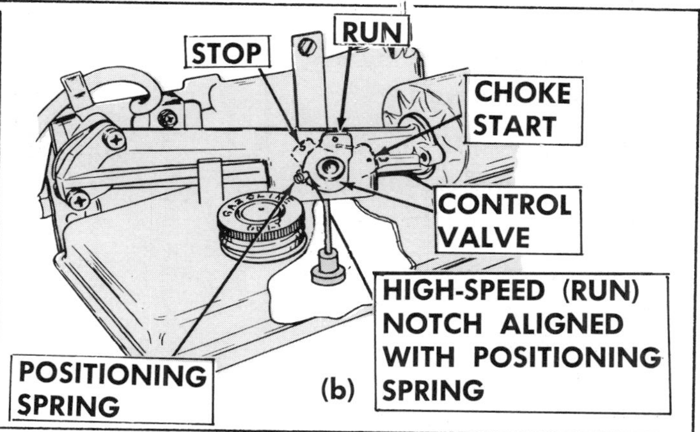

(a) 934 B & S (b) 934 TEC

GURE 151. Examples of one-tube suction-lift carburetors. (a) Carburetor equipped with a high-speed mixture adjustment and separate choke control. (b) Carburetor with a single control valve for choke, run and stop.

FIGURE 153. Types of throttle valves on suction-lift types of carburetors. (a) Rectangular valves. (b) Notched cylindrical valve used for choking, running and shutting off fuel for stopping. (c) Butterfly-type valve.

clears the retaining boss. Remove butterfly valve from shaft if your throttle valve is like the one shown in Figure 153c.

4. *Inspect "O" ring if installed (Figure 153b).*

5. *Inspect and clean air valves.*

6. *Remove fuel strainer(s) and pipe(s) (Figure 154).*

   Nylon pipes are one piece and are screwed into the carburetor body (Figure 154a).

   Brass pipes (Figures 154b and 154c) may be one or two pieces. They are usually pressed into the carburetor body. If the brass pipe is equipped with a foot valve, it is pressed onto the pipe.

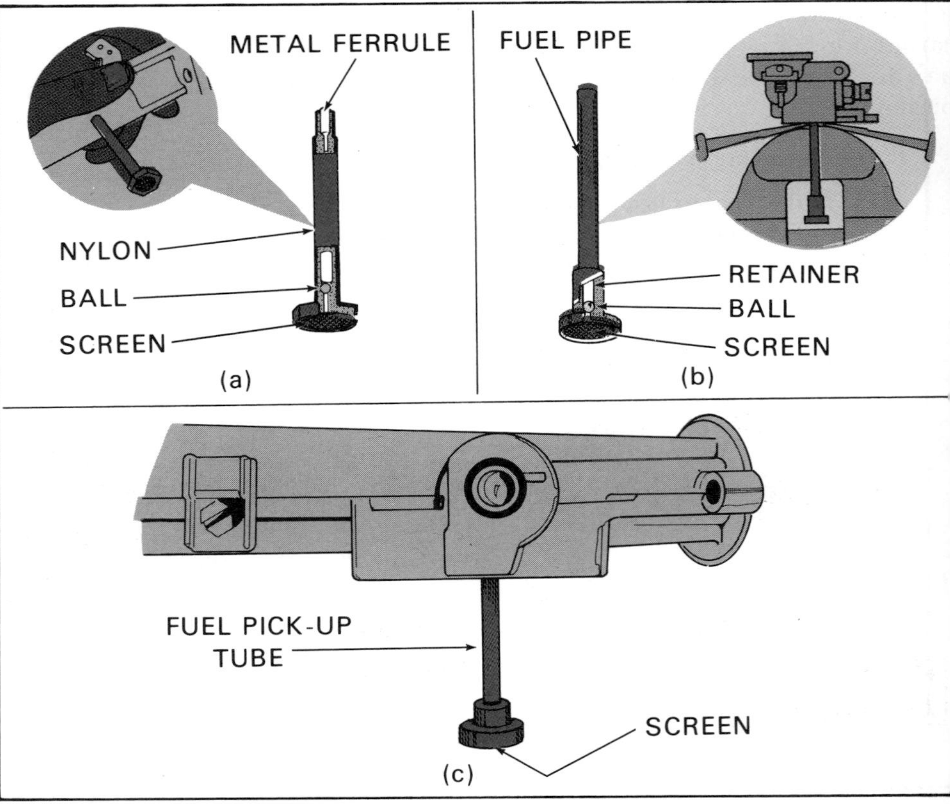

FIGURE 154. Types of fuel strainers and pipes on suction-lift type carburetors. (a) Nylon pipe. (b) and (c) Metal pipes and strainers.

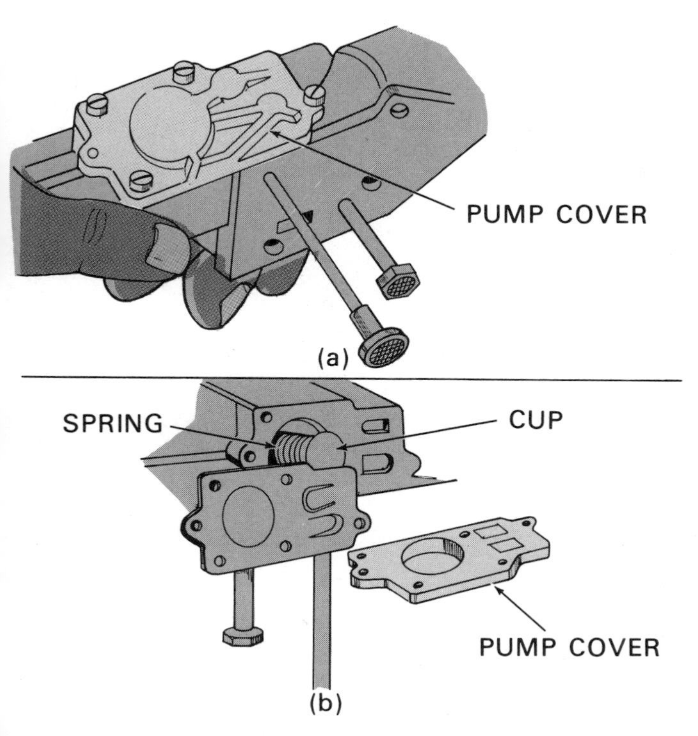

FIGURE 155. Removing the fuel pump assembly on a suction-lift carburetor. (a) Carburetor with diaphragm pump. (b) Pump disassembled.

Be sure to **check the length of the pipe** before removing it. The length of the pipe can be varied. It is adjustable by the amount inserted into the carburetor. *If it is too long,* it will jam on the bottom of the fuel tank. *If it is too short,* it will not reach into the fuel sufficiently, and there is also danger of restricting the inlet valve in the carburetor.

7. *Clean and inspect pipe and strainer.*

Do not soak nylon and other synthetic carburetor parts in carburetor cleaner more than 30 minutes. If you cannot clean the screen, replace it with a new one.

Check foot valve for freedom of operation.

8. *Remove pump assembly if installed (Figure 155).*

If your carburetor has **two suction pipes (Figure 155a),** it is equipped with a diaphragm pump. Refer to "Repairing Fuel Pumps," page 157.

If the **diaphragm is exposed to atmosphere on one side and it is punctured,** you will get a lean mixture.

If the **diaphragm is exposed to fuel on both sides and is punctured,** you will get a rich mixture.

9. *Remove and inspect the adjusting needles and seats.*

Some suction-lift carburetors have neither adjusting needle. Some have only the high-speed adjusting needle.

Some manufacturers do not recommend cleaning carburetor jets with a wire. There is danger of enlarging the hole in the jet. Some manufacturers, however, recommend using a small copper wire to clean clogged jets. Usually, soaking in carburetor solvent for several hours and blowing out with compressed air will clean them adequately.

10. *Clean and inspect the carburetor body.*

Soak the carburetor body and metal parts in Bendix, Petisol or other recommended carburetor solvent. After several hours remove the carburetor and blow it out with compressed air.

11. *Reassemble carburetor.*

Replace all gaskets, and worn and damaged parts, with new ones. Be careful about installing the fuel-strainer pipe. The length is critical (Figure 156). On some carburetors you can restrict the fuel flow by pressing the pipe too far into the carburetor. See your operator's manual for proper length.

If you have a suction-lift carburetor with a pump, replace the diaphragm also.

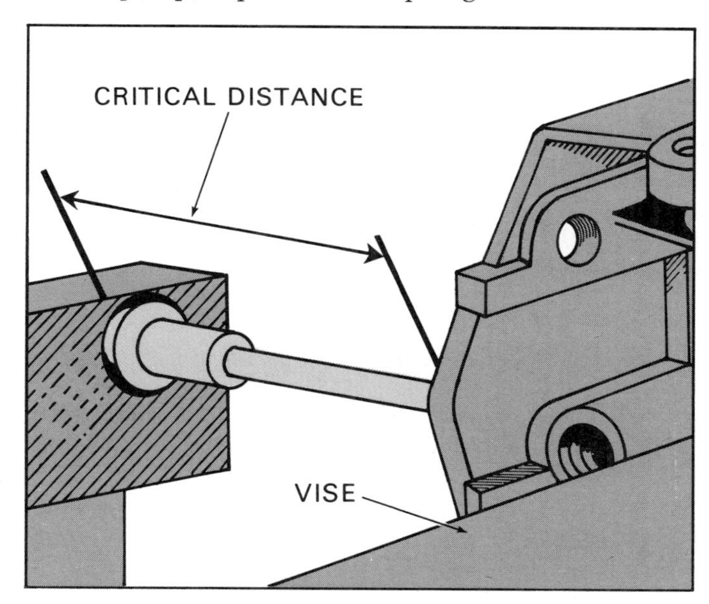

FIGURE 156. The length of the brass fuel pipe is critical.

If you have a carburetor with an **automatic choke** similar to the one in Figure 128A, refer to Figure 156A.

(1) *Assemble choke spring to diaphragm (Figure 156Aa).*

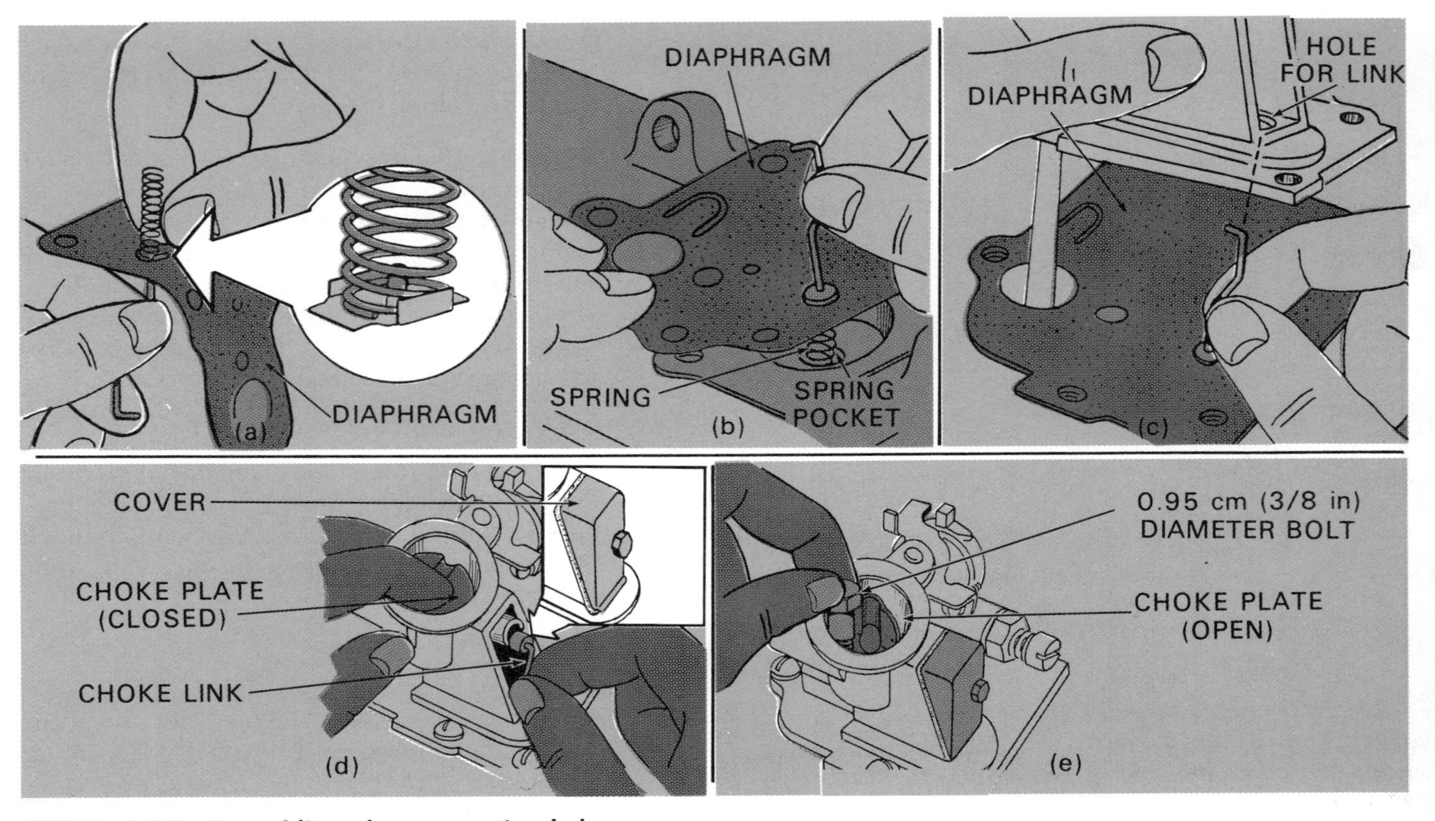

FIGURE 156A. Assembling the automatic choke.

(2) *Place diaphragm on fuel tank top (Figure 156Ab).*

(3) *Locate carburetor on diaphragm (Figure 156Ac).*

(4) *Insert choke link and cover (Figure 156Ad).*

(5) *Insert a .95 cm (3/8 in) bolt to hold the choke valve or plate open (Figure 156Ae).*

(6) *Install carburetor mounting screws and tighten alternately.*

(7) *Remove .95 cm (3/8 in) bolt.*

(8) *Check for proper operation.*

There are no adjustments on the suction-lift carburetor with no choke valve (Figure 156Ba).

12. *Readjust high-speed needle valve to setting recommended by the manufacturer.*

13. *Adjust throttle setting (single-control carburetor) (Figure 157).*

Set the throttle to high-speed run position. On some models this is done when the mark on the face of the valve is aligned with the retaining boss on the carburetor body. On other models there is a high-speed notch that aligns with a positioning spring.

If you are not sure, remove the throttle valve and observe its position when the throttle is open.

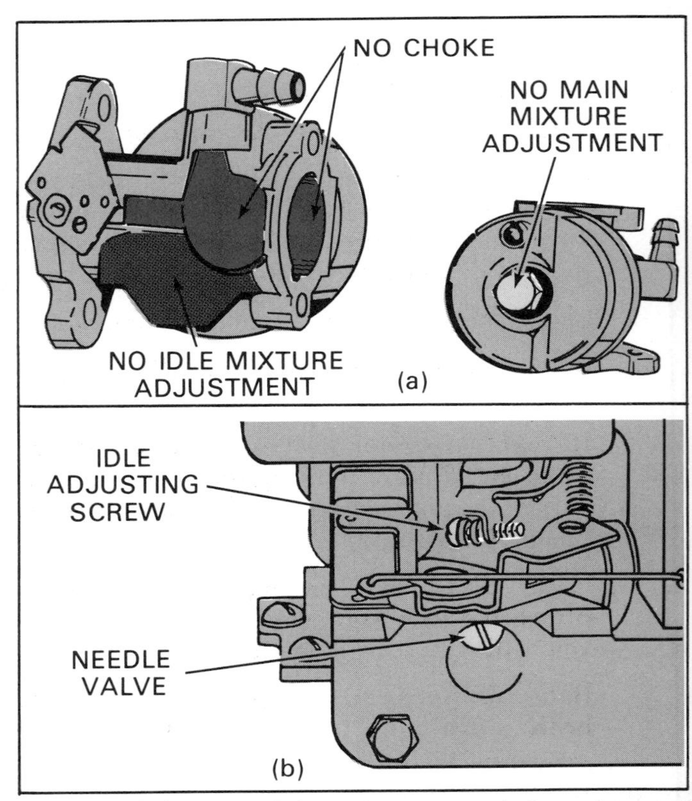

FIGURE 156B. Carburetors with (a) no choke valve and (b) automatic choke.

With the throttle valve in run position, set the remote control lever (if installed) in run position, and fasten the control wire to the throttle-valve lever.

14. *Install the carburetor.*

    See "Installing the Carburetor," next heading.

Adjustments for the carburetor in Figure 156a are shown in Figure 156b.

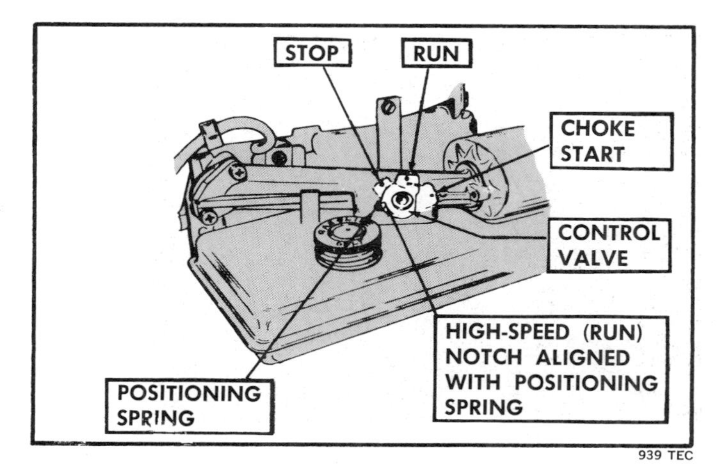

FIGURE 157. Aligning the single control type carburetor valve for the correct positions of choke, run and stop.

INSTALLING THE CARBURETOR

1. *Install gasket between carburetor and engine.*
2. *Install carburetor.*

   It may be easier, or even necessary in many cases, to connect some of the control linkages, breather return line and fuel line before securing the carburetor to the engine.
3. *Connect the governor linkage.*

   See your preliminary sketch of how to connect the governor linkage. If you are not sure of this, consult your service manual or follow procedures given under "Adjusting the Governor," page 176.
4. *Connect the choke and throttle linkage (Figure 158).*

   If your engine is equipped with a **hand-operated remote throttle-and-choke control,** check the position of the remote-control lever before attaching the linkage, or control wire, to the carburetor. Proceed as follows: For the **choke,** set the choke lever on the carburetor in closed position. Set the remote choke control to a closed position. Tighten the choke-control linkage to the carburetor linkage.

   For connecting the **throttle** control, set the remote throttle-control lever in a closed position. Set the throttle-valve control on the carburetor to be closed. Tighten the control linkage to the carburetor linkage.

   There are many variations in the carburetor linkage. No attempt is made to show all of them in this publication. If you have not made a sketch and cannot remember how your linkage should be attached, it will be necessary for you to consult the service manual for your particular engine.
5. *Connect ground ignition wire if installed.*
6. *Install the carburetor air cleaner.*
7. *Recheck for proper operation.*

   See "Checking the Carburetor for Proper Operation," Part 1.

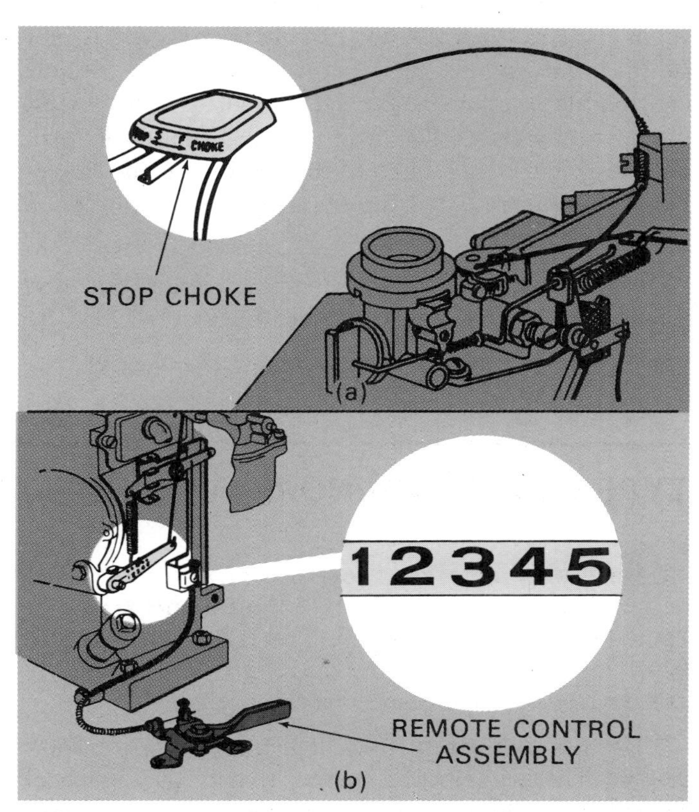

FIGURE 158. Examples of remote choke-Control linkages. (a) Combination throttle-and-choke control. (b) Throttle control.

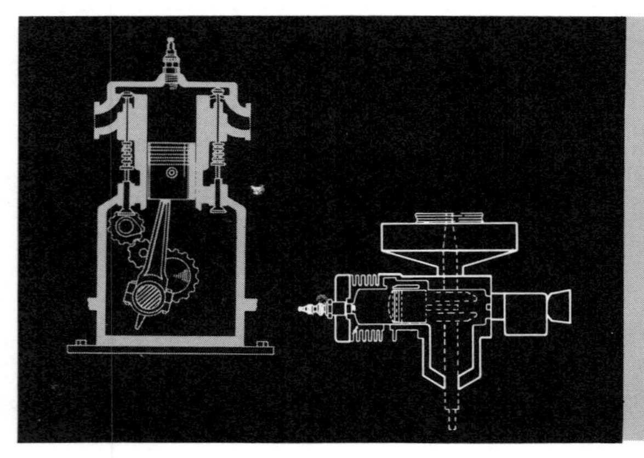

IV. Repairing Governors

The governor automatically regulates the speed of your engine at whatever setting you select on the speed control (throttle). Many people think the primary purpose of the governor is to prevent an engine from overspeeding. Most governors, however, used on small engines prevent the engine from underspeeding as well, thus maintaining a constant engine speed.

If your engine had no governor, it would choke down as load is applied. With the governor functioning properly, the engine speed remains fairly constant at variable loads as long as the load is within the horsepower capacity of the engine.

The governor regulates the engine speed by adjusting the amount of fuel-air mixture fed through the carburetor. As the load increases, the governor opens the throttle valve to provide enough more fuel so that the engine can increase its power output and maintain a constant speed. If the load decreases, the governor closes the throttle valve enough to provide less fuel to the engine, thus reducing the horsepower output and still maintaining a constant speed (Figure 159).

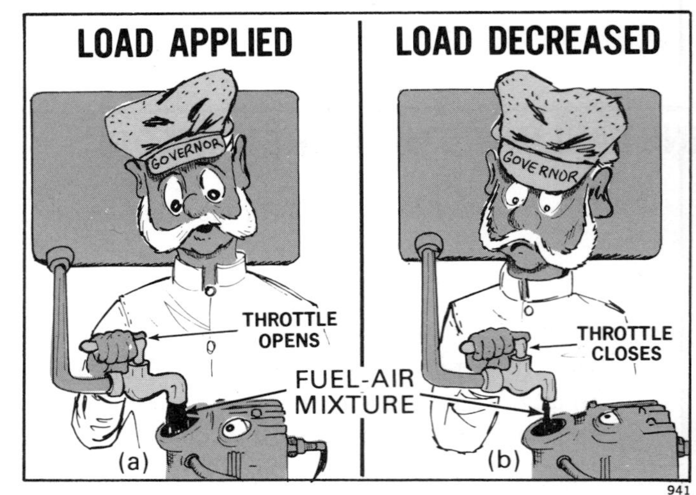

FIGURE 159. The governor on your small engine maintains a constant speed. (a) If load is applied, the governor opens the throttle and supplies enough more fuel-air mixture to maintain engine speed. (b) If the load decreases, the governor reduces the amount of fuel-air mixture and prevents the engine from overspeeding.

Maintaining and repairing governors are discussed under the following headings:

1. Types of governors and how they work.
2. Importance of proper repair.
3. Tools and materials needed.
4. Checking the governor for proper operation.
5. Adjusting the governor.
6. Repairing the governor.

TYPES OF GOVERNORS AND HOW THEY WORK

Generally there are two types of governors used on small engines: They are (1) **air-vane type,** or "pneumatic" (Figure 160); and (2) **centrifugal-type,** or "mechanical" (Figure 161). The names are derived from the speed-sensing elements.

Here is *how the air-vane governor (Figure 160) works.* The **air-vane,** located under the flywheel shroud, is in the path of the air coming from the flywheel fan. The air vane is *connected by linkage to the throttle valve* (Figure 160) in such a way that when it moves, the throttle valve opens or closes. The air vane is also connected by a spring to the throttle control. When the engine is stopped, the throttle valve is open by spring tension.

As the engine runs, air from the flywheel is directed against the air vane. The faster the flywheel turns, the greater the quantity of air directed against the air vane. This air pressure moves the air vane in the direction of air flow. **As the air vane moves,** the throttle valve closes accordingly, bringing the

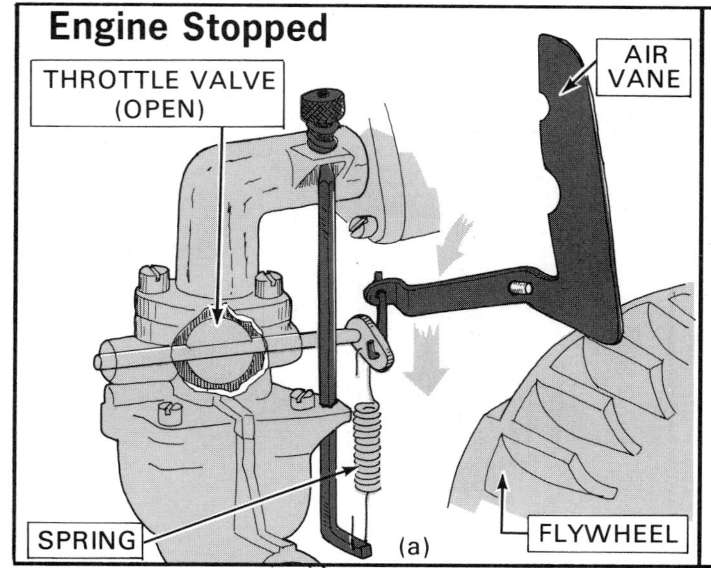

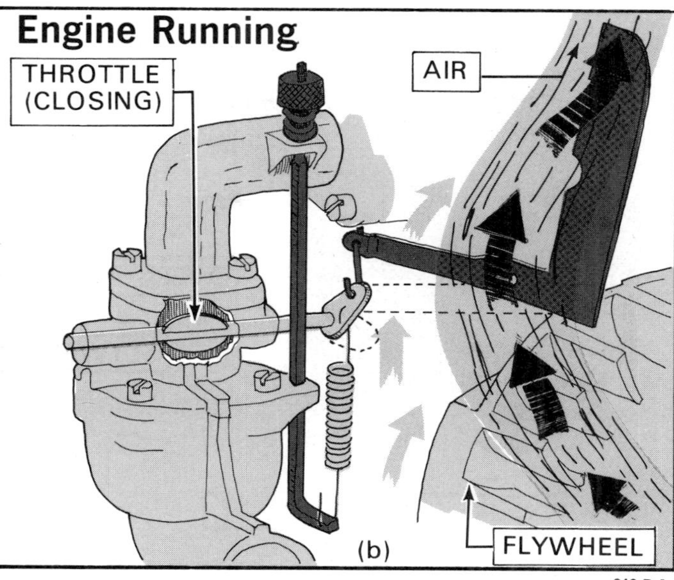

FIGURE 160. Principles of the air-vane governor. (a) With engine stopped, spring tension opens the throttle. (b) With engine running, air from the flywheel deflects the air vane and the throttle closes until the spring tension is equal to the amount of force applied to the air vane. The throttle then remains at this position unless the load changes or unless you change the throttle setting.

engine speed back in line with the speed selected by the throttle setting. At this point, the air force against the vane equals the spring tension pulling against it at the throttle. It remains in this position until you change the speed control, or the load on the engine changes.

Here is how a *centrifugal governor* (Figure 161) works. **When you adjust the throttle control** for increased speed, additional tension is exerted on the governor-control spring. The spring tension opens the throttle valve (Figure 161a) and the engine speeds up.

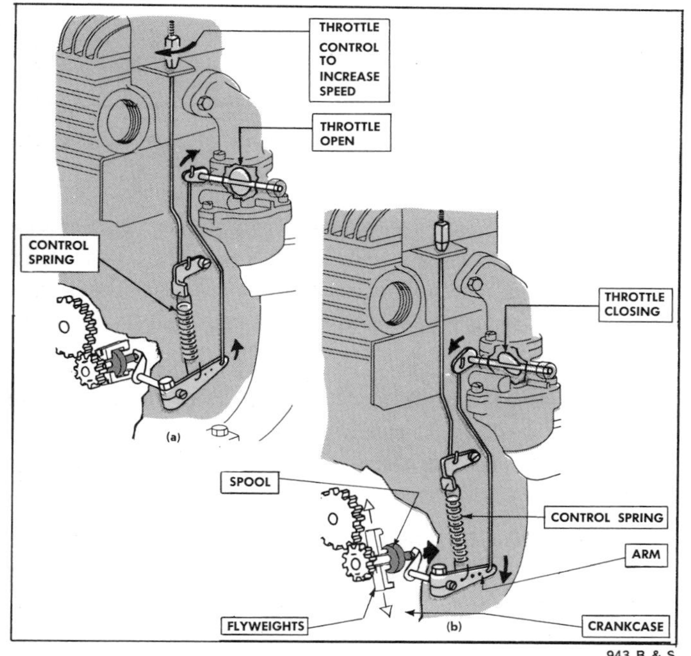

FIGURE 161. Principles of the centrifugal-type governor. (a) With engine stopped, spring tension opens the throttle. (b) With engine running, fly weights extend by centrifugal force and push the governor spool outward. This action rotates the governor arm against the spring tension and has a tendency to close the throttle.

As the engine **speeds up,** centrifugal force causes the flyweights or flyballs (Figure 162), to extend outward. As the flyweights move outward (Figure 161b), they push against a governor spool (or plate) which in turn rotates the governor arm. The rotation of the governor arm overcomes some of the spring tension and closes the throttle valve until the force against the throttle linkage equals the tension in the

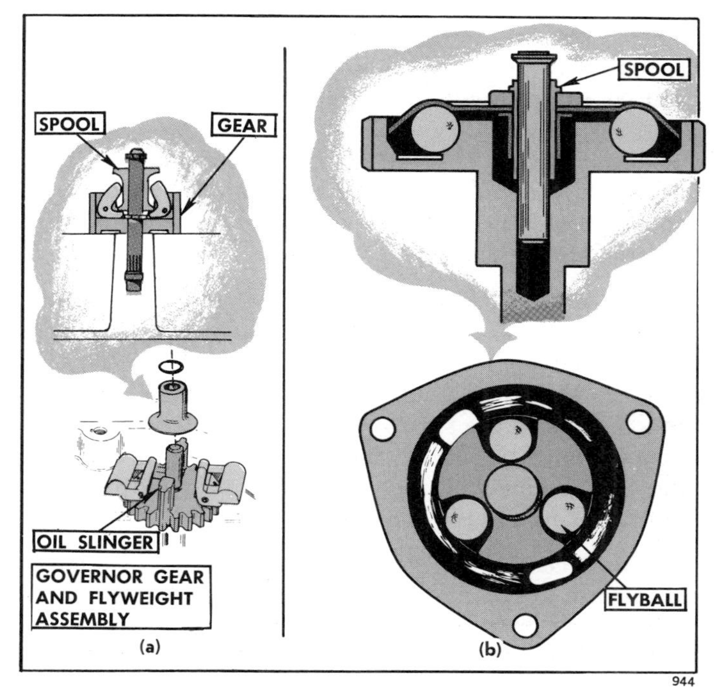

FIGURE 162. Two types of centrifugal governors. (a) Flyweights and spool. (b) Flyballs and spool.

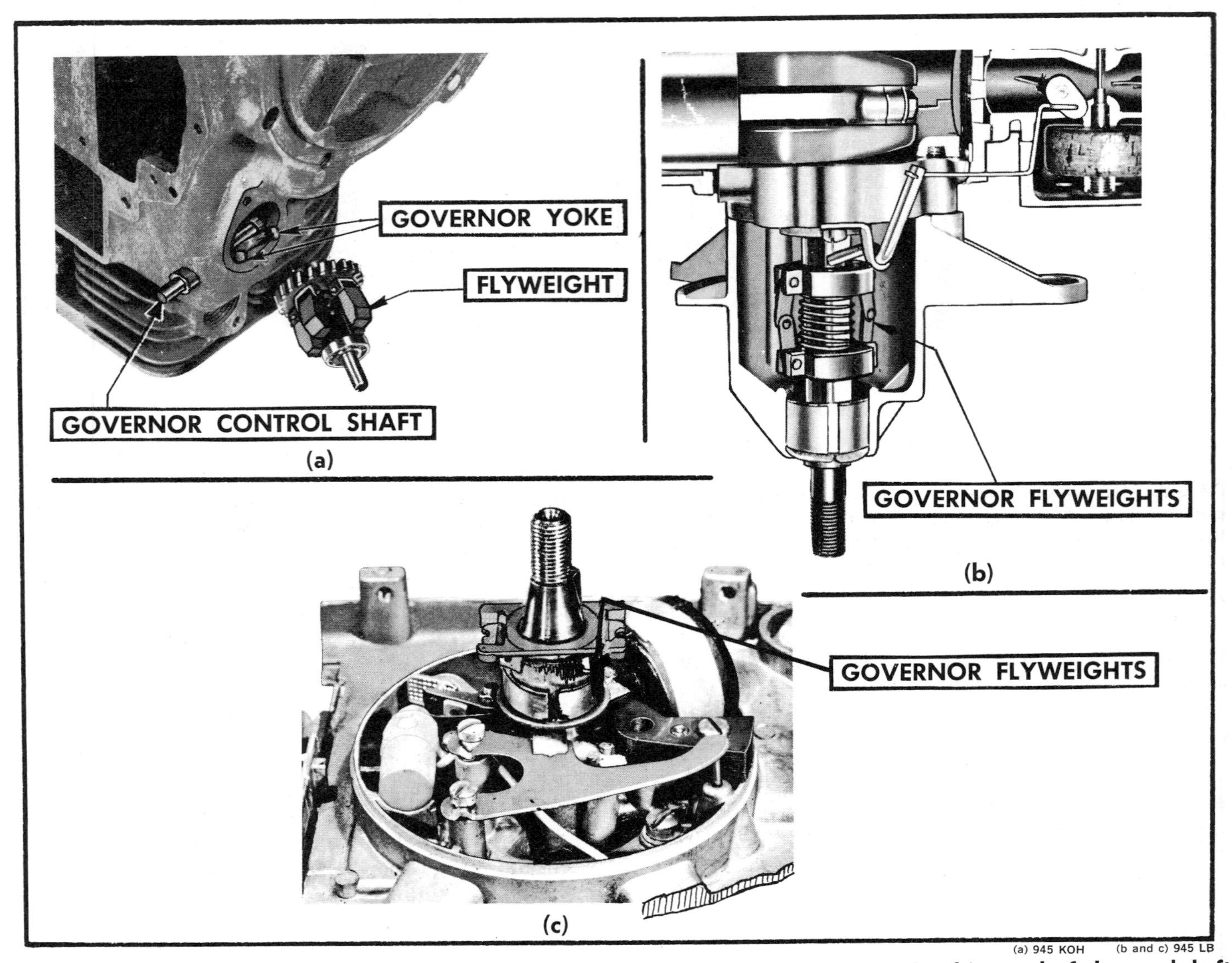

FIGURE 163. Centrifugal governors may be located (a) inside the crankcase. (b) on the drive end of the crankshaft, and (c) on the flywheel end of the crankshaft.

governor spring (Figure 161b). The amount of fuel-air mixture is reduced. In this way, the speed remains constant under variable loads, unless you change the throttle setting.

If more load is added to the engine, the engine tends to run more slowly. Centrifugal force is reduced on the flyweights and they recede. This reduces the force on the governor linkage. There is less tension in the spring, and the throttle valve opens proportionately. More fuel-air mixture then enters the engine and increases the horsepower output without reducing the engine speed.

Most centrifugal governors are located inside the crankcase (Figure 163a). They may be driven by the camshaft gear, or from the crankshaft. But some centrifugal governors are located on the drive end of the crankshaft (Figure 163b), while others are located on the flywheel end of the crankshaft (Figure 163c).

Most small engines should be operated in the high-speed range. At the high speed, your engine has the capacity to adjust to a wide range of power demands.

If the throttle setting is high enough, your engine is ready to start lugging the instant the governor action takes place. If the throttle setting is too low, there is not enough tension on the governor spring to allow the engine to adjust to full power quickly.

IMPORTANCE OF PROPER REPAIR

Engine manufacturers test their engines and select a maximum speed for each model — one that provides satisfactory power in relation to fuel consumption, wear and other factors. The speed selected varies with different makes of engines and with different models of the same make.

The *maximum idle speed* recommended for your engine is given in your operator's manual. It is the one you normally use for checking your governor.

You may find that your engine is operating either too rapidly or too slowly. *If your engine is operating too rapidly*, wear increases rapidly. *If it operates too slowly*, less power is available. The engine overloads, chokes down and overheats.

TOOLS AND MATERIALS NEEDED

1. Combination pliers — 7″
2. Long-nose pliers — 7″
3. Open-end wrenches — 3/8″ through 1/2″
4. Cleaning solvents (mineral spirits, kerosene or diesel fuel)
5. Rags

CHECKING THE GOVERNOR FOR PROPER OPERATION

The governor, when in proper condition and adjustment, will respond quickly to any change in load. It will maintain the correct engine speed without hunting (surging). There is a delicate balance between the **governor control spring** and the governor **air vane, flyballs,** or **flyweights.** If the spring loses its tension, if the governor parts wear, or if any of the control linkages become damaged or worn, the governor will not work properly.

The governor often gets blamed for poor performance when the difficulty is in the fuel system, ignition system or in the compression. Clogged cooling fins will restrict the flow of air to an air-vane type governor and cause the engine to overspeed. Before checking your governor for proper operation, see that all these components of your engine are functioning properly.

1. *Check the condition of the governor linkage.*

   Check for bent links and worn connections. Check for freedom of operation of the linkage. Check the air-vane condition and the mechanical linkage for freedom of movement.

2. *Check to see if the throttle valve is open when the engine is stopped.*

   It should be completely open when the engine is stopped.

3. *Check the speed-sensing element.*

   You will not be able to check most **centrifugal-type governor** mechanisms without removing the flywheel or disassembling the engine. Usually these parts give little trouble. But if you have checked all the other sources of trouble and still suspect the governor, follow procedures for getting to your type governor. Refer to "Removing and Checking the Flywheel," page 127, and "Removing the Crankshaft," page 254.

   If your engine has an **air-vane type of governor,** you may have to remove the engine shroud to inspect the air vane. Check to see if the air vane is parallel to the crankshaft.

4. *Find the recommended speeds for your engine in your operator's manual.*

   There are three speeds with which you are concerned when checking your engine governor. They are (1) **idle speed,** (2) **high-idle, no-load speed,** and (3) **high-idle, full-load speed.**

   Manufacturers recommended that *idle speeds* range from *800* to *2,000 r.p.m.*, usually 1,800 r.p.m.

   High-idle speeds for 4-cycle engines vary from 2,000 to 3,600 r.p.m.; for 2-cycle engines, from 2,000 to 5,000 r.p.m. Engines on chain saws and go-karts may operate as high as 9,000 r.p.m.

   Full-load speed is usually 200 to 300 r.p.m., less than the high-idle, no-load speed. Your operator's manual does not always give a *full-load speed.* It will be less, however, than the high-idle speed, even though the speed-control lever is in the same position. This decrease occurs because there is some loss of engine speed at full load that the governor does not recover. For example, an engine that has a maximum no-load speed of 3,600 r.p.m. may have a full-load speed of 3,400 r.p.m. The amount of speed loss is different with different engines.

   Some manuals will indicate a speed range such as 3,000 to 3,200 r.p.m. In this case the governor is considered to be in proper adjustment if the high-idle speed is within this range. If only one high-idle speed is given in your operator's manual, the governor is usually considered as being in satisfactory adjustment if the top engine speed is within 20 r.p.m. — faster or slower — than that speed.

5. *Start your engine and allow it to warm up.*

6. *Check the no-load, idle speed.*

   Check with the throttle control in the closed position. Use a tachometer for measuring your engine speed. Adjust the no-load idle speed by adjusting the idle stop screw. See "Adjusting the Carburetor,"Part 1.

7. *Check the no-load, high-idle engine speed.*

   Move the throttle control to open fully and check the engine r.p.m. with a tachometer. The engine speed should come up to that recommended in your operator's manual. It should remain constant.

   If the **engine does not come up to speed,** or overspeeds, proceed to "Adjusting the Governor." Your governor could be worn, sticking or out of adjustment. The governor spring may be too loose. The spring may have lost its tension. If so, replace it.

   A worn-wear block on some governors will cause a reduction in r.p.m. Adjustment for a worn-wear block is shown in Figure 163c.

   If surging occurs on changing from one load to another, the spring is too tight. If you get a substantial drop in r.p.m. when load is increased, the spring is too loose.

   If your **engine overspeeds,** the spring tension is too tight. It may be connected wrong; it may be binding or out of adjustment.

8. *Check the high-speed, full-load engine speed.*

   Use a dynamometer if possible.

   The governor should maintain the high-speed, full-load r.p.m. without fluctuating. Anything that restricts the movement of the governor mechanism and linkage can cause engine surging.

ADJUSTING THE GOVERNOR

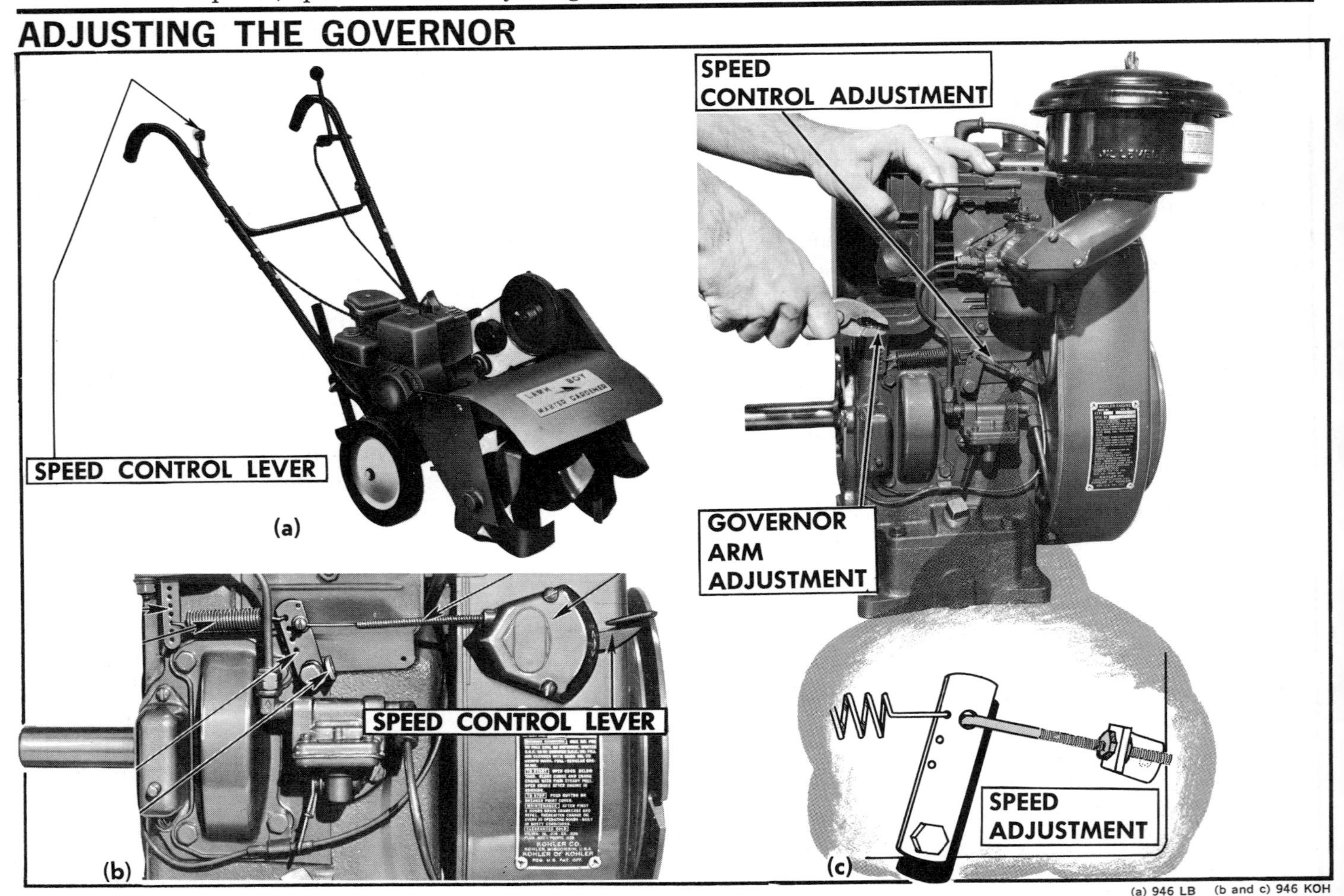

(a) 946 LB (b and c) 946 KOH

FIGURE 164. On an engine with a constant-speed governor the speed-control lever is attached to the governor spring and not the throttle valve. (a) Engine equipped with a remote speed-control lever. (b) Speed-control lever on engine. (c) Adjustment for constant speed.

Adjusting the governor consists of adjusting the tension on the governor spring. Actually that is what you do when you change the speed of your engine.

You may have a remote (throttle) control lever (Figure 164a), or the control may be mounted on the engine (Figure 164b and c).

There are many different ways provided for adjusting governors (Figure 165), but regardless of

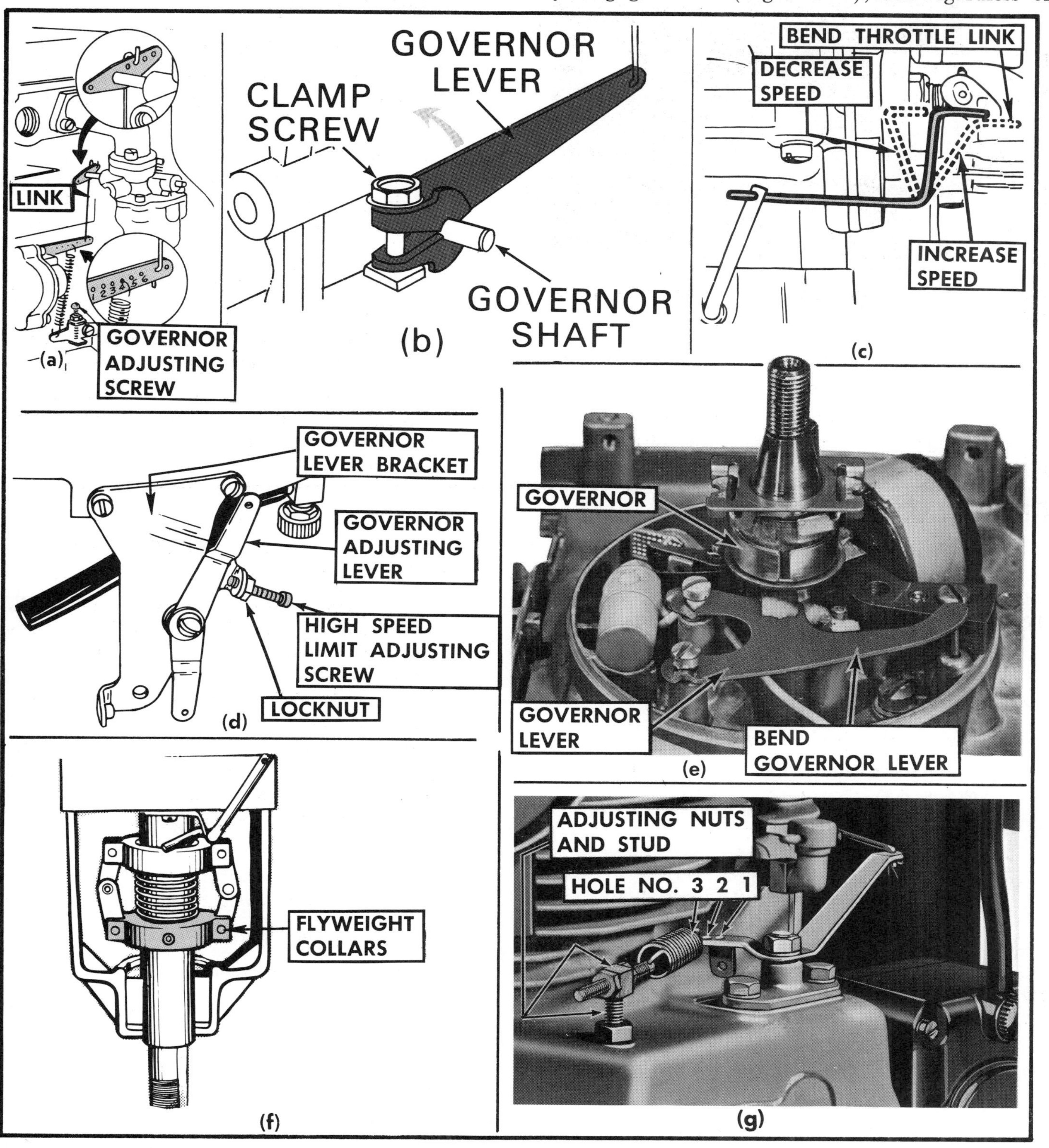

FIGURE 165. Different means for adjusting governors. (a) Changing the effective length of the throttle link. (b) Loosening clamp and rotating the governor lever. (c) Bending the throttle link. (d) Adjusting the high-speed limiting screw. (e) Bending the governor lever. (f) Repositioning the flyweight collars. (g) Changing the effective length of the governor adjusting stud.

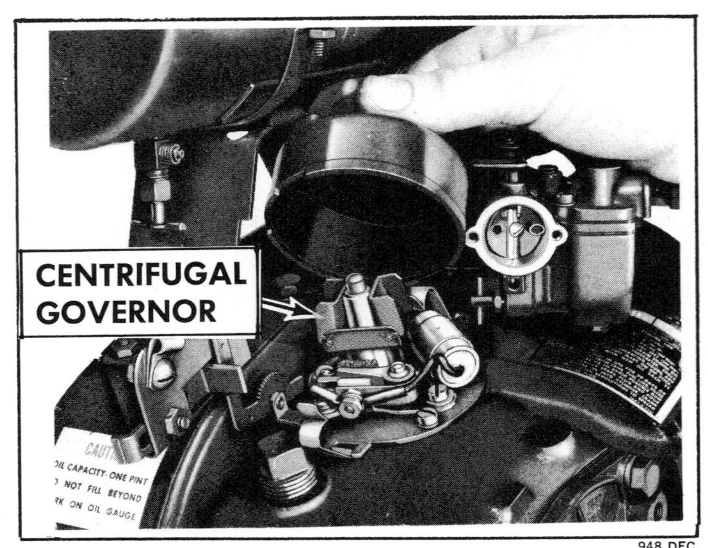

FIGURE 166. Centrifugal governor mounted on a special shaft inside the breaker box.

the method the object is to increase or decrease tension on the governor spring.

Procedures for adjusting some governors are more complicated than others (Figure 166). But the principle is the same. It may be necessary to follow procedures in your service manual, but, in general, the following procedures will apply.

1. *Adjust the governor so that the throttle valve is closed when the engine is stopped.*

   Spring tension closes the throttle valve.

2. *Start engine and operate at full speed.*

3. *Adjust the spring tension so the engine will operate steadily at the maximum recommended r.p.m.*

   If you cannot get your engine to operate properly by adjusting the spring, replace the spring with a new one. Just any spring will not do. Use the spring designed for your engine.

4. *Check and adjust high-speed stop screw if installed.*

   This should be done each time a change is made in the governor spring tension.

   On some engines a safety override stop is installed. It is a mechanical stop which limits the amount of tension you can apply to the governor spring.

REPAIRING THE GOVERNOR

From a repair standpoint there are three kinds of governors: (1) the air-vane type, (2) the centrifugal type mounted inside the crankcase, and (3) the centrifugal type mounted outside of the crankcase.

a. If your governor is the **air-vane type** (Figure 160), proceed as follows:

1. *Remove the flywheel shroud.*

   Refer to "Cleaning Engines," Part 1.

   Check the air vane for condition and position.

2. *Straighten air vane if it is bent.*

   It should be parallel to the crankshaft.

3. *Replace worn or damaged parts.*

b. If your governor is the **centrifugal type (Figure 161) and located inside the crankcase,** proceed as follows:

1. *Check and repair, or replace spring and/or linkage.*

2. *Gain access to the sensing element.*

   If the governor is **located on the flywheel end of the crankshaft (Figure 163c),** it will be necessary to remove the flywheel. Refer to "Removing and Checking the Flywheel," page 127.

   If the **governor is located inside the crankcase** (Figure 167), it likely cannot be removed without disassembling the engine. If it becomes necessary for you to get to the governor inside the crankcase, see procedures under "Removing the Crankshaft," page 254. Remember when disassembling your governor to make sure you note how to put it back together. Either draw a sketch or lay the parts out in order of their reassembly.

3. *Remove the flyweight assembly and other governor mechanisms.*

4. *Clean governor parts.*

   Use petroleum solvent.

5. *Inspect governor parts for damage and/or wear.*

6. *Reassemble governor, replace in reverse order of disassembly.*

7. *Adjust governor.*

   Follow procedures under "Adjusting the Governor" page 102.

8. *Recheck for proper operation.*

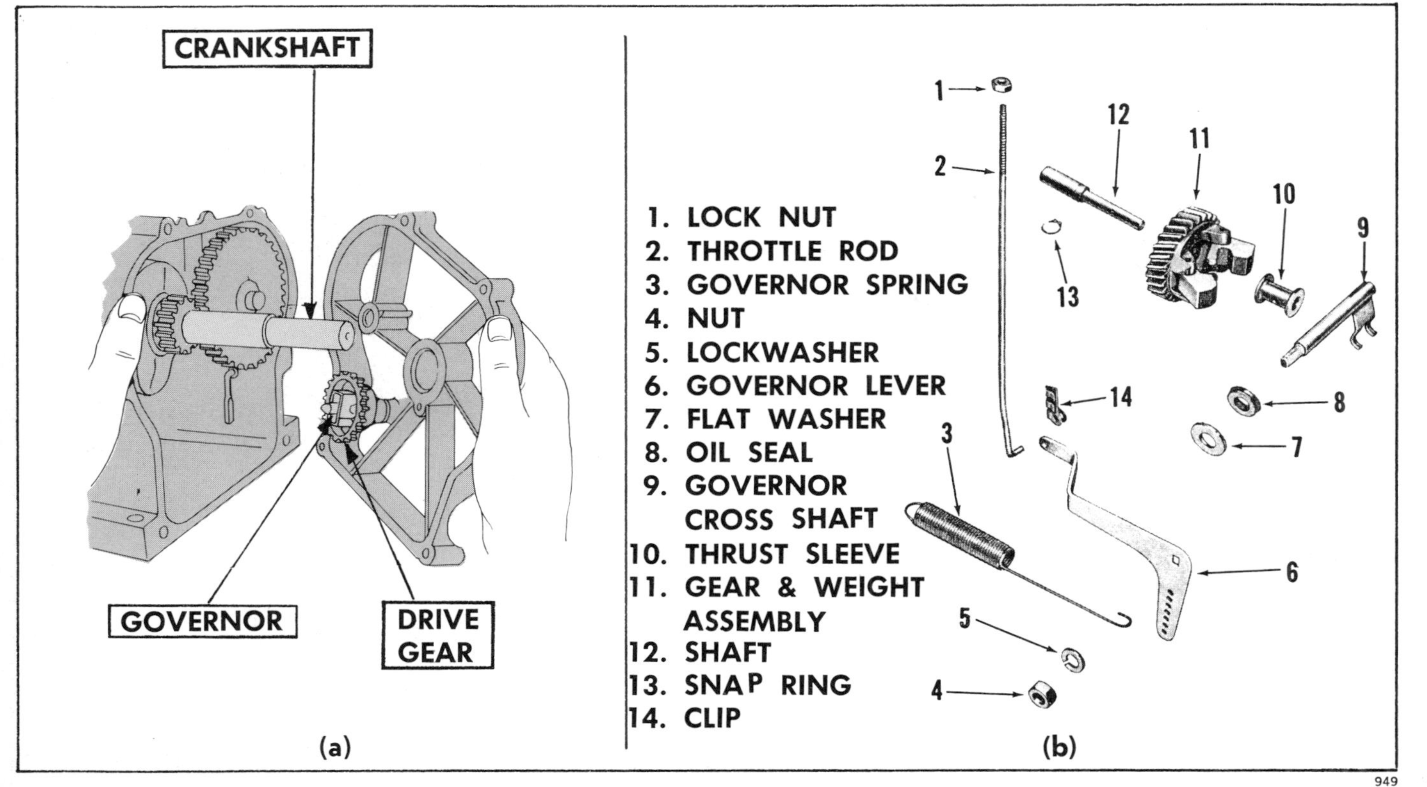

FIGURE 167. (a) Removing a crankcase type of centrifugal governor. (b) Clean and inspect governor parts for damage or wear. Lay them out in order of disassembly so you can tell how to put them back together.

c. If the governor is the **centrifugal type mounted outside of the crankcase** (Figure 168), use the procedures that follow. There are several variations of this type of governor. If the procedures given here are not sufficient for your engine, consult your service manual.

1. *Remove the flywheel.*

   Refer to "Removing and Checking the Flywheel," page 127.

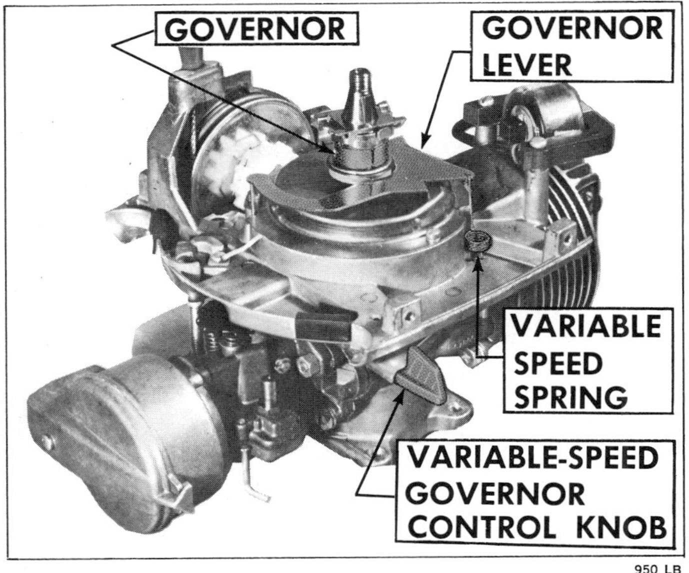

FIGURE 168. One type of centrifugal governor mounted on the flywheel end of the crankshaft outside of the crankcase.

2. *Disconnect the variable-speed spring attached to the governor lever (Figure 168).*
3. *Remove the flyweight assembly (Figure 169).*
4. *Remove the nylon thrust collar and governor lever from the crankshaft (Figure 170).*

   Look for worn or damaged parts. Replace them with new ones. Parts of this type governor are shown in Figure 171.

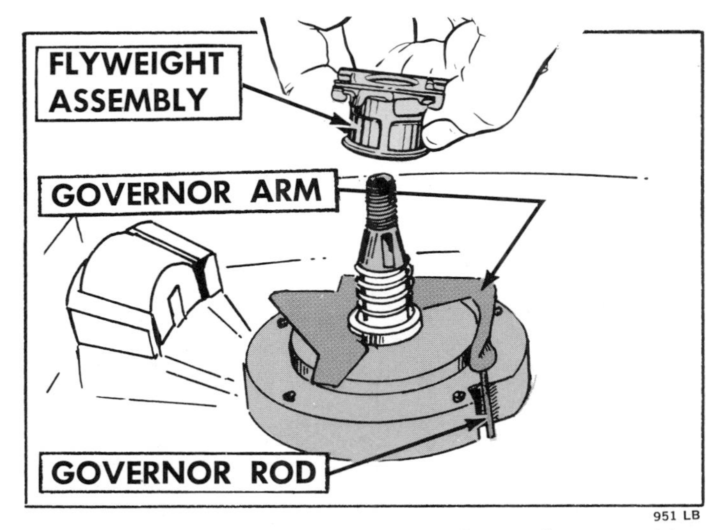

FIGURE 169. Removing governor flyweight assembly.

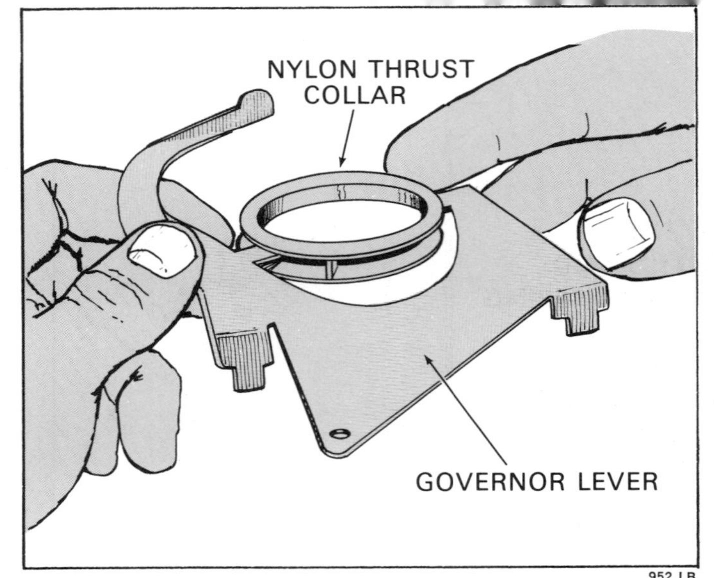

FIGURE 170. Removing the governor arm and nylon thrust collar.

5. *Install governor lever and thrust washer assembly (Figure 172a).*

   Hook the governor lever-arm hinge into slots on the breaker-box dust cover and connect the spring (Figure 172b).

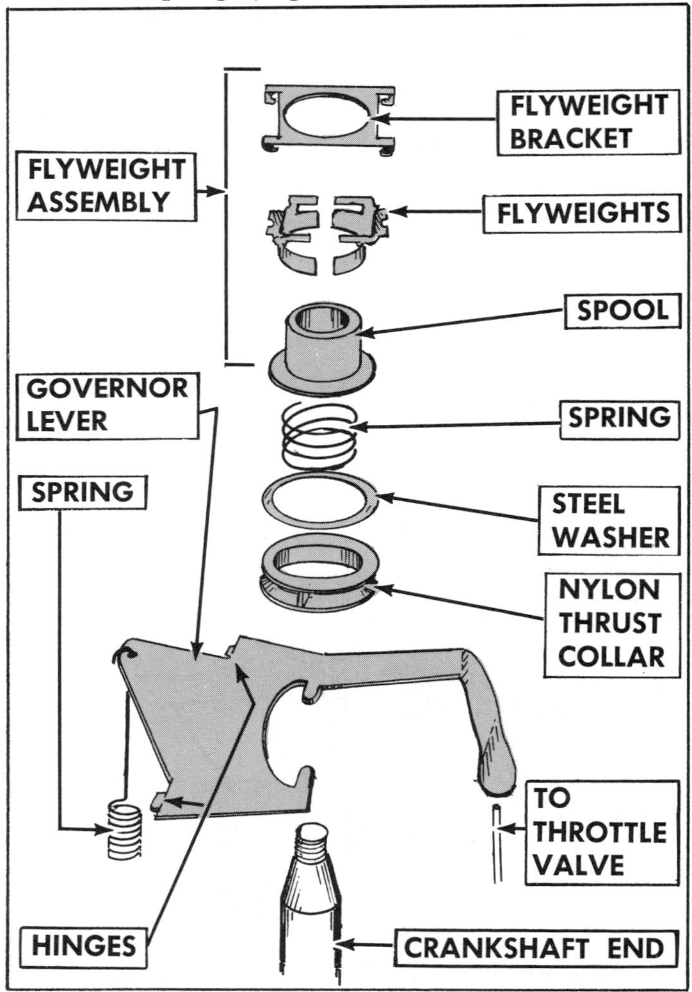

FIGURE 171. Parts of a centrifugal governor mounted on the flywheel-end of the crankshaft.

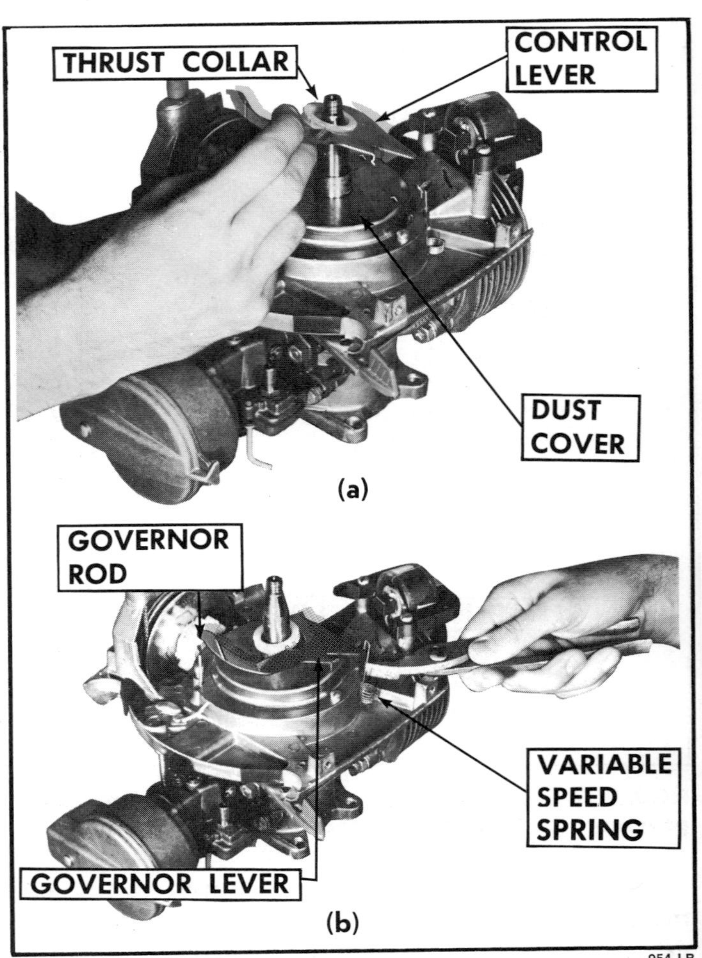

FIGURE 172. (a) Installing the governor arm and thrust collar assembly. (b) Fastening the governor spring.

6. *Install the flyweights, spring and fly-weight assembly.*

7. *Check governor lever adjustment (Figure 173).*

   Hold flyweights down with a pipe nipple. At the same time close the throttle by hand. Measure the clearance between the governor lever and the governor rod (Figure 173). It should be approximately 0.16 cm ($^1/_{16}$ in). Note in Figure 173a how the governor arm extends over the end of the governor rod. Adjust by bending the governor lever at the crease (Figure 173b).

8. *Install flywheel key.*

9. *Install flywheel (Figure 174).*

   When installing the flywheel, clean tapered end and the taper inside the flywheel with a clean cloth.

   Align the governor flyweight assembly so that it will clear the governor drive lug on the flywheel. If it does not clear, the flyweight assembly will be jammed when you tighten

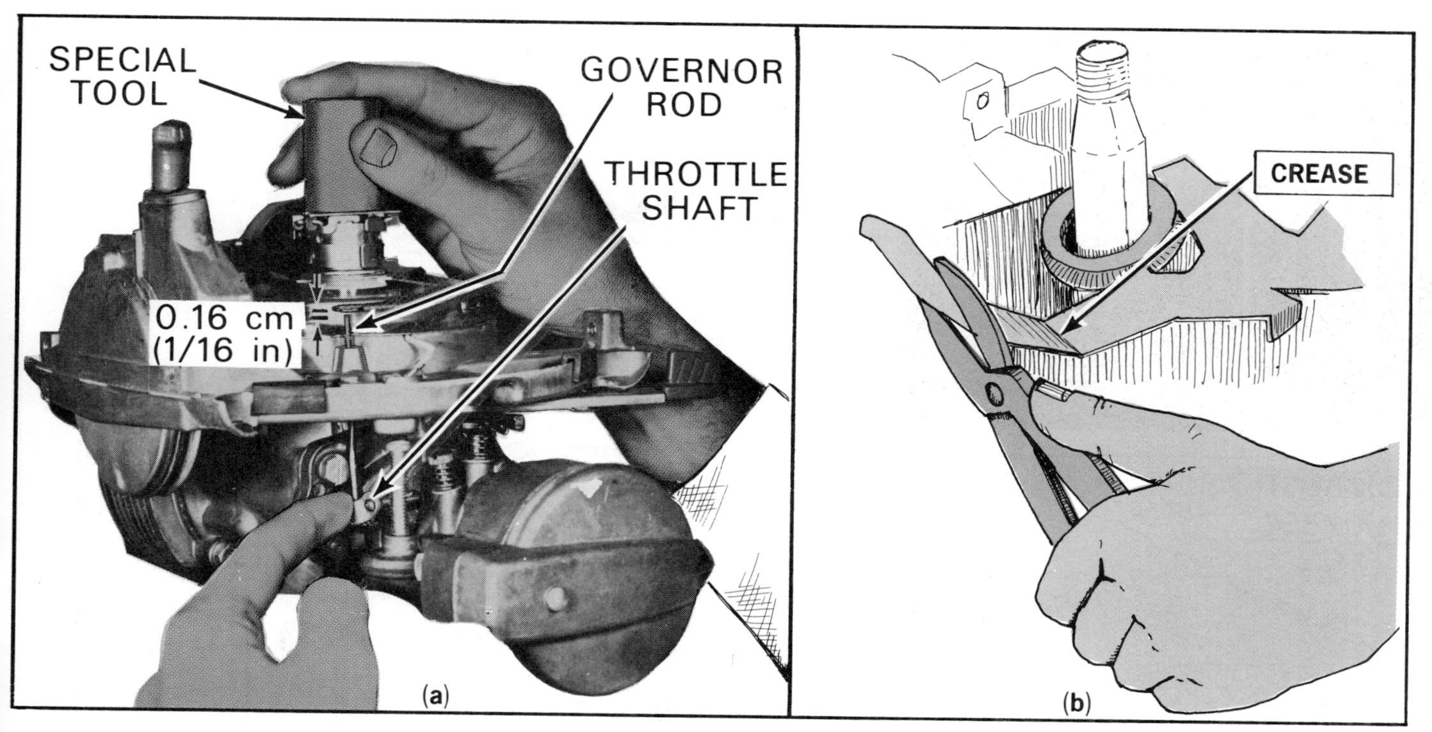

955 LB

FIGURE 173. (a) Checking the governor-lever adjustment. (b) Adjusting the governor lever for clearance between it and the governor rod.

956 LB

FIGURE 174. Align one of the straight edges of the governor with the flywheel key way.

the flywheel nut. To avoid this, some flyweight assemblies are marked "key" to indicate where the governor should be mounted in relation to the crankshaft key.

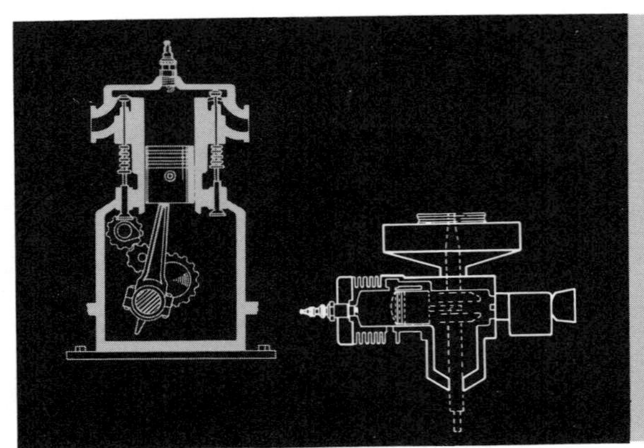

V. Repairing Valves

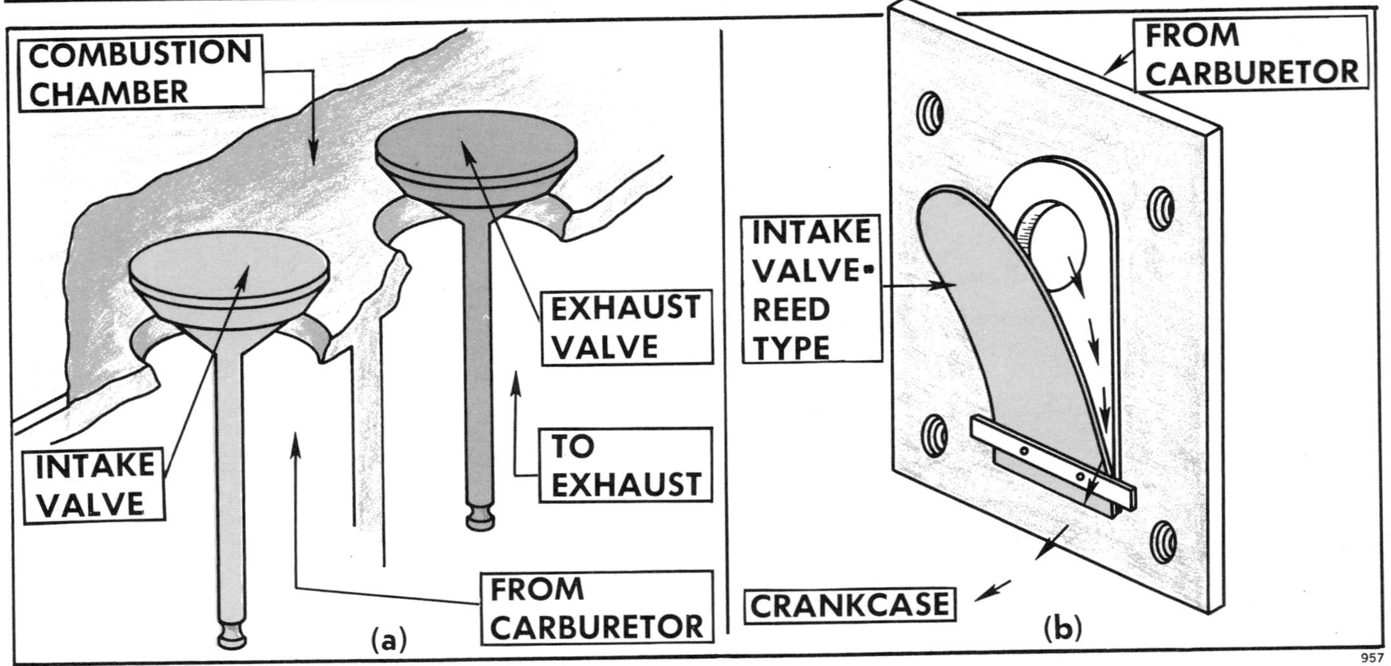

FIGURE 175. Valves on 4-cycle and 2-cycle engines. (a) Four-cycle engines have both an intake poppet valve and an exhaust poppet valve opening into the combustion chamber. (b) Two-cycle engines have only an intake valve which opens into the crankcase. It is usually a reed-type valve.

Four-cycle and 2-cycle engines are similar in many respects; but when it comes to valves, they are very different (Figure 175).

IMPORTANCE OF PROPER REPAIR

There is little preventive maintenance you can do directly to your valves. Valves, however, are seriously affected by the condition of your fuel and/or ignition system. Failure to maintain these systems properly result in valve failure.

Anything that affects valve operation will reduce the compression of your engine and result in hard starting, loss of power, and in eventual engine failure.

On **4-cycle engines,** valve failures consist of valves burning, sticking, dishing and necking (Figure 176).

Valve burning is caused by burning gases escap-

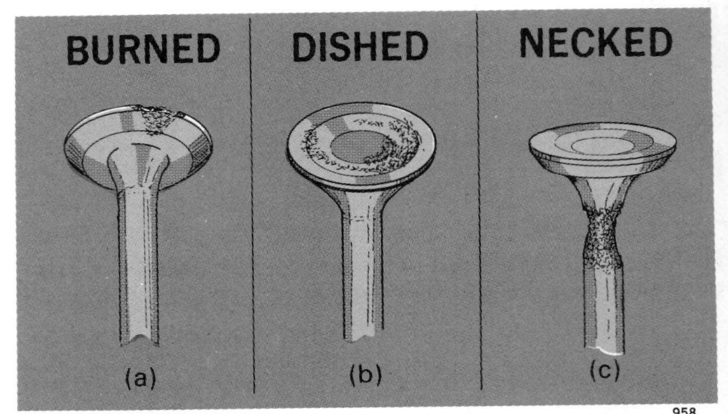

Figure 176. Improperly maintained valves may become (a) burned, (b) dished or (c) necked.

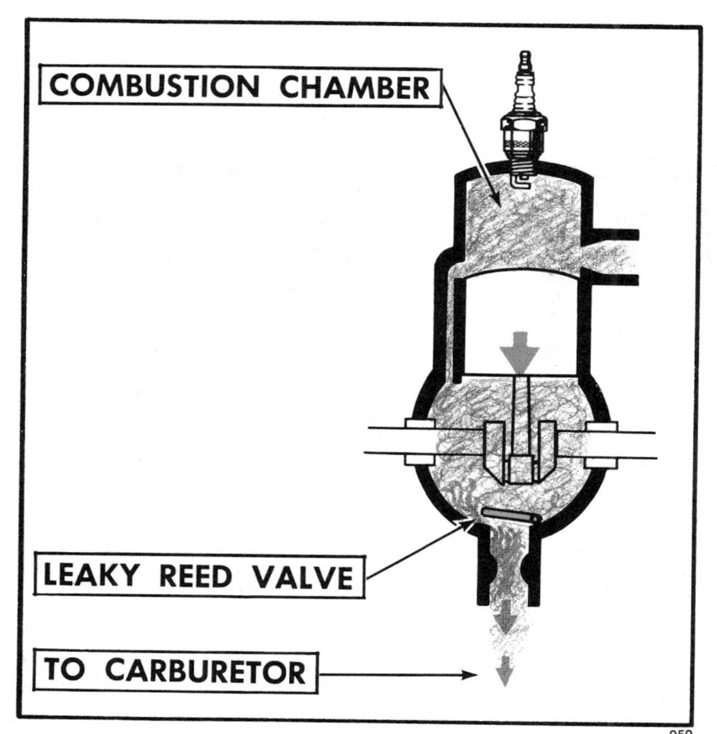

FIGURE 177. A leaky reed valve will result in loss of power.

ing by the valve. This happens when valve clearance is too small, or if the valve does not seat properly.

Valve sticking is failure of the valve to close because of the stem binding in the valve-stem guide. This sticking occurs most frequently on exhaust valves. It is caused by carbon, lead, gum and varnish. This is what happens. The hot gases escaping from the combustion chamber overheat the valve stem and guide. This overheating causes the oil on the valve stem to oxidize into varnish which holds the valve partially open. Continued operation with the valve partially open results in valve burning.

Intake-valve sticking may be caused by an excessive amount of gum in stale gasoline.

Valve dishing develops when the valve get so hot that it is cupped when it strikes the valve seat (Figure 176b). *Valve necking* results from the same conditions. The valve neck becomes red hot and elongates on impact with the valve seat (Figure 176c).

Proper valve-tappet clearance on 4-cycle engines is also essential to good engine operation. Since valve clearance on most small engines cannot be adjusted without removing them, there is not much need for checking clearance unless you are prepared to do a valve repair job.

Reed valves on **2-cycle engines** are located between the carburetor and the crankcase. They seldom give trouble; but when you suspect trouble, it is not difficult to clean, check and adjust them.

The term "reed" valve may be misleading. Reed valves are not made of bamboo, as in reed-type musical instruments. They are made of thin spring steel, plastic or phenolic materials.

A leaky reed valve will allow the fuel, air and oil mixture to be forced back into the carburetor (Figure 177), an action which decreases pressure in the crankcase. As a result of this pressure loss, the fuel charge entering the combustion chamber is lessened. The result is a loss of power.

Three common failures that result from improper engine care of reed valves are (1) **carbon build up under the reeds,** (2) **broken reeds,** and (3) **reeds out of adjustment.**

Maintaining and repairing valves are discussed under the following headings.

A. Repairing valves on 4-cycle engines.

B. Repairing valves on 2-cycle engines.

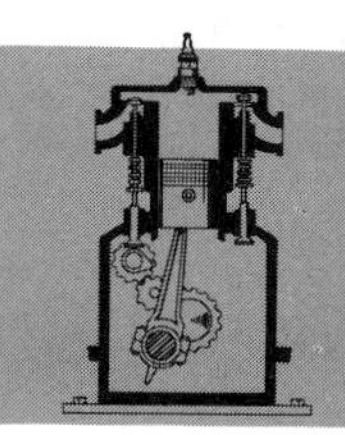

A. REPAIRING VALVES ON 4-CYCLE ENGINES

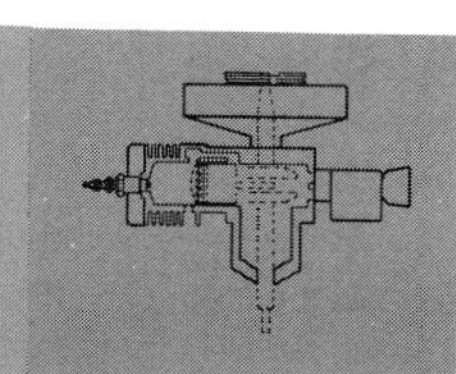

Valves on 4-cycle engines have three functions: (1) **to provide an inlet for the fuel-and-air mixture to enter the combustion chamber (Figure 178a)**, (2) **to seal the combustion chamber during the compression and power strokes (Figure 178b)**, and (3) **to allow burned gases to escape (Figure 178c).**

Maintaining and repairing valves on 4-cycle engines are discussed under the following headings:

1. Types of valves and how they work.
2. Tools and materials needed.
3. Checking valve operation by compression test.
4. Checking the valve-tappet clearance.

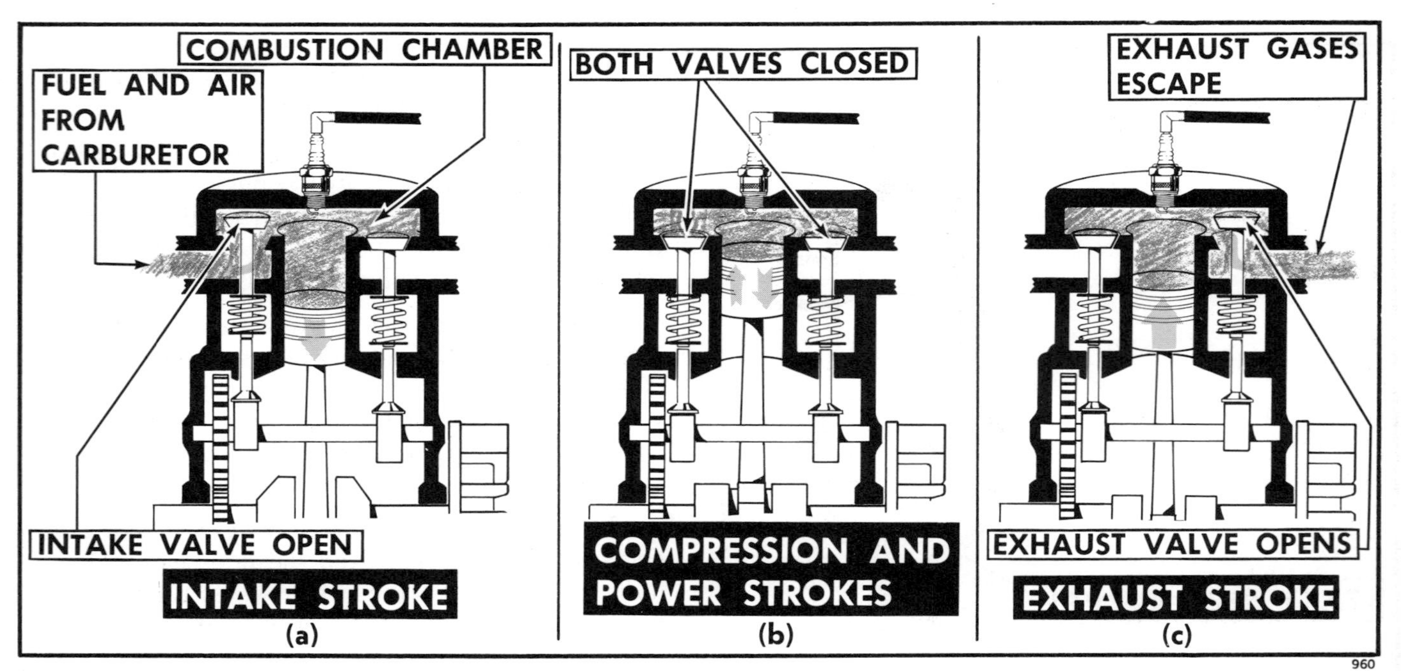

FIGURE 178. Functions of valves on 4-cycle engines. (a) Intake valve opens for fuel and air to enter the combustion chamber. (b) Both valves closed to seal the combustion chamber during the compression and power strokes. (c) Exhaust valve opens to allow burned gases to escape.

5. Removing the cylinder head.
6. Removing the valves.
7. Inspecting the valves and valve accessories.
8. Repairing valve guides.
9. Grinding (refacing) the valves.
10. Grinding (refacing) the valve seats.
11. Replacing the valve seats.
12. Lapping the valves.
13. Adjusting the valve tappet clearance.
14. Installing the cylinder head.

TYPES OF VALVES AND HOW THEY WORK

All 4-cycle engine valves are of the **poppet-type** (Figure 175a).

On 4-cycle engines there are two valves, an **intake valve** and an **exhaust valve.** *Exhaust valves*

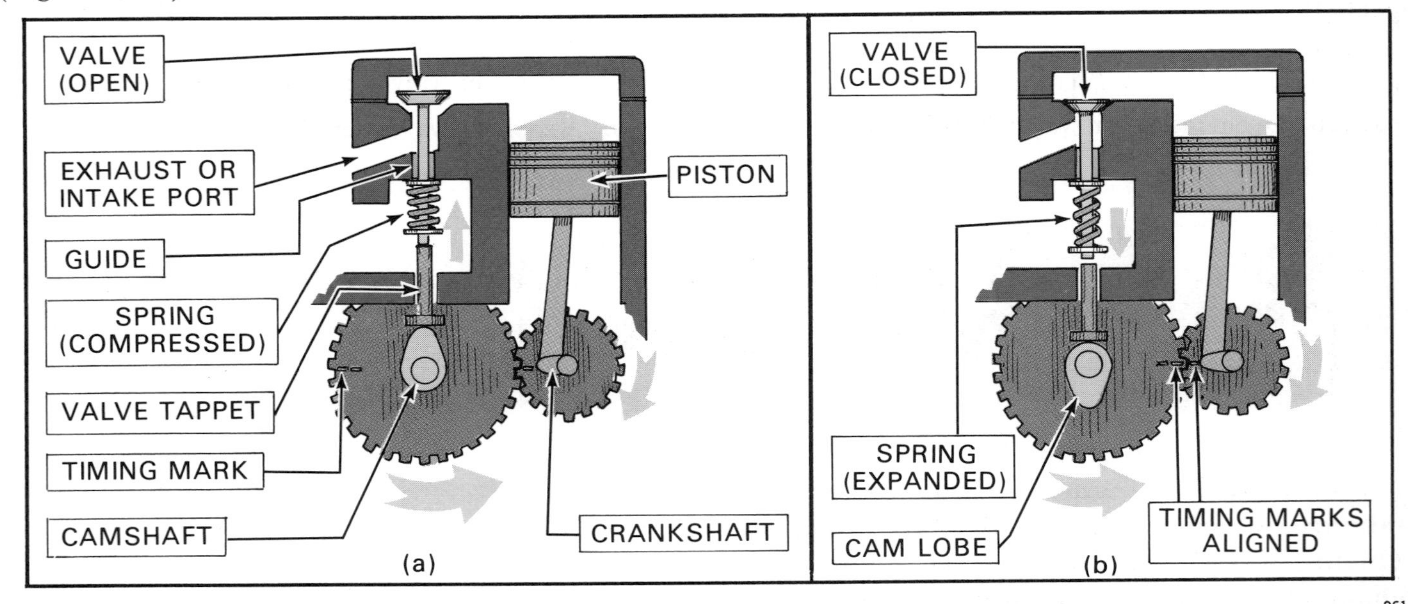

FIGURE 179. How a valve works on a 4-cycle engine. (a) Lobe on camshaft pushes valve tappet up, which opens the valve. (b) The crankshaft gear drives the camshaft gear which is connected to the camshaft. When the cam lobe moves past the valve tappet, spring compression causes the valve to close.

are made of special heat-resistant steel, and sometimes the stems are filled with a liquifying metal, such as sodium, to help them to cool faster by dissipating heat.

Here is how the valves work. Poppet valves are opened at the proper time by tappets which are driven from lobes on the camshaft (Figure 179a). They are closed by spring tension (Figure 179b). Some engines, designed for continuous operation, have *rotators* that cause the valves to turn as they operate. This helps reduce valve burning.

The **camshaft** is driven from a gear on the crankshaft (Figure 179). Note that the camshaft has twice the number of teeth as the gear on the crankshaft. This proportion enables the camshaft to turn at one half the speed of the crankshaft, since the 4-cycle engine has only one power stroke in two revolutions of the crankshaft. Refer to Figure 5, Part 1.

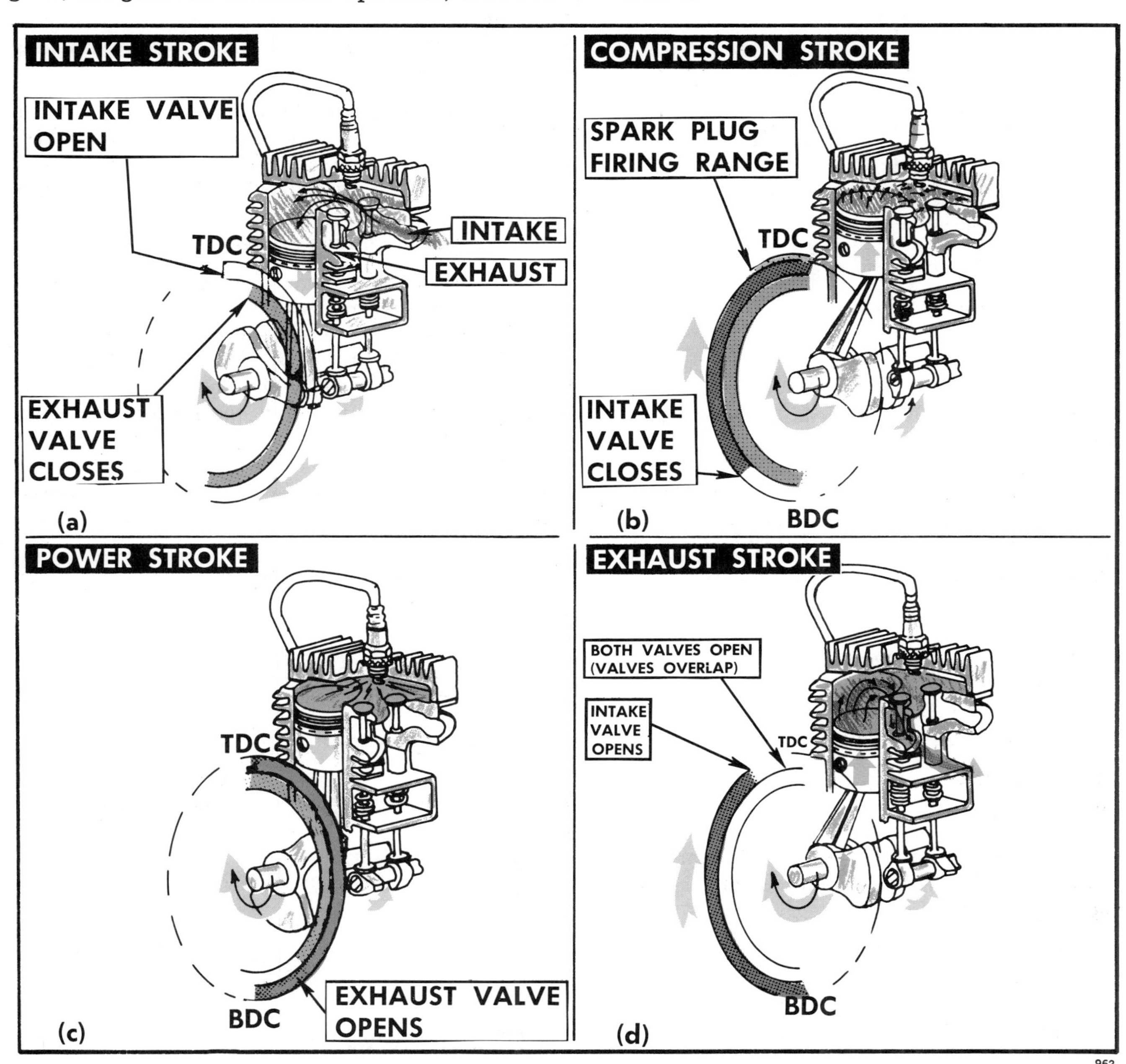

FIGURE 180. Typical positions of the crankshaft when the valves open and close. (a) Both valves are slightly open at the beginning of the intake stroke. This is called "valve overlap." Shortly afterwards the exhaust valve closes. (b) The spark plug fires at the beginning of the power stroke, or just before. The intake valve closes approximately 20° past bottom-dead-center (BDC), compression stroke, and both valves remain closed until near the end of the power stroke. (c) At approximately 35° before bottom-dead-center (BBDC) power stroke, the exhaust valve opens and remains open throughout the exhaust stroke. (d) The intake valve begins to open approximately 20° BTDC.

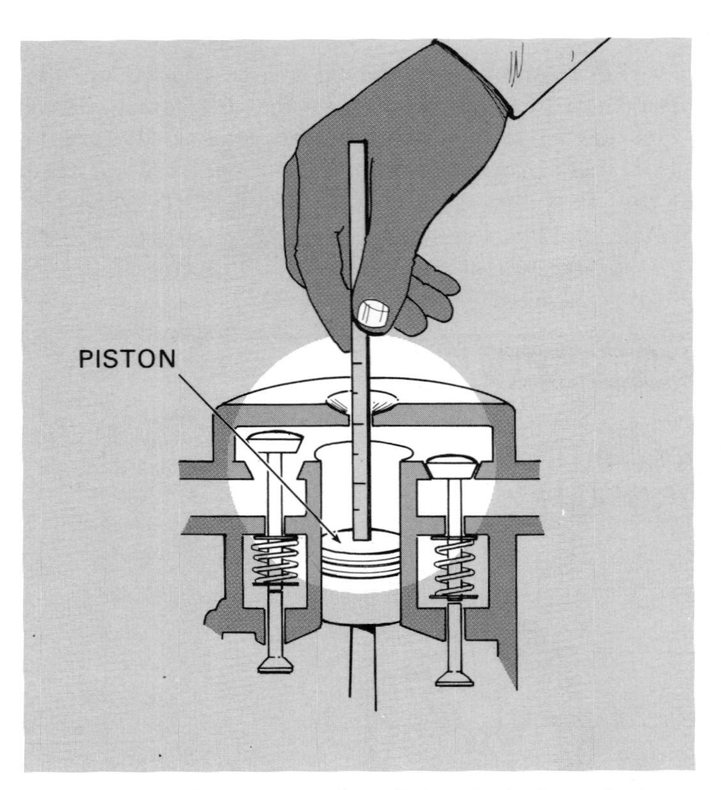

FIGURE 181. Measuring valve timing in inches of piston travel before TDC.

With your engine turning at 3,000 r.p.m., each valve must open and close in approximately 1/15 of a second. This is fast for a poppet valve, and there is not much time for the fuel charge to enter. Neither is there much time for the burned gases to escape. For this reason the cams are designed to hold the valves open as long as possible. They open a little ahead of time and close a little late (Figure 180). The time the valves are to open and to close is established by the manufacturer for each engine model. "Timing marks" are made on the camshaft gear and the crankshaft gear (Figure 179) for aligning the gears properly.

Note that the *spark-plug firing range* is also indicated in Figure 180. It ranges from 25° BTDC compression stroke to 4° ATDC. When the spark occurs before TDC, it is called "advanced-ignition timing." When the spark occurs after TDC, it is called "retarded-ignition timing."

The position of the crankshaft at which the valves open and close varies with different engines. This may be measured in degrees of turn by the crankshaft (Figure 180a), or by the measured distance of the piston from TDC (Figure 181). Manufacturers who give their timing specifications in their service manuals, may give them either in degrees of crankshaft travel or in inches or piston travel.

TOOLS AND MATERIALS NEEDED

1. Socket wrenches — 1/4″ through 9/16″
2. Wire brush
3. Valve spring compressor
4. Valve lapping tool
5. Valve seat insert remover
6. Valve grinder
7. Feeler gage
8. Torque wrench and socket set — 3/8″ through 9/16″ - 3/8″ drive, and a 3/4″ or 11/16″ spark-plug socket
9. Slot-head screwdriver — 6″ regular
10. Phillips-head screwdriver — 6″
11. Air pressure gage — 0 to 100 psi
12. Air-operating valve adapter
13. Pocket knife
14. Micrometer — 0″ to 2″ and 0″ to 6″
15. Clean rags
16. Emery cloth

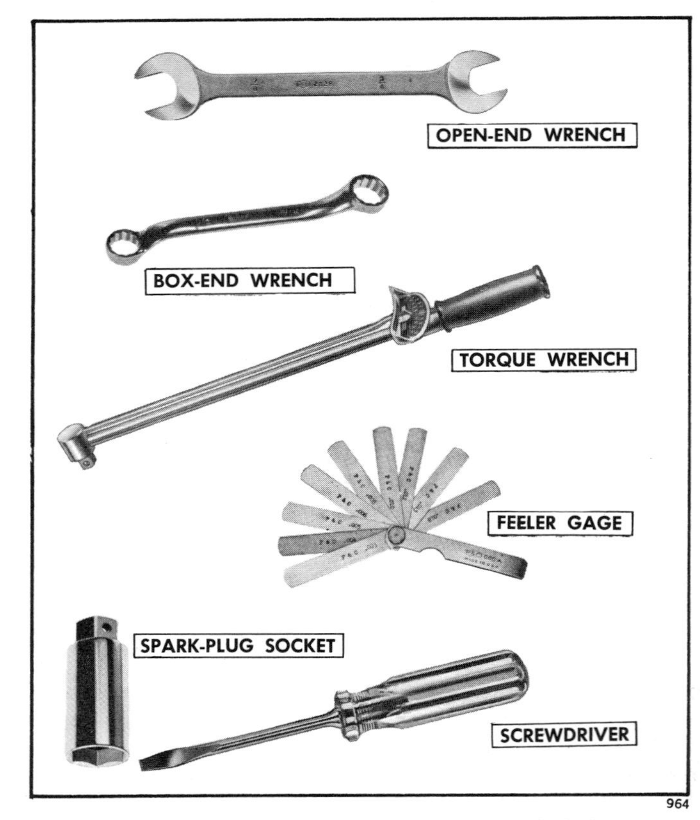

FIGURE 182. Some examples of tools needed for maintaining and repairing valves.

17. Bucket and petroleum solvent (mineral spirits, kerosene or diesel fuel)
18. Container for nuts, bolts and screws
19. Valve-lapping compound
20. Prussian blue

CHECKING VALVE OPERATION BY COMPRESSION TEST

Hard starting, rough running and loss of power are signs of poor valve operation.

Engine backfiring through the carburetor is a sign of a bad intake valve.

If you suspect valve trouble, check the compression. Improper valve operation can cause poor compression, but several other conditions can also cause it. They are as follows:

- *A dry cylinder on an engine that has not been used for some time* — the oil drains away from the rings, thus allowing air to bypass.
- *Piston rings stuck because of carbon deposits.*
- *Worn piston rings and worn cylinder.*
- *Loose or broken spark plug.*
- *Damaged cylinder head, gasket or loose cylinder head.*
- *Sticking, burned or warped valves, and/or valves out of adjustment.*

There are four methods for checking compression.

a. Turning the *crankshaft* by hand without removing the spark plug and feeling the amount of compression resistance to turning.

b. Turning the crankshaft with the starter (wind-up type) with the spark plug removed and feeling the compression.

c. Forcing compressed air into the spark-plug hole.

d. Using a compression gage in the spark-plug hole.

a. The first method **(cranking the engine by hand)** is described under "Checking Compression," Part 1. According to some manufacturers, it is sufficient.

b. The second method is a simple test **for engines with wind-up starters.** Proceed as follows:

1. *Remove the spark plug.*
2. *Wind the starter.*
3. *Place your finger over the spark-plug hole.*
4. *Release the starter.*

   Your finger should be blown away with a strong force.

c. If you use the **compressed-air method,** proceed as follows:

1. *Remove the spark plug.*
2. *Insert an "air operative valve tool" into spark-plug hole.*

   This is a special adapter for connecting an air hose to the spark-plug hole (Figure 183).
3. *Rotate the engine crankshaft to TDC, compression stroke.*

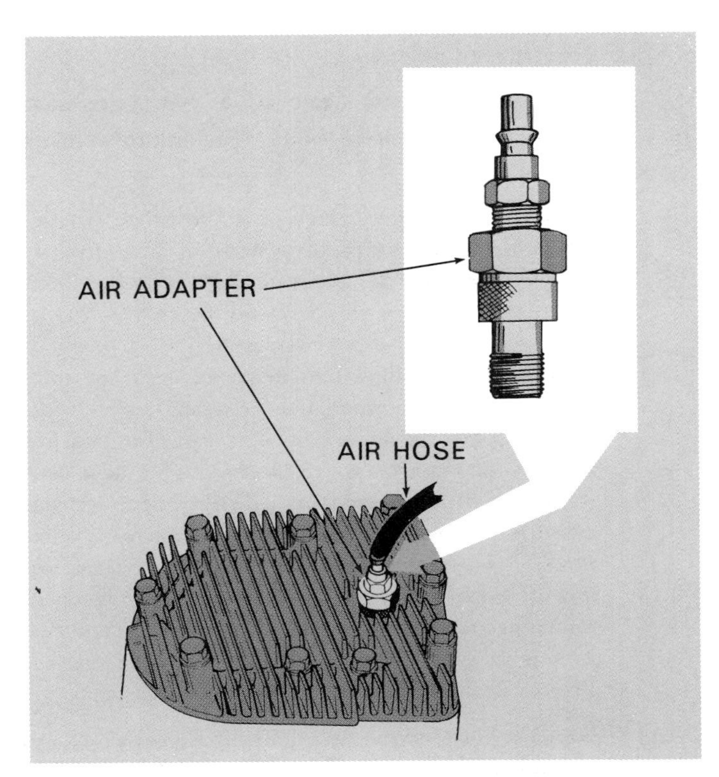

FIGURE 183. An air operative tool for connecting an air hose to the spark-plug hole.

4. *Remove the oil-filler plug.*

   Place a clean cloth over the oil-filler hole. In case there is a leak in or around the piston, this will prevent oil from being blown out.

5. *Set air compressor to supply 560 kPa (60 psi) maximum pressure.*

6. *Connect air hose to valve tool.*

7. *Listen for air leakage.*

   If you hear air through the muffler, this indicates a leak in your exhaust valve.

   If you hear air through the carburetor, this indicates a leak in the intake valve.

   If you hear air at the oil-filler hole, this indicates worn or damaged cylinder, piston or rings.

   If you find compression is low but you still are not certain of the cause, proceed with step 8.

8. *Squirt a teaspoon of oil into the cylinder through the spark-plug hole and crank the engine several times for the oil to be distributed around the piston rings.*

9. *Recheck compression.*

   If compression improves, the trouble is worn or damaged piston rings. The oil helps provide a seal which causes the compression pressure to increase.

d. The fourth method **(use of a compression gage)** is somewhat controversial. One manufacturer states:

> On single cylinder engines we think of good compression, not in terms of pounds of pressure per square inch, but in terms of horsepower output. If the engine produces the power for which it was designed, we believe the compression must be good. It is extremely difficult to make an accurate compression test on a small one cylinder engine without expensive machinery. The reasons for this are the lack of a starter to crank the engine at a constant speed and the small displacement of the cylinder. Therefore we do not publish any compression pressure figures. As a simple compression test, give the flywheel a quick spin. If the flywheel rebounds on the compression stroke, the compression is at least good enough to start the engine. *Briggs & Stratton Engine Corporation.*

If you check compression with a **compression gage,** proceed as follows:

1. *Remove the spark plug.*

966

FIGURE 184. Checking the cylinder compression with a compression gage.

2. *Insert compression gage in spark-plug hole (Figure 184).*

3. *Turn engine at least 6 to 8 revolutions at cranking speed.*

   If you have an **electric starter,** be sure the battery is fully charged. Press pressure gage tightly against spark-plug hole.

   The throttle should be fully open when making the test so as much air as possible can reach the cylinder.

4. *Observe the compression reading.*

   Compare the reading with that given in your operator's manual. For satisfactory operation the minimum gage reading should read above 560 kPa (60 psi) on most engines.

   If the pressure is **10 per cent less** than that given in your engine specifications, check for compression leaks.

If, after following one or more of the foregoing procedures, you find the compression is low, it is probably in the valves. Proceed with "Checking the Valve-Tappet Clearance," next heading.

CHECKING THE VALVE-TAPPET CLEARANCE

Proper valve-clearance adjustment is important for the following reasons:

- *Valves give longer service.*
- *Fuel is used more efficiently.*
- *Engine starting is easier.*
- *Maximum power is produced.*
- *Overheating is less probable.*
- *Engine operation is smoother.*

When a valve is properly adjusted, there is clearance between the end of the valve stem and the valve tappet when the valve is closed (Figure 185). This clearance is very small, varying from approximately .13 mm to .30 mm (.005 in to .012 in) but it is extremely important. Each manufacturer recommends the proper clearance for each model of engine. Recommendations vary depending on (1) whether the engine is hot or cold at the time of adjustment, and (2) the design of the engine—some are designed to run hotter than others.

If valves are adjusted so that there is **little or no clearance,** they are thrown out of time. This causes them to *open too early* and *close too late.* Also, the valve stems may lengthen enough from heating so that the valves do not seat completely (Figure 185). This allows hot combustion gases to leak past the valve seats, thus causing the valves, seats, stems and guides to overheat. The valves seat too briefly, too poorly or not at all. This causes loss of compression, overheating, sticking and burning of the valves.

If there is **too much valve clearance,** there is a noisy lag in valve timing which throws the engine out of balance. The fuel-air mixture is late entering the cylinder during the intake stroke. The exhaust valve closes early and prevents waste gases from being completely removed. The valves themselves become damaged because they close with heavy impact. They crack and break. (When valve clearances are correct, the camshaft lobe slows the speed of valve movement just as it closes. With too much clearance the slowing action is lost.)

To **check the valve-tappet clearance,** proceed as follows:

1. *Remove the valve cover plate (Figure 186).*

   It may be necessary to remove the carburetor to get to the valve access well. Refer to "Removing the Carburetor," page 160.

2. *Remove the crankcase breather, if necessary.*

   Some are located in the cover plate and some are located in the valve access well. Be careful not to damage the breather or drop anything into the crankcase. Refer to "Servicing Crankcase Breathers (4-cycle engines)," Part 1.

3. *Check the valves for sticking.*

   Turn the crankshaft by hand and look to see if the valves travel all the way up and down.

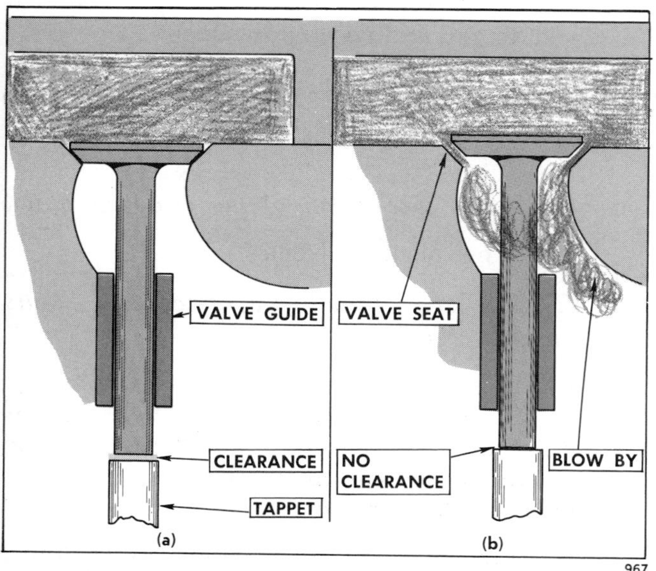

FIGURE 185. Importance of proper valve adjustment. (a) If valve-tappet clearance is correct, valve closes tightly. (b) If valve-tappet clearance is too little, valve fails to close. Exhaust gases leak by the valve seat.

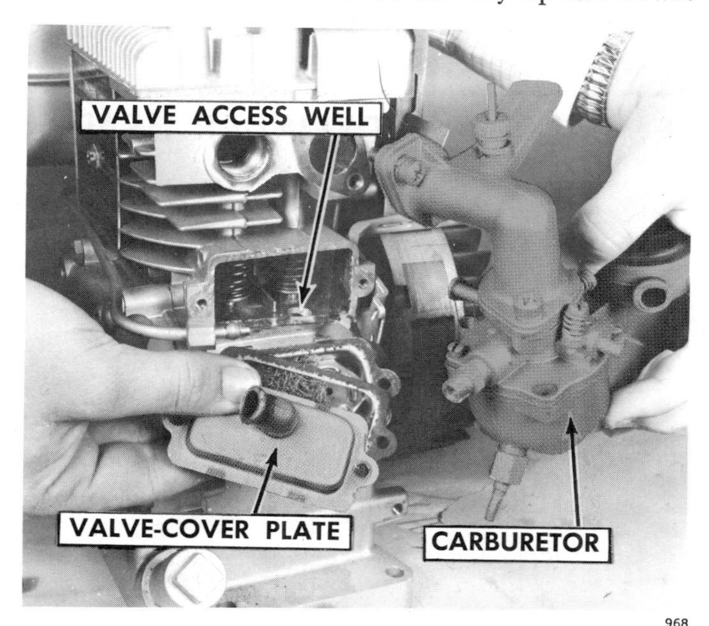

FIGURE 186. The valve access well may be located behind the carburetor.

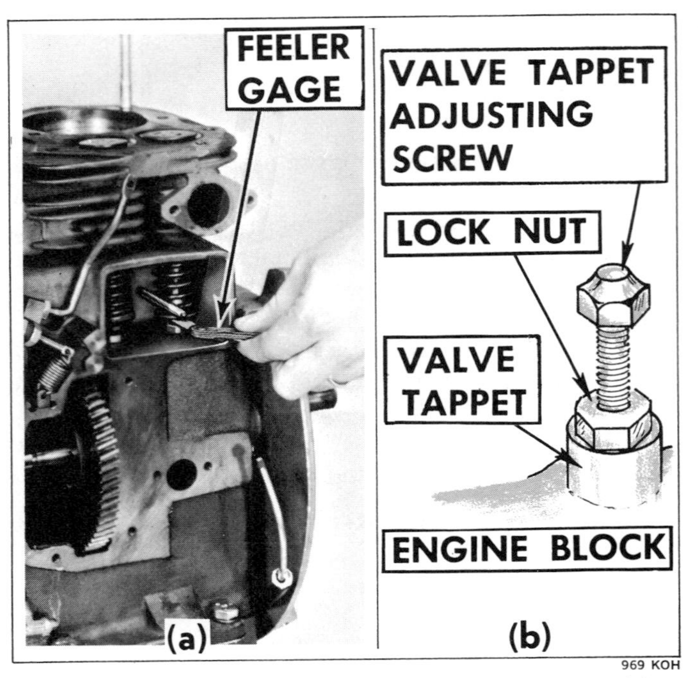

FIGURE 187. (a) Checking valve-tappet clearance with a feeler gage. (b) An adjustable tappet.

4. *Turn the crankshaft until the valve you are going to check is fully open.*
5. *Rotate the crankshaft one complete revolution.*

   This is to make sure the valve is completely closed.

   If your engine has **timing marks,** both valves will be closed when the timing marks are aligned and the piston is at TDC, compression stroke. Refer to "Timing the Magneto to the Engine," page 137.

6. *Check your operator's manual for the proper valve-tappet clearance for your engine.*

   The specifications are usually given for checking when the engine is cold.

   Intake valve-tappet clearances vary from .13 mm to .3 mm (.005 in to .012 in).

   Exhaust valve-tappet clearances vary from .23 mm to .3 mm (.009 in to .012 in).

7. *Check the valve-tappet clearance with a feeler gage (Figure 187).*

   There should be a slight drag as you move the feeler gage between the valve stem and tappet.

   If the clearance is not correct and you have **adjustable tappets,** adjust the valve clearance. Loosen the tappet with an open-end wrench, and turn it up or down until the clearance is correct (Figure 187b). Then tighten tappet.

   If the clearance is not correct and you have **no adjustable tappets,** it will be necessary for you to remove the cylinder head to correct it. Follow procedures under the next heading. Usually when you have to remove the valves for adjusting, it will be necessary to recondition them also.

   Too small a clearance is corrected by shortening the valve stem by grinding. Too much clearance is corrected by replacing the valve with a new one, then adjusting it. Reseat and/or lap and check adjustment.

REMOVING THE CYLINDER HEAD

1. *Remove the spark-plug wire from the spark plug.*
2. *Remove the spark plug.*

   Use a spark-plug wrench to avoid breaking the spark plug.

3. *Remove the carburetor if not already removed (Figure 188).*

   See "Removing the Carburetor," page 160.

4. *Remove the cowling if necessary to get to the cylinder head (Figure 189).*
5. *Remove the valve-cover plate if it is not already removed (Figure 186).*
6. *Clean and inspect all of the removed parts.*

   Use a petroleum solvent.

7. *Loosen and remove cylinder-head bolts (Figure 190).*

   Use a socket wrench with a ratchet handle or "T" handle. Sometimes the bolts are of different lengths. Note any difference in the lengths so you can properly reassemble them in the right positions.

   If you should use a **bolt which is too short,** it may not reach enough threads to hold; and it will strip out.

   If you use a **bolt that is too long,** it may

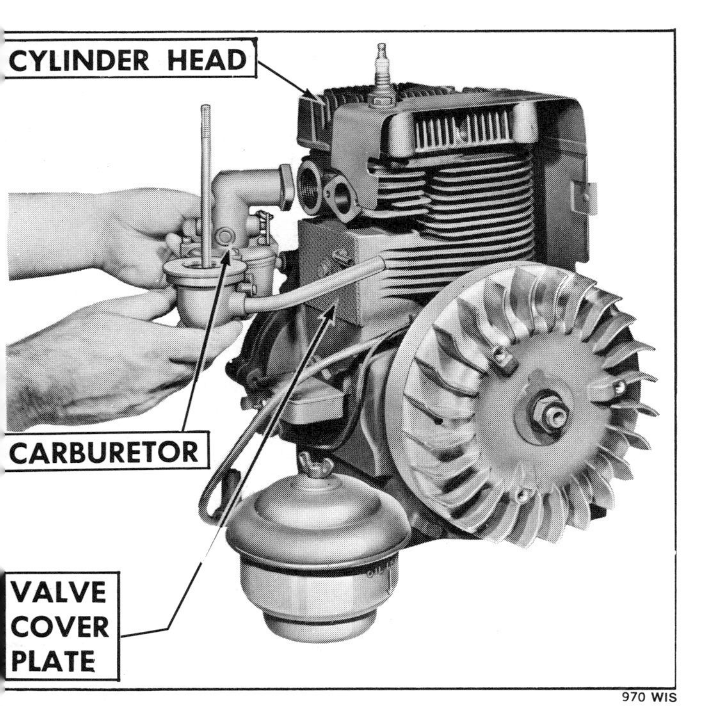

FIGURE 188. It may be necessary to remove the carburetor to get to the valve-cover plate.

bottom on a cooling fin and break the fin; or it may not tighten the head sufficiently.

Some bolts are long because braces, supports and/or brackets are anchored to them.

8. *Remove the cylinder head and cylinder-head gasket.*

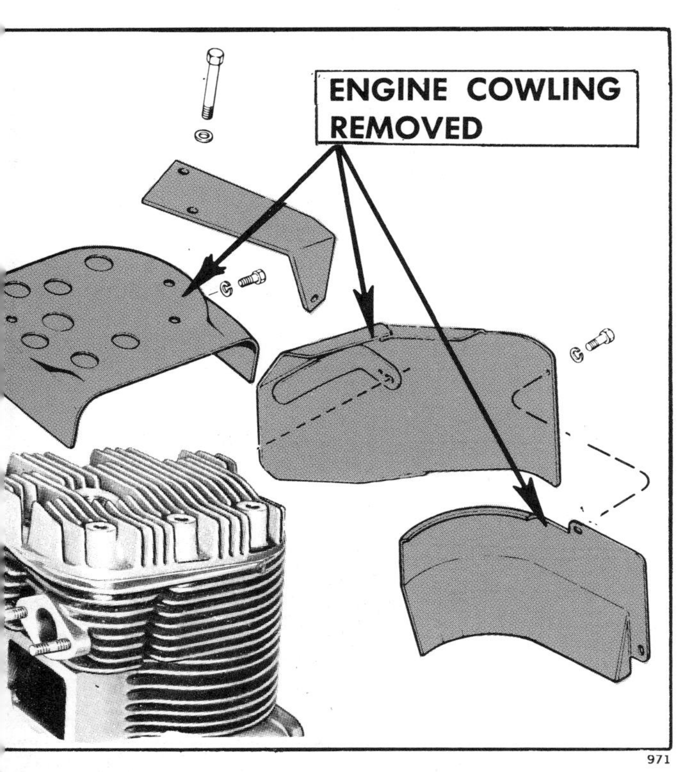

FIGURE 189. Remove cowling from around the cylinder and cylinder head.

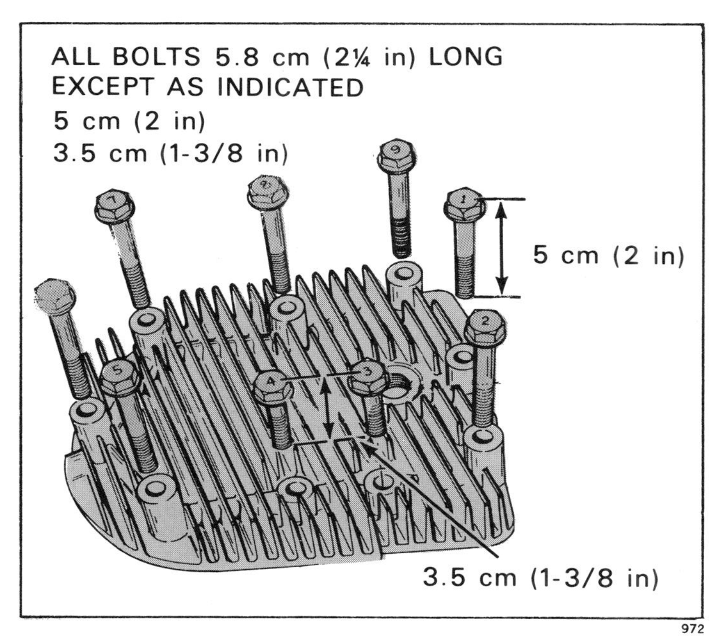

FIGURE 190. Cylinder-head bolts may be of different lengths for different positions.

To loosen the cylinder head, tap it sharply on the edge with a soft hammer.

The gasket may stick to either the cylinder head or the engine block or both.

Generally the gasket needs to be replaced with a new one.

9. *Clean the head, the piston and the area around the valve seats (Figure 191).*

Remove carbon and any gasket material that may remain. Use a metal scraper. If the

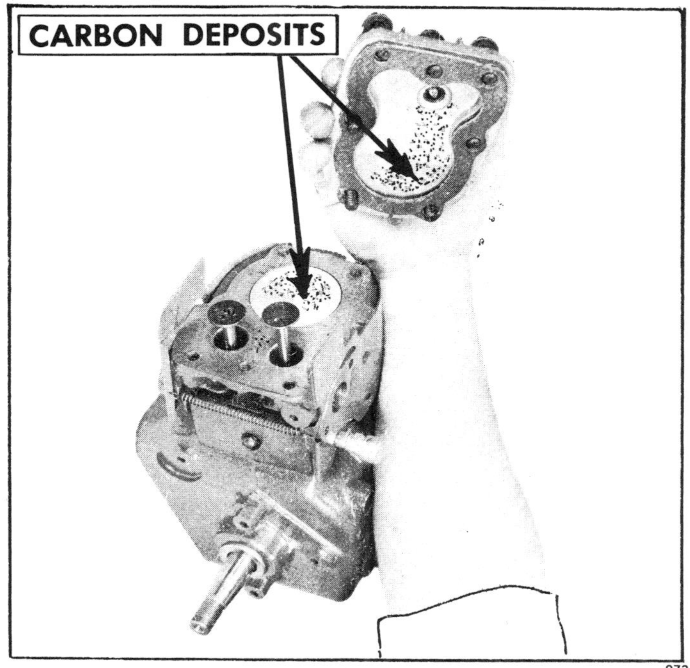

FIGURE 191. Clean the carbon from the cylinder head, from the piston and from around the valve seats.

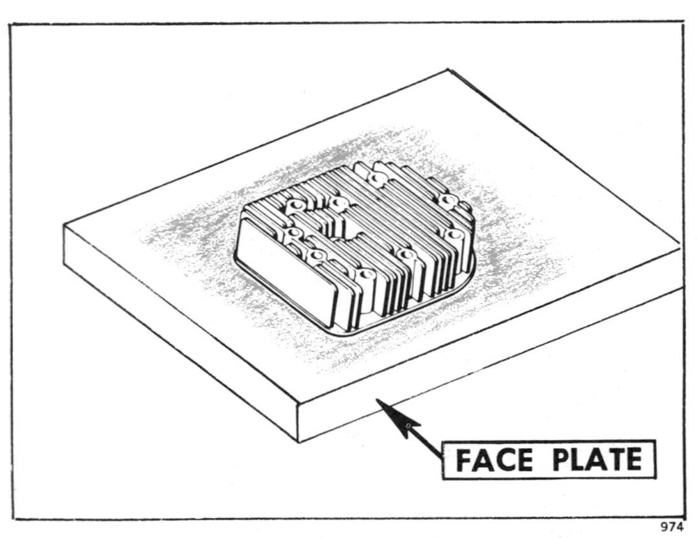

FIGURE 192. Checking the cylinder head for warpage.

metal you are cleaning is soft, like aluminum, be careful to avoid scratching it.

10. *Check the cylinder head for warpage (Figure 192).*

    Use a face plate. A face plate is a thick metal plate with a perfectly flat surface. Apply thin layer of prussian blue to the face plate, and lay the head (gasket-side) down on the face plate. Slide the head back and forth a couple of times. Then remove it and check to see that bluing has transferred to all contact surfaces of the cylinder head. If not, the cylinder head is warped and must be replaced with a new one.

11. *Remove prussian blue from head and face plate.*

    Use a clean dry cloth.

12. *Place the parts in a clean, well-protected place and in an orderly manner.*

REMOVING THE VALVES

1. *Compress the valve spring (Figure 193).*

A valve-spring compressor is almost essential

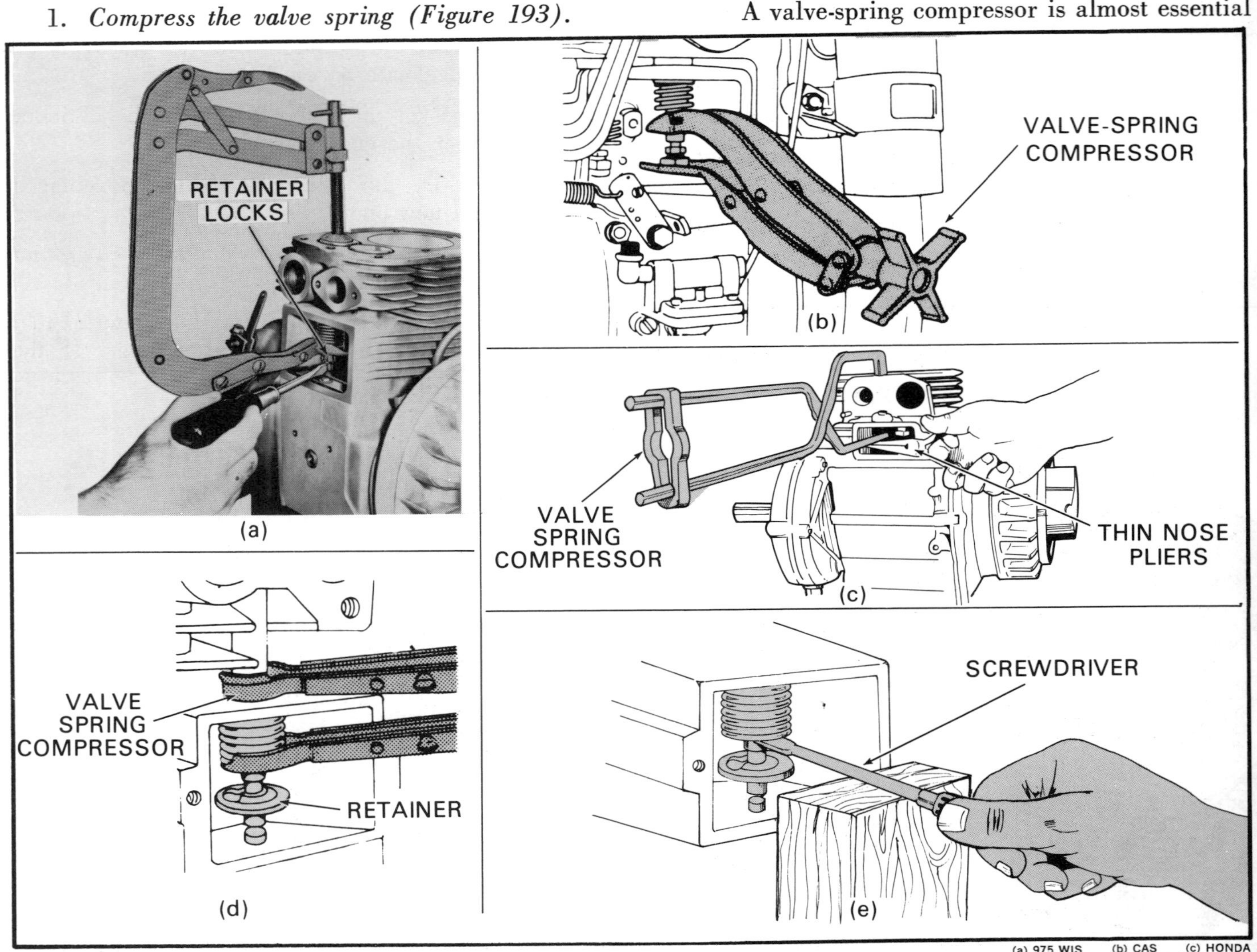

FIGURE 193. Compressing the valve springs. (a, b, c and d) Types of commercial valve-spring compressors. (e) Using a screwdriver for compressing the valve spring.

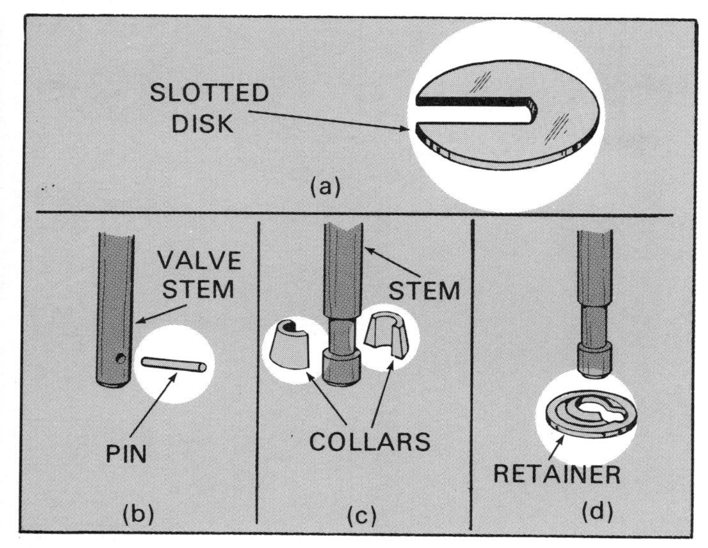

FIGURE 194. Types of valve spring retainer locks: (a) slotted disc, (b) pin, (c) split collar and (d) combination retainer and lock.

to releasing the valve. It is possible, however, to compress the spring with a screwdriver enough to remove the valve spring and retainer lock (Figure 193e).

2. *Hold the valve closed with your thumb, if the valve is not held by valve compressor.*
3. *Remove the valve-spring retainer and/or retainer lock (Figure 194).*

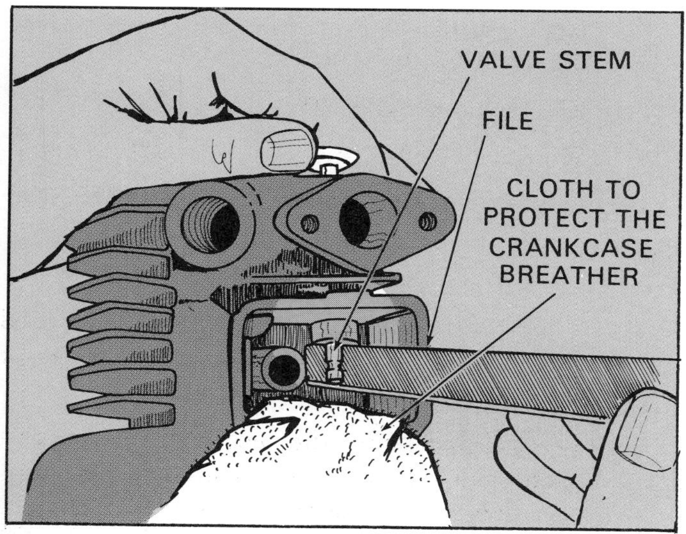

FIGURE 195. Filing away burrs from a valve stem before removal of the valve.

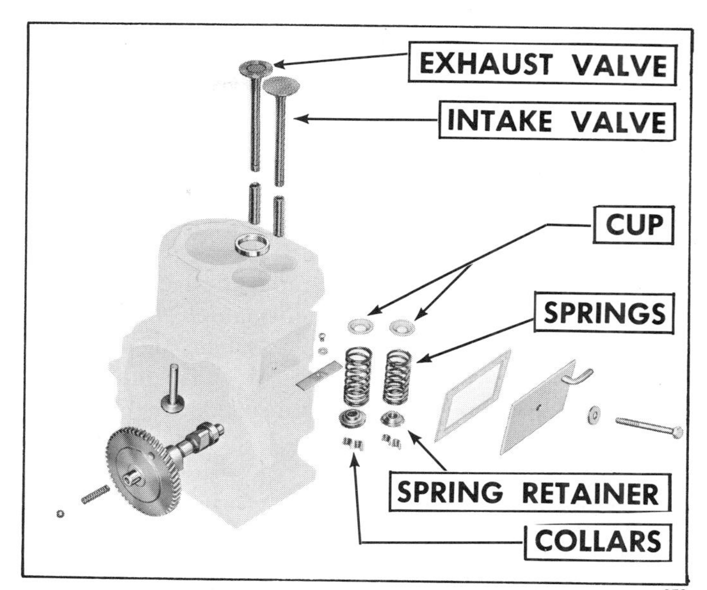

FIGURE 196. Valve assembly removed from the engine.

4. *Remove the valves.*

   It may be necessary to file away burrs that have developed on the valve-stem lock and groove, before removing the valve, to prevent damage to the valve guide (Figure 195). Use a fine, flat file. Do not over file.

   Do not let the filings drop into the crankcase breather hole, unless you plan to do a complete engine overhaul. Stuff a rag in the valve access well.

   Be careful not to damage the crankcase breather plate in the bottom of the valve access well, if it is installed.

   Tag the valves and the springs so you will know how to replace them in the same positions. On some engines the exhaust spring is heavier than the intake spring.

5. *Remove the valve springs and cups (Figure 196).*
6. *Clean valves and accessories.*

   Carburetor cleaner, such as Petisol or Bendix, may be used to remove gum. For normal cleaning, use petroleum solvent.

INSPECTING THE VALVES AND VALVE ACCESSORIES

Before going too far with valve repair, check the cylinder and bearings. The engine may be worn too much to repair. See "Repairing Cylinders and Piston-and-Rod Assemblies," page 212. Also see "Repairing Crankshaft Assemblies," page 248.

The condition of the valves tell you much about the condition of your engine.

Intake valves are cooled by the fuel-and-air mixture going into the cylinder. Therefore, intake valve trouble is usually caused by the use of stale gasoline or a poor grade of gasoline. Gum builds up on the valve neck and burns, thus leaving a hard deposit of carbon material under the valve head. This deposit causes the valve to stick and/or burn.

Another cause of intake valve sticking and burning is leaking valve guides — guides that are worn and out of round. Oil seeps into the ports — the upper ends of the valve guides — and builds up a soft oily deposit under the valve head.

Exhaust valves operate under extremely high temperatures (red hot) under normal engine operation. For cooling they depend upon heat dissipation through the valve seat and the valve guides and then out onto the cooling fins. So anything that causes the engine to run only slightly hotter than normal will affect the exhaust-valve life. For this reason exhaust valves are more likely to fail than intake valves. This is the reason most of them are made of a heat-resistant steel.

You can tell by looking at the condition of the valves what may have caused the failure. Examine the valves; determine their condition; then refer to the following guide to find out what caused the trouble. The guide indicates what steps to take to prevent the reocurrence of the trouble.

VALVE CONDITIONS AND PROBABLE CAUSE:

Gum and Varnish Deposits:

Use of stale or low quality gasoline

Use of poor quality oil, or insufficient lubrication

Worn valve guides

Loose guide, or guide extending too far into the exhaust port

Dirty cooling fins

Overspeeding or overloading the engine

Operating with an excessively rich fuel-air mixture

Poor crankcase ventilation (clogged crankcase breather)

Evidence of Valve Burning:

Sticking valve stem

Leaking valve seat

Insufficient valve-tappet clearance

Weak valve spring

Excessive exhaust back pressure (clogged muffler)

Excessively lean fuel-air mixture

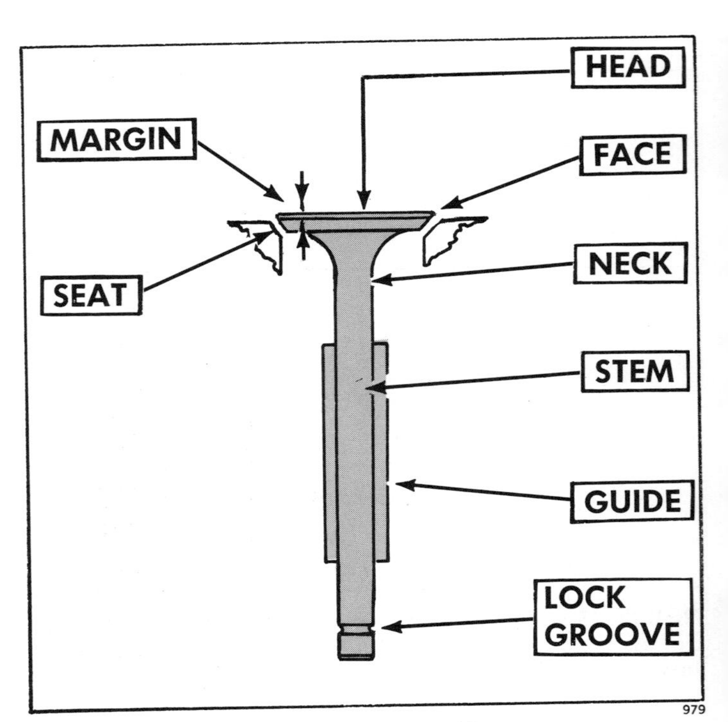

FIGURE 197. Wearing parts of a valve.

Warped valve

Deposits on the valve face

Necked (stretched) valve stem

Loose valve seat insert or warped valve seat

Carbon Deposits on the Valve Face:

High seating impact from excessive valve-tappet clearance

Excessive spring tension

Worn valve retainers and retainer grooves

To **inspect the valves and valve accessories,** proceed as follows:

1. *Inspect the valve head (Figure 197).*

   Look for warped valve heads; bent, necked or worn stems; valve face burned, cracked or out of round.

   You seldom know whether a valve is reusable until you see how much grinding it is going to require to smooth the face. This is discussed under "Grinding (Refacing) the Valves," page 198.

2. *Check condition of valve stem (Figure 198).*

   Polish the stem with an emery cloth; then check the stem diameter with a micrometer.*

*If you do not know how to use a micrometer, get a copy of "Micrometers and Related Measuring Tools," a publication available from Vocational Agriculture Service, University of Illinois, Urbana, Illinois.

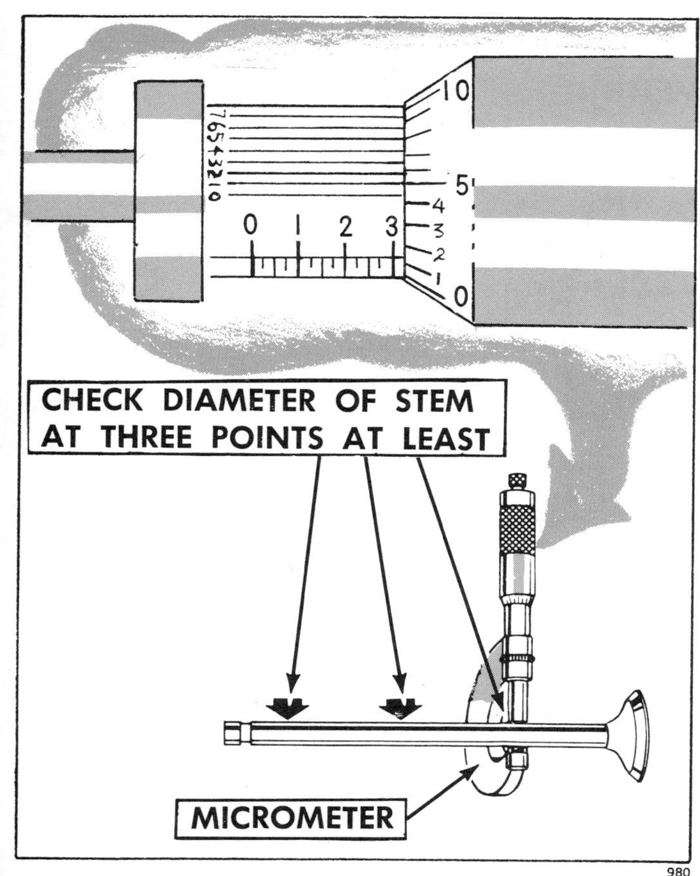

FIGURE 198. Checking the valve-stem diameter with a micrometer.

If the stem is worn more than .013 mm (.005 in), replace the valve. See your service manual for the original diameter of the valve-stem.

3. *Inspect the valve springs (Figure 199).*

Inspect valve spring for squareness at the top and bottom. Also inspect for distortion and for pits or cracks.

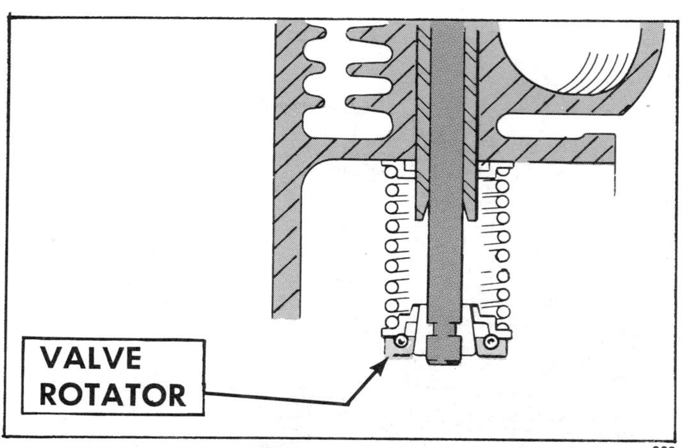

FIGURE 200. Valve rotators are used on engines under constant use, to rotate the valves as the engine runs. This helps prevent valve burning.

Some manufacturers give specifications for the expanded length of the valve spring so you can measure them (Figure 199a). Others give the required spring tension. This is measured with a spring tension tester (Figure 199b).

4. *Check valve-spring retainer and retainer locks for wear (Figure 196).*

Also check valve rotator if installed. They are installed in the valve-spring retainer (Figure 200).

5. *Check the valve seats for cracking and pitting.*

On some engines you can replace the valve seats with a new insert (Figure 206b).

Some have only the exhaust valve seat replaceable. On others, neither valve seat is replaceable (Figure 206a).

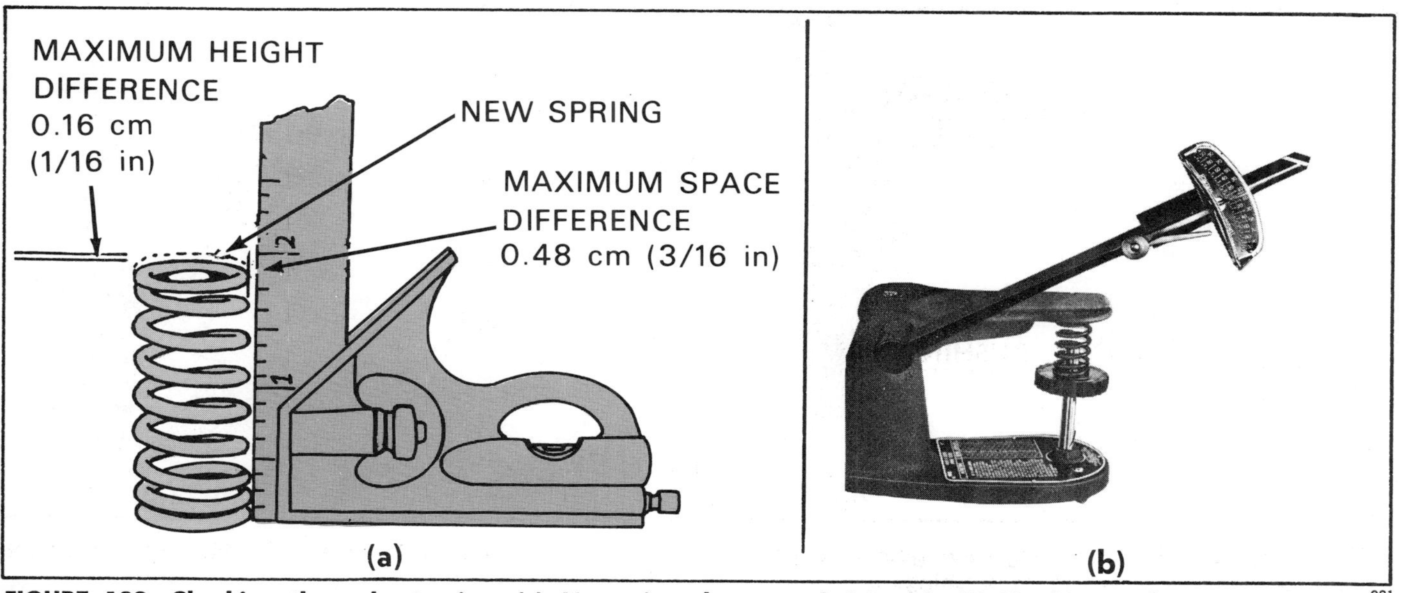

FIGURE 199. Checking the valve-tension. (a) Measuring the expanded length. (b) Checking with a tension tester.

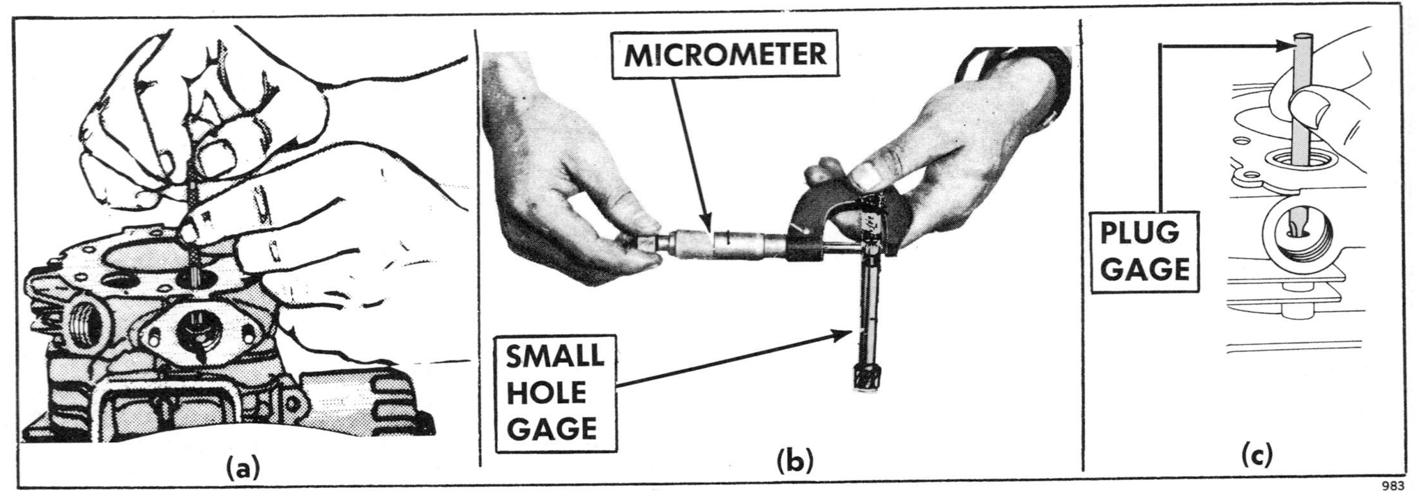

FIGURE 201. Check the valve guide for wear. (a and b) Use a small-hole gage and micrometer, or (c) Use a special go and no-go gage supplied by your manufacturer.

If the valve seat in your engine is damaged and is not replacable, it will need to be refaced. If it can not be refaced properly, it will be necessary to replace the engine block.

6. *Checking the valve-guide bushings (Figure 201).*

   Clean valve guides thoroughly and check for wear. Use a small-hole gage and a micrometer (Figure 201a and b). Or you may use a special go and no-go gage (Figure 201c).

   Check the bushing at the top and at the bottom. Turn the gage $\frac{1}{4}$ turn and recheck.

   See your service manual for specifications.

   As a rule, there should be no more than .005″ difference in the valve stem and bushing diameter. Clearance greater than this allows the valve to move about on the seat and not seal properly.

   On some engines you can get over-sized valve stems which will allow you to ream worn guides to oversize. If you are replacing the valve anyway, get a new standard-size valve guide.

REPAIRING THE VALVE GUIDES

Before repairing the valve guides, you will have to determine the type of valve guide you have and determine which method to use for repairing it (Figure 202). There are three alternatives:

a. Some aluminum engines **have no guide bushing inserts.** If the guide is worn, rebore the valve guide and install replacement inserts (Figure 202a

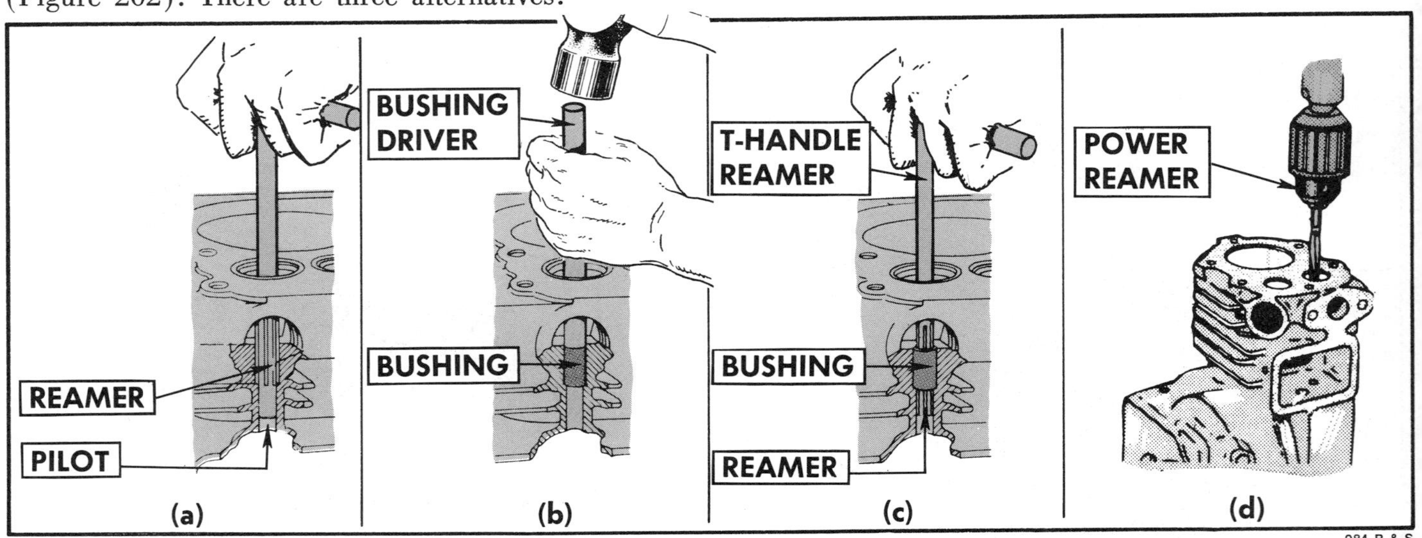

FIGURE 202. Methods of repairing valve guides. (a) Reaming the cylinder block where no bushings are used. (b) Installing new bushings. (c) Reaming the bushing with a T-handle reamer. (d) Reaming bushing with a power-driven reamer.

and b). Then ream the insert to fit standard-size valve stems (Figure 202c and d).

b. If your engine comes equipped **with guide bushing inserts,** they can be replaced when they become worn. Remove the old guide, install a new one, and ream it to fit standard-size valve stems.

c. If your engine is equipped **with or without guide bushing inserts,** it is possible to ream the valve guide bushing to *oversize* and install new valves with oversize stems to fit.

a. If the engine **has no guide bushing and you elect to install a bushing,** proceed as follows:

1. *Protect the engine from metal cuttings.*

   Unless you plan to disassemble completely and clean the engine, do not allow metal chips to fall into the crankcase. Place a *clean cloth* in the valve access well to catch the cuttings.

2. *Select the proper size reamer.*

   Follow the manufacturer's recommendation. Most manufacturers supply a special reamer already sized. They may be purchased by part number.

   The bushing is a press fit, so the hole in the engine block should be .025 mm (.001 in) smaller than the bushing.

3. *Check to see how valve-guide bushing is mounted in the block.*

   On some engines the hole for the valve guide is reamed all the way through to the valve access well. On other engines there is a limit to the depth you should rebore for the valve-guide bushing (Figure 202b). Check your service manual and be sure not to bore further than recommended. The cylinder block wall may get thin below a certain point.

4. *Ream the cylinder block to accommodate the valve-guide bushing (Figure 202a).*

   Make a positive stroke with the reamer. Rotate reamer clockwise with each up and down stroke. Never turn the reamer backwards — you will score the cylinder block. Continue to turn it in a clockwise direction as you remove it. Do not cut any more than necessary when removing the reamer.

5. *Remove metal clippings.*

   Use compressed air and then remove cloth.

CAUTION! Wear protective goggles when you use compressed air on the engine. There is a danger of blowing metal and dirt into your eyes.

6. *Chill the guide bushing in a home-type freezer for an hour.*

   This contracts the size of the guide so it is more easily inserted in the cylinder block. It will become tight when it reaches the temperature of the block.

7. *Drive new guide bushing into cylinder block.*

   Use a brass rod as a punch and guide. If the engine has a flat base and the valve guide is at a right angle to the base, use an arbor press. Otherwise, drive the bushing out with a hammer.

8. *Ream new valve-guide bushings to fit valve stem (Figure 202c and d).*

   Use a reamer that will provide .05 mm to .04 mm (.002 in to .004 in) clearance between the valve stem and the valve guide.

   Make only one cut through with the reamer. Otherwise, you will cut too much.

   If you use a power reamer, operate it no faster than 600 r.p.m. There is danger of overheating.

9. *Check the dimension and the condition of the newly reamed guide (Figure 201).*

   See that there are no burrs or pieces of metal left.

10. *Lubricate the valve-guide bushing with crankcase oil.*

b. If the engine comes **equipped with guide bushings and you wish to replace them,** proceed as follows:

1. *Remove old valve-guide bushings (Figure 203).*

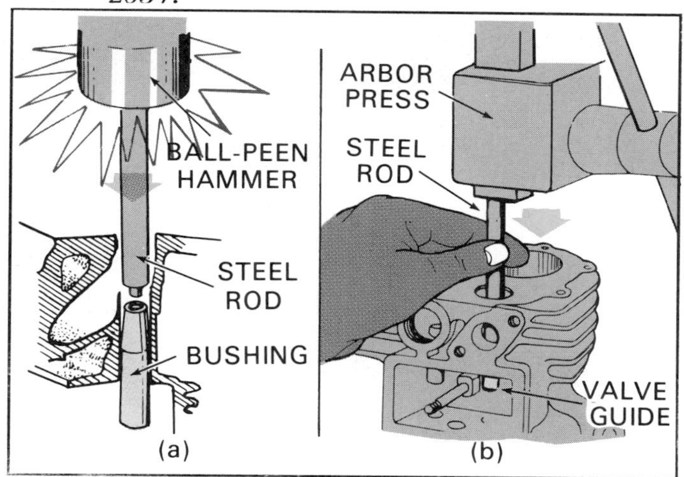

FIGURE 203. Removing valve-guide bushings. (a) Driving valve-guide bushing out with a special punch. (b) Pressing valve-guide bushing out with an arbor press and steel rod.

You may use a steel rod as shown in Figure 203a or an arbor press if your engine has a flat base and if the valve guide is at a right angle to the base (Figure 203b).

Some valve guides are too long to be removed in one piece through the valve access well. In this case, use a press designed to press the bushing toward the cylinder head.

Or, if you have a **cast-iron engine,** you can drive the bushing down part way and break it off. Use a sharp cold chisel to aid in breaking the bushing. Be careful not to damage the cylinder block.

2. *Install new bushings.*

   Follow procedures given in steps 4 through 10, under "a" preceding.

c. If you elect to **ream your old valve guides, or valve-guide bushings, to oversize and install valves with oversize stems,** proceed as follows:

1. *Select reamer of proper size for the oversize valve you have selected.*
2. *Follow steps 8 through 10 in "a" preceding.*

GRINDING (REFACING) THE VALVES

With old valve, first grind and then lap both the valve faces and valve seats (Figure 204). With new valves only lapping is necessary. Lapping is matching the valve face to the valve seat by rubbing the two together with a grinding paste.

1. *Dress the valve grinding wheel if needed.*

   If the grinding wheel is glazed or irregular, dress it according to directions by the grinder manufacturer.

2. *Adjust the valve grinder for grinding the proper face angle (Figure 205).*

   Most valves on small engines are ground at a 45° angle. Some older engines had valves which were ground at a 30° angle. Check the engine specifications before doing any machine work.

3. *Place the valve in the grinder chuck and tighten it securely.*

   Make sure it is straight and does not wobble.

4. *Check to see that the valve face is in proper position to contact the grinding wheel.*

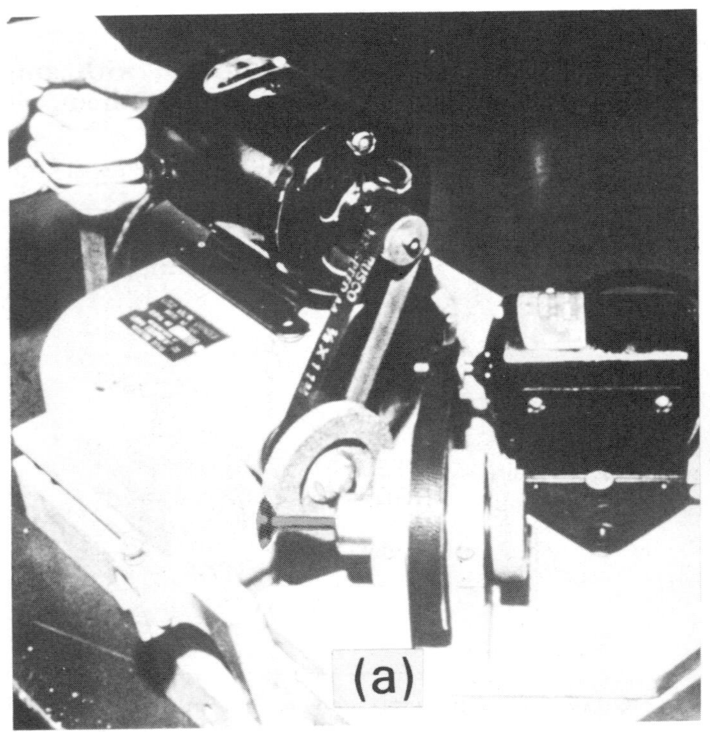

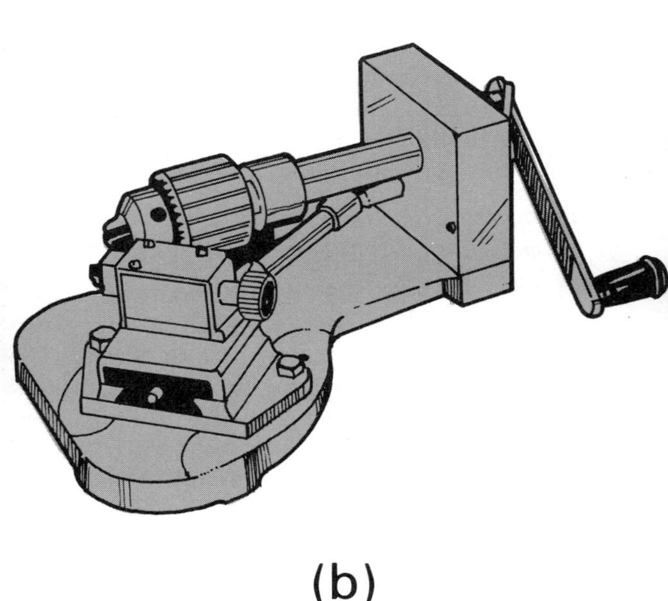

FIGURE 204. Valve grinding is done with a special valve-grinding machine.

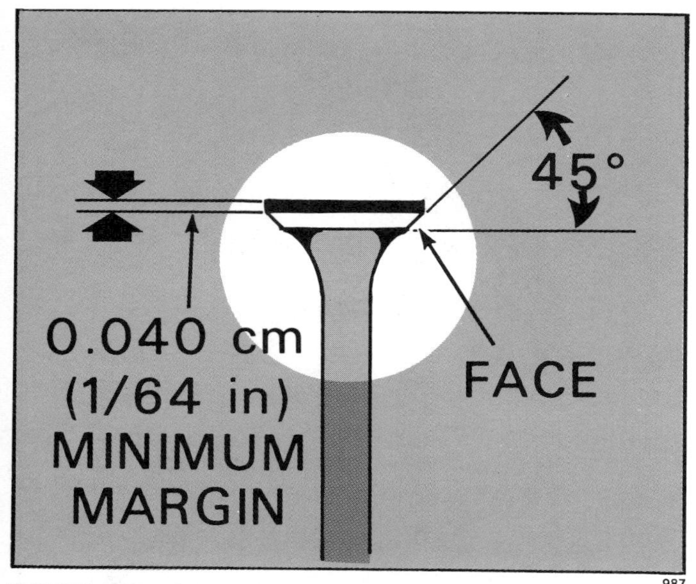

FIGURE 205. Typical specifications for grinding a valve.

Move the valve in and out before starting the grinder while observing its position. Set the machine so the wheel will not contact the valve stem.

5. *Start grinder.*

   Make sure the cutting oil is flowing properly.

6. *Move the valve face up to the wheel and start grinding lightly.*

   Take light cuts and continue until the grinding sound is regular. Do not try to finish the grinding without checking your progress often.

7. *Remove the valve from the grinding wheel.*
8. *Stop the grinder.*
9. *Check the valve face width (Figure 205).*

   You may not be able to grind the valve face adequately before grinding the margin too thin. If the margin (Figure 205) is less than 1/64″, or less than 1/2 the thickness of a new valve margin, discard the valve.

   Check your manufacturer's specifications. If you use a refaced valve with the margin too thin, it is likely to crack and burn.

10. *Clean valve with petroleum solvent.*

GRINDING (REFACING) THE VALVE SEATS

To insure a *tight seal between the valve and valve seat,* it is necessary to grind (or reface) the valve seats whenever you grind the valves. Some valve seats are machined in the cylinder block (Figure 206a) and others are separate inserts (Figure 206b).

If you have a **cast-iron cylinder block,** you may find the *intake-valve* seat is machined in the cylinder block and the *exhaust-valve* seat is a moly-nickel, chrome-steel (Stellite) insert.

It is possible on some cast-iron engines without inserts to install them if the seats become badly worn.

When the *valves and valve seats have been reworked a number of times,* the valve rests too deeply in the block. This condition requires that too much metal be removed from the end of valve stem to get the proper valve-tappet clearance. Also the valve spring tension is weakened by the lengthening of the stem. A good gage of when a valve seat has been cut too

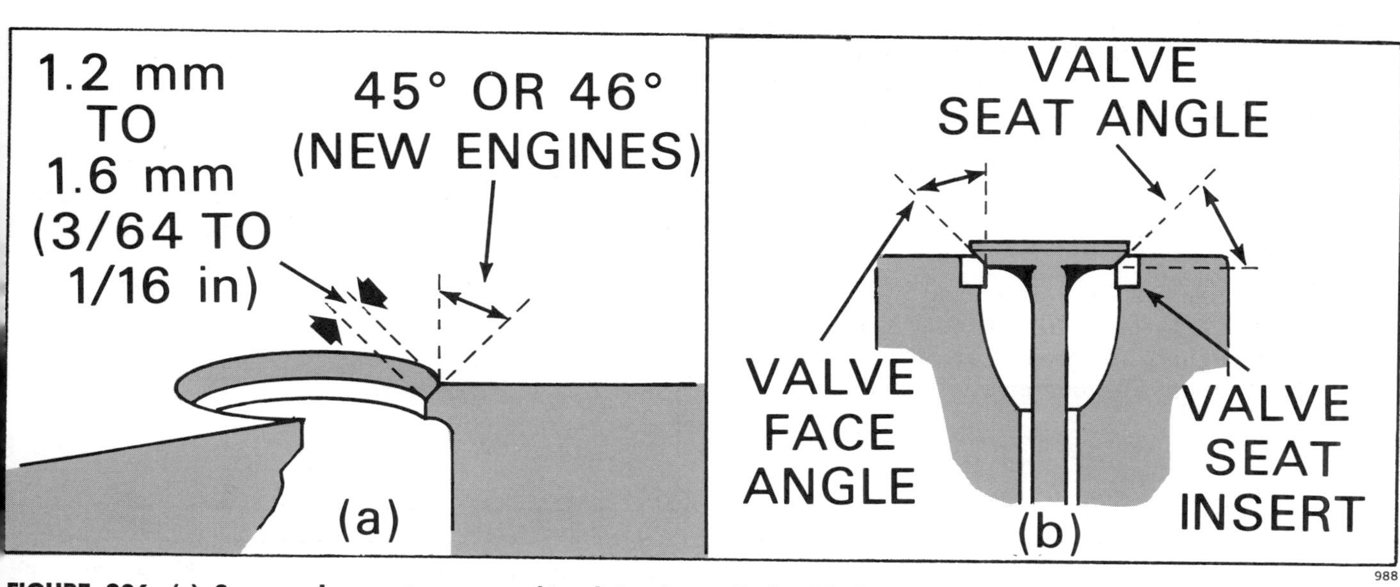

FIGURE 206. (a) Some valve seats are machined in the cylinder block. (b) Others are separate inserts.

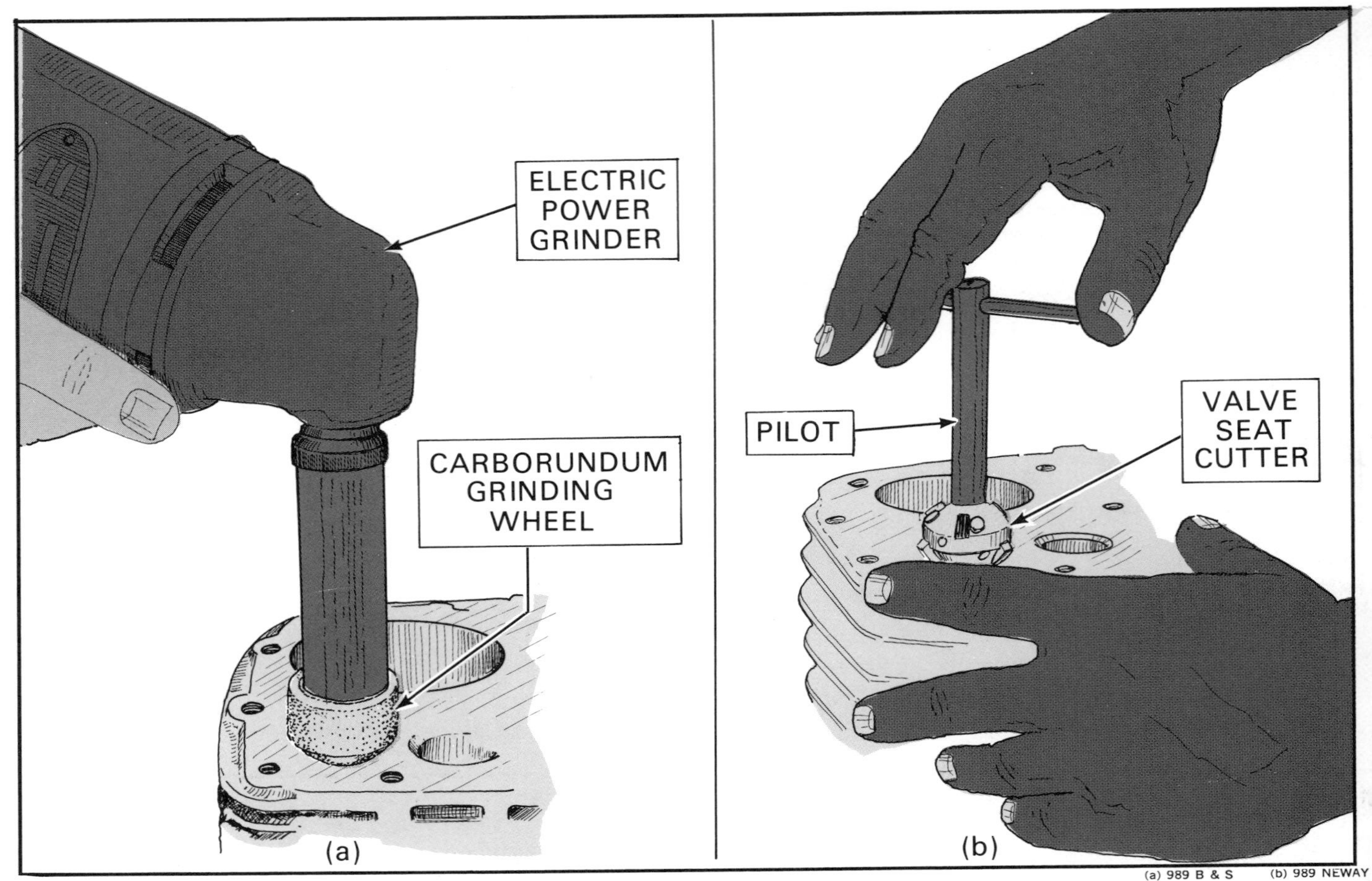

FIGURE 207. Two methods for refacing valve seats. (a) Portable electric grinder equipped with carborundum grinding wheel and pilot. (b) Special valve-seat cutter, pilot and T-handle.

deeply into the block is to compare the length of the valve stem with a new one. If over half of the metal from the lock groove to the end of the valve has been removed to get the proper clearance between the tappet and the valve stem, install a new valve-seat insert.

There are two tools used for servicing valve seats. They are (1) **electric-power grinder** with a carborundum grinding wheel (Figure 207a), and (2) a **special valve-seat cutter** (Figure 207b).

a. If you use the **electric-power grinder,** proceed as follows:

1. *Select a grinding wheel with the proper angle.*

   Most valve seats are ground at a 45° angle. Some 44°; some 46°. Others vary. Some older engines are ground at 30°. Check the specifications in your engine manual.

2. *Be sure the wheel is dressed properly.*

3. *Install the pilot in the valve guide.*

   This is why it is important that the valve guides be serviced before grinding the valve seats.

4. *Grind the valve seats.*

   Cut away all oxidized (discolored) metal. Grinding action is fast; so be careful not to do too much. Keep the grinder straight.

5. *Check width of valve seat (Figure 208).*

   It should not be over .16 cm ($^{1}/_{16}$ in) wide, or

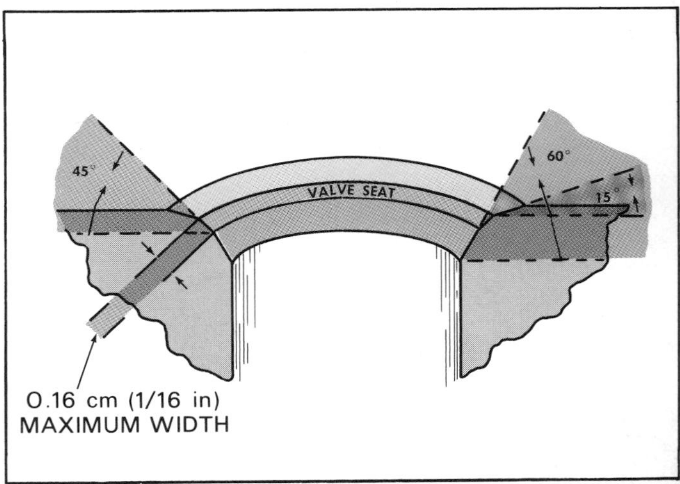

FIGURE 208. The seat width can be reduced by grinding both at the top and the bottom with the proper type grinding wheels.

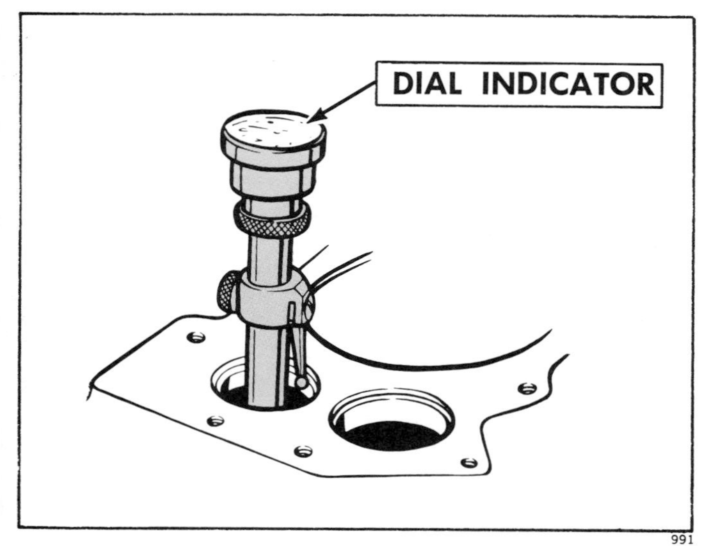

FIGURE 209. Checking the valve-seat trueness with a dial indicator.

the valve will not seat properly. The minimum seat width is from .08 cm to .12 cm ($^1/_{32}$ in to $^3/_{64}$ in). If the seat angle is 45° and the seat width is over .16 cm ($^1/_{16}$ in) wide, narrow it down with a 15° to 30° angle wheel at the top, and a 60° to 75° angle wheel at the bottom (Figure 208).

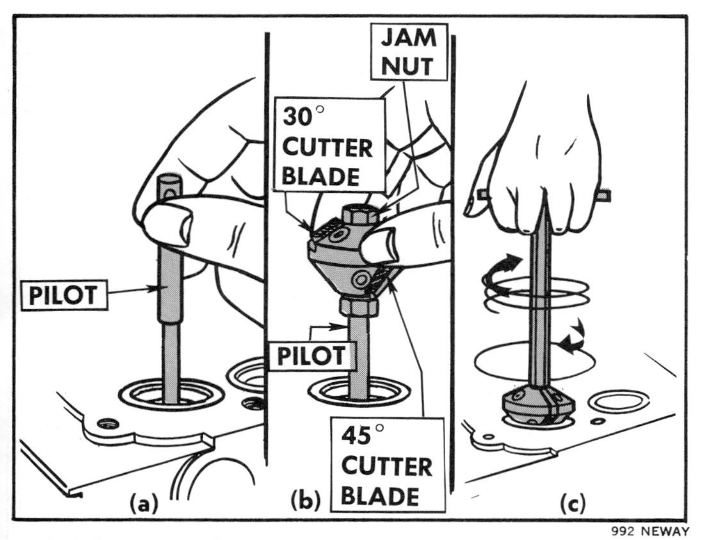

FIGURE 210. Refacing valve seats with a special cutter. (a) Installing the cutter pilot in the valve guide. (b) Installing the cutter head on the pilot. It is held by a jam nut. (c) Cutting the valve seat.

6. *Smooth burred edge with the original seat-grinder wheel.*
7. *Check the valve in the seat.*

   Apply prussian blue to the valve, and insert it in the engine. Turn the valve with a slight pressure of your thumb on the valve head ¼ turn. Remove the valve, and check to see if the blue has transferred to all areas of the valve seat. A *slight imperfection* may be taken out by lapping.

   If you are not able to shape the seat properly, replace the valve seat with a new insert if possible. See "Replacing Valve Seats," next heading, for procedures.

   It is possible to check trueness of the valve seat with a dial indicator (Figure 209).
8. *Lap the valve to the seat.*

   Refer to "Lapping the Valves," page 204.

b. If you use a **valve-seat cutter,** proceed as follows:

1. *Determine the angle of your valve seat.*

   See the engine specifications.
2. *Install the pilot in the valve guide (Figure 210a).*
3. *Install the cutter head on the pilot (Figure 210b).*
4. *Install the T-handle on the cutter head and cut the valve seat (Figure 210c).*

   Press down on T-handle and turn in clockwise direction.
5. *Check valve seat for condition and width.*

   If it requires over .16 cm ($^1/_{16}$ in) of an inch width (Figure 208) to recondition the valve seat, turn the valve cutter over and trim the width with the 30° angle cutter (45° seats). The minimum seat width is from .08 cm to .12 cm ($^1/_{32}$ in to $^3/_{64}$ in).
6. *Lap the valve to the seat.*

   Refer to "Lapping the Valves," page 204.

REPLACING THE VALVE SEATS

If your engine is aluminum and you plan to rebore the cylinder, do it after replacing the valve-seat inserts. There is a danger of distorting the rebored cylinder wall when installing the valve-seat insert.

Most engines have provisions for replacing valve seats if they cannot be reground properly. All **aluminum engines** have **valve-seat inserts.**

Cast-iron engines that do not have inserts can be counterbored (bored larger) and inserts installed. Some cast-iron engines have an exhaust-valve seat insert, and the intake-valve seat machined in the cylinder block.

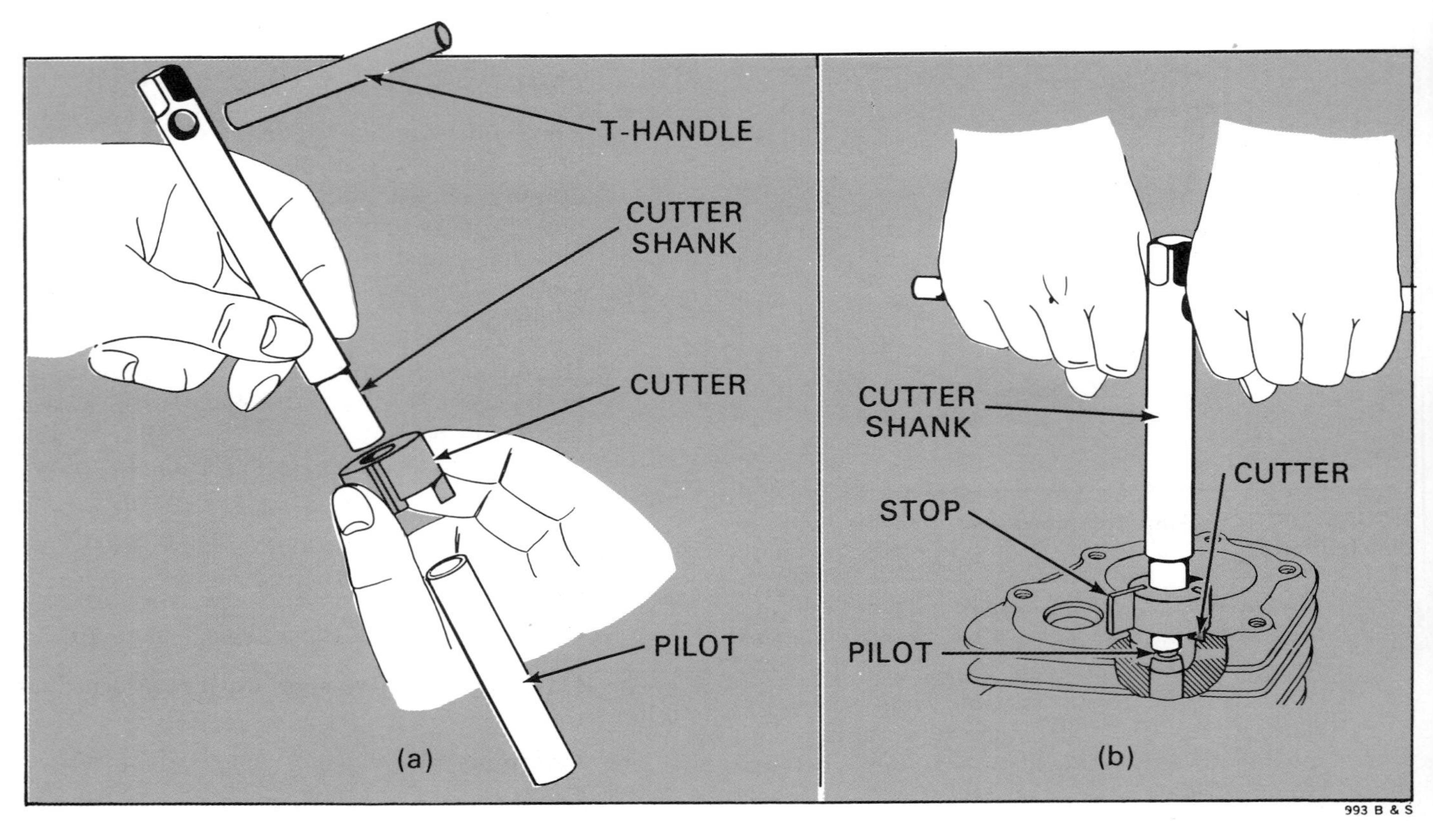

FIGURE 211. Counterboring for installing cast iron engine valve-seat inserts. (a) Tools required. (b) Counterboring.

a. If the **valve seat is machined in the cylinder block** (cast-iron engine), proceed as follows:

1. *Select the proper size of seat insert, cutter shank, cutter and pilot for your engine (Figure 211a).*

   Special tools are prescribed by your manufacturer. The counterbored hole is slightly smaller .025 mm to .05 mm (.001 in to .002 in) than the outside diameter of the new insert.

2. *Insert the pilot in the valve guide.*

   The pilot must be snug in the guide or the cutter will cut oversize.

3. *Protect the engine from metal cuttings.*

   Place cloth in valve-access well. Be certain openings to crankcase are covered.

4. *Counterbore to the depth of the stop on the cutter or according to specifications for your engine.*

   There is a critical boring depth for the new insert. You will have to get this from the engine specifications.

   If a drill press or a hand drill is used, keep the speed down to 300 r.p.m. for proper cutting and to prevent overheating.

   Be very careful not to allow the cutter to wobble, or the hole will be too large for the insert.

5. *Clean away all metal cuttings.*

   Use compressed air. If cleaning is done before removing the pilot, this will prevent metal chips from getting into the valve guide.

6. *Remove boring tools.*

7. *Measure counterbore depth.*

   Compare depth with the depth of the valve seat.

   The top of the valve insert should seat just below the cylinder-head gasket level.

8. *Chill the insert in a freezer for an hour.*

   This chilling is to reduce the insert size for ease of installation and to help insure a tight fit when both the cylinder block and valve insert reach the same temperature.

9. *Install the insert.*

   Install with a valve-seat insert driver (Figure 212a).

   Be sure the seat is right side up.

   Do not pound heavily on the insert. There is danger of damaging the block.

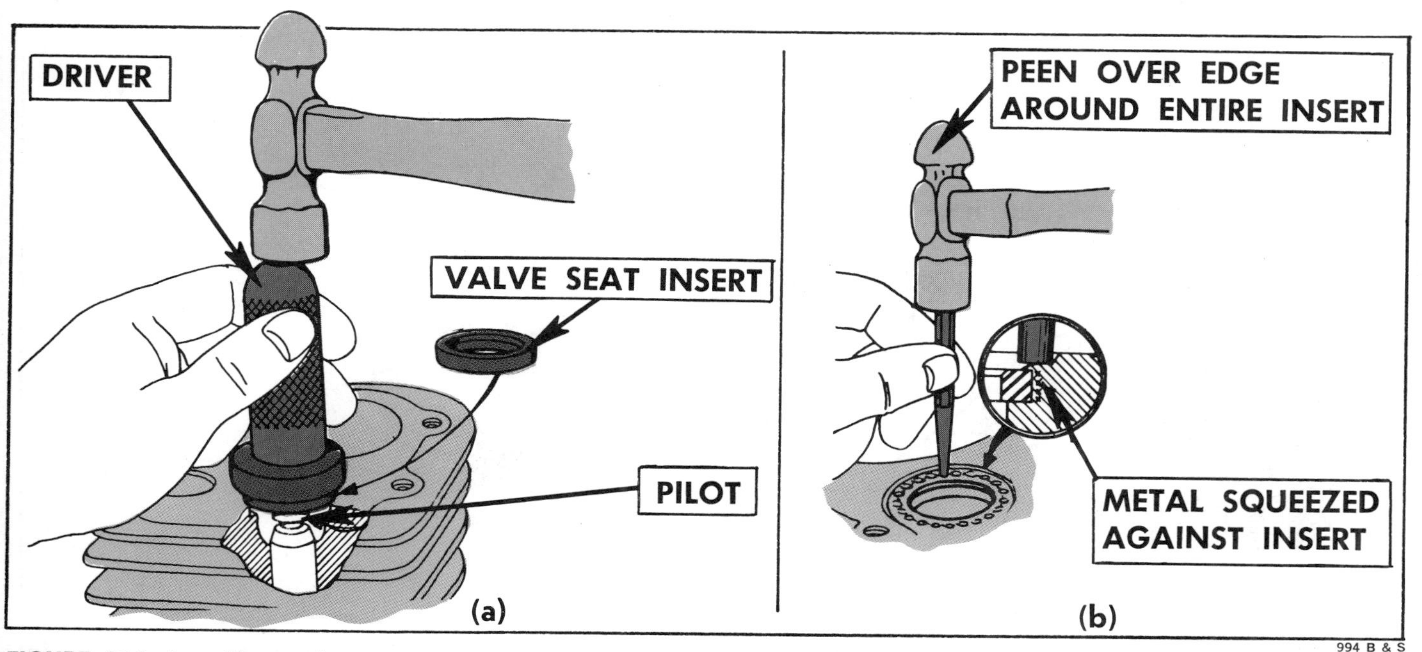

FIGURE 212. Installing valve-seat inserts. (a) Driving the insert into the cylinder block. (b) Peening the cylinder block over the valve-seat insert.

Do not pound the pilot onto the valve tappet. You can damage the camshaft gear.

10. *Peen the metal over the insert (Figure 212b).*

    This is to wedge the insert so as to further tighten it. Alternate from side to side when peening to prevent shifting the insert to one side.

    NOTE: This procedure is not recommended by some manufacturers. They say that peening pushes the seat away from the block and causes it to overheat.

11. *Grind insert lightly.*

    Refer to "Grinding (Refacing) the Valve Seats," page 199.

b. If your **engine already has valve-seat inserts** and they must be replaced, proceed as follows:

1. *Remove the peened or rolled metal lip from over the top of the insert if necessary.*

   Use counterbore cutter (Figure 211).

2. *Remove the old insert.*

   Note three methods shown in Figure 213.

   Take care not to damage the cylinder block.

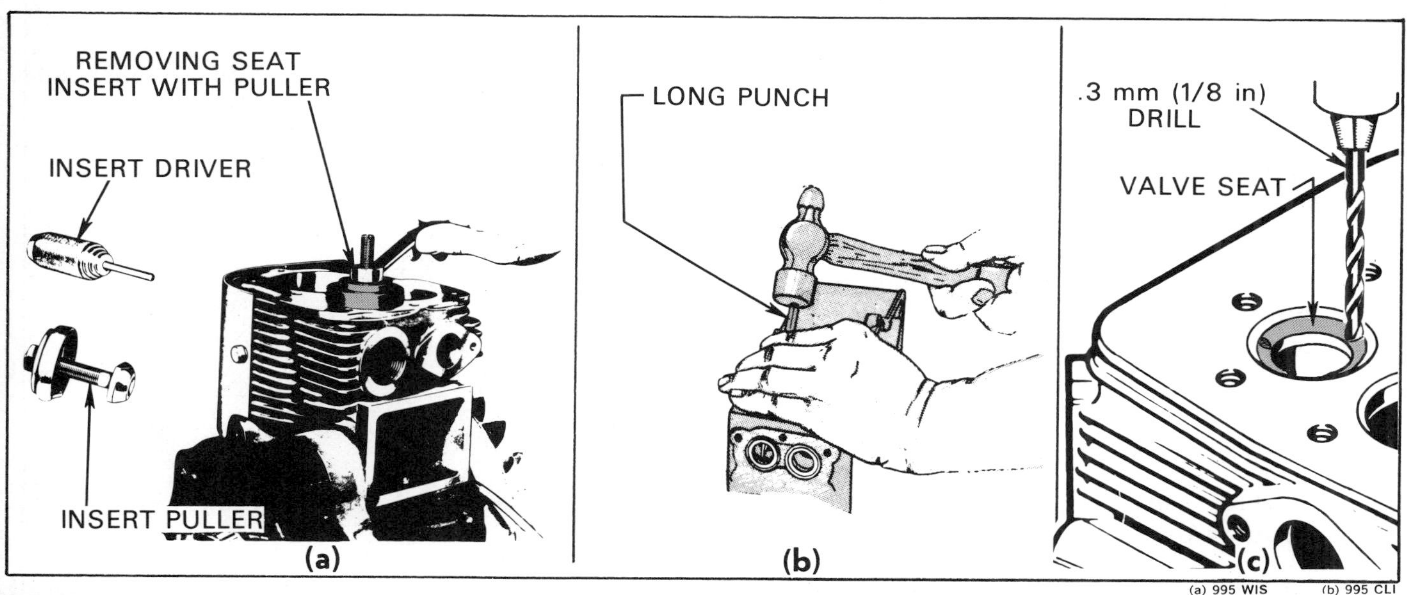

FIGURE 213. Removing old valve-seat inserts. (a) Using a special puller. (b) Using a punch. (c) Drilling holes in insert for breaking it out.

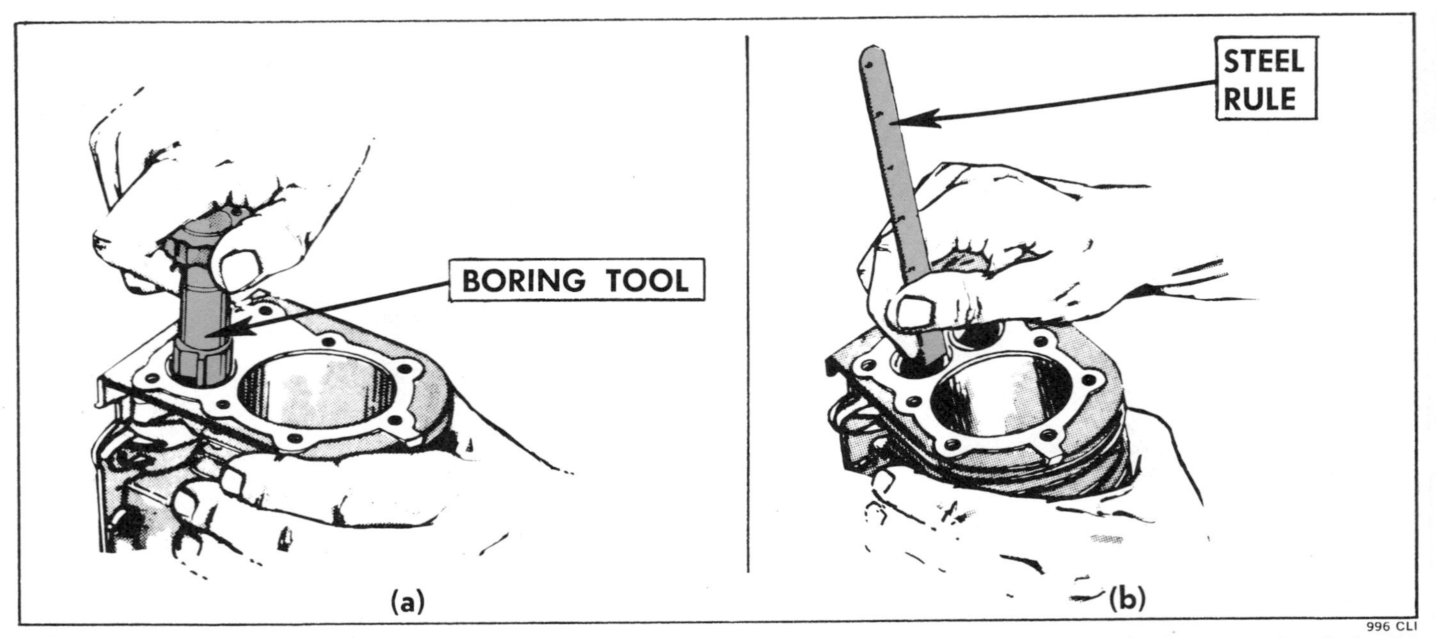

FIGURE 214. Preparing cylinder block for installing of a new valve-seat insert. (a) Counterboring oversize hole (b) Measuring the depth of the counterbore.

3. *Clean counterbore.*

   Remove dirt and metal chips. Use rag dampened with petroleum solvent. Scrape away deposits with a knife.

4. *Check hole diameter.*

   It should be .025 mm to .05 mm (.001 in to .002 in) smaller than the outside diameter of the insert.

   Over size inserts are available.

5. *Rebore oversize hole if necessary.*

6. *Install new insert.*

   Follow steps 8 through 11, under "a" preceding.

LAPPING THE VALVES

Some manufacturers recommend that you "lap" the valves to the seats anytime you have the valves out of your engine. To assure a good fit, lapping is recommended by some for new valves and seats and even for valves that appear to be in good condition. Proceed as follows:

1. *Apply thin coat of lapping compound (carborundum paste) to the valve face.*

   If you *have serviced the valve seats,* or if you *are installing new valves,* use **coarse-grain lapping compound** for fast action; then finish with a **fine-grain paste.** Wipe away the coarse compound before finishing with the fine-grained paste.

2. *Insert valve into the cylinder block.*

   Be sure to use the proper valve. Intake and exhaust valves are of different sizes. The intake valve is usually larger than the exhaust valve.

3. *Attach special valve-lapping tool.*

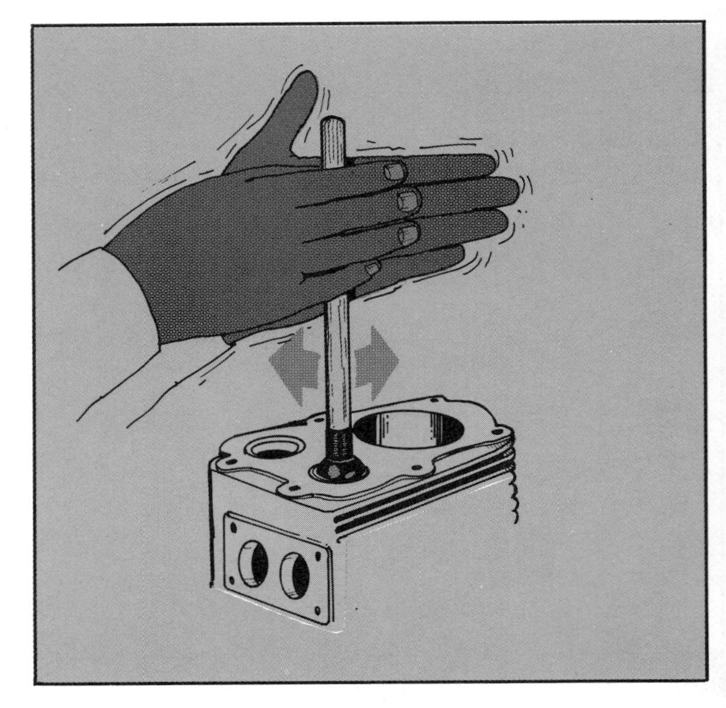

FIGURE 215. Lapping valves.

It has a suction cup to hold the valve and a spring-loaded cap; so you can apply constant and uniform pressure.

4. *Rotate valve while applying pressure (Figure 215).*

5. *Work back and forth and generally to the right (Figure 216).*

   Clean away all lapping compound with petroleum solvent. Apply prussian blue to the valve face.

   Install the valve and turn it on the seat with your thumb. Remove the valve and observe how the bluing is distributed on the valve seat. When properly lapped, the prussian blue should be distributed evenly around the center of the valve seat. This means the valve is seating at all points.

   An alternate method for checking the valve seat for an unmounted engine is as follows: (1) Install the valve and valve spring. (2) Adjust the valve-tappet clearance. (3) Turn the cylinder block upside down, pour gasoline into the valve port, and see if any gasoline leaks out. (4) If the valve leaks, remove and relap.

6. *Clean all compound from the valve and the engine.*

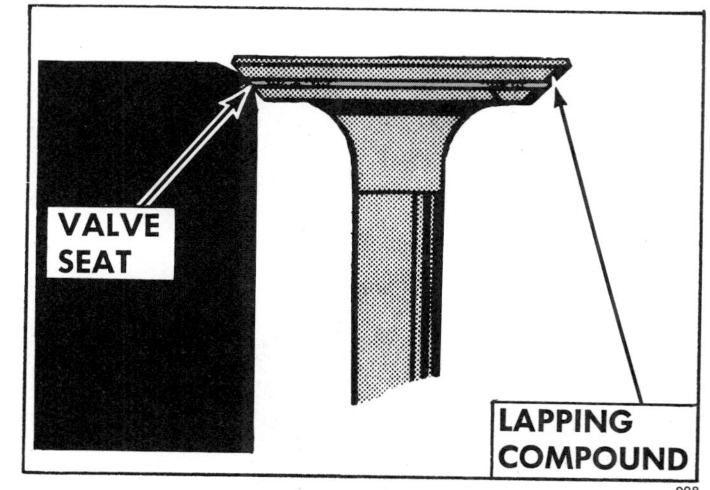

FIGURE 216. Valve face should contact the seat evenly and at the center.

   If the lapping compound is a **petroleum-solvent base,** clean with petroleum solvent.

   If the lapping compound is **water base,** use water.

   Only a small amount of lapping compound left in your engine will get into the lubricating oil and grind away vital parts.

7. *Install valve.*

   See procedures under next heading.

ADJUSTING THE VALVE-TAPPET CLEARANCE

There are two methods for checking and adjusting the valve clearance: (1) Grinding a small amount from the end of the valve stem, or (2) using adjustable tappets.

Some manufacturers recommend that you adjust the valves before installing the valve springs. Others object because they say you cannot get an accurate check without spring tension on the valve.

a. If your engine has **no valve-tappet adjusting screw,** proceed as follows:

1. *Apply crankcase oil to the valve stems, valve guides, springs and spring retainers.*

   Be sure they are clean and have no dirt or grinding compound on them. The oil makes installation easier, and provides lubrication when the engine is started.

2. *Place the valve in the cylinder block.*

3. *Turn the crankshaft to top-dead-center, compression stroke (piston at the top and both valves closed.)*

   You may turn the crankshaft until the valve in its highest position (open); then make one complete revolution.

4. *Check for the recommended valve-tappet clearance in your operator's manual.*

   See "Checking the Valve-Tappet Clearance," page 189.

5. *Install valve spring.*

   If the manufacturer of your engine recommends installing the valve spring before checking the tappet clearance, add these steps:

   (1) *Place the valve spring, valve-spring seat and retainer in the valve access well.*

   On some engines the exhaust-valve spring is heavier than the intake-valve spring. Be sure to install the proper spring.

   (2) *Insert the proper valve in its position.*

   (3) *Compress the valve spring (Figure 217).*

   See Figure 193 for other types of valve-spring compressors.

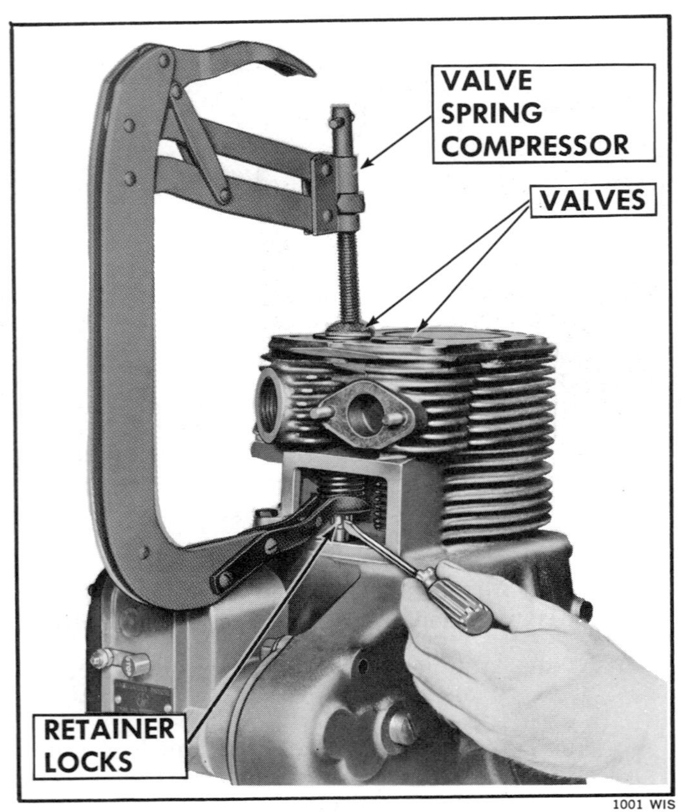

FIGURE 217. Installing a valve with a valve-spring compressor.

(4) *Install the valve-spring retainer locks.*

Use needle-nose pliers to install pin-type locks (Figure 194b). A little multi-purpose grease applied to split-type (Figure 194c) retainer locks will help you install them.

Be sure the retainer-combination lock (Figure 194d) is centered on the valve if you have this type.

6. *Check valve-tappet clearance (Figure 218).*

Use the proper thickness gage. Press the valve down firmly and insert the gage (Figure 219).

Note the amount needed to be removed from the valve stem before removing the valve. For example, if you have .05 mm (.002 in) clearance and you need .18 mm (.007 in) clearance, you will need to remove .13 mm (.005 in) from the valve stem.

7. *Remove the valve.*

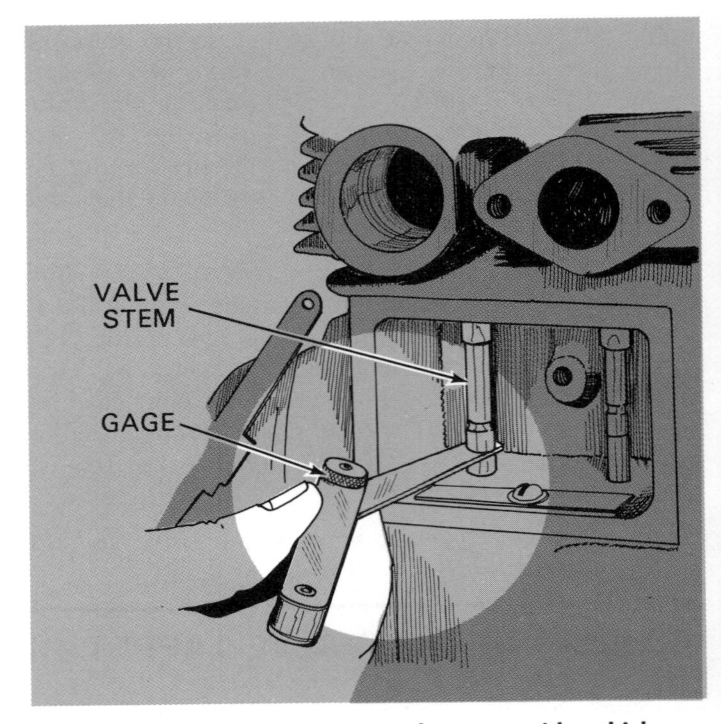

FIGURE 219. Check valve-tappet clearance with a thickness gage.

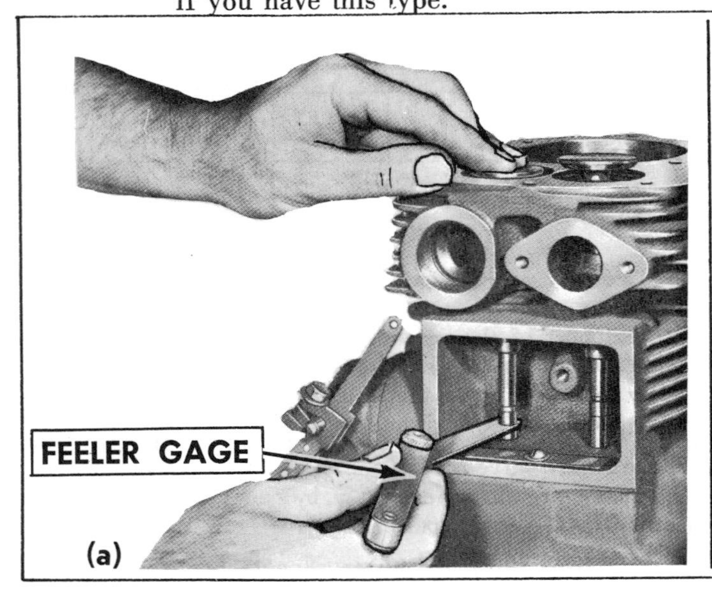

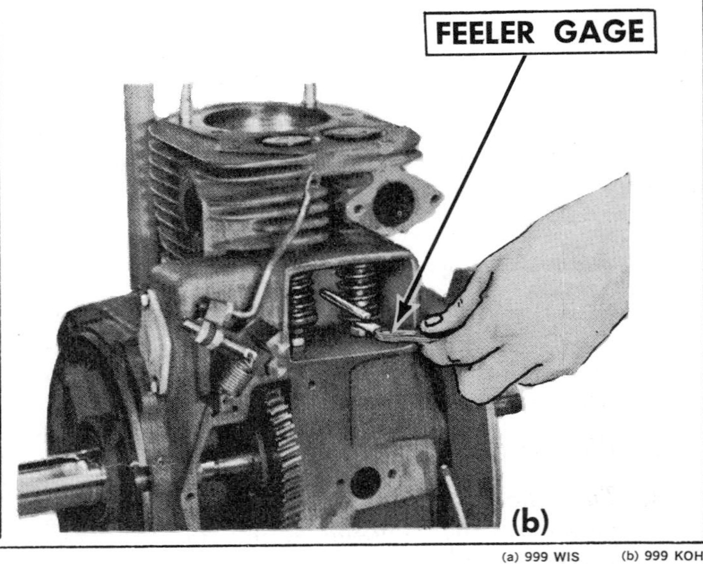

FIGURE 218. Checking the valve-tappet clearance. (a) Checking the clearance before installing the valve spring. (b) Checking the clearance after installing the valve spring.

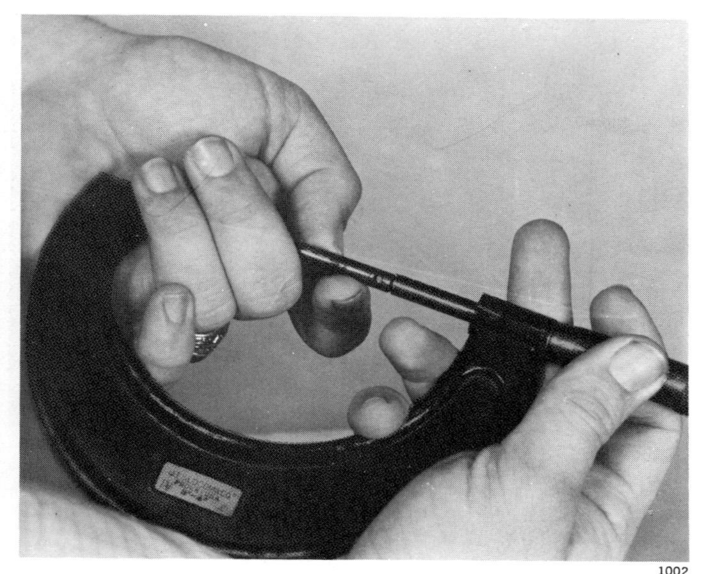

1002

FIGURE 220. Measuring valve length with a micrometer.

8. *Measure the length of the valve with a large micrometer (Figure 220).*

   With the micrometer you can determine how much to grind away without replacing the valve assembly each time for a measurement. If your valve is 7.64 cm (3.010 in) and you need to remove .13 mm (.005 in), the final length of the valve should be 7.63 cm (3.005 in).

9. *Grind required amount from valve stem.*

   Grind it square. Grind a little at a time. Do not overheat.

10. *Reinstall valve assembly in cylinder block.*

    Note steps 2 through 5.

11. *Recheck clearance.*

    If too short, reface and lap.

b. If your engine has **valve-tappet adjusting screws,** proceed as follows:

1. *Position the camshaft for valve adjustment.*

   Follow steps 1 through 3 under "a" preceding.

2. *Install valve-tappet adjusting screws (Figure 187b).*
3. *Install the valve assembly.*

   Follow steps 2 through 5 under "a" preceding.

4. *Adjust valve-tappet clearance by turning the valve-tappet adjusting screws.*
5. *Tighten jam nuts.*

INSTALLING THE CYLINDER HEAD

There are two points to keep in mind when tightening the cylinder-head cap screws. They are (1) **the sequence in which the cap screws are to be tightened,** and (2) **the amount of torque to apply.** The recommendations vary so much among different makes and models of engines that it is impossible to give a general recommendation.

The recommended *sequence of cap screw tightening* is given to prevent warping the cylinder head by uneven tightening. Usually you are directed to tighten alternate cap screws back and forth across the cylinder head (Figure 221).

The *amount of torque to apply* depends primarily on the size of cylinder-head cap screws, the metal composition of the cylinder block (aluminum or cast iron), the depth of the threads, and the number of cap screws used. Most manufacturers give torque in inch-pounds (Figure 222). The torque may be as low

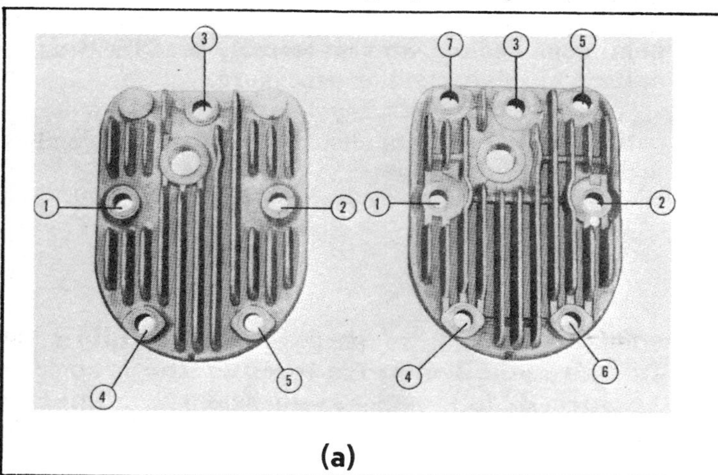

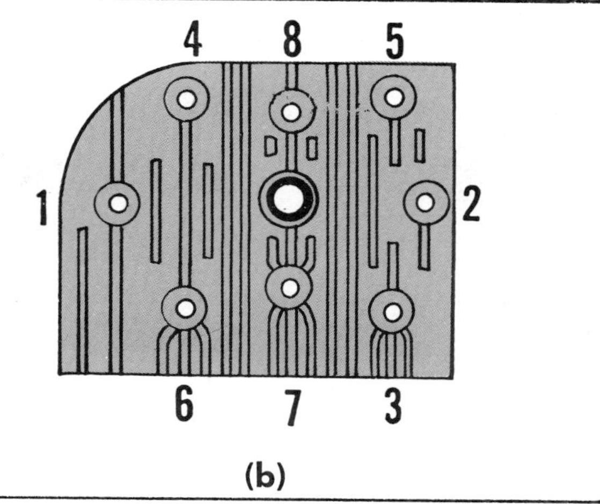

(a) 1003 CLI (b) 1003 B & S

FIGURE 221. Three examples of recommended sequence for tightening cap screws on cylinder heads.

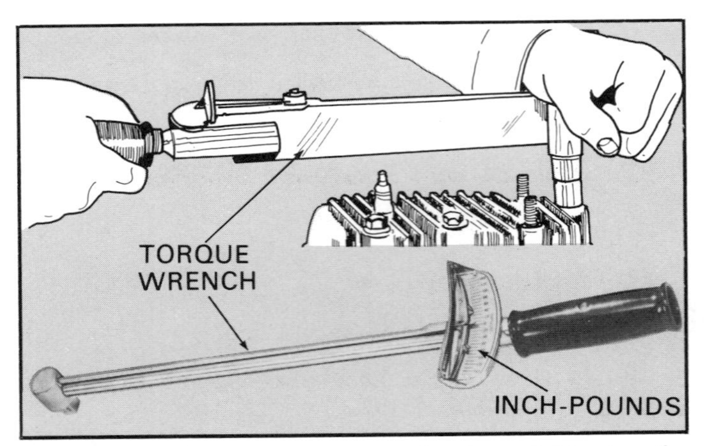

FIGURE 222. Use a torque wrench when tightening the cylinder-head screws.

as .18 N·m (50 in·lbs) on some 2-cycle engines and .5 N·m to 1.7 N·m (150 in·lbs to 480 in·lbs) on 4-cycle engines. When you use a torque wrench, tighten the cap screws in at least three stages.

Some manufacturers do not give torque recommendations but instead recommend tightening the cap screws by feel. For example, tighten the cap screws finger tight; then turn each 1/4th turn in the recommended sequence. The general procedure is as follows:

1. *Check piston, cylinder, valves and cylinder head for cleanliness.*
2. *Install new head gasket.*

   Be sure the gasket is the proper one for your engine.
3. *Place cylinder head on cylinder block (Figure 223).*
4. *Install all cylinder-head cap screws snugly. Do not tighten!*

   Remember that some cylinder-head cap screws are longer than others. Replace them in their original positions.

   Clean threads on cap screws and lubricate before installing.

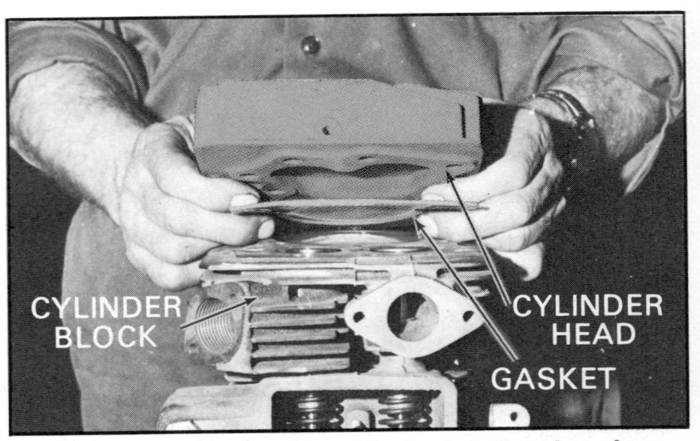

FIGURE 223. Installing the cylinder head and gasket.

If **threads are damaged,** you can replace them with a Heli-coil thread (Figure 223A).

5. *Tighten the cap screws according to procedures given by your manufacturer.*

   Follow the proper sequence and torque properly. If the cap screws are not torqued properly, the cylinder head will warp.

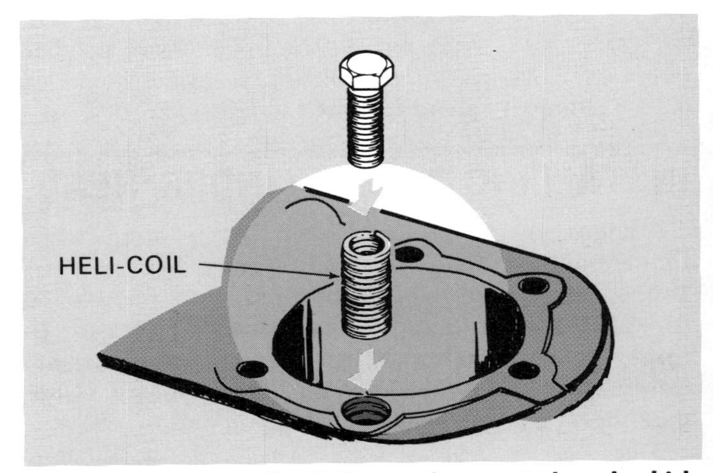

FIGURE 223A. A Heli-coil is a replacement thread which comes in various sizes. Drill the damaged thread out, retap the hole and install the Heli-coil. Follow instructions from Heli-coil for the proper size drill, tap, and coil.

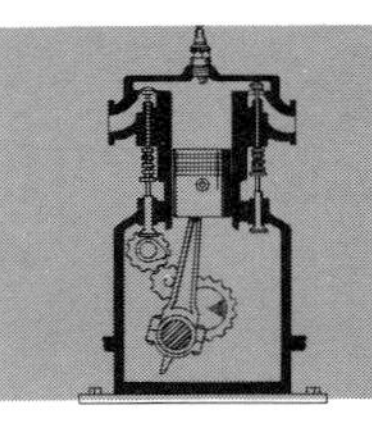

B. REPAIRING VALVES ON 2-CYCLE ENGINES

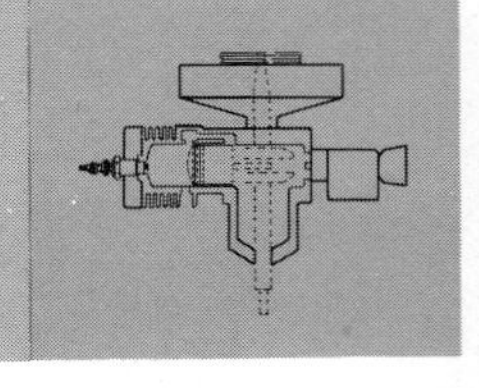

Two-cycle engines have no valves in the combustion chamber. The intake and exhaust ports are located in the cylinder wall, and their opening and closing are controlled by the piston.

There is a valve between the carburetor and the crankcase. It has two purposes: (1) **to allow the fuel-air-and-oil mixture to enter the crankcase (Figure 224a)**, and (2) **to seal the crankcase on the downward stroke of the piston (Figure 224b).**

The fuel mixture that is displaced by the piston

on the downward stroke enters the combustion chamber under pressure.

Maintaining and repairing valves on 2-cycle engines are discussed under the following headings:

1. Types of valves and how they work.
2. Tool and materials required.
3. Checking the reed valve for proper operation.
4. Removing and replacing the reed-valve assembly.

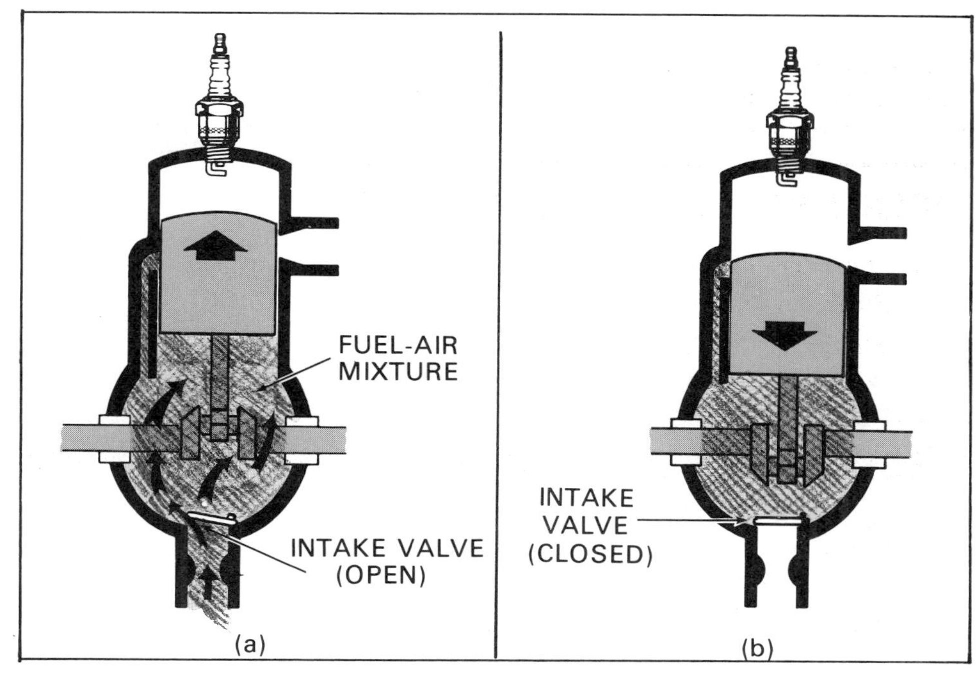

FIGURE 224. Functions of a fuel valve on a 2-cycle engine. (a) Valve opens to allow fuel-air-and-oil mixture to enter the crankcase. (b) Valve closes on the downward stroke of the piston to seal the crankcase.

TYPES OF VALVES AND HOW THEY WORK

Most valves used on small, 2-cycle engines are of the **reed type** (Figure 225). They are made of **spring steel, phenolic or plastic.**

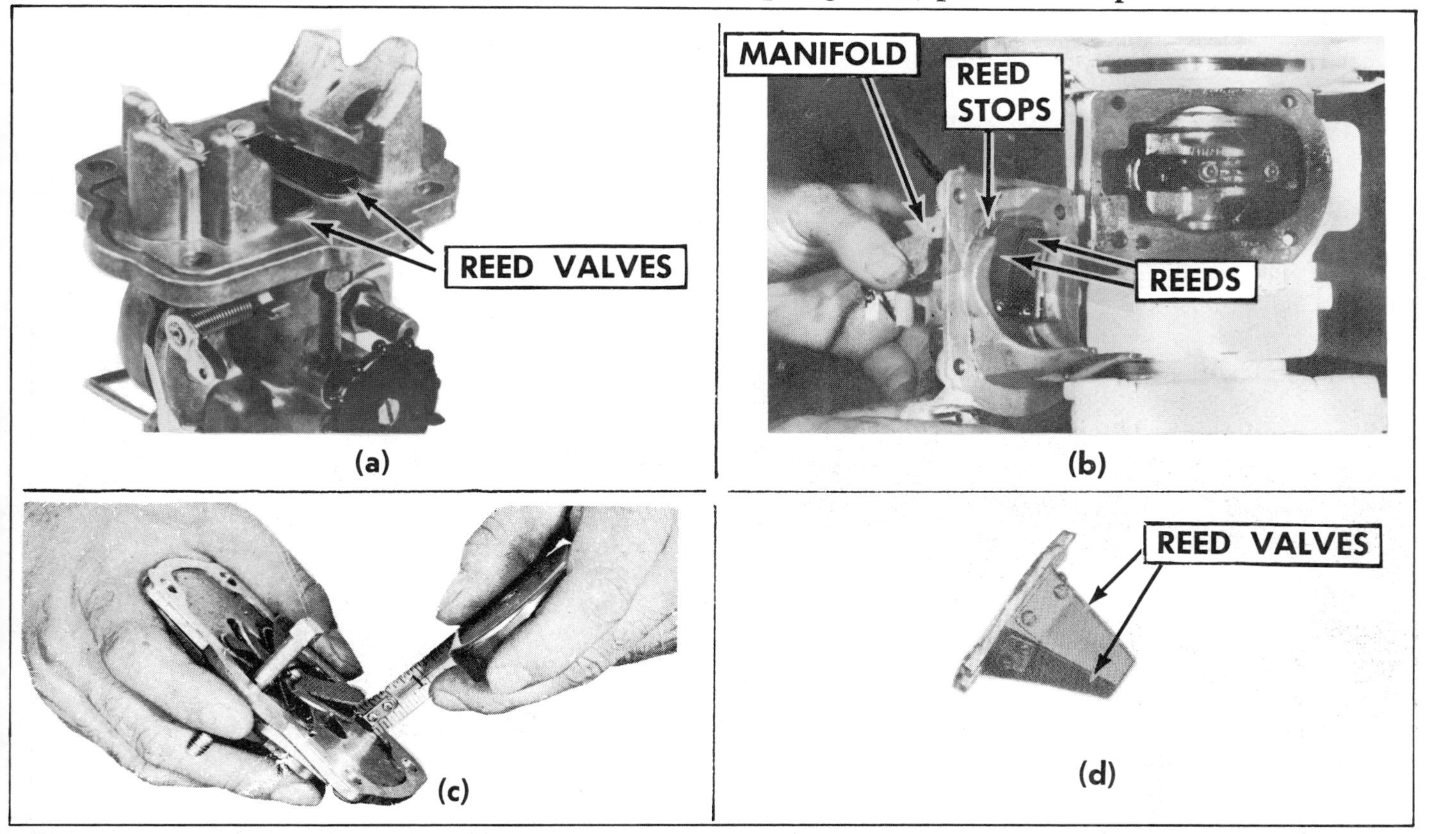

FIGURE 225. Types of reed-valve arrangements used on 2-cycle engines. (a and b) Twin reeds. (c) A 4-reed cluster. (d) Pyramid-type 4-reed cluster.

The valves are opened by the partial vacuum created in the crankcase on the upward stroke of the piston (Figure 224a). They are closed by the built-in spring tension in the reed and also by the pressure the crankcase developed by the downward stroke of the piston.

TOOLS AND MATERIALS NEEDED

1. Open-end wrenches — 3/8″ to 9/16″
2. Steel rule — 6″
3. Slot-head screwdriver — 6″
4. Punch — 5/16″
5. Carburetor cleaning solvent
6. Clean rags
7. Wooden scraper

CHECKING THE REED VALVE FOR PROPER OPERATION

If your engine is **hard to start and spits back through the carburetor when starting,** check the reed valve. Proceed as follows:

1. *Remove the carburetor air cleaner.*

   Check for signs of blowback in the carburetor. This will show as carbon deposits in the carburetor manifold.

2. *Turn the engine by hand quickly and listen to the carburetor.*

   The engine must be turning at cranking speed or above for enough compression to build up in the crankcase to close the reed valves. If you hear air passing backwards through the carburetor, it indicates a leaking valve. The reed may be bent, broken or out of adjustment, or the reed-valve seat may need cleaning.

REMOVING AND REPLACING THE REED-VALVE ASSEMBLY

1. *Remove the reed plate.*

   On some engines it is a simple matter to get to the reed plate. All you have to do is to remove it along with the carburetor (Figure 226a).

   On others it is necessary to remove the carburetor first (Figure 226b).

2. *Disassemble the reed-valve assembly (Figure 227).*

   Check reed for cracks and smoothness.

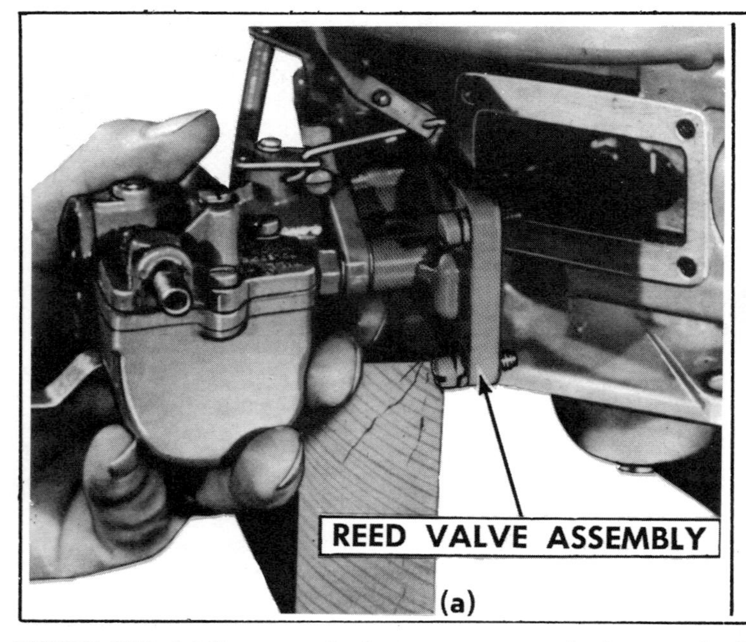

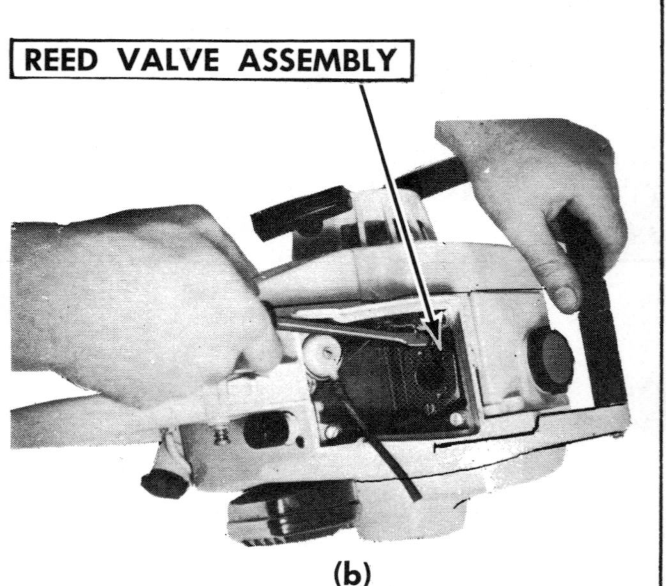

(a) 1008 JAC

FIGURE 226. (a) Some reed plates are removed along with the carburetor. (b) Others cannot be removed until after the carburetor is removed.

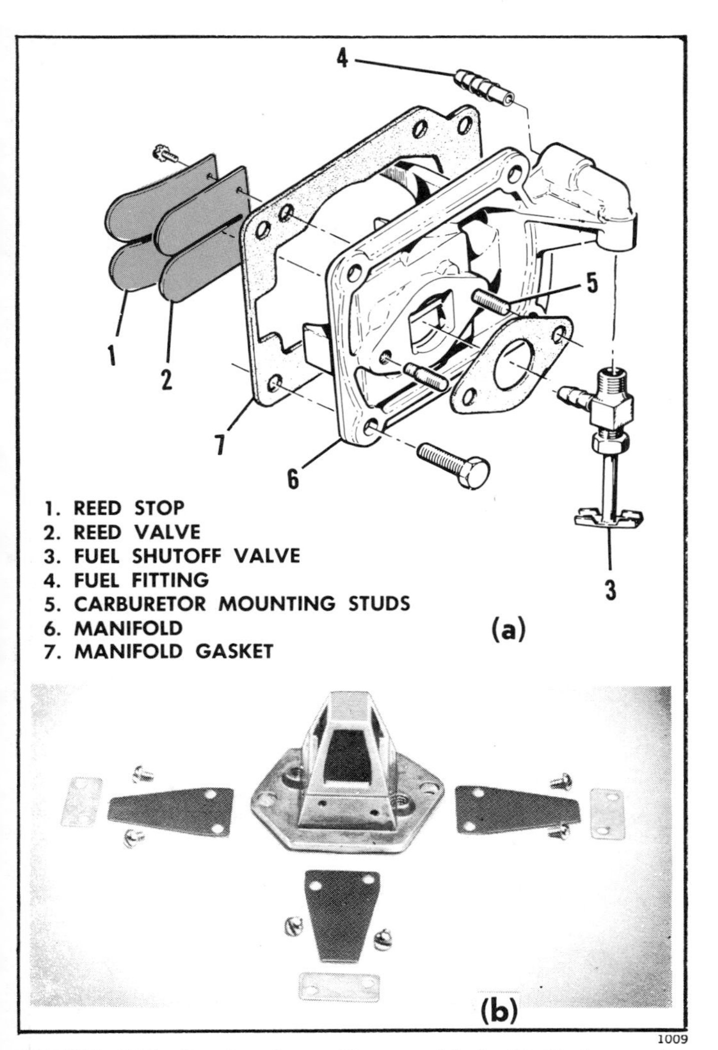

FIGURE 227. Reed valves disassembled. (a) Twin reeds. (b) Four-reed cluster.

3. *Clean and inspect parts (Figure 228).*

4. *Check reed-stop plate adjustment if installed (Figure 229).*

   Some leaf-type reeds have a stop plate to prevent the reed from opening too far. If they are bent or distorted, they should be replaced.

5. *Reassemble reed-valve assembly.*

6. *Install reed plate.*

   Use new gasket.

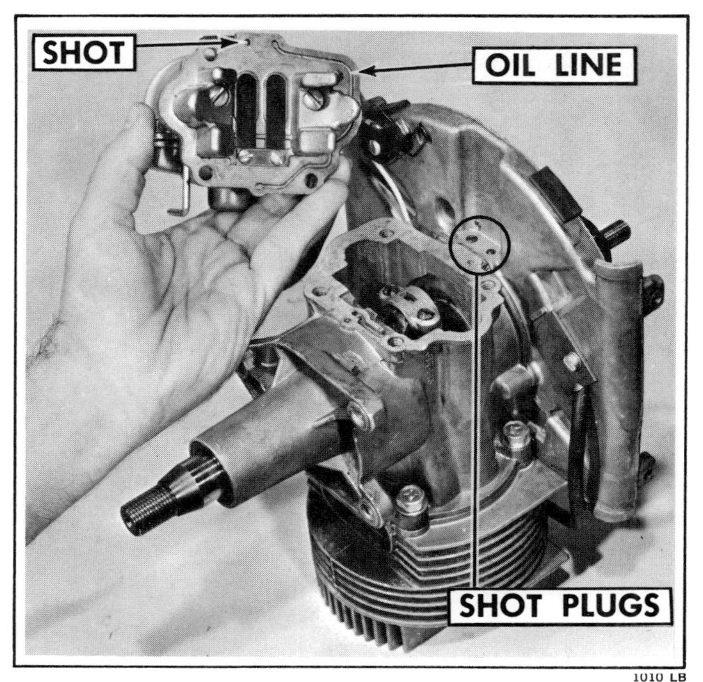

FIGURE 228. Check shot plugs and oil lines if present on your assembly.

7. *Install carburetor.*

8. *Check for proper operation.*

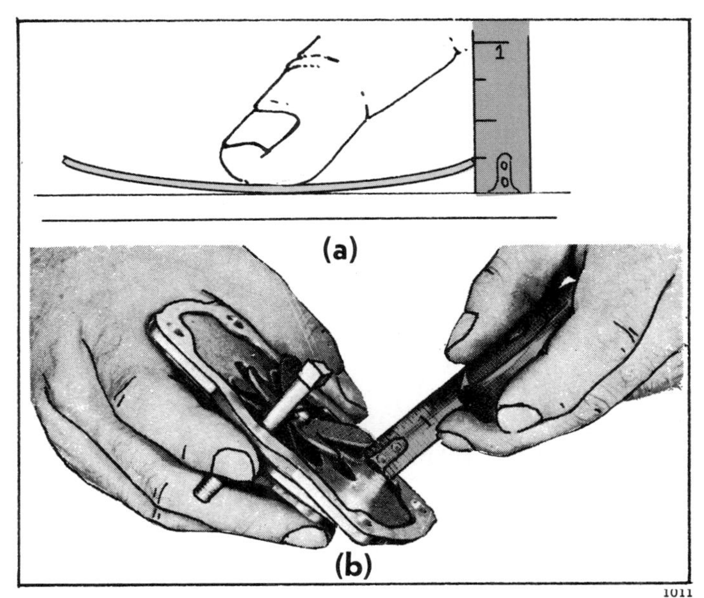

FIGURE 229. Checking reed-stop plate adjustment. (a) Some reed-stop plates are removed for checking. (b) Others may be checked while assembled. If the specified clearance is too small to measure with a rule, use a thickness gage.

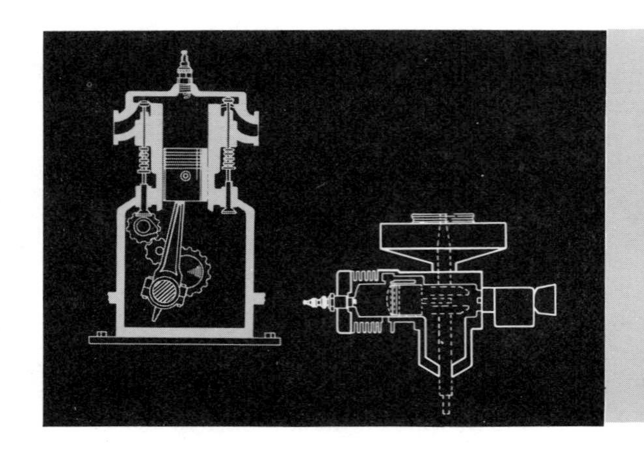

VI. Repairing Cylinders & Piston-and-Rod Assemblies

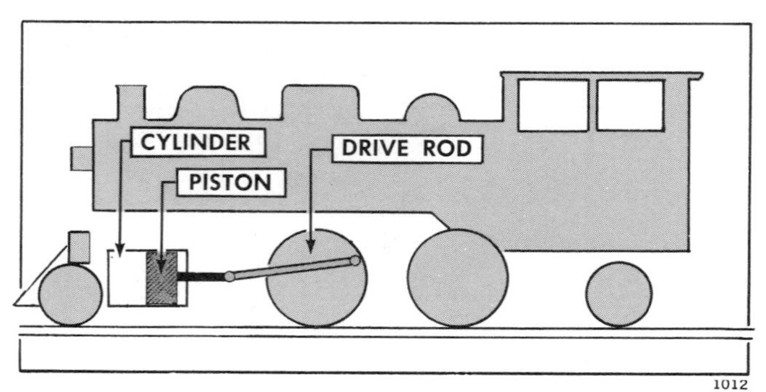

FIGURE 230. The drive rod transfers power from the piston and cylinder to the drive wheel on a steam locomotive.

The function of the **cylinder and piston-and-rod assembly** in a small engine is similar to that of the cylinder, piston and drive rod on steam locomotives (Figure 230).

It is through the action of the engine piston-and-rod assembly that chemical energy of the gasoline is converted to mechanical energy. The rod and crankshaft change the motion of the energy from reciprocating motion (back and forth) to rotating motion.

Figure 231a illustrates how the power is transferred from the piston to the crankshaft in your engine. Power is developed in the cylinder and transferred through the piston and piston rod to the crankshaft. The piston acts as a pump on the intake, compression and exhaust strokes. On the power stroke, however, the piston is driven by the rapidly expanding gases that developed when the fuel burned in the combustion chamber. This is how the power is developed. For each complete cycle (up and down motion) of the piston, the crankshaft makes one revolution.

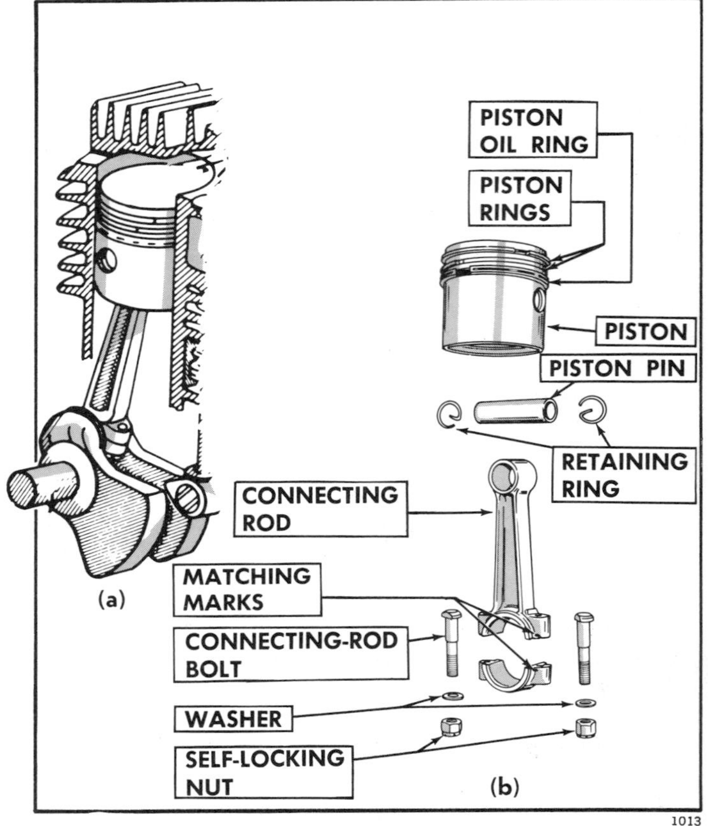

FIGURE 231. (a) Power is transferred from the piston to the crankshaft. (b) The cylinder and piston-and-rod assembly consist of a cylinder, piston, piston rings and a connecting rod.

Again you will note that on **4-cycle engines** each second downward stroke of the piston is a power stroke—or, every second revolution of the crankshaft. On **2-cycle engines** each downward stroke, or every revolution of the crankshaft, is a power stroke.

IMPORTANCE OF PROPER REPAIR

If the cylinder and piston-and-rod assembly are to perform effectively, there are some critical conditions and tolerances which must be met. The rings must be fitted properly. The compression rings must seal combustion gases out of the crankcase, thus preventing blow-by. The oil ring must prevent the

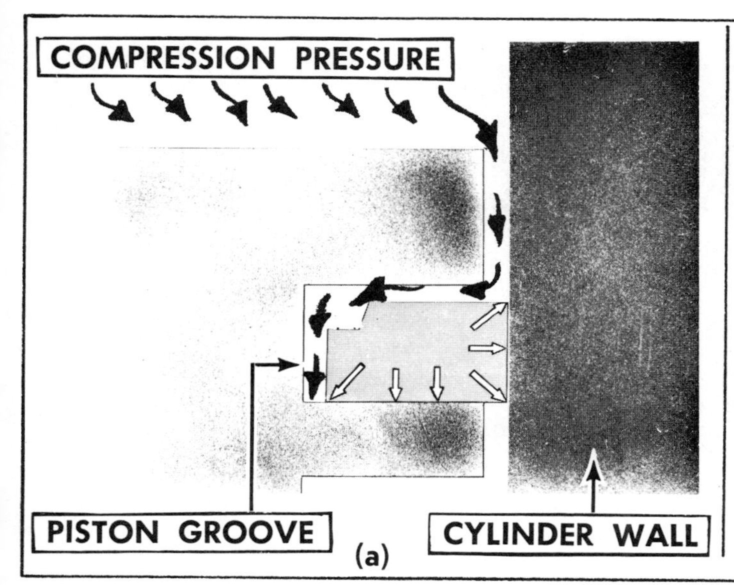

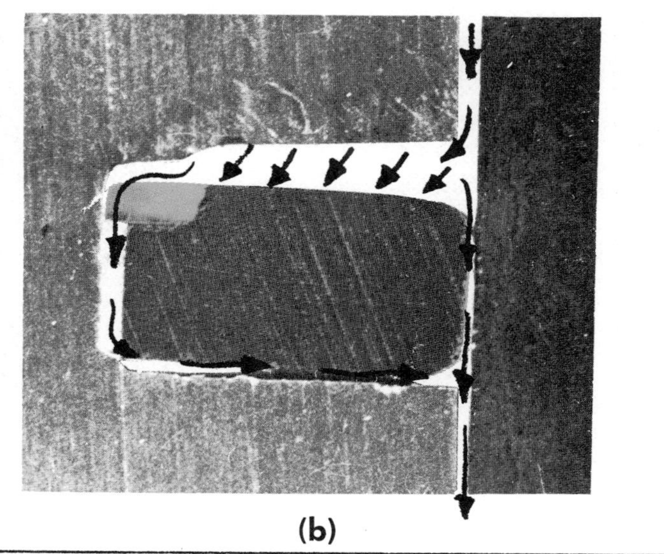

FIGURE 232. (a) On compression stroke the compression ring seats against the cylinder wall and the bottom of the piston-ring groove. (b) A ring cannot seal properly in a worn ring groove.

oil in the crankcase from entering the combustion chamber, thus preventing oil consumption.

The seal between the piston ring and the cylinder wall takes place at two areas of contact (Figure 232): (1) **the cylinder wall** and (2) the **piston groove** (either top or bottom depending upon the direction of pressure). When the piston is being forced downward, the contact points are as shown in Figure 232. When the piston is moving downward on the intake stroke, the seal shifts to the top of the piston groove.

There is a top and a bottom to most piston rings. If you install them upside down, the engine will use oil. New rings cannot seat properly if the ring grooves are worn badly (Figure 232b). Also, they are likely to develop a fluttering and break.

Rings which are **too tight** in the cylinder will *overheat, score* and eventually *seize.*

Rings which are **too loose** in the grooves will *not seat properly* and the ring *may break* during operation.

Other critical areas in the repair of your cylinder, and piston-and-rod assembly are the bearing surfaces between the rod and piston pin, and the bearing surfaces between rod and crankshaft. The bearing surfaces must have smooth finishes and proper clearances.

There is a **critical clearance between the piston and the cylinder wall.** If there is *too little clearance,* the parts will overheat and score — and perhaps seize because of friction. If there is *too much clearance,* it will wear unevenly in the cylinder and the rings cannot perform properly.

Cylinders which are worn excessively may be rebored, and oversized pistons and rings may be installed. But if the cylinder walls are not honed (ground slightly) properly, the rings will not seat properly, and the engine will use oil (Figure 276).

It is most important that you **keep the parts clean** when repairing your engine. Clean all the parts to be reused, and protect your engine from dirt and dust when repairing it. A small amount of dirt left in the engine will wear out bearings and rings in a very short time. Your efforts at repair will have been wasted.

Checking and repairing the cylinder and piston-and-rod assemblies are discussed under the following headings:

A. Checking the cylinder and piston-and-rod assembly for proper operation.

B. Repairing pistons, rods and rings.

C. Repairing cylinders.

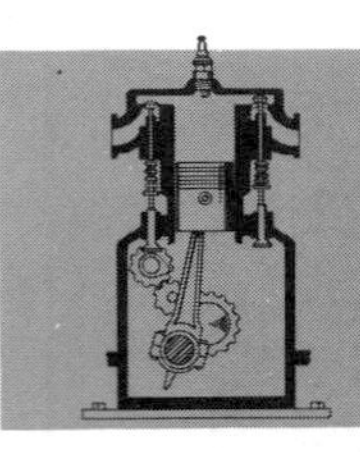

A. CHECKING THE CYLINDER AND PISTON-AND-ROD ASSEMBLY FOR PROPER OPERATION

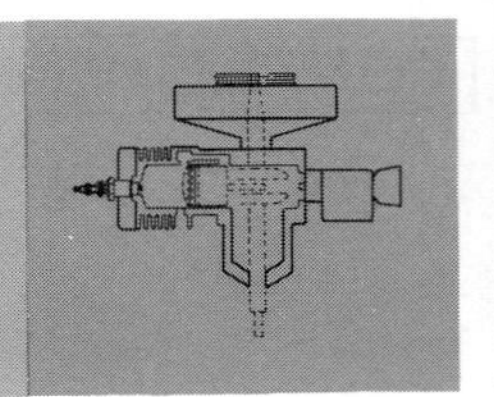

Indications of improper operation of the cylinder and piston-and-rod assembly are (1) poor compression, (2) excessive oil consumption and (3) noisy operation.

It is difficult to tell just what may be causing the trouble until you disassemble the piston-and-rod assembly and inspect the parts. Inspection of the parts will give you a clue as to what caused the trouble. The following procedures for checking before disassembly, however, may give you an indication as to the source of your trouble.

1. *Check the compression.*

   For procedures see "Checking the Valve Operation by Compression Test," page 187. This will give you an indication as to the condition of the cylinder and piston rings.

2. *Check for oil consumption.*

   Puffs of blue smoke from the exhaust indicate oil consumption which may be due to worn or seized piston rings. Another indication, of course, is necessity for adding oil frequently.

3. *Check for noisy operation.*

   Noisy operation may be due to worn piston-rod bearings — either at the piston pin or at the crankshaft.

Before repairing the cylinder and piston-and-rod assembly, it is a good idea to check the crankshaft, bearings and valves. If they are worn also, you may decide to purchase a "short block" assembly rather than to overhaul the engine. "Short blocks" consist of a cylinder block, crankcase, crankshaft, camshaft assembly, piston assembly and valves.

The cost is approximately one half of the cost of a new engine.

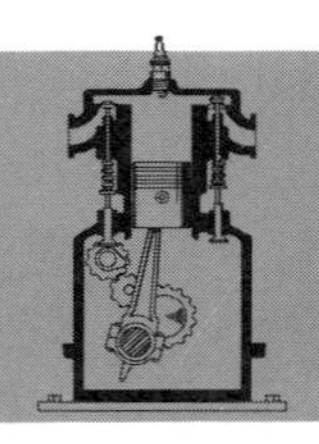

B. REPAIRING PISTONS, RODS AND RINGS

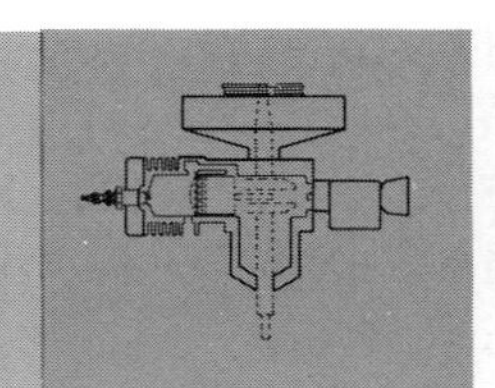

Repairing pistons, rods and rings is discussed under the following headings.

1. Types of pistons and how they work.
2. Types of piston rings and how they work.
3. Tools and materials needed.
4. Preparing to remove the piston-and-rod assembly.
5. Removing the piston-and-rod assembly.
6. Disassembling the piston-and-rod assembly.
7. Cleaning and inspecting the piston and rings.
8. Inspecting the piston rod and bearings.
9. Installing the piston rings.
10. Installing the piston pin.
11. Installing the piston-and-rod assembly.

TYPES OF PISTONS AND HOW THEY WORK

Pistons in **4-cycle engines** appear to be the same diameter from top to bottom (Figure 233). If you check them closely, you will usually find they are tapered slightly toward the top. The piston top is flat.

Pistons on **2-cycle engines** may be tapered or they may have the same diameter from top to bottom. Some may have an odd-shaped piston top, depending on the scavenging system for which it is designed (Figure 234).

Pistons may be made of aluminum alloy or steel. The piston is drilled for a **piston pin.** The pin connects the piston to the connecting rod. During operation the piston floats — moves from side to side — on the piston pin. This reduces the possibility of the piston binding against the cylinder wall.

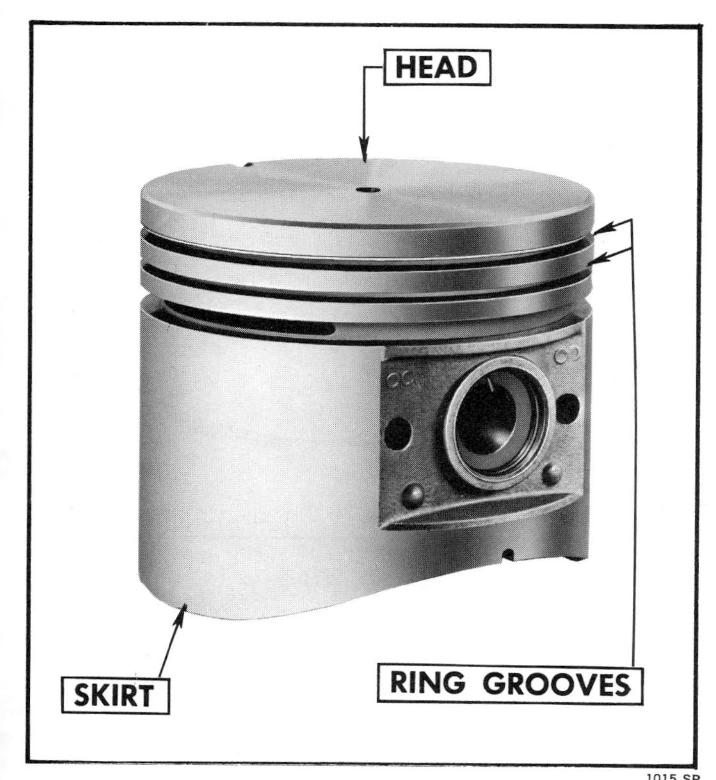

1015 SP

FIGURE 233. A typical piston used in small 4-cycle gasoline engines. Some are slightly tapered from bottom to top.

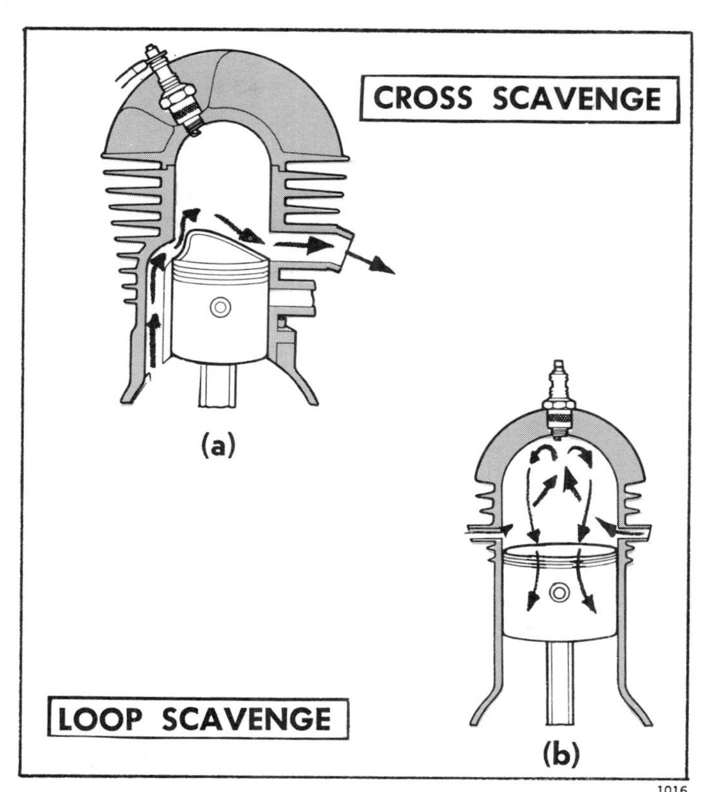

1016

FIGURE 234. Two types of pistons used on 2-cycle engines. (a) A piston with a baffle on top for deflecting the fuel-air mixture going into the cylinder. This reduces the amount of unburned fuel mixture escaping through the exhaust port. (b) A flat-top piston used for loop scavenging. The fuel-air mixture enters through two intake ports on opposite sides of the cylinder. It is directed up and over by the dome-shaped combustion chamber and out through the exhaust ports which are located between the intake ports.

Pistons are slightly smaller than the cylinder .13 mm to .25 mm (.005 in to .010 in). After being sized and installed properly, they barely touch the cylinder walls when operating.

TYPES OF PISTON RINGS AND HOW THEY WORK

All pistons are grooved for **rings** (Figure 235). Most pistons on **4-cycle engines** have three rings. Pistons on **2-cycle engines** generally have two rings.

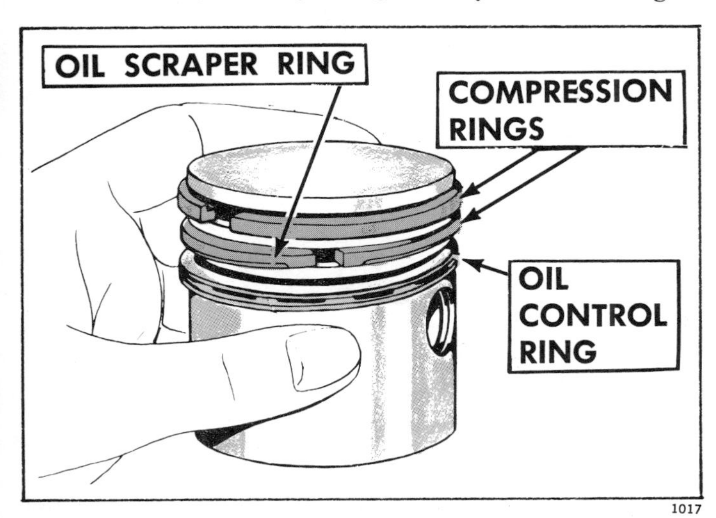

1017

FIGURE 235. Typical piston and piston rings.

Each ring is designed for a specific purpose. They are described as follows:

- **Compression rings** (Figure 236a). These rings are designed to make a tight seal against the cylinder wall. This is particularly important during the power stroke to prevent combustion gases from being forced by the piston, thus resulting in a loss of power.

  Compression ring(s) are always placed in the top groove(s) of the piston on **4-cycle engines.** On **2-cycle engines** compression rings are the only ones used.

- **Oil-wiper rings** (Figure 236b). These are designed to wipe the excess oil off the cylinder wall as the piston moves down. This prevents the oil from bypassing the ring, entering the combustion chamber and being burned. Compression rings also provide for some oil wiping.

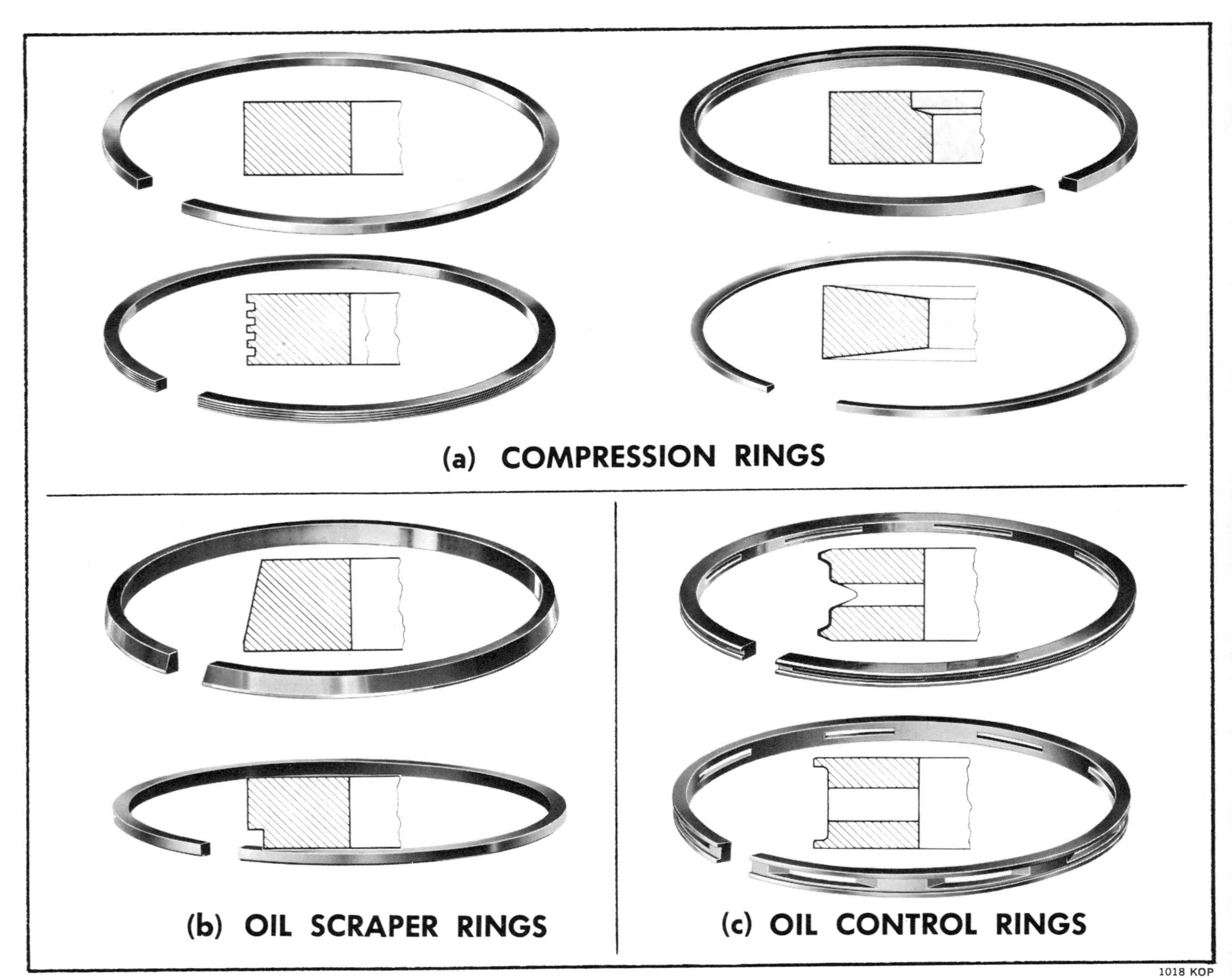

FIGURE 236. Types of piston rings used on small gasoline engines. (a) Compression rings. (b) Oil-wiper rings. (c) Oil-control rings. These types shown in (b) and (c) are used only on 4-cycle engines.

- **Oil-control rings** (Figure 236c). These are located below the compression rings. They regulate the amount of oil going to the compression rings and cylinder walls. Sometimes an additional expander spring is placed under the oil-control ring to help maintain its tension as it wears (Figure 237). The expander spring also helps the oil ring to take the shape of a slightly distorted cylinder wall.

  Two-cycle engines have no oil-control rings since the oil is mixed with the fuel.

Most piston rings are made of cast iron. Some have a chrome-plated face, or are specially heat treated, to harden the surface and reduce the wear. Oil rings do not run so hot as compression rings; consequently, they can be made of steel without annealing while running. Therefore, many oil rings

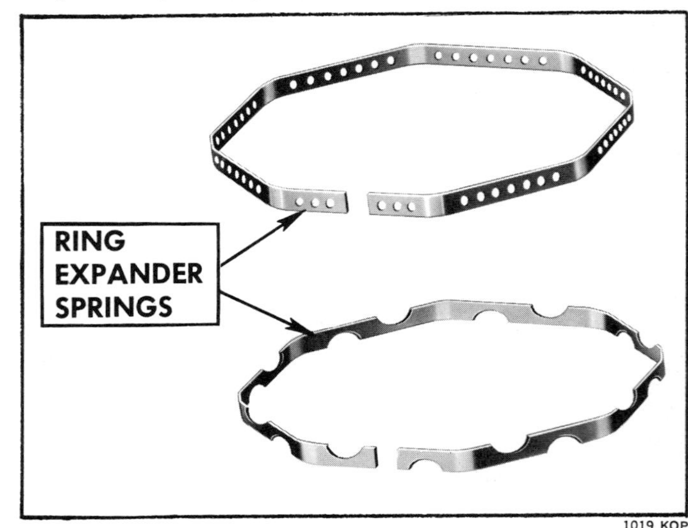

FIGURE 237. Expander springs can be placed under oil-control rings.

are designed with two narrow steel rails that contact the cylinder wall.

Some manufacturers now have a replaceable ring which they claim will take the shape of a worn, uneven or out-of-round cylinder. You can use them in cylinders that have not worn to the extent that they must be rebored.

All piston rings provide enough spring action to press against the cylinder wall. This spring action provides a seal between the piston and the cylinder.

TOOLS AND MATERIALS NEEDED

1. Long-nose pliers — 7″
2. Piston ring expander
3. Reamer for piston pin
4. Micrometer (outside and inside)
5. Ring-groove gage
6. Open-end wrenches — 3/8″ through 9/16″
7. Sockets, ratchet handle — 3/8″ through 9/16″ — 3/8″ drive
8. Brass punch — 1/2″ x 6″
9. Piston solvent — John Deere, Stoddard or equivalent
10. Clean rags
11. Crankcase oil

PREPARING TO REMOVE THE PISTON-AND-ROD ASSEMBLY

All piston-rod assemblies are removed by disconnecting the rod (Figure 238) from the crankshaft. Most of them are pushed out through the top of the cylinder (Figure 239a). Some (on 2-cycle engines) are removed from the bottom of the cylinder (Figure 239b).

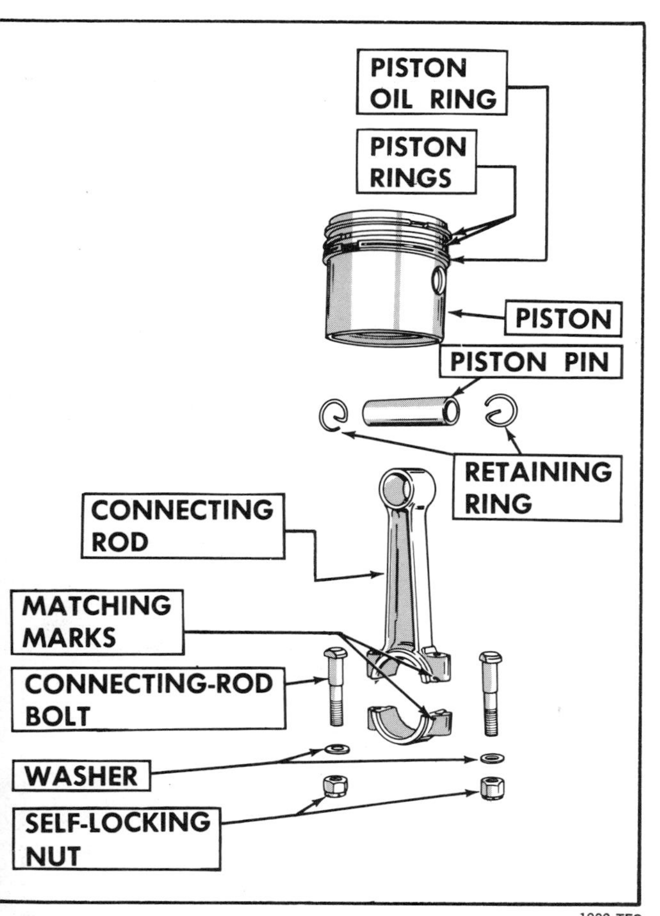

FIGURE 238. Parts of the piston-and-rod assembly.

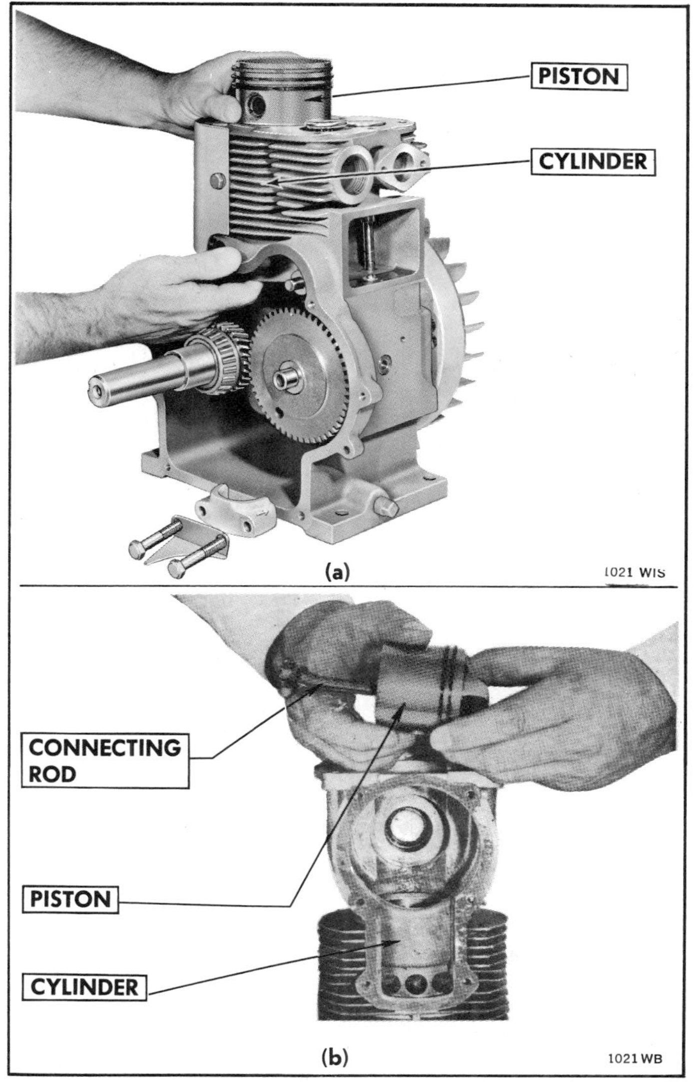

FIGURE 239. Removing the piston-and-rod assembly. (a) Most piston-and-rod assemblies are removed from the top of the cylinder. (b) Some are removed from the bottom of the cylinder.

There are several steps you have to complete before removing the piston assembly, since it is connected inside the crankcase. Proceed as follows:

1. *Clean the engine and equipment.*

   See "Cleaning Small Engines," Part 1.

2. *Drain the oil (4-cycle engines).*

   See "Changing the Crankcase Oil (4-Cycle Engines)," Part 1.

3. *Remove the engine from the equipment if convenient.*

   A major job of repairing your engine will be easier if you can remove it from the tractor, mower, chain saw or the equipment to which it may be attached.

4. *Remove the fuel tank.*

   See "Repairing Fuel Tank Assemblies," page 154.

5. *Remove the starter (if one is installed).*

   See "Repairing Starters," page 78.

6. *Remove the flywheel shroud and cylinder baffles.*

   See "Removing and Checking the Flywheel," page 127.

7. *Remove the flywheel if necessary.*

   On some engines it is not necessary to remove the flywheel, unless you are going to remove the crankshaft.

8. *Remove the spark plug.*

   See "Servicing Spark Plugs," Part 1.

9. *Remove the carburetor air cleaner.*

   See "Servicing Carburetor Air Cleaners," Part 1.

10. *Remove the carburetor.*

    See "Repairing Carburetors," page 160.

11. *Remove the muffler (if one is installed.)*

    See "Cleaning Small Engines," Part 1.

12. *Remove the cylinder head if you plan to overhaul the engine.*

    See "Removing the Cylinder Head," page 190.

13. *Remove the valves if they are to be serviced.*

    See "Removing the Valves," page 192.

REMOVING THE PISTON-AND-ROD ASSEMBLY

The piston rod is attached to the crankshaft by a **rod-bearing cap** (Figure 240). To remove the piston-and-rod assembly, you must remove the rod-bearing cap which is located inside the crankcase. Proceed as follows:

1. *Remove the cylinder head — if it is removable.*

   If your **cylinder head is bolted on to the cylinder block,** the piston-and-rod assembly is removed from the top, or cylinder-head end of the cylinder.

   If the **cylinder head and cylinder are cast in one piece** and bolted to the crankcase (Figure 249), proceed to step 2.

   If the **cylinder head, cylinder and crankcase are cast in one piece** (Figure 250), proceed to step 2.

2. *Gain access to the rod-bearing cap.*

   This is a matter of removing some part of the crankcase. On **4-cycle engines** remove the oil sump (Figure 241a) or a PTO bearing plate (Figure 241b and c).

   On **2-cycle engines** remove the carburetor and reed plate (Figure 242a) or a rod-bearing access plate (Figure 242b).

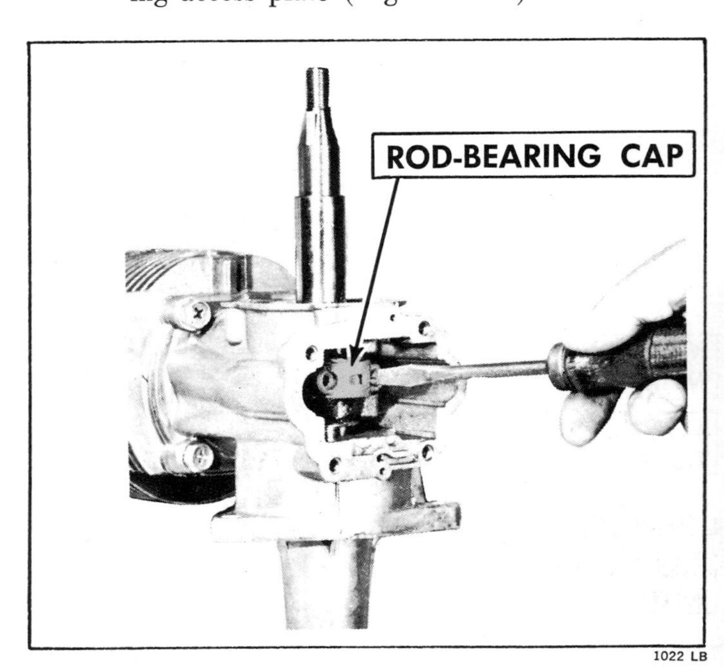

FIGURE 240. Removing the rod-bearing cap.

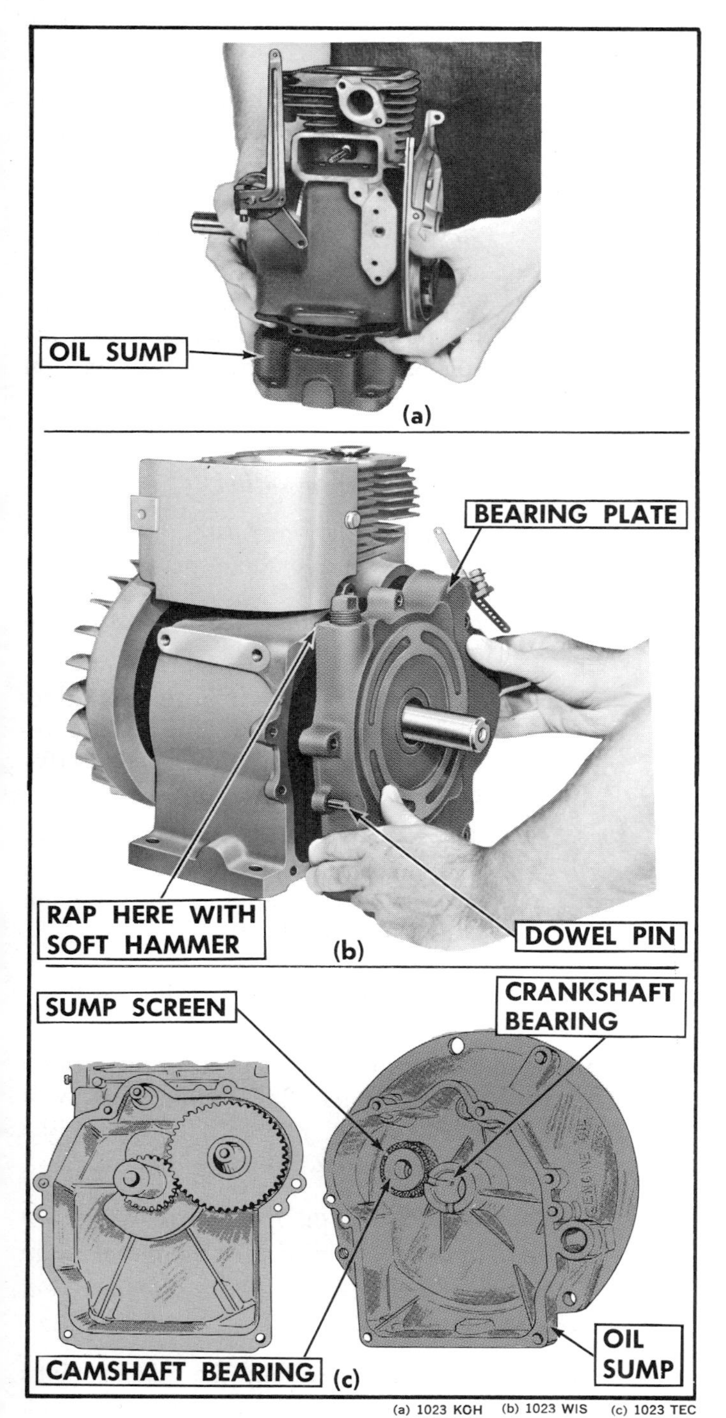

FIGURE 241. Gaining access to the rod-bearing cap on the crankshaft of different types of 4-cyle engines. (a) Removing the oil sump. (b) Removing the bearing plate. (c) Removing the oil sump and bearing plate.

a. If you are removing from the crankcase **a part which does not serve as a bearing support,** such as an oil sump (Figure 241a), or a carburetor and reed plate (Figure 242a), or a rod-bearing access plate (Figure 242b), proceed as follows:

(1) *Remove nuts, or capscrews, that hold the part to be removed from the crankcase.*

Some parts have alignment dowels that must be removed also.

If you are removing the carburetor, refer to "Removing the Carburetor," page 160, for procedures.

(2) *Jar the part loose with a soft hammer*

(3) *Remove the part from the crankcase.*

b. If you must remove a **part that supports a bearing,** such as a base plate on a horizontal cylinder engine (Figure 241b), or a bearing plate on a vertical cylinder engine (Figure 241c), proceed as follows:

(1) *Clean rust and burrs from the power*

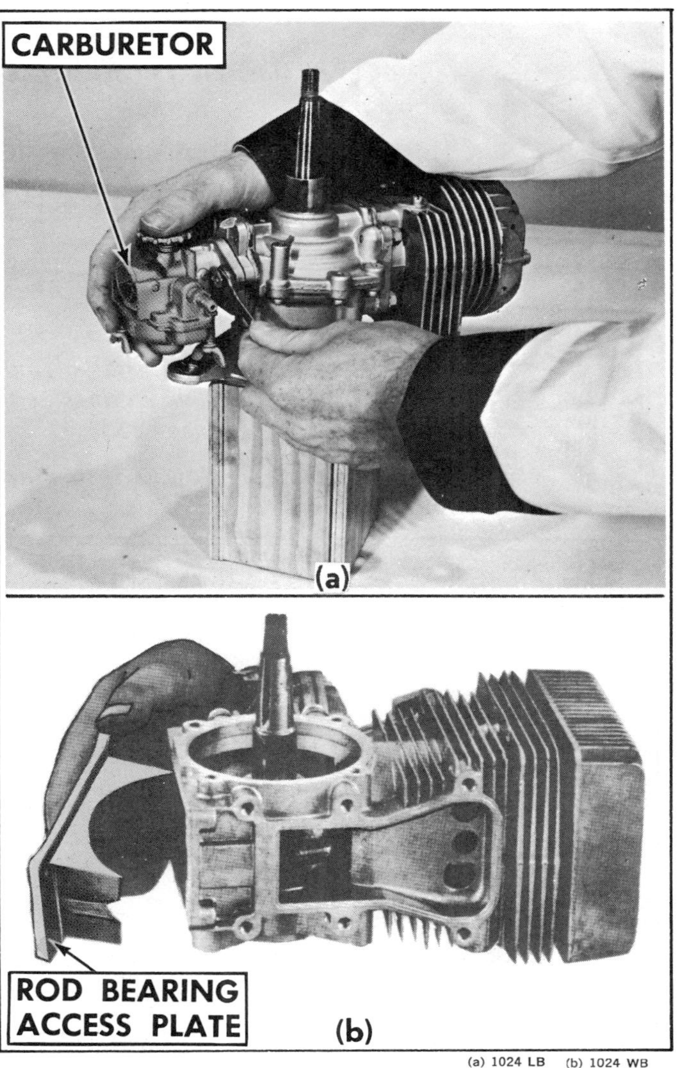

FIGURE 242. Two methods for gaining access to the rod-bearing cap on 2-cycle engines. (a) Removing the carburetor and reed plate. (b) Removing a special rod-bearing access plate.

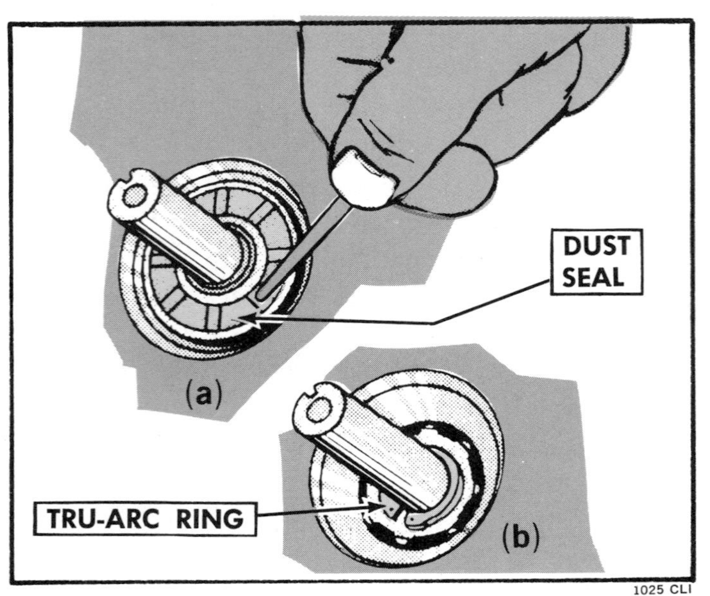

FIGURE 243. (a) Removing metal dust cover. (b) Tru Arc lock ring.

take-off (PTO) end of the crankshaft.

This is to prevent damage to the main bearing and oil seal when the bearing plate is removed.

On engines with clutch or speed-reduction units, these must be removed first.

(2) *Remove dust cover if installed and Tru Arc lock ring on engines with ball bearings (Figure 243).*

(3) *Remove alignment dowels if they are the removable type.*

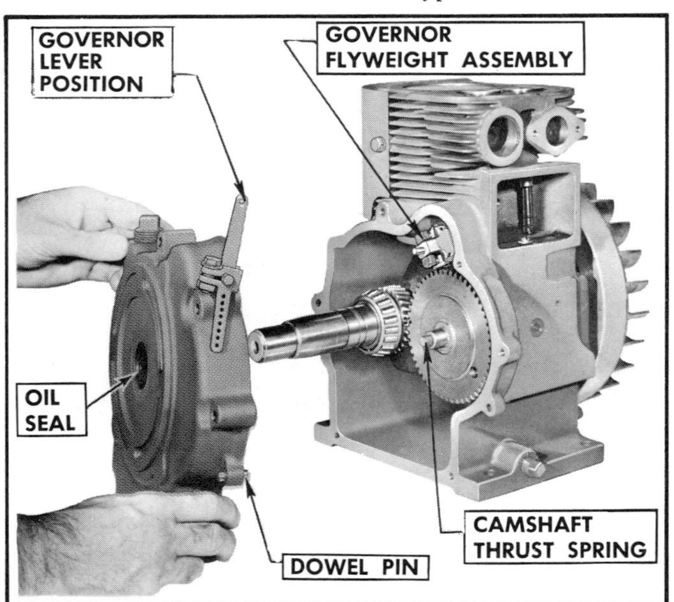

FIGURE 244. Removing the bearing plate.

(4) *Remove base plate, or bearing plate (Figure 244).*

Tap with a soft hammer to loosen. Be careful not to damage oil seal and bearings.

Some engines have a *steel ball for end thrust* at the end of the camshaft. Watch for this because it may fall out when you remove the bearing plate. There is also an end-thrust spring on the end of the camshaft on this type.

(5) *Remove the governor and oil pump assembly if installed.*

Note how they are assembled so you can reassemble them properly.

(6) *Remove the camshaft if necessary to get to the rod-bearing cap.*

Turn the crankshaft until the valve-timing marks are aligned. This relieves the cam pressure on the valve tappets.

The valve tappets will fall out when the camshaft is removed. Mark them so you will be able to install them in the original location.

3. *Inspect the crankshaft bearings that are removed.*

Refer to "Cleaning and Inspecting the Crankshaft Bearings," page 257. This check will help you determine whether to repair the engine or buy a new "short block."

4. *Scrape the carbon ring from around the top of the cylinder and/or remove the metal ridge from top of cylinder on removable-head engines (Figure 245).*

This is to avoid breaking the piston rings and/or damaging the piston when removing the piston-and-rod assembly from the top. If left, it may also break the new rings.

Remove metal ridge with a boring bar, cylinder hone or a "ridge reamer" (Figure 245b).

5. *Turn the crankshaft so the piston will be at bottom-dead-center.*

The rod-bearing cap will be at its lowest point so that you will have better access to it.

6. *Turn the engine upside down or on its side (Figure 246).*

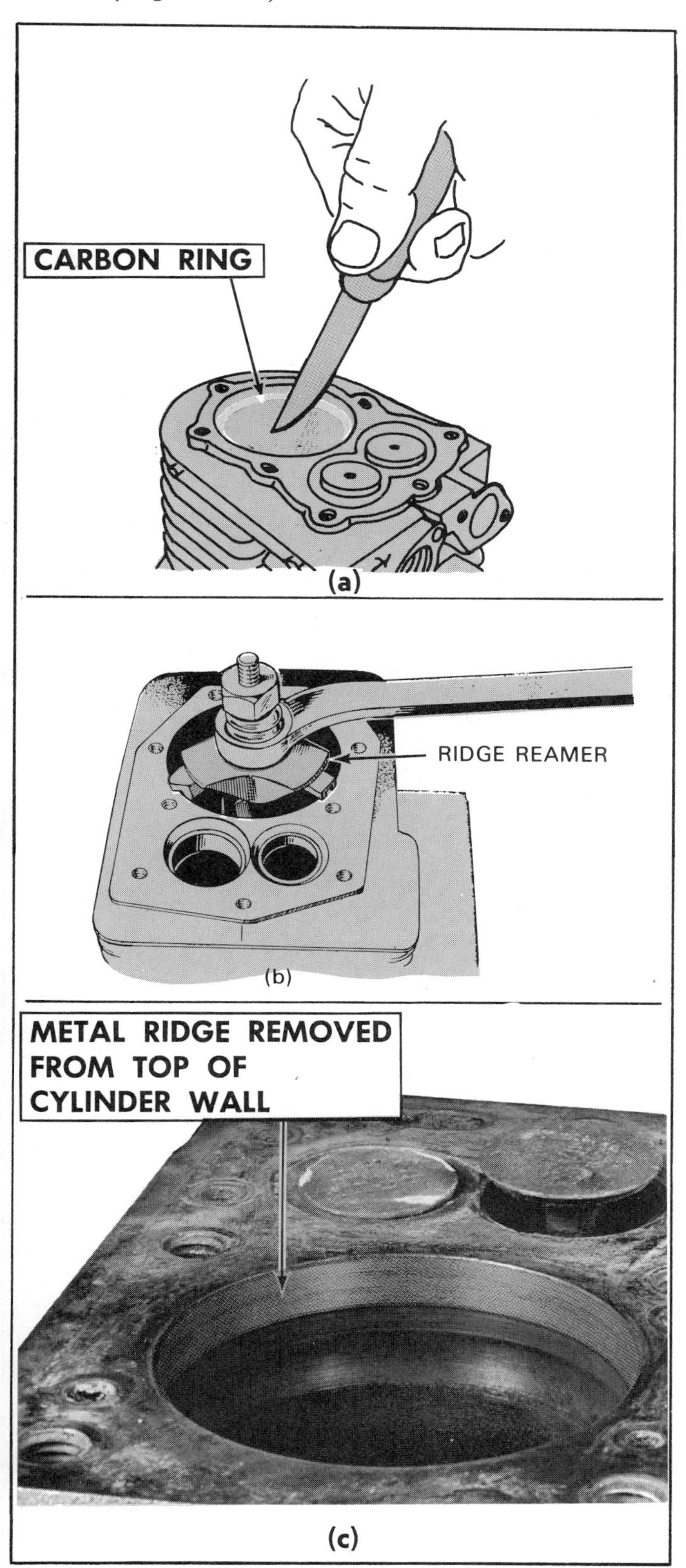

FIGURE 245. (a) Scraping the carbon ring from around the top of the cylinder. (b) Removing the metal ridge from the top of the cylinder with a ridge reamer. (c) Ridge removed.

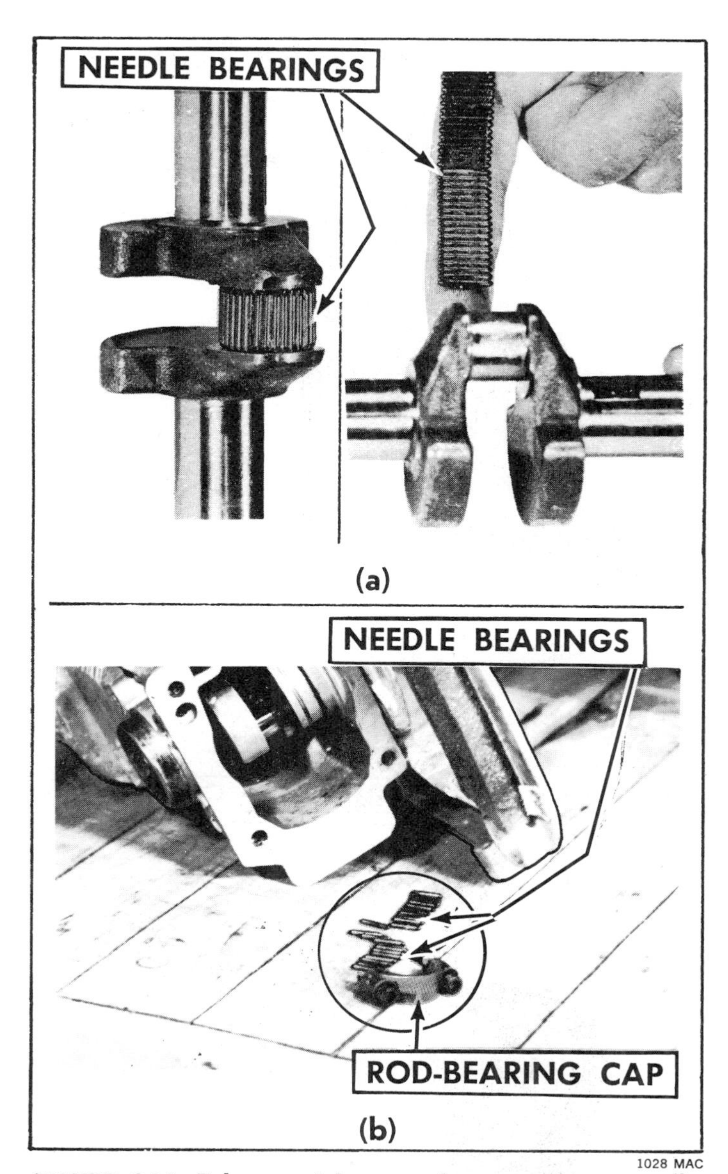

FIGURE 246. Take special care when removing needle bearings. (a) Removing needle bearings. (b) Needle bearings and rod-bearing cap removed.

This provides better access to the rod-bearing cap.

7. *Remove the nut or capscrews from the rod-bearing cap.*

Some capscrews are held by a lock which is bent along the edge of the screw head. If so, it must be flattened first. Use a .080 cm (5/16 in) punch.

8. *Note positions of cap and rod.*

Before removing the rod-bearing cap, take note of its position. It is usually marked with matching notches or bosses so that you can replace them in the same position. The bearings have been fitted with the cap in this position. If you turn the cap around when you reinstall it, you may not get a good fit.

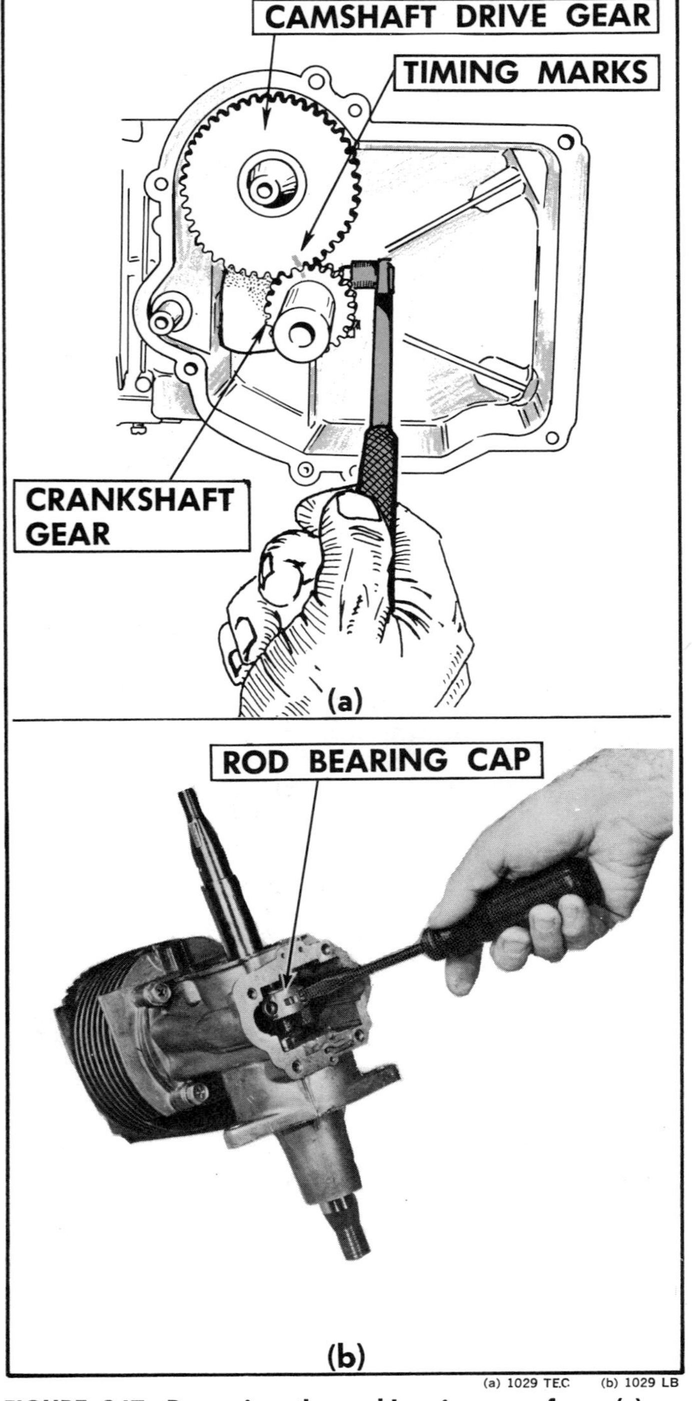

FIGURE 247. Removing the rod-bearing cap from (a) a 4-cycle engine, and (b) a 2-cycle engine.

Another reason for noting the position of the cap is that some **caps have an oil dipper** which must be positioned a certain way to sling oil over the crankcase.

Also **note the position of the connecting rod.** Some rods work satisfactorily in either position, but most of them will not. If the cap is not marked, mark it and the rod with a center punch. Some rods have needle bearings. If so, these bearings may fall out when you remove the rod-bearing cap. Take special care not to lose the bearings and bearing races (Figure 246.)

9. *Remove the rod-bearing cap and bearings, if installed (Figure 247).*
10. *Remove the piston-and-rod assembly from the cylinder.*

    If your engine has a **removable head,** the piston is to be removed from the top. Push the piston out with a wooden stick (Figure 248).

    If the **cylinder and cylinder head are cast in one piece** and the casting is removable,

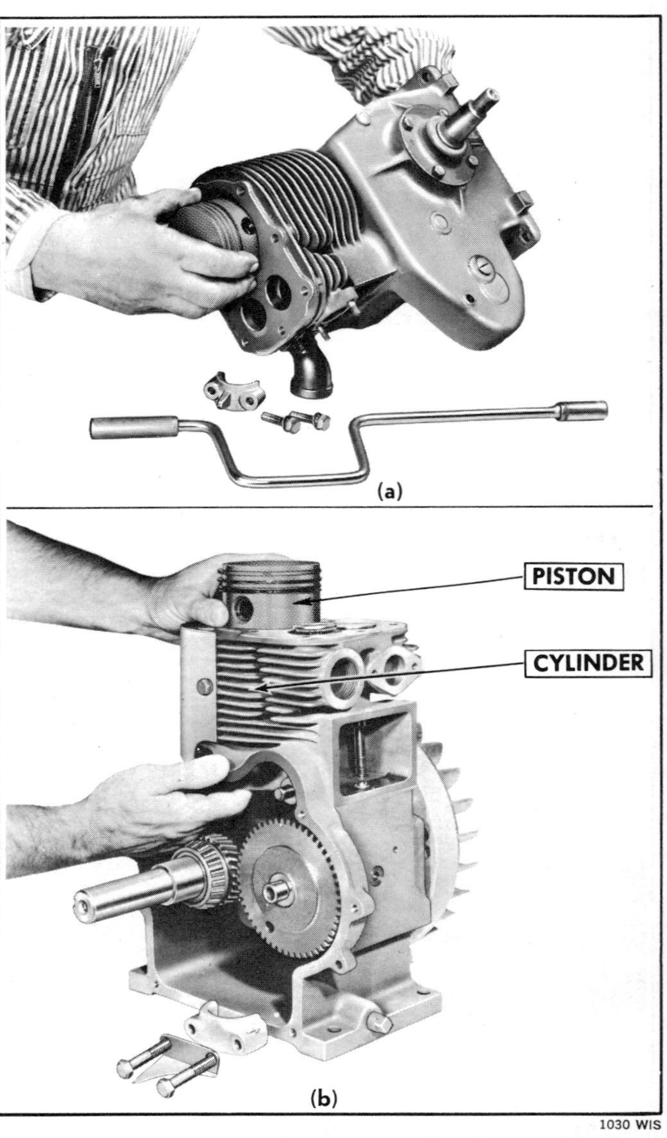

FIGURE 248. Removing the piston-and-rod assembly from the cylinder (a) with the oil sump removed and (b) with the bearing plate removed.

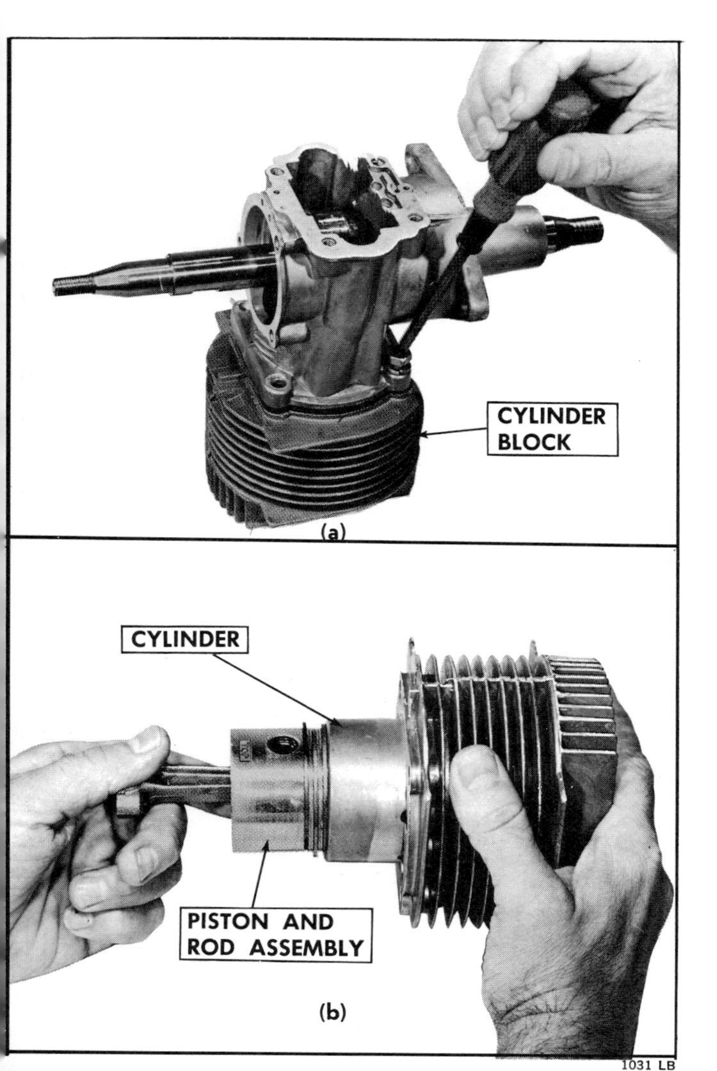

FIGURE 249. (a) Removing the 1-piece cylinder block and head from one model of 2-cycle engine. (b) Removing the piston-and-rod assembly from the cylinder.

it will be necessary to remove the casting from the crankcase (Figure 249a).

If the **cylinder, cylinder head and crankcase are cast in one piece** (Figure 242b), it will be necessary for you to remove the crankshaft and then remove the piston through the crankcase (Figure 250). Proceed as follows:

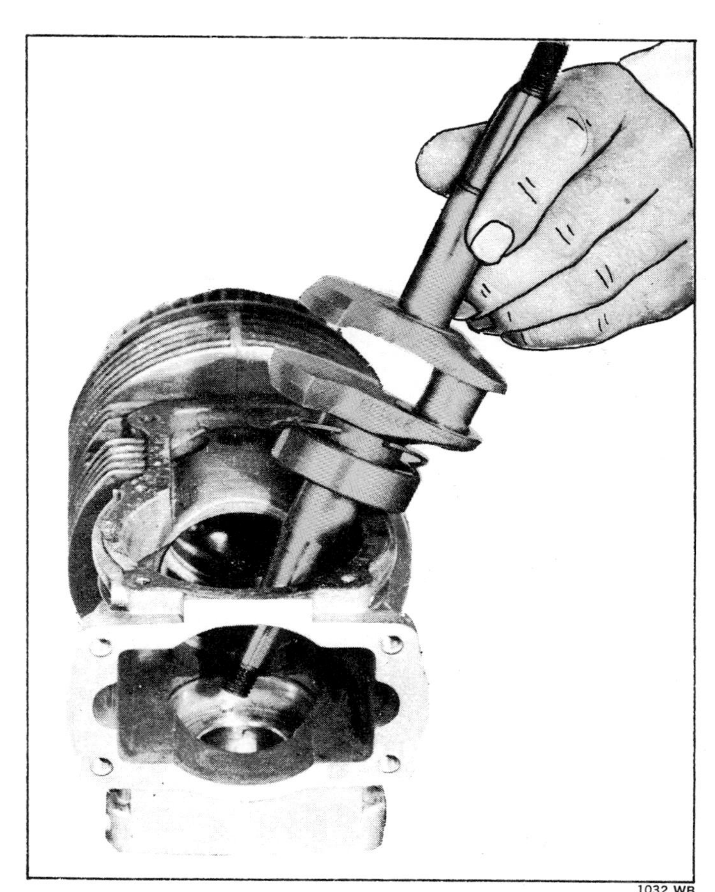

FIGURE 250. Removing the crankshaft from a 2-cycle engine that has the head, cylinder block and crankcase cast in one piece.

(1) *Push the piston up into the cylinder to provide for clearance for removing the crankshaft.*

(2) *Remove the crankshaft retaining ring if installed.*
Some require special tools.

(3) *Rotate the crankshaft so that the counterweights will clear the connecting rod.*

(4) *Remove the crankshaft (Figure 250).*

(5) *Remove the piston-and-rod assembly from the crankcase.*

Pull the piston out through the crankcase by the rod.

DISASSEMBLING THE PISTON-AND-ROD ASSEMBLY

Before going any further with piston repair, inspect the cylinder to determine if the engine is worth repairing. Refer to "Inspecting the Cylinder," page 157.

If the **cylinder must be rebored,** you will need a new, oversized piston.

If the **cylinder dimensions are all right,** proceed as follows:

1. *Remove the two retaining rings that hold the piston pin in the piston (Figure 251).*

Special tools are available, or you can pry the ring out with a small screwdriver. Be careful not to damage the piston.

Mark the relative position of the piston and piston pin on the rod so that you can reassemble it in the same position.

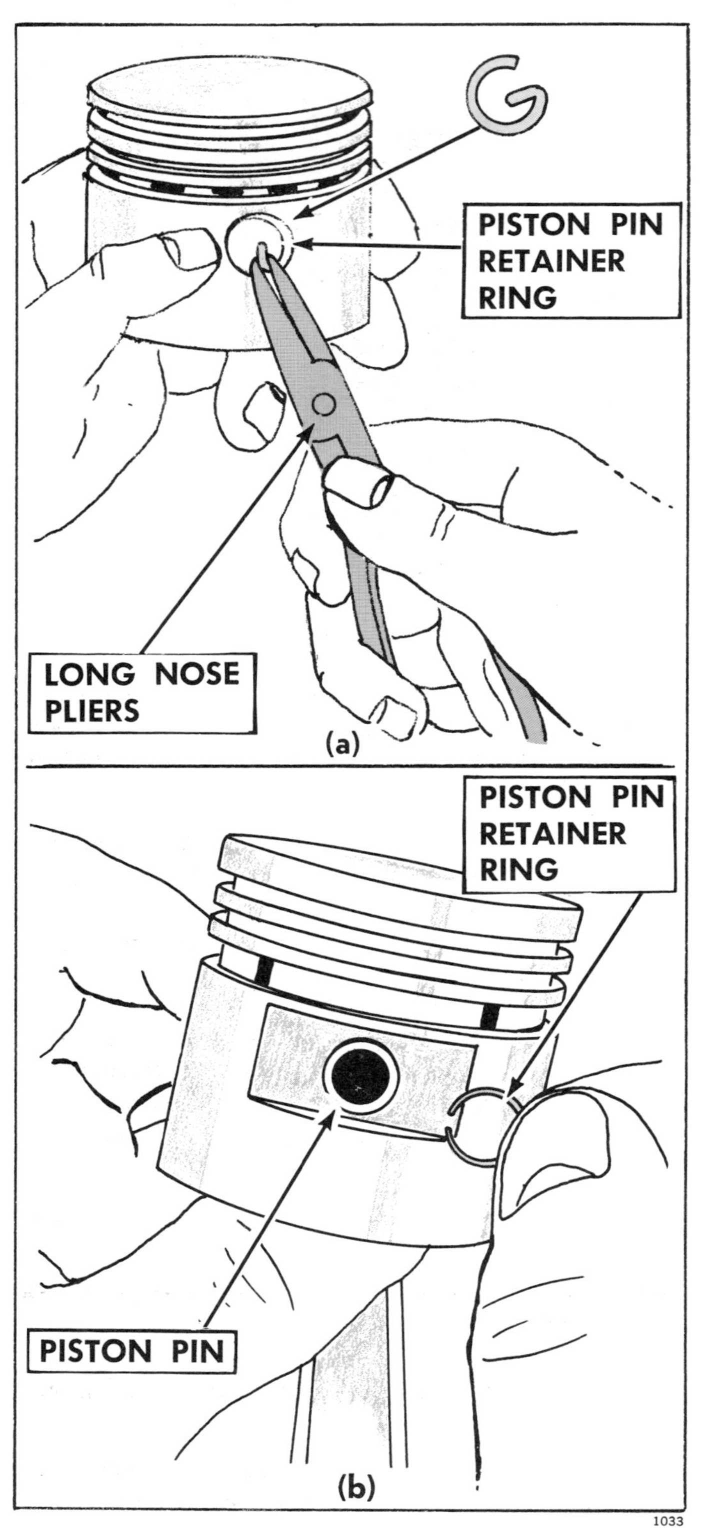

FIGURE 251. Removing piston retainer rings. (a) Some rings can be removed with long-nose pliers. (b) Others have no provision for removal with pliers and must be pried out with a special tool or a small screwdriver.

FIGURE 252. Piston with needle bearings between the pin and piston.

2. *Remove pin from piston.*

   On some 2-cycle engines there are needle bearings in the piston end of the rod also. Watch for these. Some have needle bearings between the rod and the pin. Others have needle bearings between the pin and the piston (Figure 252). Some piston pins are installed with a slight pressure fit. If yours is tight, heat the piston with a light bulb and drive the pin out with a punch (Figure 253).

3. *Remove the rings (Figure 254).*

   Use a piston-ring expander to avoid breaking

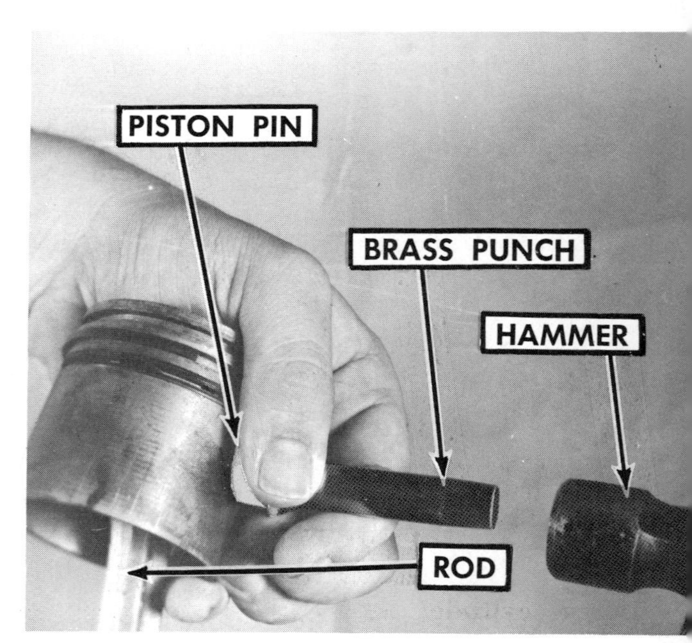

FIGURE 253. Removing the piston pin.

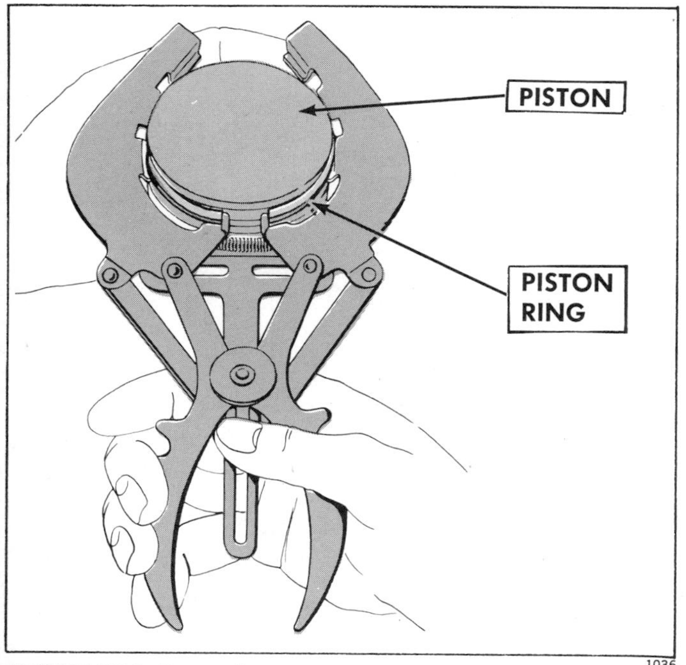

FIGURE 254. Removing piston rings with a ring expander.

the rings. Make a note describing the position and the top and bottom of the piston rings.

Otherwise, you may not know how to reinstall them or to install new ones.

CLEANING AND INSPECTING THE PISTON AND RINGS

1. *Soak the piston in a special carbon solvent.*

   Some manufacturers supply special piston solvents. Commercial brands, such as John Deere and Stoddard solvents, are available for this purpose.

2. *Scrape the carbon from the ring groove with a broken piston ring if available.*

   The broken ring fits the ring groove. Do not scratch the piston. Special scraping tools are also available (Figure 255).

3. *Check the condition of the piston and rings (Figure 256).*

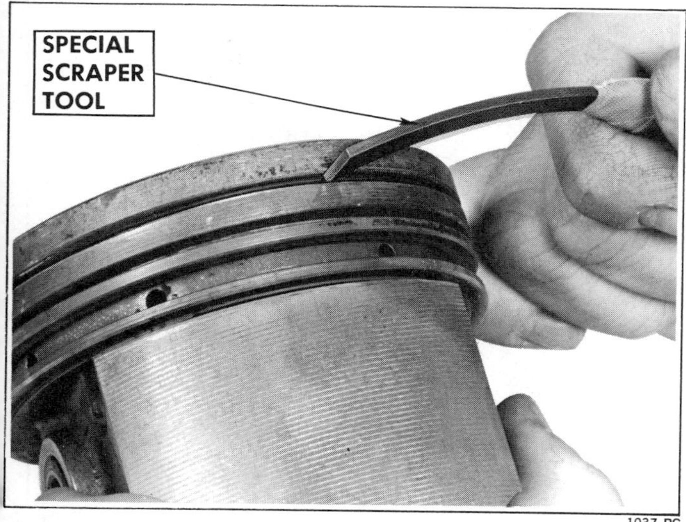

FIGURE 255. Cleaning the ring grooves with a special tool.

FIGURE 256. Examples of piston failures. (a) Top land burned and melted away. (b) Hole blown in the top of the piston from preignition. (c) Piston scuffed because of lack of proper lubrication.

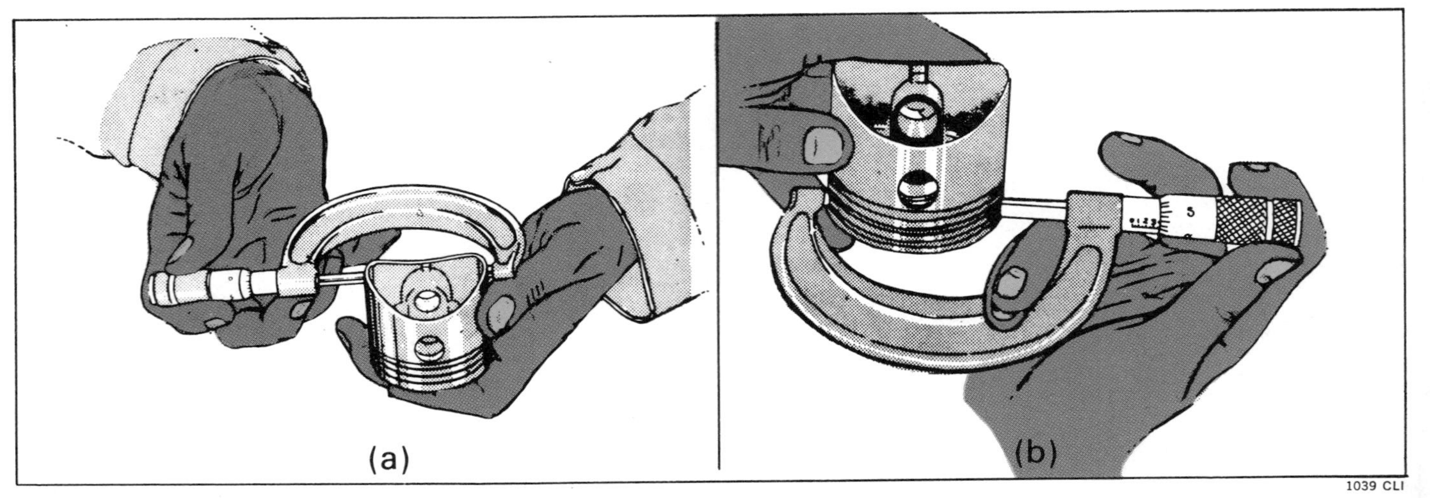

FIGURE 257. Measuring the diameter of the piston. (a) Measuring the skirt diameter. (b) Measuring the diameter at the lands.

Check for burning, top groove wear and broken or stuck rings. Such conditions are caused by detonation and preignition. Detonation causes extremely high pressure against the top ring which is resisted by the second land of the piston. As a result it is usually the second and sometimes lower lands that are broken by detonation. For an explanation of detonation and preignition, refer to "Refueling Small Engines," Part 1.

Check for scuffing and scoring. (Figure 256b and c). Scoring is the result of metals getting so hot because of lack of lubrication that they reach the melting point and tend to fuse. Scuffing is a lighter form of scoring.

Check for scratches on the piston and rings. Vertical scratches and excessive wear indicate that too much dirt and other abrasive materials are entering the engine. This is most likely the result of your letting dirt get into the engine through the air cleaner, intake manifold or the oil system. Abrasives from the use of a spark plug sand blast may also cause piston scratches.

4. *Measure the piston for wear (Figure 257).*

Refer to your repair manual for the maximum allowable wear. It gives the tolerances.

Whether or not you will be able to use your old piston may also depend on the diameter of the cylinder. There is a critical clearance between the piston and the cylinder. These clearances range from 0.13 mm to .25 mm (.005 in to .010 in). If it is too much, the piston will "slap" against the cylinder wall and cause more wear. Note discussion under "Checking and Repairing Cylinders," page 230.

5. *Check the ring-groove wear (Figure 258).*

Use a ring-groove wear gage (Figure 258a),

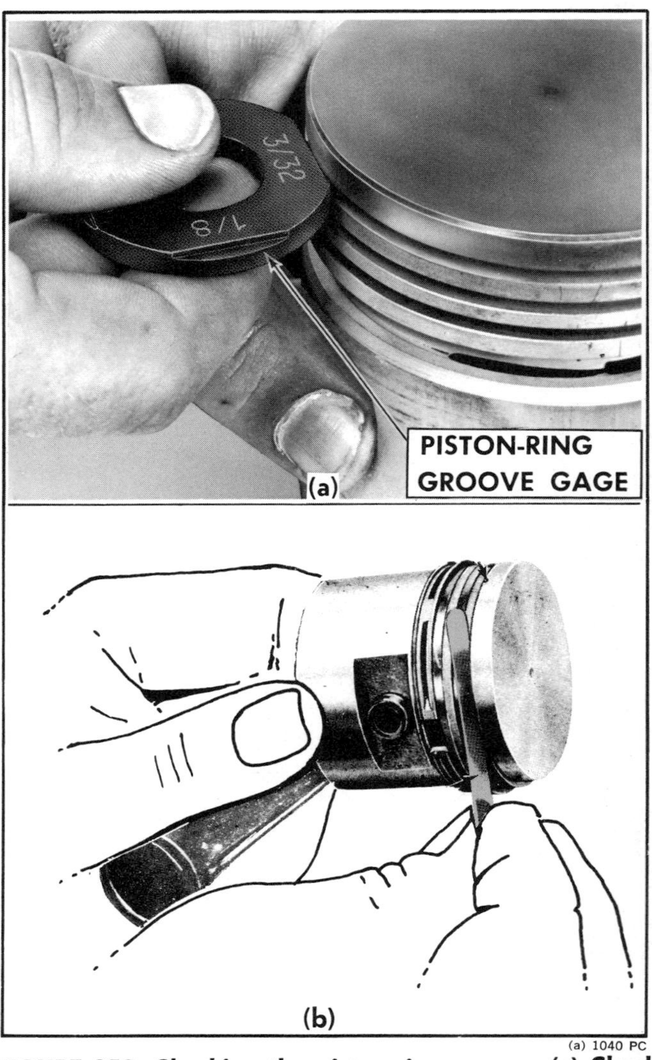

FIGURE 258. Checking the piston ring groove. (a) Checking the width, or (b) checking the clearance.

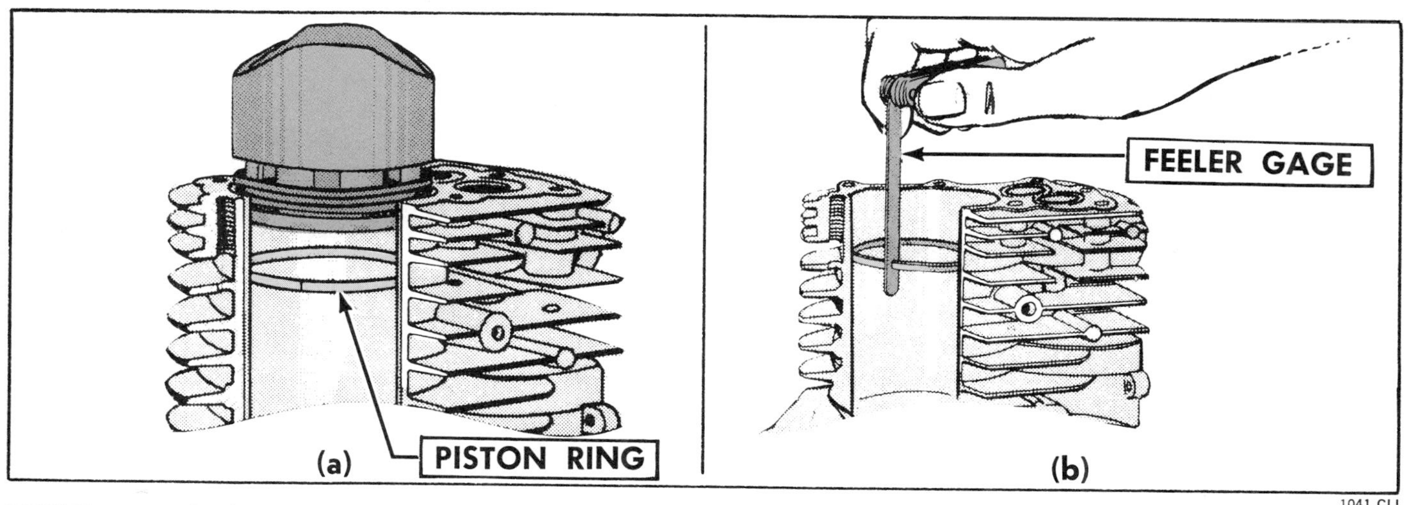

FIGURE 259. Checking the piston ring gap. (a) Insert the ring with a piston. (b) Check the gap with a feeler gage.

or install new rings and use a thickness gage (Figure 258b). See your repair manual for the proper clearance. It may range from .05 mm to .18 mm (.002 in to .007 in).

6. *Check the rings for wear (Figure 259a).*

   Insert the rings one at the time into the cylinder. Push the ring in with a piston (rings removed) so that it will be in its natural position in relation to the cylinder wall (Figure 259a). Measure the ring gap with a feeler gage (Figure 259b).

   Refer to your repair manual for the proper ring gap. It ranges from .18 mm to 1.2 mm (.007 to .045 in). If the *ring gap is too much* and your cylinder is not worn too much, replace the rings with a new set of standard-size rings. Special chrome rings are available for warped cylinders.

INSPECTING THE PISTON ROD AND BEARINGS

1. *Check the condition of the rod.*

   Inspect for straightness and for cracks.

2. *Install rod temporarily and check for side clearance between the rod and the crankshaft.*

   Check with a feeler gage.

   Refer to your repair manual for the proper tolerance. Too much clearance indicates excessive wear. If left in that condition, the rod will "slap" during operation and increase the wear rate.

3. *Remove rod from crankshaft.*

4. *Place a piece of plastigage across the full width of the bearing cap (Figure 260a).*

7. *Check the piston pin and bore.*

   Refer to your repair manual for the proper dimensions. Use an inside gage and micrometer for the piston-pin bore and an outside micrometer for the pin.

   Some manufacturers make oversize piston pins. If they are available, you can hone the piston and install the oversize pin. Others supply new connecting rods and bearing to use when the holes in the piston are worn. Then a standard-size pin can be used. Installing oversize pins and bearings is a delicate operation and should be done at a shop which is well equipped for this job.

If your piston has **needle bearings** that show flat spots or do not fit snugly in position, replace them.

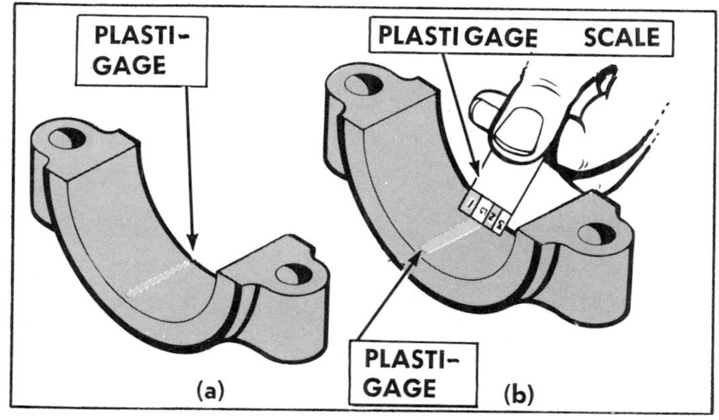

FIGURE 260. Checking rod-bearing clearance with plastigage. (a) Plastigage is placed in the bearing cap. (b) Checking the width of the expanded plastigage after the rod-bearing cap has been installed and tightened to the specified torque and then removed.

"Plastigage" is a plastic ribbon that is designed to expand in proportion to the amount of pressure applied to it.

5. *Install the rod and cap on the crankshaft.*
6. *Tighten the cap nuts or screws to the torque given in your repair manual.*
7. *Remove the rod cap and measure the flattened plastigage at its widest point with the plastigage scale (Figure 260b).*
   The number corresponding to the width of the flattened plastigage is the bearing clearance in thousandths of an inch. If the **clearance is more than allowed** in your repair manual, replace the rod or rod bearing. Some manufacturers, however, recommend filing away some of the rod-bearing-cap flange. If you do, be sure to take the same amount off of both sides. Very little filing will be needed.

   Too much filing will cause the bearing to be out of round and will result in very rapid wear.

INSTALLING THE PISTON RINGS

1. *Check the end gap of each new ring (Figure 259b).*

   If the **end gap is too small,** file the end of the ring squarely to the proper gap. If the **end gap is too large** and the cylinder is the proper size, discard the ring.

2. *Apply crankcase oil to the rings and pistons.*
3. *Install the rings on the piston (Figures 259 and 261).*

   Begin with the bottom ring. If you use expander springs under the lower oil ring, install them first.

   Use a piston ring expander if available. If not available, you can install them by hand. Start open end of ring in slot first; then push it over top of piston into position.

   Be sure you install them right side up. Some are marked "top"; others are not marked. If not labeled, you cannot tell by looking at the rings how they should go. Refer to the notes you made when removing the rings and/or to your operator's manual. Most manuals give specific instructions (Figure 262).

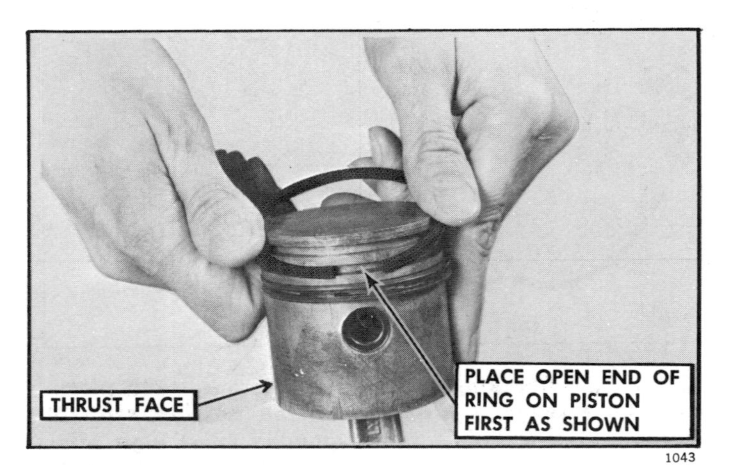

FIGURE 261. Installing rings without a ring expander.

4. *Check the rings for freedom of movement in the grooves.*

   If they are not free to move in the grooves, there is a danger of their scoring the cylinder wall and breaking.

5. *Stagger the ring gaps.*

   Some claim that if the ring gaps are in line on a 4-cycle engine, the engine will lose compression and use oil. Others dispute this claim.

   With **2-cycle engines,** however, it is agreed you should have the ring gaps on any two adjacent rings 180° apart. Some have stops at the ring-gap position to insure this.

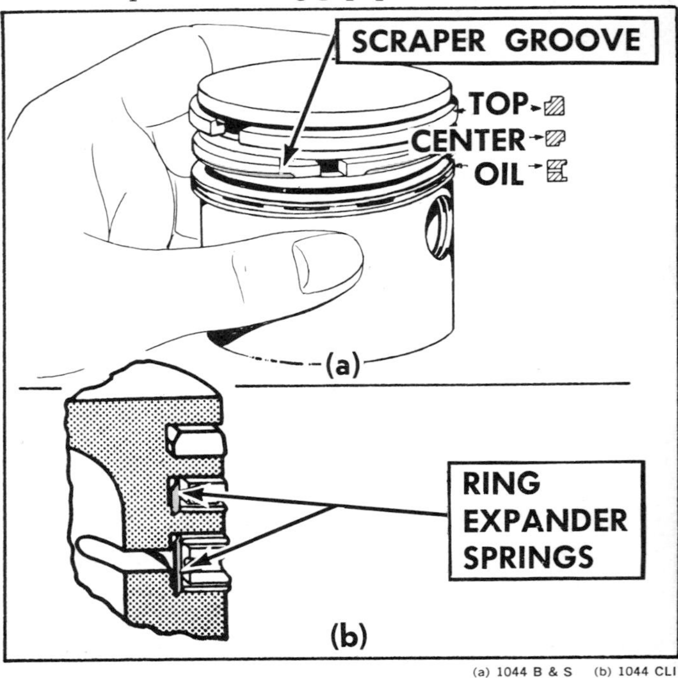

FIGURE 262. Rings must be installed right side up and in the proper grooves. (a) One example of the relative location of a standard set of rings. (b) Cross section of a chrome re-ring set for out-of-round cylinders using expander springs back of the two lower rings (See Figure 237).

INSTALLING THE PISTON PIN

1. *Install the needle bearings if used.*

   Coat them with multi-purpose grease so they will stay in place.

2. *Install a lock ring on one side of the piston.*

3. *Apply oil to the rod bearing and piston bore.*

4. *Align the rod bearing with the piston bore and insert the piston pin.*

   Insert solid ends first on hollow pins. Some are easy to install with a handpress fit. Others are "sweat fitted."

   If the *pin appears to be too large,* it probably requires sweat fitting. You will have to heat the piston to expand the hole, and shrink the pin by cooling with dry ice before installation.

   Pistons on cross-scavenged, **2-cycle engines** must have the abrupt side of the piston head facing the fuel intake.

   Some pistons have hollow pins with one closed end. With these the closed end should be turned toward the exhaust port.

5. *Install the second piston-pin lock ring.*

INSTALLING THE PISTON-AND-ROD ASSEMBLY

If you are doing a complete overhaul job, before installing the piston and rod, proceed with the steps under "Repairing Crankshaft Assemblies," page 248.

To install the piston-and-rod assembly, proceed as follows:

1. *Oil the piston rings and cylinder.*
   Oiling makes the installation easier and supplies lubrication when the engine is started — before normal lubrication can be supplied.

2. *Install ring compressor on rings (Figure 263).*
   Compress the rings completely; then release slightly.

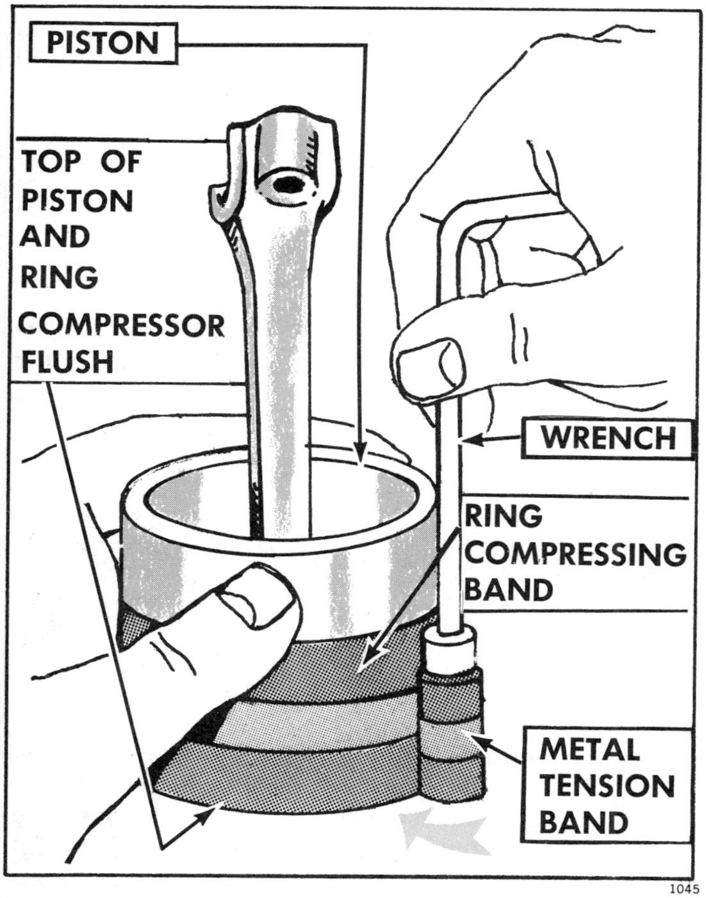

PISTON
TOP OF PISTON AND RING COMPRESSOR FLUSH
WRENCH
RING COMPRESSING BAND
METAL TENSION BAND
1045

FIGURE 263. Attaching a piston-ring compressor.

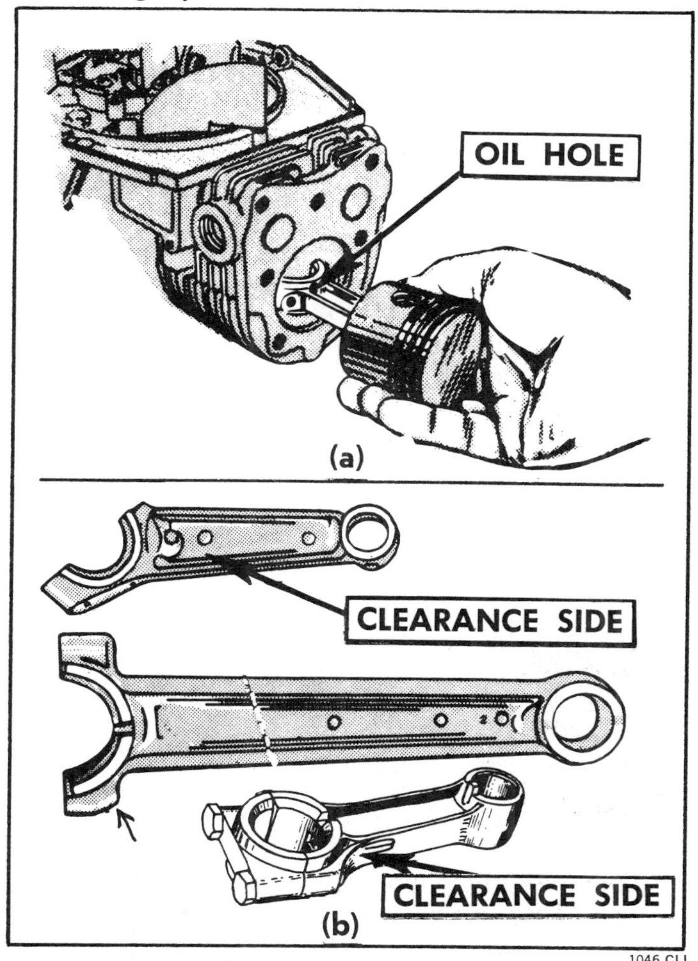

FIGURE 264. Some rods must be installed in a certain position. (a) Rod with a special oil hole. (b) Special-clearance rods.

1047 KOH

FIGURE 265. Tapping the piston into the cylinder.

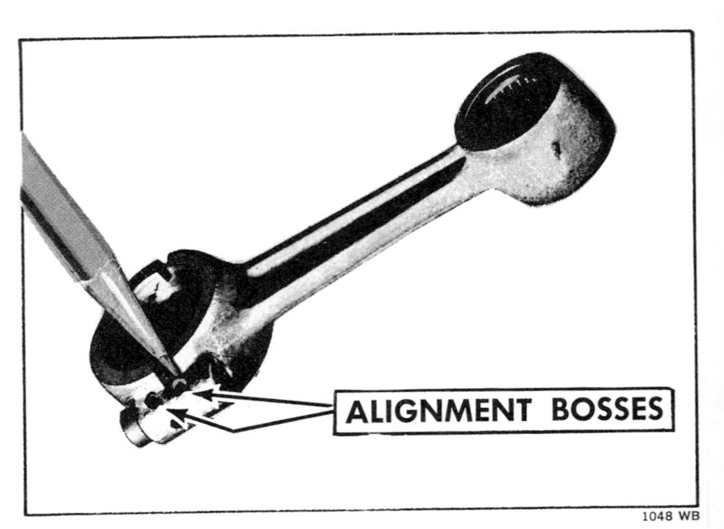

1048 WB

FIGURE 266. Match marks embossed on rod and rod-bearing cap.

3. *Place the piston and rod over the cylinder (Figure 264a).*

   Some engines have a certain way the rod fits on the crankshaft (Figure 264).

   For example, some engines with horizontal cylinders have a special oil hole in the rod aligned with the crankshaft. This oil hole must be on the top side so that oil will flow to the bearing (Figure 264a). Others have a special clearance provision to bypass the bottom edge of the cylinder as the crankshaft turns (Figure 264b).

4. *Center ring compressor over the cylinder, and tap the piston and rings into the cylinder (Figure 265).*

5. *Install the rod-bearing cap.*

   Install the oil dipper if it is attached to the rod-bearing cap. There should be a boss or some marking for aligning the cap to the rod (Figure 266). If not, refer to the notes you made during removal. The rod and cap are made in two separate pieces and then tightened together and bored. If you turn the cap around, you may change the bore alignment.

6. *Install the nuts or cap screws and apply the proper torque.*

   Use a torque wrench and tighten to specifications given in your repair manual. These specifications vary with the size of the engine. They range from 38 N·m to over 440 N·m (35 lb f·in to over 400 lb f·in).

   When the rod bearing was bored, the cap was tightened to a certain torque. This is why it is necessary for you to apply the same torque that was applied when the rod was bored so that the bearing will not be distorted.

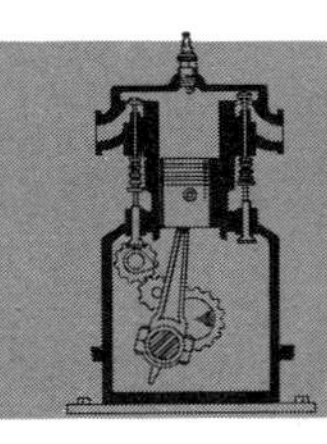

A. CHECKING AND REPAIRING CYLINDERS

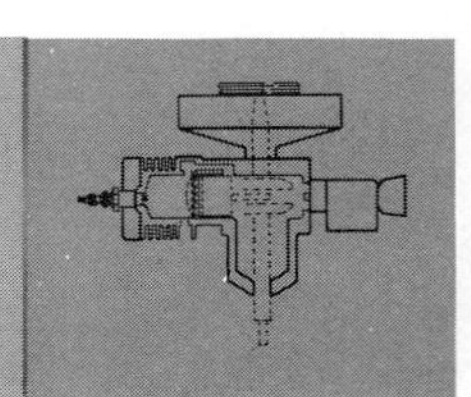

There are three types of cylinders used on small engines. They are as follows: (1) **cast iron** — the cylinder is part of the cast-iron block; (2) **aluminum** — the cylinder is part of the aluminum cylinder block; and (3) **steel sleeves** in an aluminum cylinder block.

Even though aluminum is soft, compared to cast iron or steel, satisfactory wear has been experienced by using aluminum cylinders when proper rings are used. One experiment showed only a .10 mm (.004 in) total ring-and-bore wear after 1,000 hours of operation.

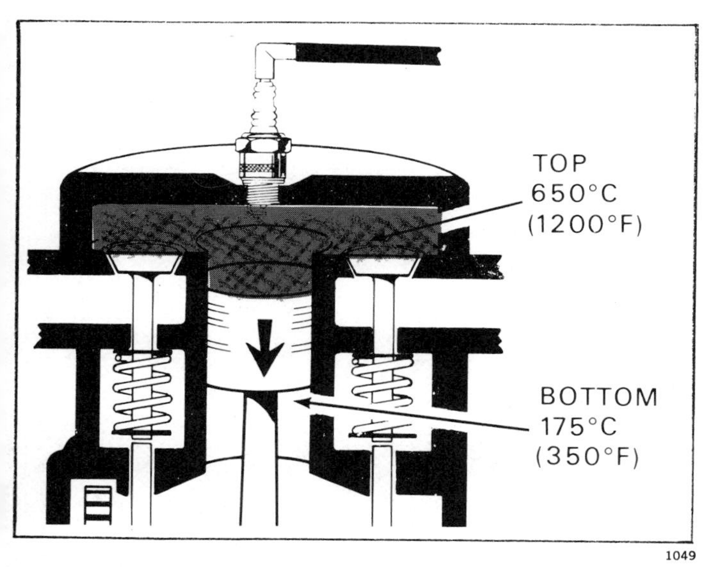

FIGURE 267. The cylinder is subjected to extreme temperature differences.

Cylinders are subjected to extreme temperature differentials. The top of the cylinder is next to the combustion chamber, where the temperature may reach 650°C (1200°F). The bottom of the cylinder is in the crankcase (Figure 267) where temperatures are comparatively cool. The cylinder walls on all air-cooled engines take the heat from the piston and rings and conduct it to the cooling fins. This is why you should always keep the cooling fins free of dust, trash and oil.

Cylinders wear mostly at the top (Figure 268). This is the point of maximum stress, where the greatest force is exerted against the piston because of combustion. If the cylinder is worn very much, there is a ridge left at the top of the cylinder which should be removed before you attempt to remove the piston. Otherwise, there is a danger of breaking the rings and damaging the piston. See Figure 245. Of course, if you rebore the cylinder, this will be bored out.

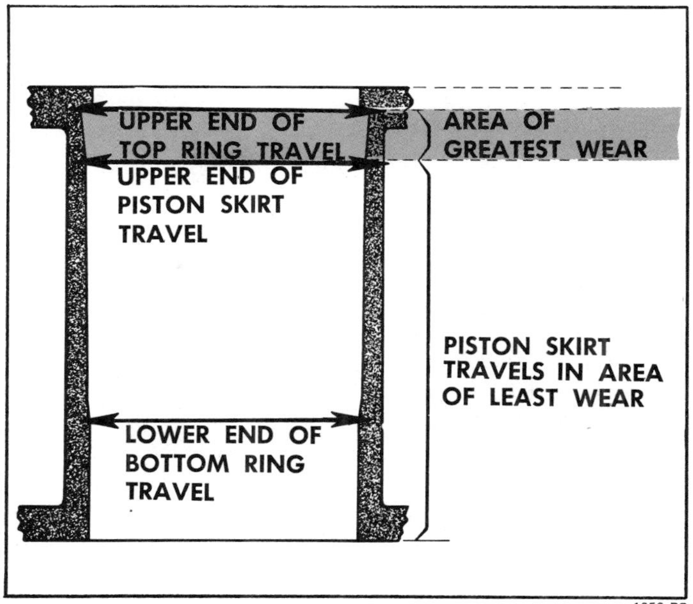

FIGURE 268. Cylinder cross-section showing the area of greatest wear.

Repairing cylinders is discussed under the following headings:

1. Tools and materials needed.
2. Inspecting the cylinder.
3. Reboring the cylinder.

TOOLS AND MATERIALS NEEDED

1. Inside micrometer or cylinder gage
2. Hone and/or boring bar (coarse and fine honing stones)
3. Clean rags
4. Cleaning solvent (mineral spirits, kerosene or diesel fuel)
5. Soap and water
6. Paper and pencil
7. Cutting fluid

INSPECTING THE CYLINDER

There are three reasons for measuring the cylinder diameter: (1) to determine if it is **worn too much for a new standard ring** to seat properly—tolerances range from .08 mm to 20 mm (.003 in to .008 in); (2) to determine if the cylinder is **warped or out of round**—tolerances range from .04 mm to .08 mm (.0015 in to .003 in); (3) to determine if the cylinder has **too much taper**—tolerances range from .10 mm to .20 mm (.004 in to .008 in).

If you are unable to repair the cylinder by resizing without exceeding any of these tolerances, you will have to replace the cylinder block.

1. *Clean the cylinder inside and out with petroleum solvent.*

   Clean all dirt from the cooling fins and all carbon from the cylinder head.

2. *Inspect the cylinder block for cracks, stripped bolts threads, broken fins and scored cylinder walls.*

   Scored cylinder walls are a sign of overheating and/or lack of proper lubrication.

1 CYLINDER DIMENSION	2 Actual Size (Present) cm (in)	3 Standard Size (New) cm (in)	4 Actual Wear cm (in)	5 Maximum Wear Allowed cm (in)	6 Reboring Required	7 Minimum Oversize cm (in)	8 Rebore to Oversize Dimension cm (in)
Top	5.093 (2.0050)	5.08 (2.0000)	.013 (.0050)	.007 (.0030)	Yes	.025 (.0100)	5.105 (2.0100)
Center	5.087 (2.0030)	5.08 (2.0000)	.007 (.0030)	.007 (.0030)	No		5.105 (2.0100)
Bottom	5.083 (2.0010)	5.08 (2.0000)	.003 (.0010)	.007 (.0030)	No		5.105 (2.0100)
ROTATE 90° TO ABOVE MEASUREMENTS							
Top	5.098 (2.0070)	5.08 (2.0000)		.007 (.0030)			
Center		5.08 (2.0000)		.007 (.0030)			
Bottom		5.08 (2.0000)		.007 (.0030)			

FIGURE 269. Chart for determining cylinder wear. The figures shown are examples only.

Dirt causes uniform wear, not scoring.

3. *Prepare chart similar to the one shown in Figure 269.*

   Record the standard size dimension of your cylinder in column 2 and the maximum wear allowable in column 5 from your repair manual.

4. *Measure and record the diameter of the cylinder for wear (Figure 270).*

   Take *measurements* in six places (Figure 271) and record in column 2 (Figure 269). (1) Start *at the top* of the piston-ring travel, then (2) *at the center* of the piston-ring travel, and then (3) *at the bottom* of the piston-ring travel. Rotate gage 90 degrees and repeat measurements.

5. *Determine the amount of cylinder wear for each position.*

   Subtract the "Standard Size" dimensions (Column 3) from the "Actual Size" dimensions (Column 2). This gives the amount of actual wear. Record the amount of wear in column 4.

   Example: 5.0930 cm − 5.0800 cm = .130 cm (2.0050 in − 2.0000 in = .0050 in)

6. *Compare the amount of "Actual Wear" (column 4) with "Maximum Wear Allowed," column 5 (Figure 269).*

   If the actual wear is greater than the maximum wear allowed at any of the six positions, resize the cylinder and get a new oversize piston and rings to fit it. Cylinders that cannot be rebored within the oversize limits must be replaced. Most manufacturers make oversize pistons and rings in increments of .025 cm, .050 cm and .075 cm (.010 in, .020 in and .030 in) oversize.

7. *Determine the oversize dimensions of the cylinder and record them in the chart.*

   Example: 5.0800 cm + .0215 cm = 5.1050 cm (2.000 in + .0100 in = 2.0100 in)

   Resize to the smallest oversize that will even up the cylinder. See column 8, Figure 262.

8. *Check to see if the cylinder is out of round.*

   Prepare a chart similar to the one shown in Figure 272 and record the measurements from columns 2 and 3 (Figure 269) into column 2 (Figure 272).

   Subtract the dimensions at each of 3 measurements from the measurements taken at 90° at the same location in the cylinder.

   Example: 5.0980 cm − 5.0930 cm = .0050 cm (2.0070 in − 2.0050 in = .0020 in)

   If the difference at any of the three locations is greater than the maximum allowable out-of-round dimensions, you must replace the block or resize the cylinder. **If the actual out-of-round in column 4 is less**

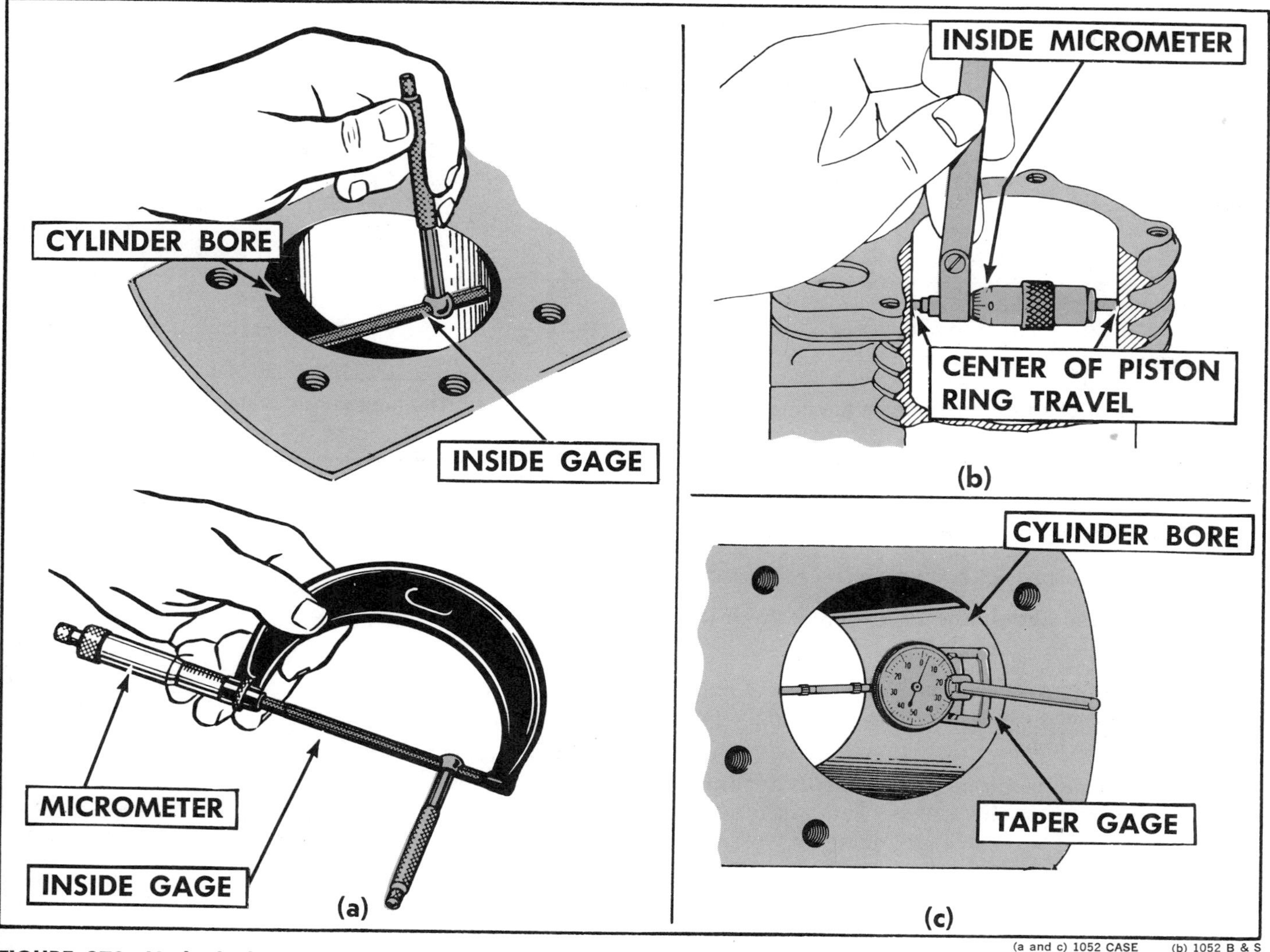

FIGURE 270. Methods for measuring the diameter of a cylinder. (a) Adjustable inside gage and micrometer for measuring gage length. (b) Inside micrometer, (c) Special gage and dial indicator for measuring cylinder taper.

than the maximum out-of-round allowance in column 5, proceed to step 9.

9. *Check the cylinder for taper.*

Subtract the measured dimensions at the bottom of the cylinder from the measured dimensions at the top (Figure 272).

If the difference exceeds the maximum allowable given in your repair manual, replace the block or resize the cylinder.

NOTE: If any one dimension in the cylinder exceeds the factory allowance, the cylinder will require resizing; and there is no need to measure all the other dimensions.

If the taper is within allowable tolerance, hone the cylinder lightly and reinstall your standard-size piston with new standard-size rings.

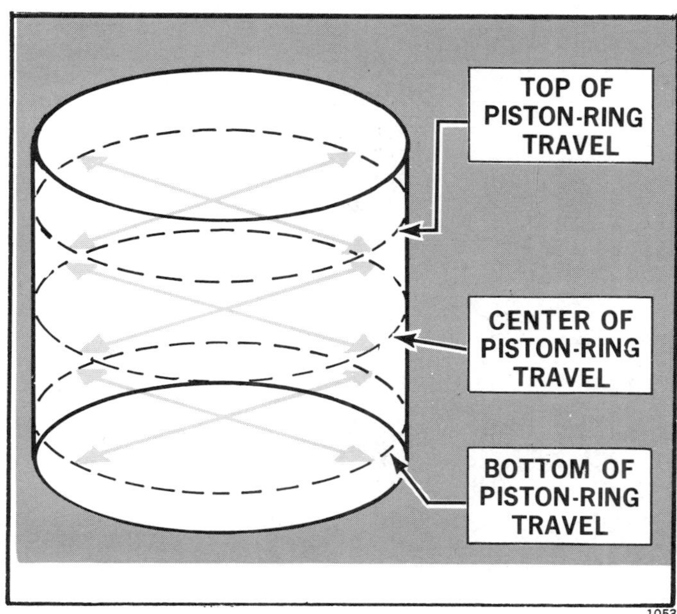

FIGURE 271. Measure the cylinder diameter in six different positions.

1	2 Cylinder Dimension cm (in)	3 Cylinder Dimension at 90° cm (in)	4 Actual Out of Round Computed cm (in)	5 Maximum Allowance Out of Round cm (in)	6 Maximum Allowance for Taper cm (in)	7 Reboring Required
Top	5.093 (2.0050)	5.098 (2.0070)	0.005 (.0020)	0.0033 (.0013)		Yes
Center						
Bottom						

FIGURE 272. A chart for determining the amount of cylinder is out-of-round. The figures are examples only.

RESIZING THE CYLINDER

1. *Anchor the cylinder block (Figure 273).*
2. *Select the proper boring tools (Figure 274).*

   Hones for aluminum are different from those of cast iron or steel.

   A boring bar uses a high carbon steel cutting tool. Procedures given here apply only to the use of a hone.

   Most manufacturers recommend a **"hone"** for resizing small engines. This machine uses coarse honing stones for removing most of the bore and fine honing stones for finishing.

   A **boring bar** (Figure 274c) is necessary

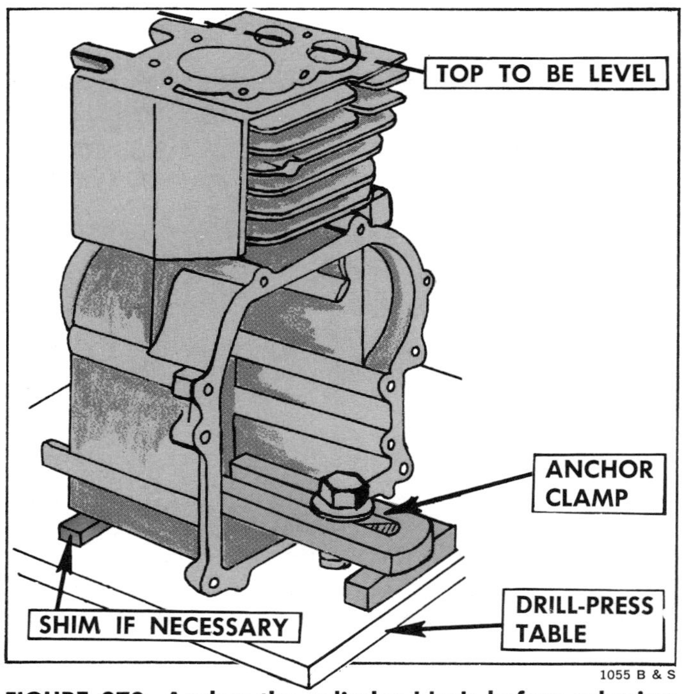

FIGURE 273. Anchor the cylinder block before reboring the cylinder.

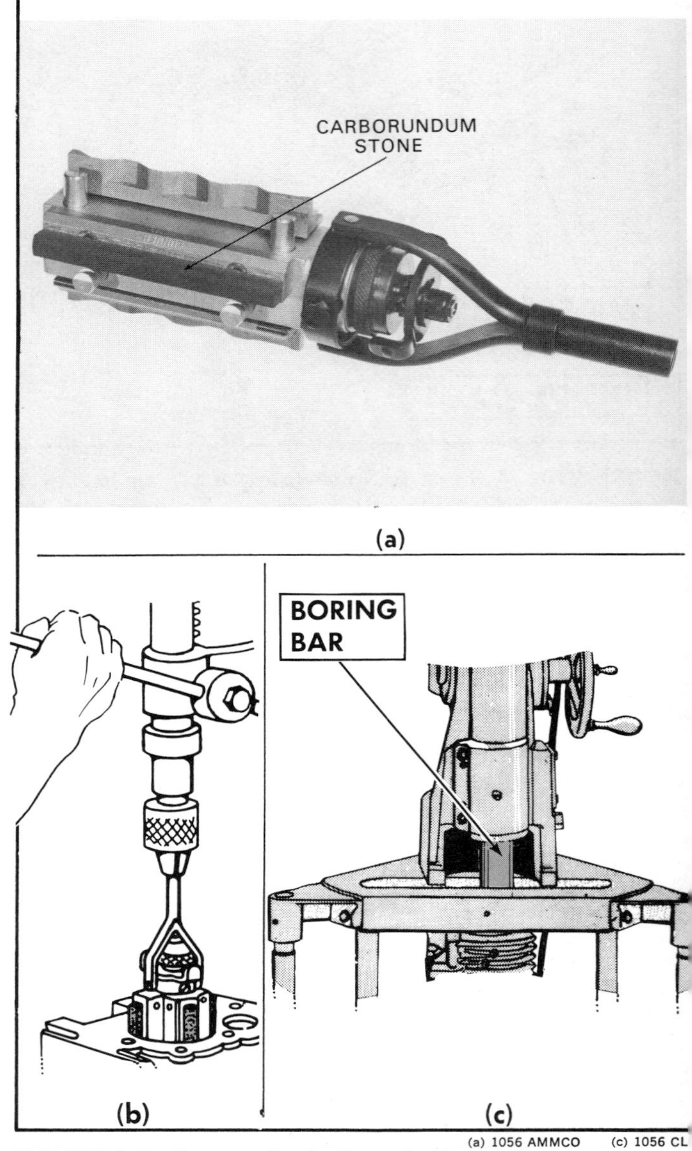

FIGURE 274. Two methods for reboring cylinders are as follows: (a and b) hone with silver carbide abrasives or stones, and (c) boring bar with a cutting tool.

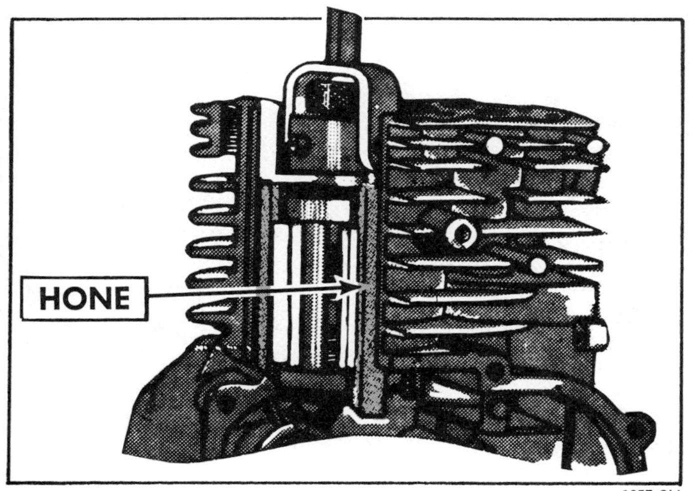

FIGURE 275. Allow the stones to protrude 1.3 cm to 2.5 cm (½ in to 1 in) below and above the cylinder when honing.

for reboring some cylinders on 2-cycle engines—ones that do not have a removable head. Some manufacturers recommend replacing rather than reboring this type of cylinder.

Be sure the hone is in good condition and the stones are not worn to much.

3. *Set the drill press to operate from 450 to 700 r.p.m.*

4. *Lower the hone to the point where the lower end protudes 1.3 cm to 2 cm (½ in to ¾ in) past the end of the cylinder (Figure 275).*

5. *Rotate the adjusting nut on the hone until the stones come in contact with the cylinder wall at the narrowest point.*

   The narrowest point is normally on the crankcase end of the cylinder.

6. *Turn the hone by hand.*

   If you cannot turn it, it is too tight. Loosen it until it can be turned by hand.

7. *Start the drill.*

8. *Move the hone up and down in the cylinder approximately 40 cycles per minute.*

   Usually it is necessary to work out the bottom of the cylinder first because it is smaller. Then when the cylinder begins to take a uniform diameter, move the hone up and down all the way through the cylinder. Follow the hone manufacturer's recommendation. Some require oil, and some will not work with even a small amount of oil on the cylinder wall.

FIGURE 276. To aid in seating the new rings, the cylinder should have a cross-hatched surface.

9. *Check the diameter of the cylinder regularly during honing.*

   Stop the drill before measuring, and remove the hone from the cylinder.

10. *Change the stone grit size.*

    When the cylinder is approximately .0050 cm (.0020 in) within the desired bore, change to fine stones and finish the bore. Finish should not be perfectly smooth. You should have a cross hatch pattern similar to the one illustrated in Figure 276.

11. *Clean the cylinder block thoroughly.*

    Use soap, water and clean rags. Clean the cylinder wall for a "white glove" inspection. You should not be able to soil a clean white rag on the cylinder wall. Do not use solvent or gasoline. They wash all the oil from the cylinder but do not remove metal particles produced during honing. These cause wear.

12. *Dry the cylinder and coat it with crankcase oil.*

    Cover the cylinder block if you are not ready to install a new piston and rings.

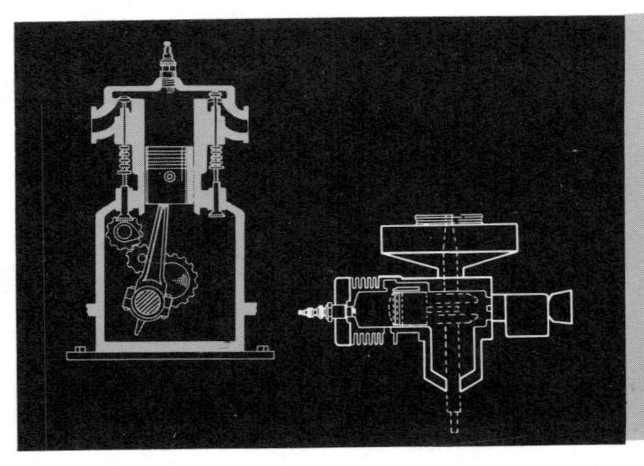

VII. Repairing Lubricating Mechanisms In 4-Cycle Engines

Without oil, the metal rubbing parts of your engine would become overheated from friction and finally weld themselves together.

Wherever two metal parts move, one against the other, there must be some form of cushion between them. Petroleum companies have developed high-quality crankcase oils that provide this cushion as needed for different engines. (See "Lubricating Small Engines," Part 1.) The problem encountered by engine manufacturers is to design a mechanism that will furnish a **constant** and **adequate supply of oil** to all bearing surfaces.

Repairing lubricating mechanisms in 4-cycle engines is discussed under the following headings:

1. Types of lubricating mechanisms and how they work.
2. Importance of proper repair.
3. Tools and materials needed.
4. Removing and servicing the oil pump.

TYPES OF LUBRICATING MECHANISMS AND HOW THEY WORK

Lubrication is provided in **2-cycle engines** by the oil which is mixed with the gasoline. It is fed first into the crankcase and pressurized. Small droplets of oil suspended in the fuel mixture collect on the bearing journals and gradually feed into the bearing surfaces.

Four-cycle engine lubrication is different. Oil is supplied to the bearings by means of a dipper, scoop, slinger, gear pump or barrel-and-plunger type of pump (Figure 277).

How these five types of oil distribution mechanisms work is as follows:

- **The dipper type rotates with the crankshaft throw.** On the lower end of the stroke the dipper dips into oil, picks it up, and throws it against the sides of the crankcase. Some of it bounces off the crankcase walls and falls onto the crankshaft, thus lubricating the bearings. It is also splashed onto the cylinder walls and lubricates the piston and rings. Oil is splashed onto the cam gear, camshaft bearing and cam lobes. It is very important that you maintain the proper oil level in your engine for this type oiler. This is especially true if your engine does not have an oil trough. If the oil level gets too low, there will not be enough lubrication of the engine. The engine in Figure 277a has a plunger pump which is operated by an eccentric on the camshaft. It maintains a constant oil level for the dipper. Many small engines with splash-type systems have no such provision. An oil strainer is used wherever a pump is used.
- **An oil scoop type on the cam gear** (Figure 277b) operates in a similar manner to the dipper. It is used on vertical crankshaft engines. As the crankshaft turns, the scoop agitates the oil and throws it onto all parts requiring lubrication.
- **The oil slinger type** (Figure 277c) is partially submerged in the oil at all times. It picks up oil as it turns and slings it around inside the crankcase to supply lubrication to all parts.
- **The gear-type oil pump** (Figures 277d and 278) is different from the pumps previously mentioned. It is a positive acting type of pump and requires a relief valve. Some engines provide for regulating the oil pressure by adjusting the relief valve. Others have the desired oil pressure designed into the pump,

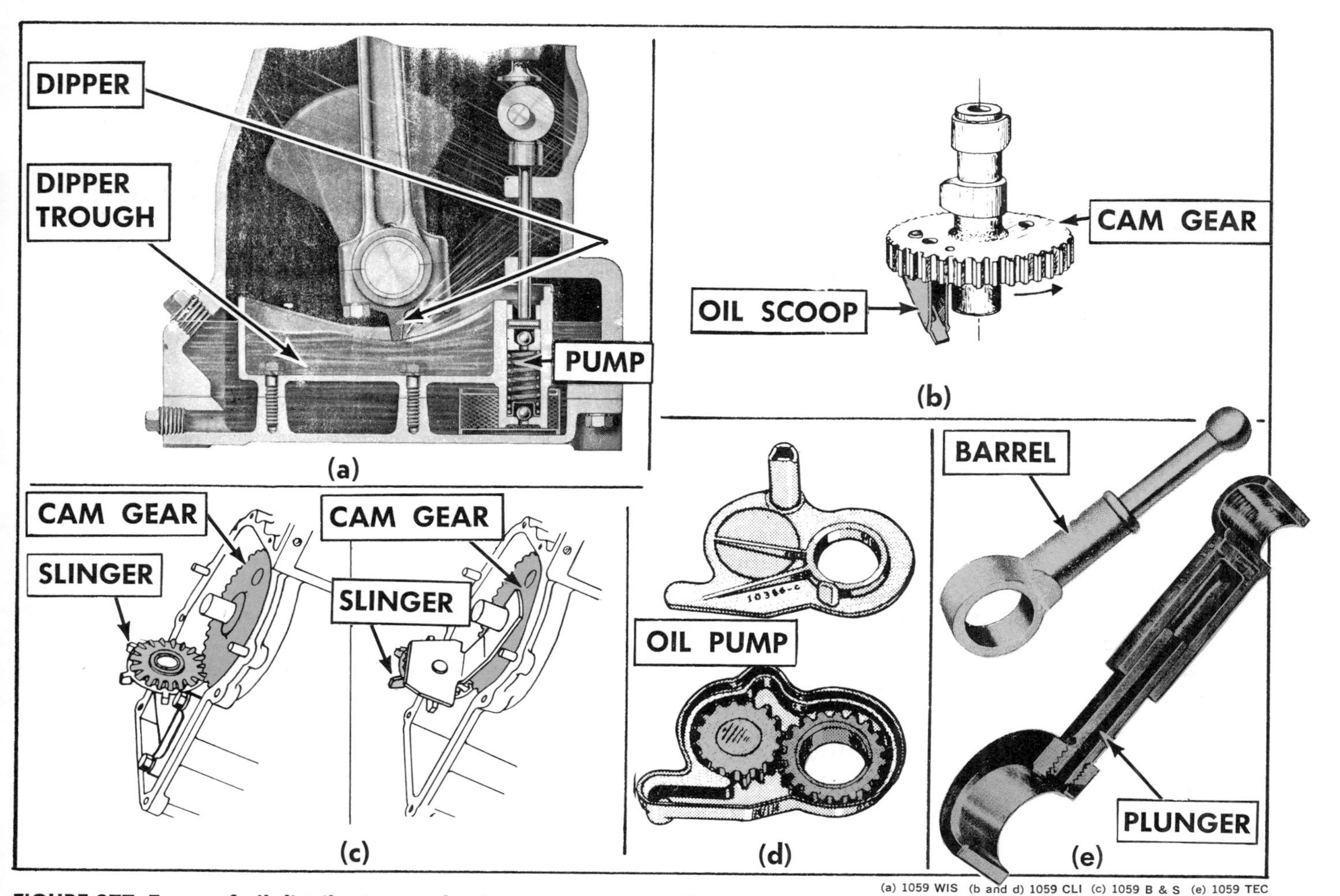

(a) 1059 WIS (b and d) 1059 CLI (c) 1059 B & S (e) 1059 TEC

FIGURE 277. Types of oil distribution mechanisms used on 4-cycle engines. (a) Dipper on rod-bearing cap picks up oil and slings it over the crankcase. A pump is also supplied with this engine to maintain an oil level in the dipper trough. (b) A scoop on the cam gear. (c) A rotary oil slinger driven from the cam gear. (d) Gear-type pump driven from the cam gear. (e) Barrel-and-plunger type of pump operated from an eccentric drive located on the camshaft.

and the relief valve is only a safety measure in case something happens to cause the pressure to build up excessively in the crankcase. See Figure 280.

Oil is picked up from the oil sump and is either sprayed or pumped through drilled passageways to the bearings.

- **Barrel-and-plunger type of pump** (Figures 277e and 279) is located in the oil sump. This type is also a positive-acting pump. It picks up oil from the sump and pumps it into a drilled passageway through the camshaft and crankshaft and into the bearings. The piston and cylinder are lubricated by the oil splashed on the cylinder walls and on the piston by the splashing action from the crankshaft.

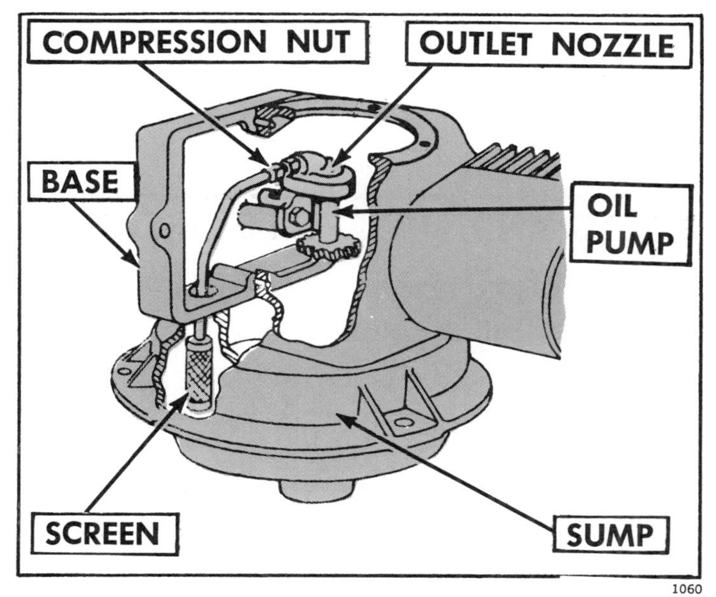

1060

FIGURE 278. Gear-type oil pump.

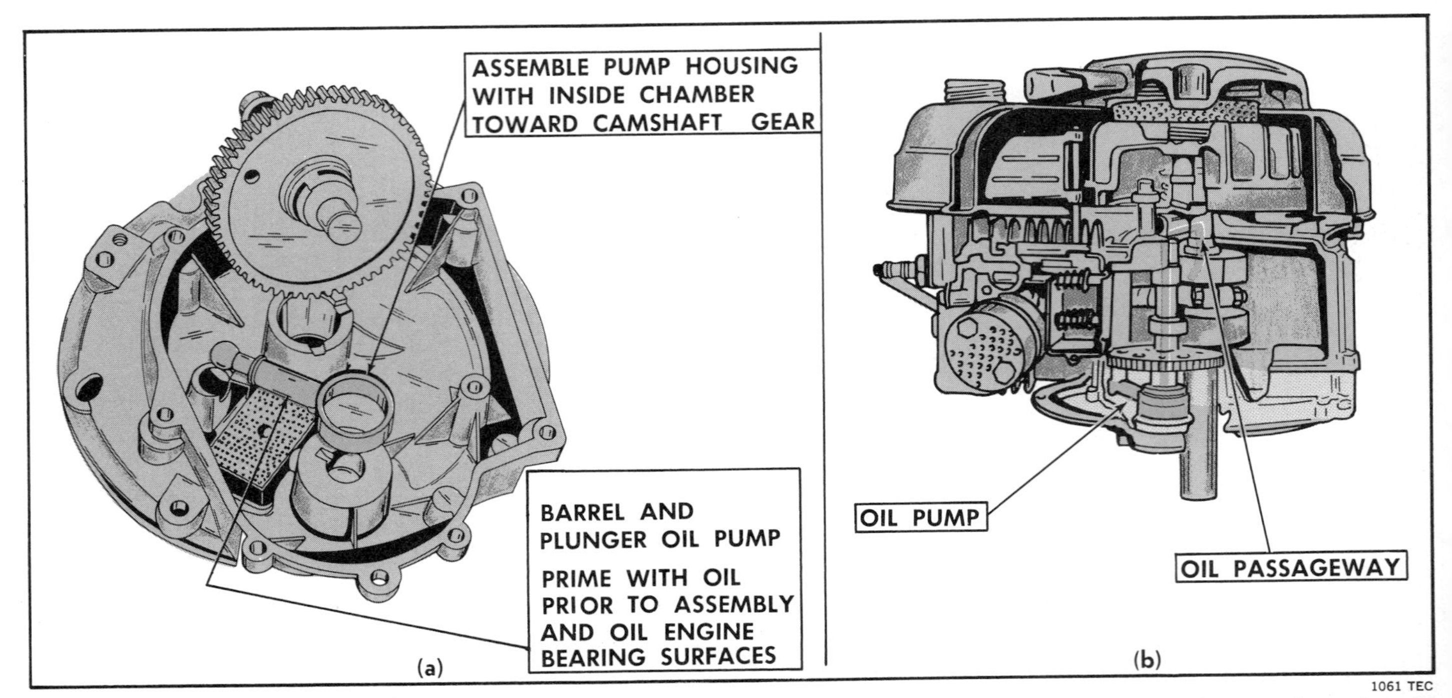

FIGURE 279. Barrel-and-plunger type of pump. (a) Pump is located on the oil sump. (b) It pumps oil through the camshaft and crankshaft to the camshaft and crankshaft bearings.

IMPORTANCE OF PROPER REPAIR

If your engine is equipped with a **dipper, scoop or slinger,** your main concern is to see that they are installed in the proper position for distributing oil. They require no further servicing.

Oil pumps and relief valves give little trouble and are not generally reparable.

You can check most pumps for proper operation by placing the suction end of the pump in oil and operating it.

A relief valve, provided on some engines for regulating the oil pressure, can usually be adjusted without removing the pump (Figure 280).

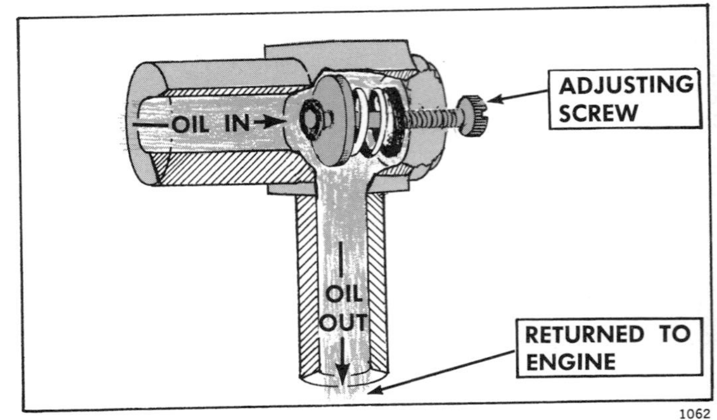

FIGURE 280. Principle of an oil-pressure regulating valve. Oil pressure pushes against the spring-loaded valve while the engine operates. When pressure becomes excessive, the valve lifts and oil discharges back into the crankcase.

TOOLS AND MATERIALS NEEDED

1. Open-end wrenches — 3/8″ - 9/16″
2. Slot-head screwdrivers — 6″
3. Phillips-head screwdriver — 6″
4. Crankcase oil
5. Squirt can
6. Clean rags
7. Petroleum solvent (mineral spirits, kerosene or diesel fuel)

REMOVING AND SERVICING THE OIL PUMP

Figures 281, 282, 283, and 284 show different types of oil-pump assemblies. If you have an oil pump on your engine, it may look different from any of these; but the operating principle is the

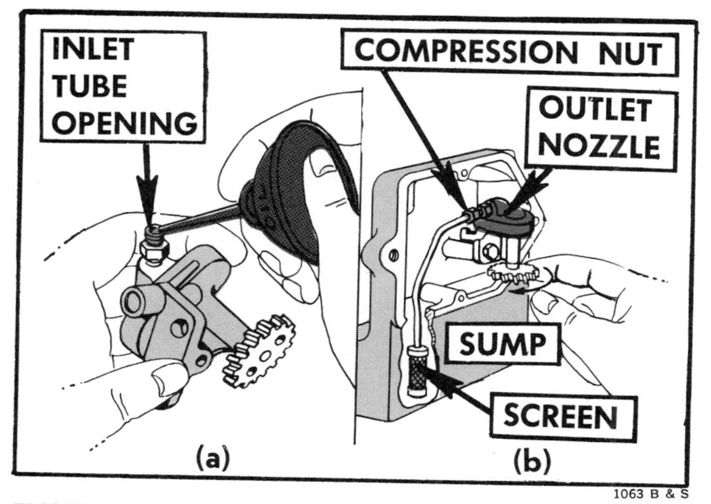

FIGURE 281. One type of removable gear pump. (a) Priming after cleaning. (b) Checking the pump for proper operation.

same for each one of them. Since you cannot see the oil pumps, it will be necessary for you to refer to your service manual to determine which type of oil pump is used in your engine.

a. If your oil-pump unit is the **removable gear type** (Figure 278), proceed as follows:

1. *Open the crankcase as explained under "Removing the Piston and Rod Assembly," page 218.*
2. *Remove the oil pump.*

   Most oil-pump housings are attached to the crankcase by capscrews. Observe carefully how the unit is assembled so you can reassemble it in the same way.
3. *Disconnect the pipe at the pump and remove the pipe and screen.*
4. *Flush the pump with a cleaning solvent.*

   Turn the pump while it is submerged in cleaning solvent; then blow solvent out with compressed air.
5. *Clean the screen and pipe with petroleum solvent.*
6. *Fill the pump with oil for priming (Figure 281a).*
7. *Connect the pipe to the pump but leave the nut loose.*

   This is for easy installation.
8. *Attach the pump to the housing.*

   Use cap screws.
9. *Center the oil pipe and strainer over the sump opening and tighten the compression unit.*
10. *Check the pump for proper operation (Figure 281b).*

    Place enough oil in the sump to cover the screen and spin the pump. Oil should flow freely.

b. If your oil pump unit is a **gear design of the non-removable type** — pump housing part of casting (Figure 282), — proceed as follows:

1. *Open the crankcase as explained under "Removing the Piston and Rod Assembly," page 218.*

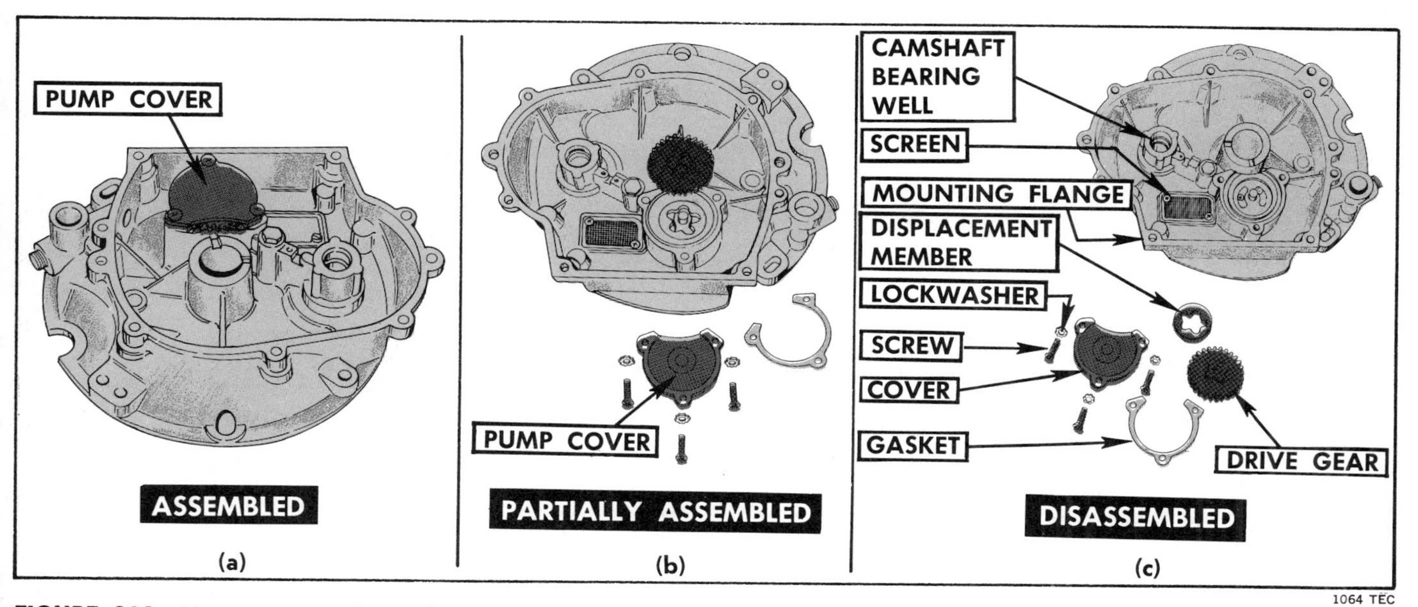

FIGURE 282. A gear pump located in the oil sump. (a) Pump with cover. (b) Pump partially diassembled. (c) Pump disassembled — except screen.

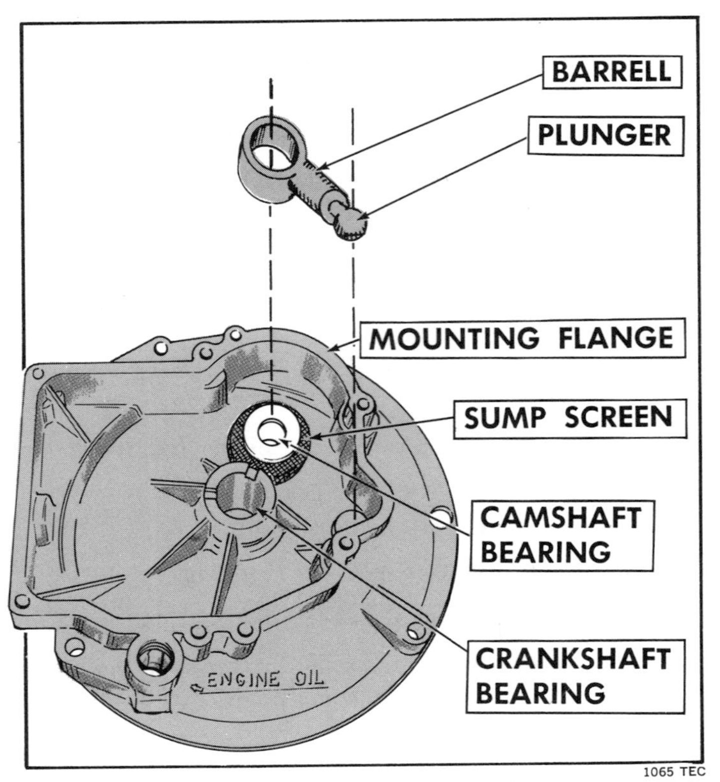

FIGURE 283. A removable barrel-and-plunger type of pump.

2. *Remove the cap screws that hold the pump cover.*
3. *Remove the cover (Figure 282a).*
4. *Lift out the drive gear and piston (Figure 282b and c).*
5. *Remove the oil screen.*
6. *Clean and inspect all the parts.*
7. *Reassemble the pump.*
8. *Put a new gasket on the pump cover and replace it.*
9. *Install the oil screen.*
10. *Check for proper operation.*

    Fill the oil sump (over the screen) with oil and spin the pump. Oil should flow freely into the camshaft-bearing well.

c. If your oil pump is a **removable barrel-and-plunger type** (Figure 283), proceed as follows:

1. *Open the crankcase as explained under "Removing the Piston and Rod Assembly," page 218.*
2. *Remove the pump (Figure 283).*

   This is a simple-type plunger pump. It is anchored to the camshaft by a ring and to the crankcase by a ball joint or a pivot post.

   When you remove the base plate from your engine, the pump will most likely remain on the camshaft. All you have to do is lift it off.
3. *Disassemble the pump.*

   Remove the plunger from barrel.
4. *Clean and inspect the parts, screen and sump.*

   Use petroleum solvent. It is difficult to remove the screen without damaging it. So try to clean it without removing it.
5. *Reassemble the pump.*
6. *Check for proper operation.*

   Submerge the pump in oil and operate. Place your finger over the pump outlet and observe operation.
7. *Install the pump properly.*

   Some have a bevel inside the camshaft ring on one side only. If yours is like this, put the bevel side next to the camshaft.

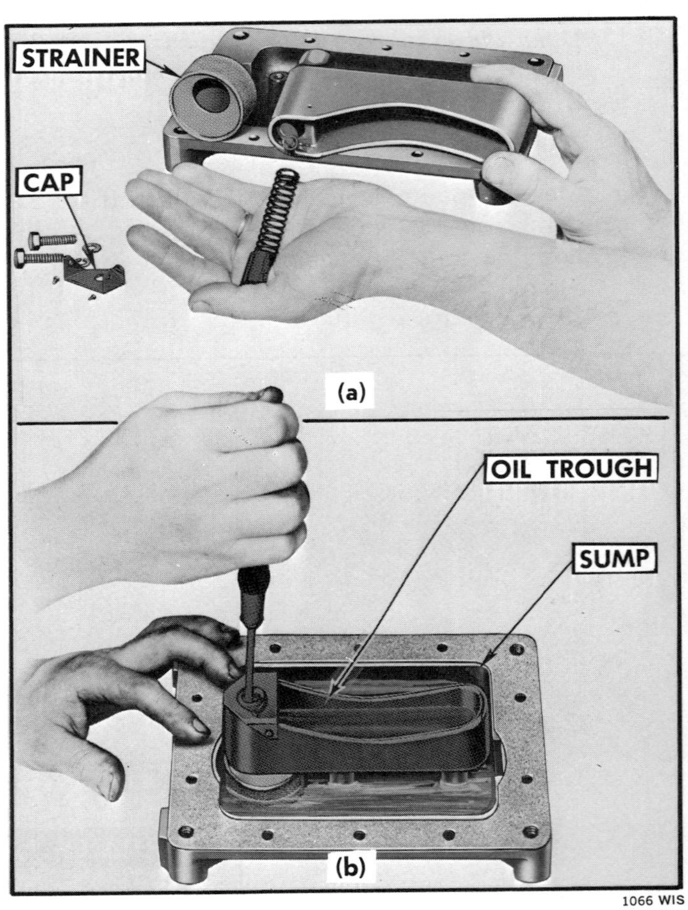

FIGURE 284. A combination oil trough for a dipper and plunger pump. (a) Pump disassembled. (b) Checking the pump for proper operation.

d. If your oil pump unit is a **non-removable barrel-and-plunger type** (Figure 284), proceed as follows:

1. *Open the crankcase as explained under "Removing the Piston and Rod Assembly,"* *page* 218.
2. *Remove the oil trough and strainer from the engine base.*
3. *Remove the cap from the top of the oil pump (Figure 284).*
4. *Remove the pump parts.*
5. *Clean in petroleum solvent and inspect.*
6. *Reassemble in reverse order.*
7. *Check for proper operation (Figure 284b).*

   Fill the sump with oil over the screen. Use a small screwdriver and operate the plunger.

   The trough should fill with oil.

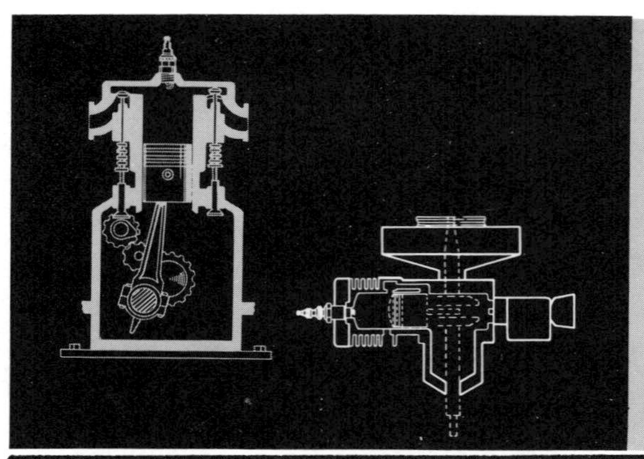

VIII. Repairing Camshaft Assemblies In 4-Cycle Engines

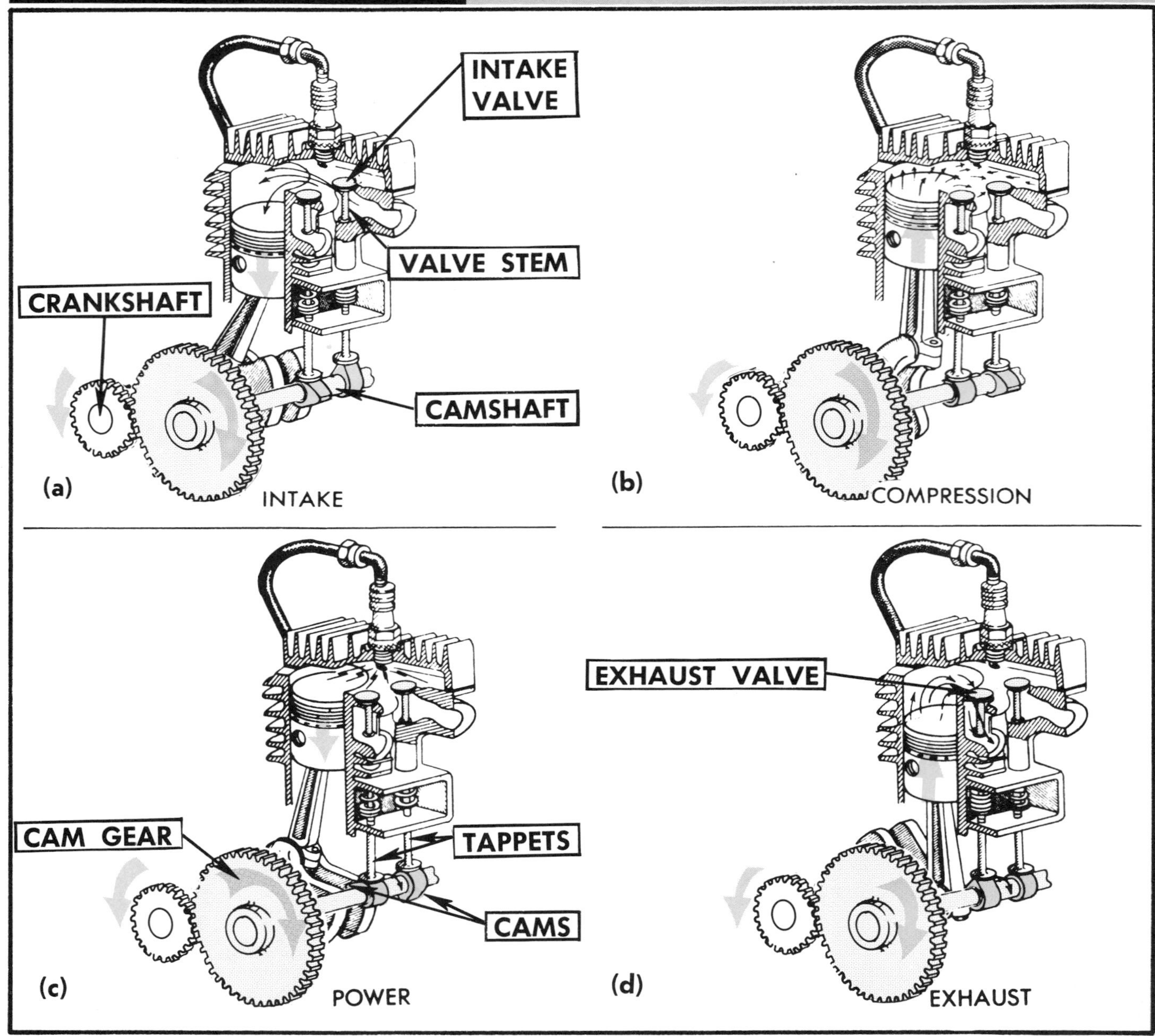

FIGURE 285. The primary function of the camshaft assembly is to open and close the valves at exactly the right time. (a) Intake stroke — cam pushes tappet against the valve stem and forces the intake valve open. There is no cam action against the exhaust valve tappet in this stroke. It remains closed. (b) Compression stroke — no cam action against either valve tappet. Both valves are closed. (c) Power stroke — both valves remain closed. (d) Exhaust stroke — cam pushes tappet against exhaust-valve stem and opens the exhaust valve.

The primary function of the camshaft assembly in your 4-cycle engine is to open and close the valves at the proper time (Figure 285) It is used only with 4-cycle engines. Two-cycle engines have no poppet valves.

The camshaft has two cams machined on the shaft. They are located so that when the shaft turns, the valves open and close at the proper time. The valves are closed by spring tension.

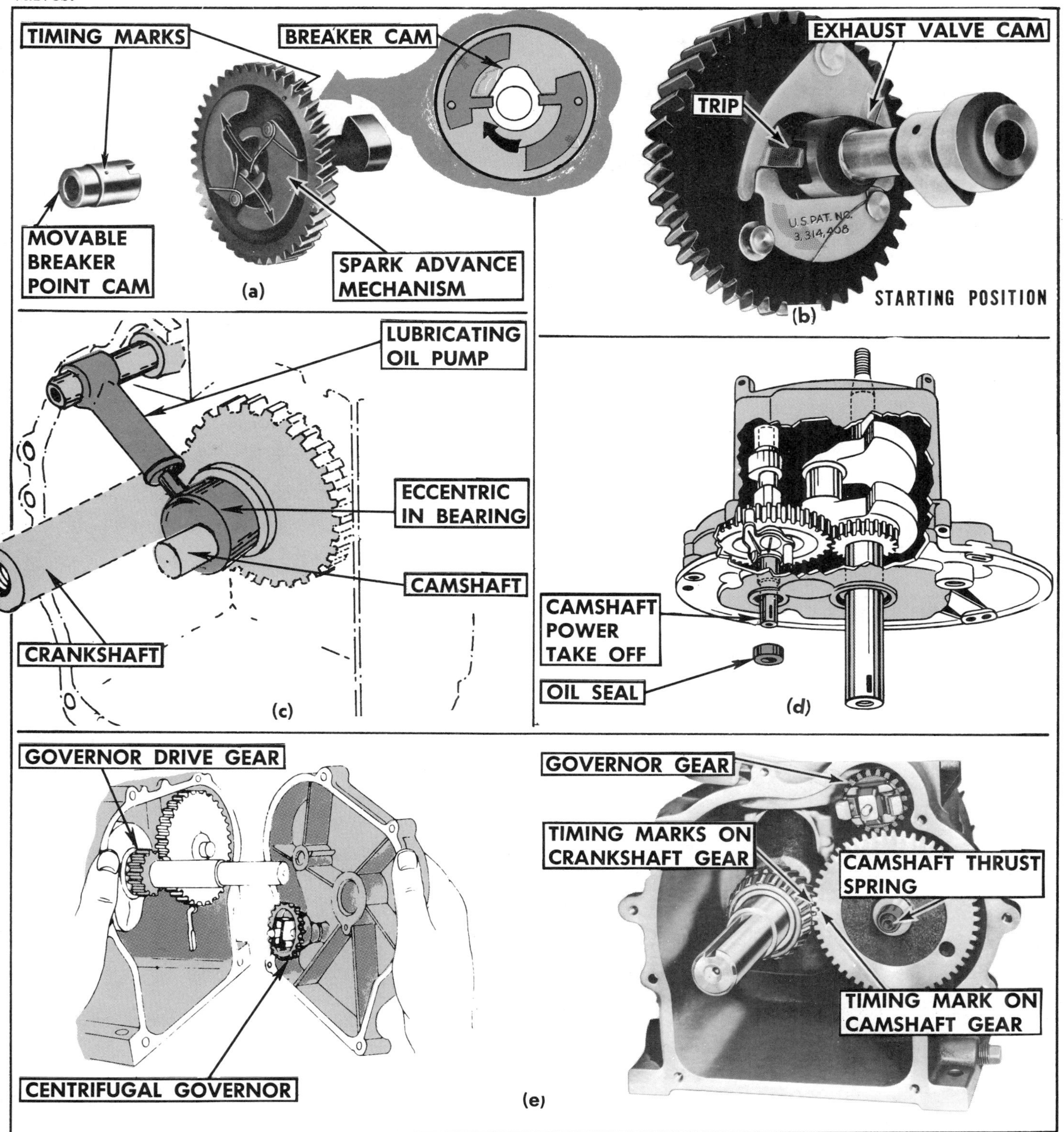

FIGURE 286. The camshaft assembly, in addition to opening and closing valves may also have: (a) a breaker-point cam and a spark-advance mechanism, (b) an automatic compression-release mechanism for easier starting, (c) an extra cam for operating an oil pump, or a gear-type oil pump driven from the cam gear, (d) a camshaft extending through the housing to provide for additional source of power, and (e) a cam gear which turns a centrifugal-type speed-control governor.

The camshaft is driven by the crankshaft through gears (Figure 285). Since there is only one power stroke for each two revolutions of the crankshaft, the gear ratio is designed so that the camshaft will turn at only half the speed of the crankshaft. This is done by making twice the number of teeth in the camshaft as in the crankshaft (Figure 287).

For more information on valve operation refer to "Types of Valves and How They Work," page 184.

In addition to opening and closing the valve, the camshaft assembly may also operate:

- **An (ignition) breaker-point cam and spark-advance mechanism (Figure 286a).**

  The spark-advance mechanism is operated by a centrifugal device. It advances or retards the breaker point cam. When the engine is being cranked, the centrifugal device retards the breaker-point opening time and makes starting easier. As engine speed increases, the centrifugal device opens the breaker points earlier. This provides higher efficiency at high speeds.

- **A compression release mechanism (Figure 286b) for easy starting** — At slow speeds the flyweights remain closed and the exhaust-valve trip lifts the exhaust valve slightly on compression stroke to relieve the compression before starting. At high speeds the flyweights move out because of centrifugal force, thus allowing the trip to drop in the slot back of the exhaust valve cam for normal operation.

  Some engine manufacturers accomplish the same action by shaping the intake valve cam so that the intake valve closes late.

- **An oil-pump drive cam (Figure 286c).** An eccentric inside of the bearing journal provides pumping action to the plunger.

- **An external power source (Figure 286d).** On some engines, the camshaft is extended to provide a power connection for supplying power to operate equipment. It operates at ½ of crankshaft speed.

Repairing the camshaft assembly is discussed under the following headings:

1. Tools and materials needed.
2. Removing the camshaft assembly.
3. Inspecting the camshaft assembly.
4. Installing the camshaft assembly.

TOOLS AND MATERIALS NEEDED

1. Open-end wrenches — 3/8″ through 9/16″
2. Socket wrenches and handle — 3/8″ through 3/4″ — 3/8″ drive
3. Slot-head screwdrivers — 6″
4. Phillips-head screwdriver — 6″
5. Combination pliers — 7″
6. Needle-nose pliers — 7″
7. Micrometer — 1/2″ to 1″
8. Petroleum solvent (mineral spirits, kerosene or diesel fuel
9. Clean rags

REMOVING THE CAMSHAFT ASSEMBLY

There are one of two methods of removing the camshaft, depending on the design of the crankcase.

If your *engine has no separate oil sump, or base plate (Figure 241b)*, or if it *has a vertical crankshaft (Figure 241c)*, you can usually remove the camshaft by removing the base plate or gear cover (Figure 287a) because they are bearing supports. If your engine **has a base plate or oil sump and no removable bearing support** (Figure 287b), it will be necessary for you to remove the crankshaft before you can completely remove the camshaft.

a. If your engine has a **removable bearing support,** remove the camshaft as follows:

1. *Remove the gear cover or mounting base.*

   See procedures under "Removing the Piston Rod Assembly," page 218.

2. *Note the position of the timing marks on the camshaft and crankshaft gears (Figure 288).*

   Timing marks are made in several different ways — punch marks, bosses, slots. Some gears are not marked. If yours are not, mark both the camshaft gear and the crankshaft gear with a center punch before removing the camshaft.

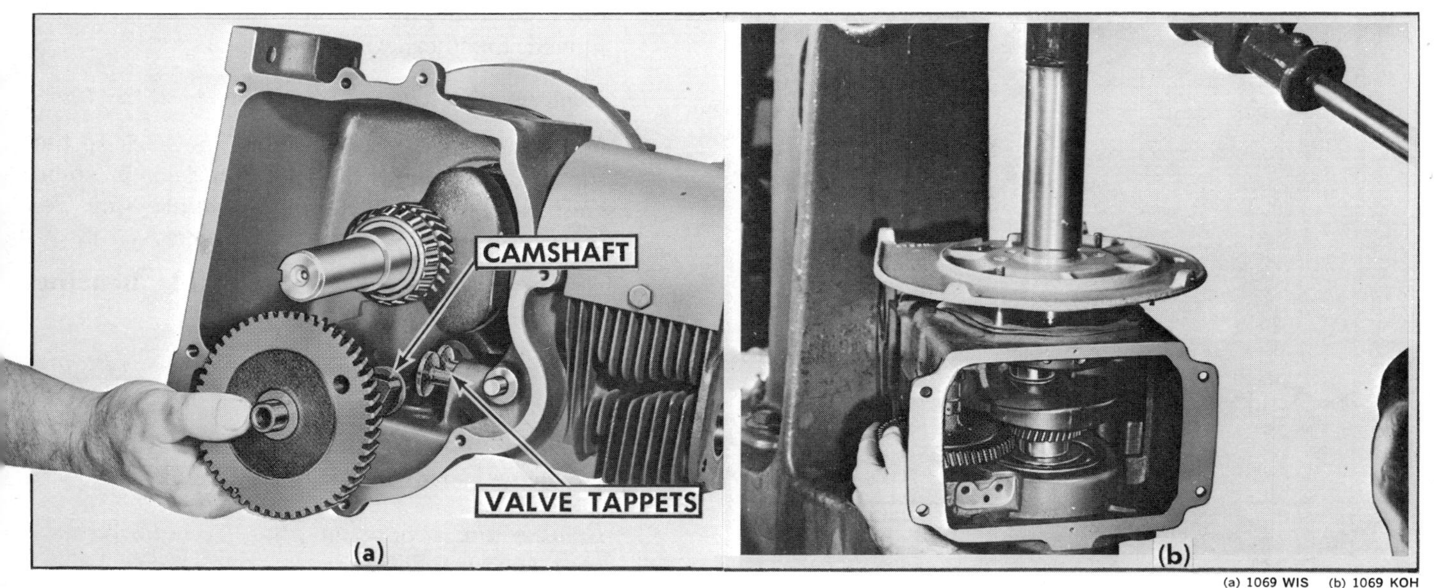

FIGURE 287. Removing the camshaft. (a) The camshaft is easily removed on some engines after you remove the gear cover or mounting base. (b) It is necessary to remove the crankshaft on some engines before you can completely remove the camshaft.

3. *Remove the thrust bearing and spring if installed.*

   Some engines have a camshaft end-thrust ball bearing and spring. Sometimes the ball will fall out when you remove the gear case. Most camshafts do not have this end-thrust bearing.

4. *Turn the crankshaft until the timing marks are aligned on compression stroke — both valves closed.*

   This action will relieve the pressure from the tappets and make removal of the camshaft easy.

5. *Turn the crankcase on its side or upside down.*

   This prevents the tappets from falling out.

6. *Remove the camshaft.*

   On some engines with ball bearings, both

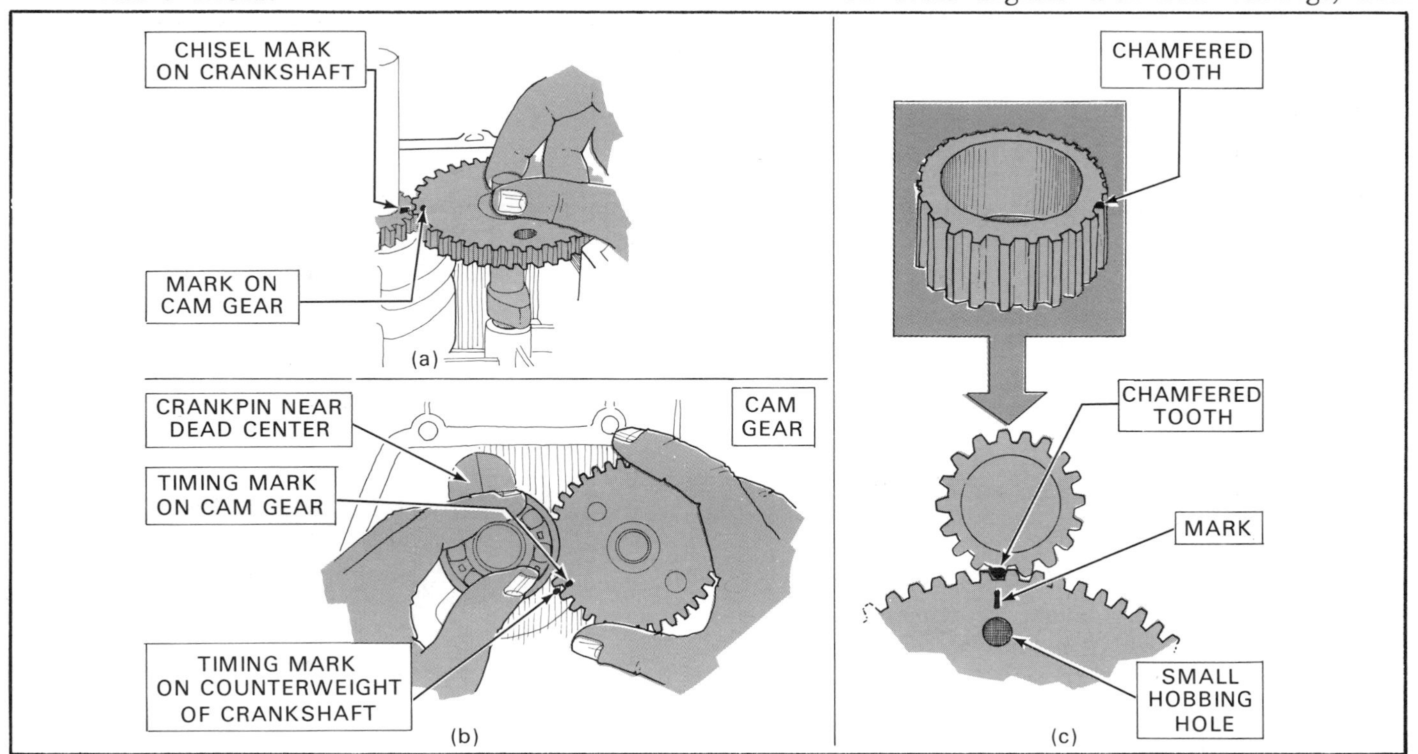

FIGURE 288. Valve timing marks are shown in different ways. (a) Chisel or punch marks. (b) Punch marks. (c) Chisel marks and chamfered tooth.

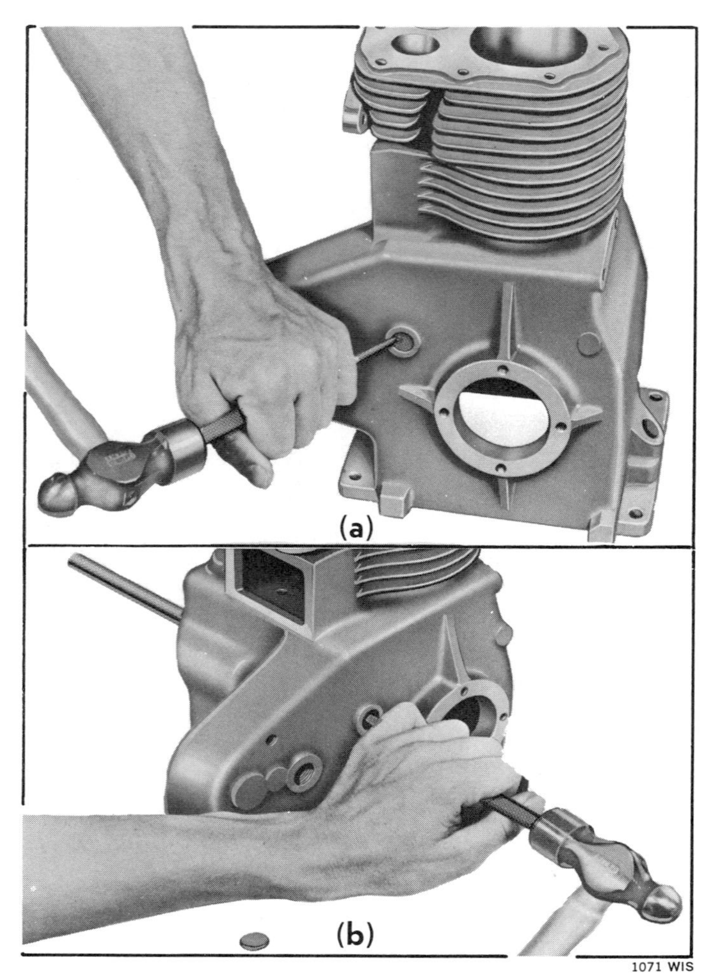

FIGURE 289. Removing the camshaft pin. (a) Remove the expansion plug. (b) Drive out the pin.

the camshaft and the crankshaft must be removed together.

7. *Remove the valve tappets one at a time.*

   Mark the tappets. They should go back in the same holes from which they came. On some engines with an automatic compression release, the exhaust tappet is shorter.

b. If your engine has **no removable bearing support,** proceed as follows:

1. *Remove the camshaft if possible (Figure 287).*

   On some engines this is not possible until you have performed steps 2 and 3.

2. *Remove the expansion plug from the crankcase (Figure 289a).*

   Some shafts are removed from the flywheel end and others from the power take-off end. For removal of the crankshaft see "Removing the Crankshaft," page 254.

3. *Drive the camshaft pin out (Figure 282b).*

   It will be easy after you have driven it past the housing.

4. *Remove the camshaft assembly.*

   Watch for endplay, adjusting washer if installed.

INSPECTING THE CAMSHAFT ASSEMBLY

1. *Clean the parts in petroleum solvent.*
2. *Inspect gear teeth for wear and nicks.*
3. *Check automatic spark advance if installed (Figure 286a).*

   Place the cam gear in normal position with the movable weight down. Press the weight down and release it. The spring should lift the weight. If it does not, check for binding. If the spring is weak, replace it.

   Some spark-advance assemblies do not have replaceable parts, The entire assembly has to be replaced.

4. *Check the tappets for wear.*
5. *Check the automatic compression-release mechanism for freedom of operation and excessive wear, if installed.*
6. *Clean out the oil passages, if the camshaft is drilled for oil passageways.*

   Use compressed air.

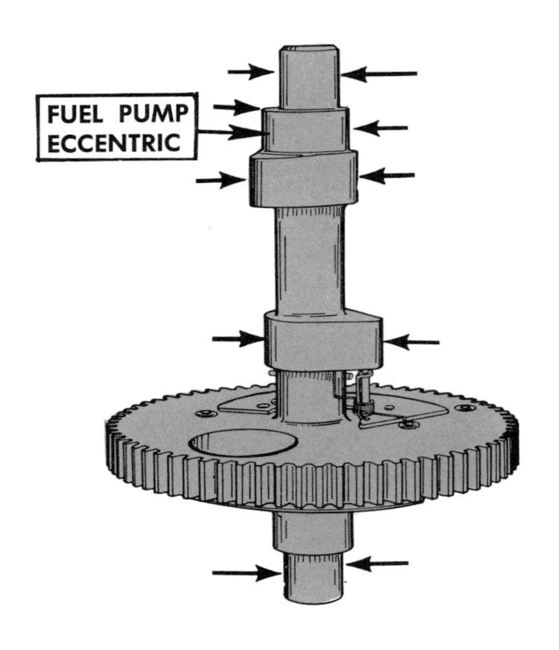

FIGURE 290. Points at which the camshaft should be measured.

7. *Measure the camshaft dimensions (Figure 290) and check against the tolerances given in your repair manual.*

   Use an outside micrometer.

8. *Measure the height of the cams with a micrometer.*

   Check the tolerances given in your repair manual.

INSTALLING THE CAMSHAFT ASSEMBLY

1. *Assemble the automatic compression-release and spark-advance mechanism, if you have had them disassembled.*

2. *Oil all parts.*

   Make sure they are free from dirt.

3. *Turn the crankshaft so that the piston is at top-dead-center, compression stroke.*

4. *Turn the cylinder block upside down.*

5. *Install the valve tappets.*

   Be sure you install the tappets in the guides from which they came. If you do not, you may change the valve-tappet clearance. Also, the tappets in some engines are of different lengths.

6. *Install the camshaft.*

   If your engine is similar to the one shown in Figure 287b, install the crankshaft before driving in the camshaft pin. This is so you can move the camshaft out of the way in order to install the crankshaft.

7. *Install the end-thrust spring and bearing, if applicable.*

   If you have an **external fuel pump, governor arm** or **breaker points operated by the camshaft assembly,** be sure the drive mechanisms are set in the proper position so that they will operate properly.

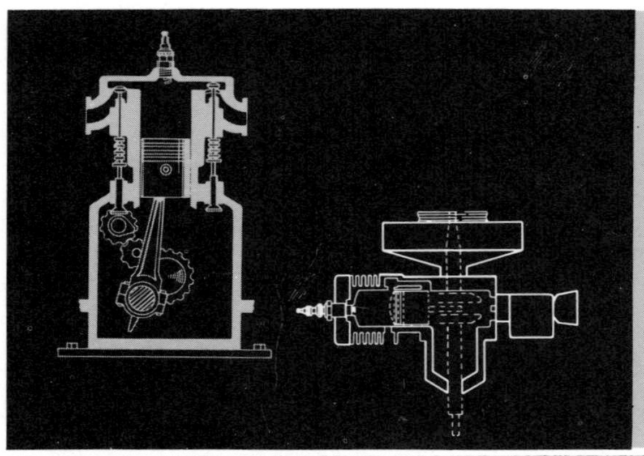

IX. Repairing Crankshaft Assemblies

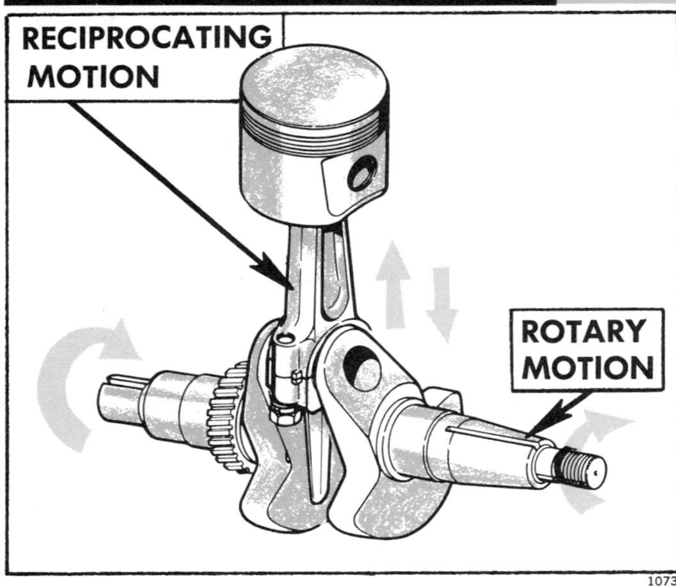

FIGURE 291. The crankshaft converts the reciprocating (up and down) motion of the piston to rotary motion.

The primary function of the crankshaft assembly is to convert the reciprocating motion of the piston to rotary motion (Figure 291). Other functions of the crankshaft include:

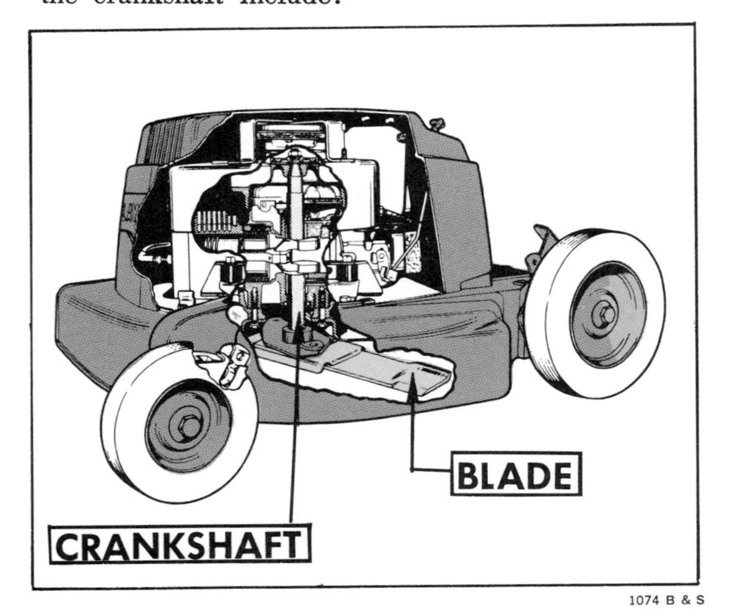

FIGURE 292. Mower blade attached to the end of a crankshaft.

- **Providing power for driving equipment such as a saw, mower blade, wheels on a self-propelled unit, pump or other units (Figure 292).**
- **Driving the camshaft (Figure 284e).**
- **Turning the flywheel (Figure 293).**

The flywheel gives balance and stability to the engine.

Some small engines are equipped with crankshaft balances. These are called "counter-

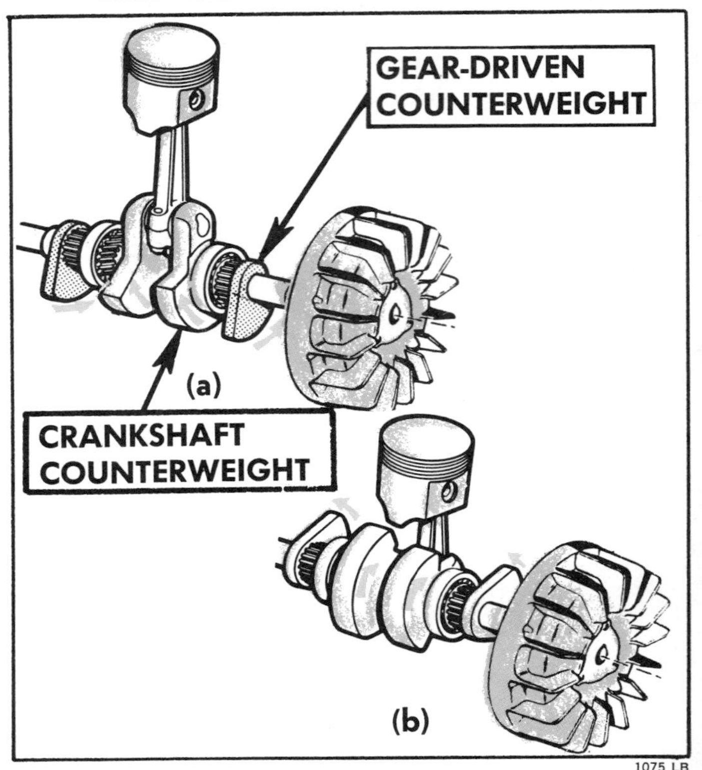

FIGURE 293. How counter-balanced crankshaft engines offset vibrations developed by the up-and-down motion of the piston. Counterbalances at each end of the crankshaft are geared to rotate in a direction opposite to that of the crankshaft counterweights. (a) In the horizontal position the counterbalances are positioned to offset the unbalanced weight at the crankshaft. (b) In the vertical position the synchronized counterbalances combine with the unbalanced weight of the crankshaft to balance the weight of the piston.

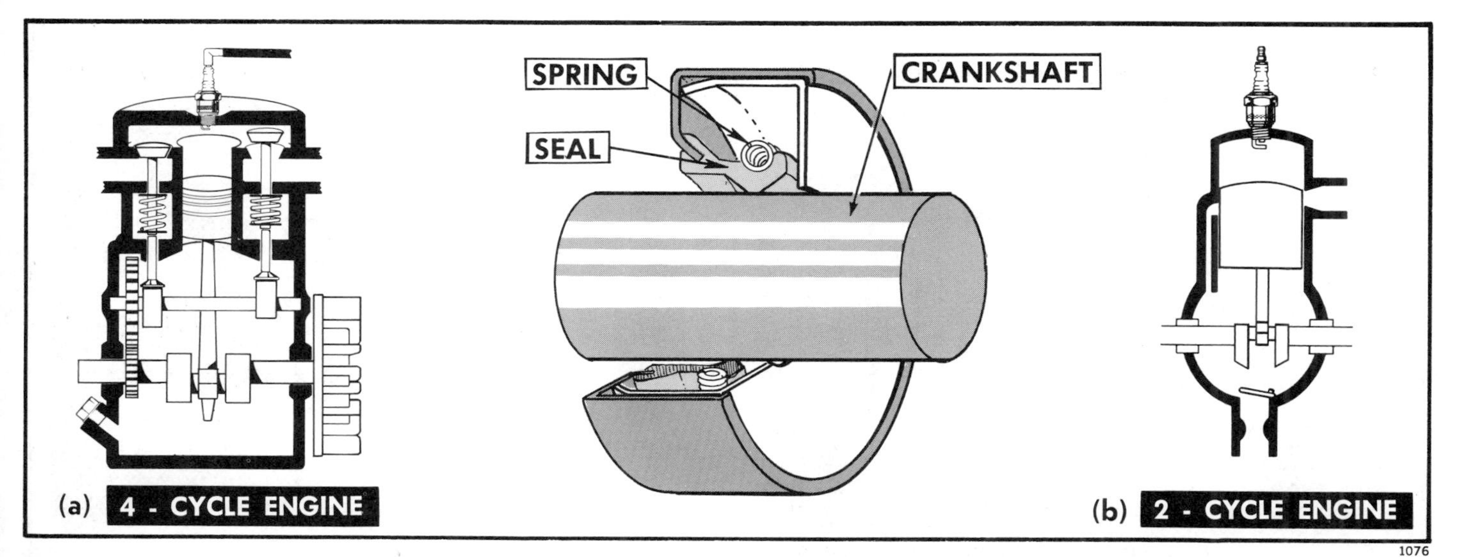

FIGURE 294. Seals are used at both ends of the crankshaft. (a) They prevent oil from leaking from the crankcase on 4-cycle engines. (b) They prevent fuel, air, oil and pressure leakage from the crankcase on 2-cycle engines.

balances." They overcome much of the vibration caused by the reciprocating motion of the piston and the unbalanced weight of the crankshaft. Figure 293 explains how they work.

- **Driving a cam for opening the breaker points on some engines.** On flywheel magnetos the spark occurs on each revolution of the crankshaft, even though with 4-cycle engines it is not needed except on every second revolution. On most 4-cycle engines, the points are operated from the crankshaft.

- Providing **seals to prevent leakage of pressure** from within the crankcase (2-cycle engines, Figure 294a) and oil from within the crankcase (4-cycle engines, Figure 294b).

  Crankcases on 4-cycle engines may operate at either slightly negative pressure or at a slightly positive pressure. The seal is installed the same for either operating condition.

- Providing **power for the speed-control governor** on some engines (Figure 295).

- **Driving gear-type oil pump** on some engines (Figure 296).

Repairing crankshaft assemblies is discussed under the following headings:

1. Importance of proper repair.
2. Tools and materials needed.
3. Checking the crankshaft for proper operation.
4. Removing and replacing crankshaft oil seals (with shaft in place).

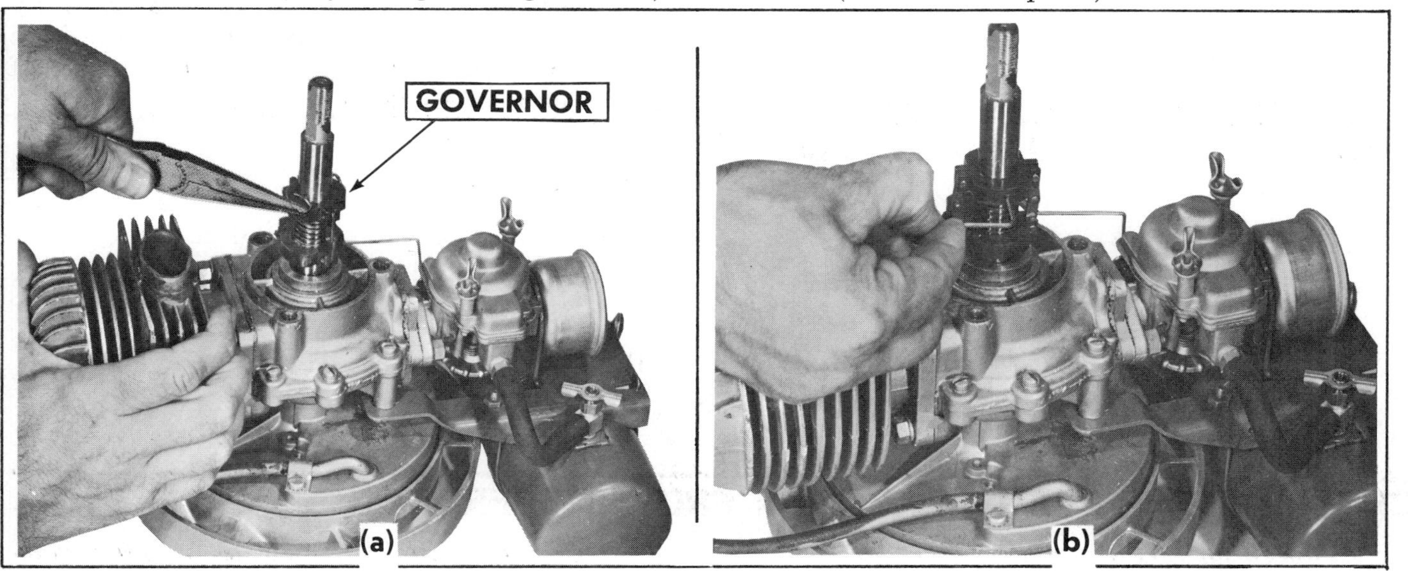

FIGURE 295. Governor attached directly to the crankshaft of a 2-cycle engine.

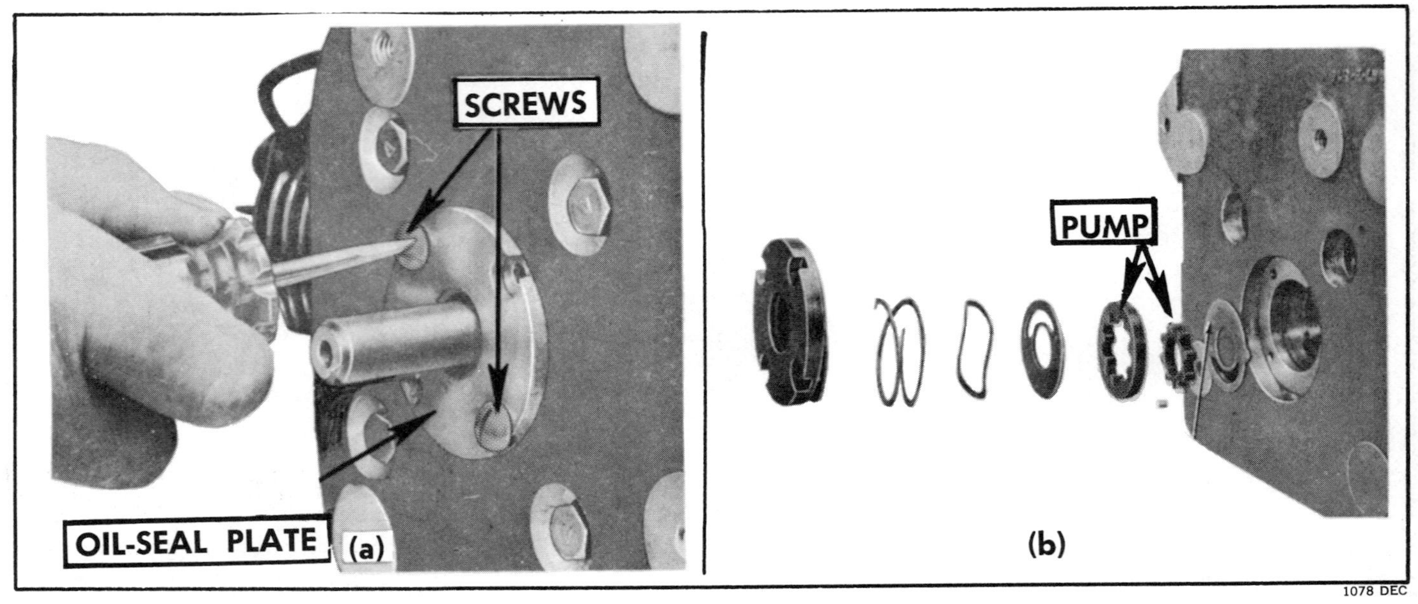

FIGURE 296. Oil-seal plate and crankshaft-driven gear-type oil pump. (a) Removing the oil seal. (b) Oil seal plate-and-pump assembly.

5. Removing the crankshaft.
6. Cleaning and inspecting the crankshaft.
7. Cleaning and inspecting crankshaft bearings.
8. Removing and replacing bearings.
9. Installing the crankshaft.

IMPORTANCE OF PROPER REPAIR

If you expect to get much service from your engine after overhaul, you will have to recognize crankshaft troubles and do the crankshaft repair job correctly. A **worn crankshaft will allow too much play** between the crankshaft and the bearings. The oil seal will leak. With **4-cycle engines,** which operate with a slight crankcase vacuum, dirt will be drawn inside the crankcase which increases bearing wear. With **2-cycle engines,** worn bearings and seals will prevent them from starting easily and running properly.

Since most spark-ignition breaker points operate from the crankshaft, **a worn crankshaft will cause the points to open and close at the wrong time.**

A partially sheared keyway or twisted shaft will also throw the timing off.

A bent crankshaft will cause the engine to vibrate excessively.

If you find very much crankshaft damage, it will be desirable to purchase a short block assembly which includes the cylinder block, crankcase assembly, pistons, rings and valves if one is available for your engine. A new shaft costs more than half as much as a new short-block assembly.

TOOLS AND MATERIALS NEEDED

1. Long nose pliers — 7″
2. Slot-head screwdriver — 8″
3. End wrenches — 3/8″ through 3/4″
4. Clean rags
5. Micrometers — inside and outside
6. Arbor press
7. Bearing puller
8. Special tools — gages and reamers
9. Petroleum solvent (mineral spirits, kerosene or diesel fuel)
10. Emery cloth
11. Gasket cement
12. Gaskets

CHECKING THE CRANKSHAFT FOR PROPER OPERATION

1. *Check for excessive vibration.*

Excessive vibration indicates a bent crankshaft. If you have a mower blade attached to the crankshaft, remove it and check the blade for balance. An unbalanced mower blade will cause vibration.

A bent crankshaft also can cause excessive vibration.

2. *Check for noisy operation.*

   Worn bearings, especially rod bearings, can be noisy and indicate trouble.

3. *Check for leaking crankshaft oil seals.*

   Oil leaks in **4-cycle engines** indicate a bent crankshaft, bad oil seals, and/or worn bearings.

   If you have an *oil leak and excessive vibration,* this indicates a bent crankshaft. See step 4.

   If you have an *oil leak and there is no evidence of a bent crankshaft or worn bearings,* replace the oil seal. Follow procedures under "Removing and Replacing Crankshaft Oil Seals (With Shaft in Place)," page 252.

   If you have a **2-cycle engine** and it is hard to start, it may be leaking at the crankshaft oil seals. Put some crankcase oil around the seal and observe to see if the oil is drawn into the crankcase as you turn the starter.

   Another method is to use a special tool or kit for this purpose. The kit includes adapter plates that are bolted over the intake and exhaust ports to seal them off. A tube, in one adapter, accepts a hose to which is attached a pressure gage and a pump. A check valve is used in the hose to prevent leakage through the hose. Pressure is applied, by the pump, through the hose and into the engine and is registered on the pressure gage.

   If there is a seal leak or other leak in the crankcase, this will show as the pressure reading on the gage drops. This same tool can be used to detect carburetor or fuel line leaks.

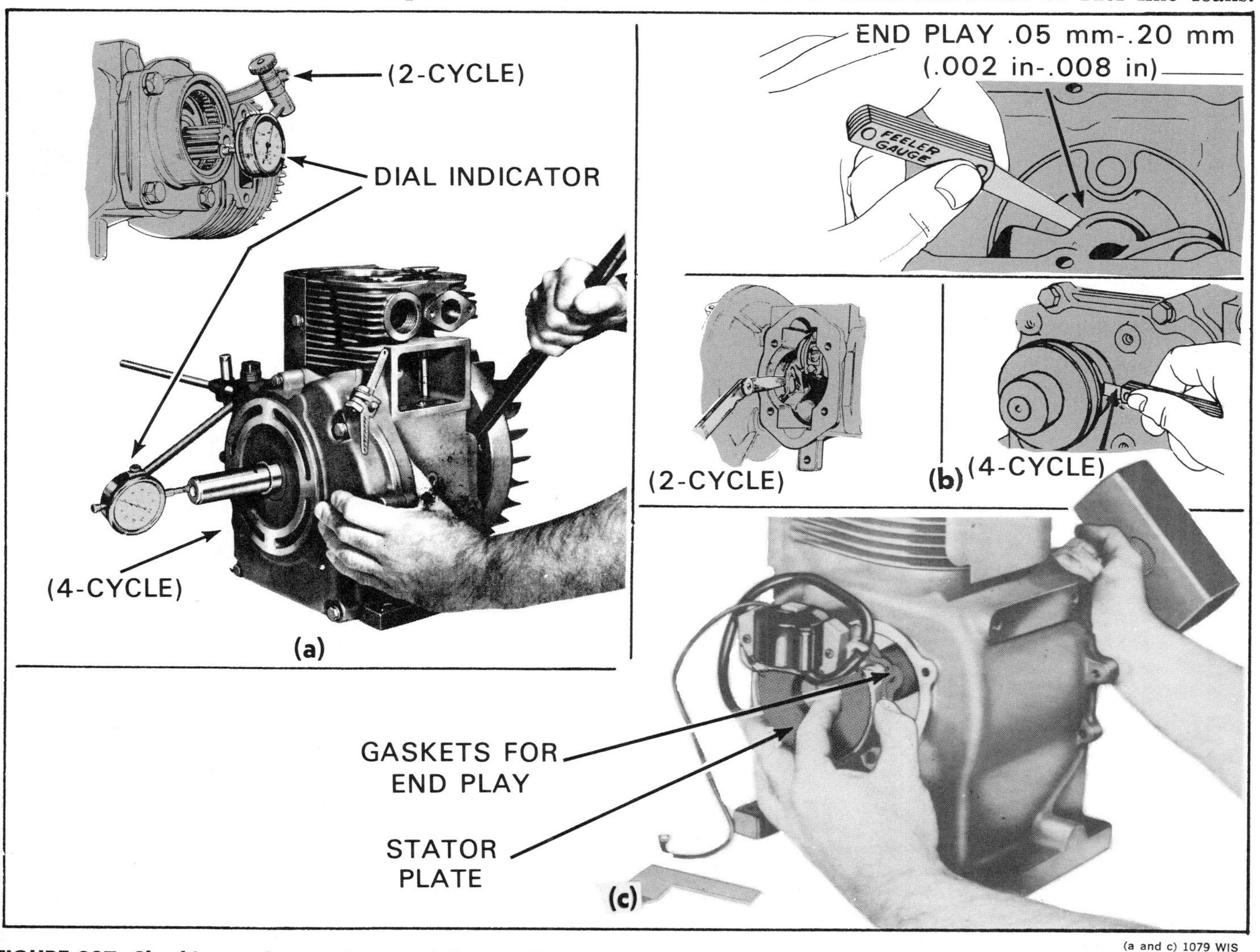

(a and c) 1079 WIS

FIGURE 297. Checking and correcting crankshaft endplay. (a) Checking the crankshaft endplay with a dial indicator. (b) Checking the crankshaft endplay with a feeler gage. (c) One method for correcting crankshaft endplay is by inserting gaskets under the stator plate.

Hard starting may also be caused by a pin hole in the crankcase. Check it the same way.

4. *Check for bent crankshaft or warped housing.*

   If you cannot turn the shaft by hand either way, this binding indicates a bent crankshaft or seized bearing.

   If the crankshaft will turn, you can check with a dial indicator to see if it is bent before removing it from the engine.

   If you have already installed a new crankshaft and it still binds, this binding indicates the housing is warped.

5. *Check the crankshaft endplay (Figure 297).*

   Before removing the crankshaft check the endplay so you will know how many, and what size gaskets to install to correct it when reassembling the crankcase. Move the crankshaft back and forth and check the amount of travel.

   Too much endplay will allow the shaft to move back and forth, thus causing excessive pounding on the bearings.

   Too little endplay will cause excessive pressure against the bearings.

   Endplay is corrected by adding or removing the gaskets or shims under the bearing plate or stator plate (Figure 297c).

REMOVING AND REPLACING CRANKSHAFT OIL SEALS WITH SHAFT IN PLACE

If you have a leaking oil seal, you can usually replace it without removing the crankshaft or bearing support. Proceed as follows:

1. *Remove the flywheel if the seal is to be removed from the flywheel end.*

   Follow procedures under "Removing and Checking the Flywheel," page 127.

2. *Remove the dust cover if one is installed (Figure 298).*

3. *Determine if the oil seal can be removed without removing the crankshaft.*

   If the **seal is accessible,** it can be replaced without removing the crankshaft. Proceed to step 3.

4. *Remove the oil seal (Figure 299).*

   Some oil seals are **1-piece.** The neoprene seal is encased in a metal housing (Figure 294). This type is pressed into place around the crankshaft.

   Others are supplied in **3-piece sets** with the seal and retainer held by a lock ring (Figure 300).

   a. If you have the **1-piece type** (Figure 299), proceed as follows:

      (1) *Drill or punch .30 cm (1/8 in) holes in the seal on opposite sides.*

      Do not drill past the seal, or you will damage the crankcase or bearings.

FIGURE 298. Removing the oil-seal dust cover.

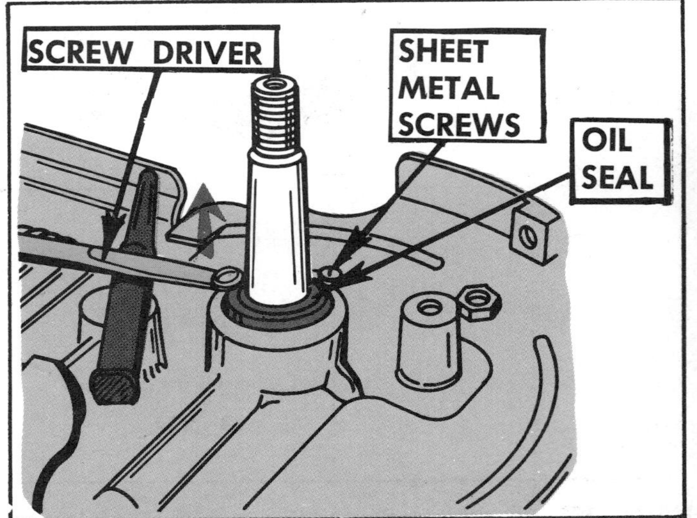

FIGURE 299. Removing a 1-piece oil seal.

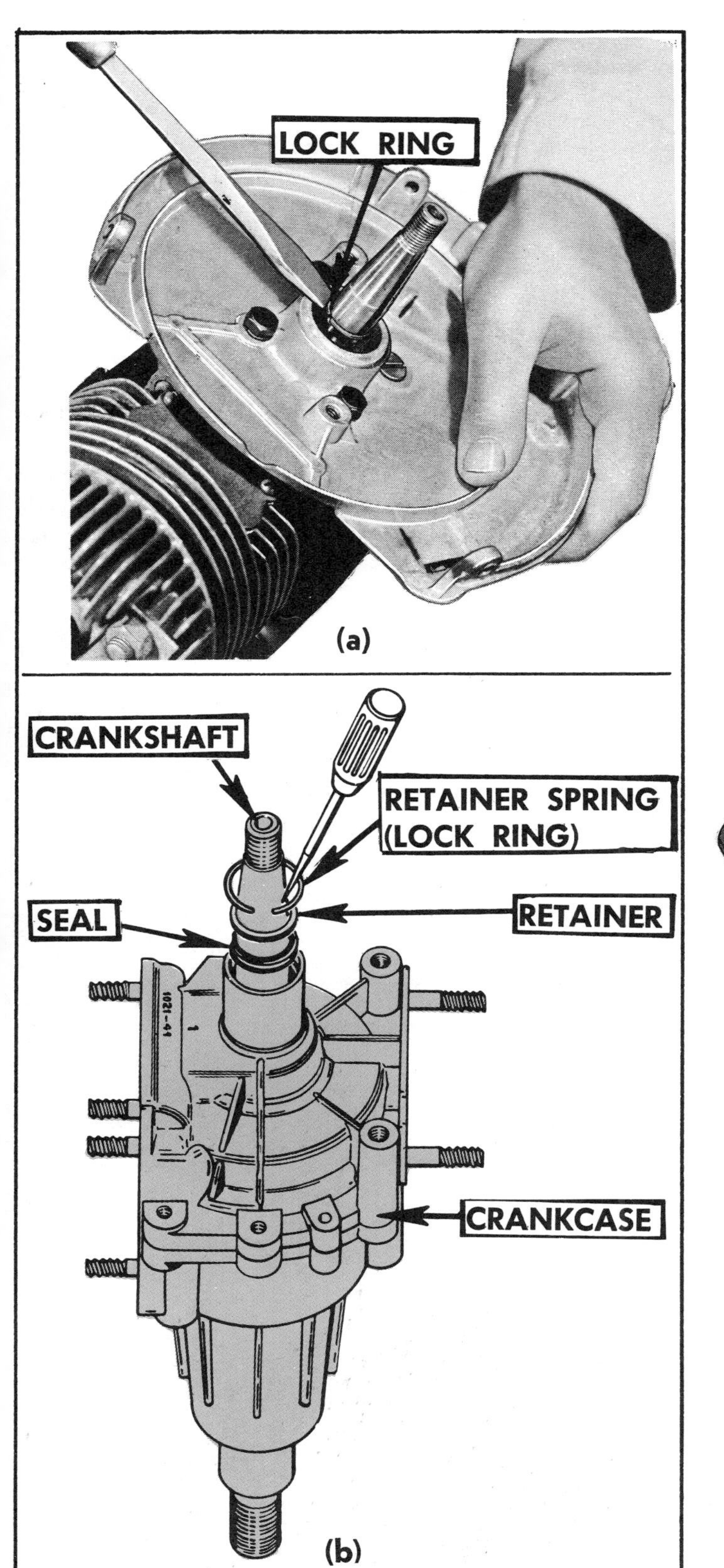

FIGURE 300. Removing a 3-piece oil seal. (a) Removing the lock ring. (b) Lock ring, retainer and seal removed.

(2) *Install .50 cm ($^3/_{16}$ in) sheet-metal screws in each hole part way.*

(3) *Pry the oil seal out with a screwdriver (Figure 299).*

A special tool is available for removing this type seal.

b. If yours is a **3-piece oil seal** (Figure 300), proceed as follows:

(1) *Remove the lock ring (Figure 300).*

Use a prick punch.

(2) *Remove the seal retainer.*

Turn the engine so the end of the crankshaft is down, and try to jar the retainer out by pounding the end of the crankshaft with a soft hammer. If the retainer does not fall out, follow procedures under "a" preceding; then proceed with step 3.

(3) *Remove the seal with a prick punch.*

(4) *Clean the oil-seal recess.*

Wipe dry with a clean cloth.

(5) *Apply gasket cement on the outside rim of the oil seal.*

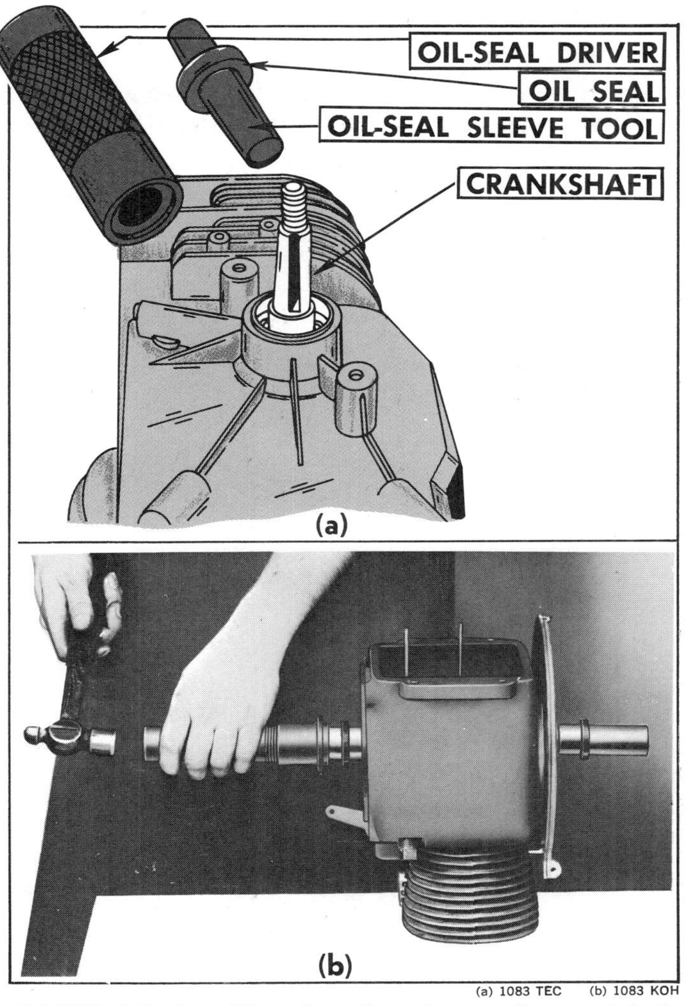

FIGURE 301. Installing the oil seal over the crankshaft. (a) Special oil-seal installation tools. (b) Installing the seal.

The cement seals the side next to the crankcase so that it is oil tight.

(6) *Apply crankcase oil to the inside of the seal and to the crankshaft.*

(7) *Install the oil seal over the oil-seal-sleeve tool (Figure 301a), or wrap the crankshaft with thin cardboard.*

This precaution protects the seal from being damaged while passing over the crankshaft.

(8) *Position the seal over the recess.*

(9) *Drive the seal in place (Figure 301b).*

(10) *Install dust cap if provided.*

REMOVING THE CRANKSHAFT

1. *Prepare the engine for removal of the crankshaft.*
   The parts to be removed are listed under "Preparing to Remove the Piston-and-Rod Assembly," page 217.

2. *Remove the flywheel.*

   For procedures refer to "Removing and Checking the Flywheel," page 127.

3. *Remove the magneto breaker-point assembly if installed.*

   Follow procedures under "Removing and Replacing the Stator Plate and/or Coil," page 145.

4. *Remove the magneto-bearing plate or PTO-bearing support and gear cover.*

   A bearing plate on the flywheel end, and/or a *bearing support on the PTO end* will have to be removed. The bearing plate on the flywheel end usually must be removed with a bearing puller (Figure 302a). Some bearing supports on the PTO end are removed along with the crankshaft, after the bearing plate on the opposite end is removed (Figure 302b).

5. *Release the ball bearing if it is held by cam locks (Figure 303).*
   Loosen lock nuts and turn one quar'er turn with a screwdriver.

6. *Remove the oil pump, camshaft and governor from inside the engine if installed.*

7. *Remove crankshaft retaining ring from opposite end of crankshaft, if installed.*

   Some 2-cycle engines have a retainer ring inside the crankcase.

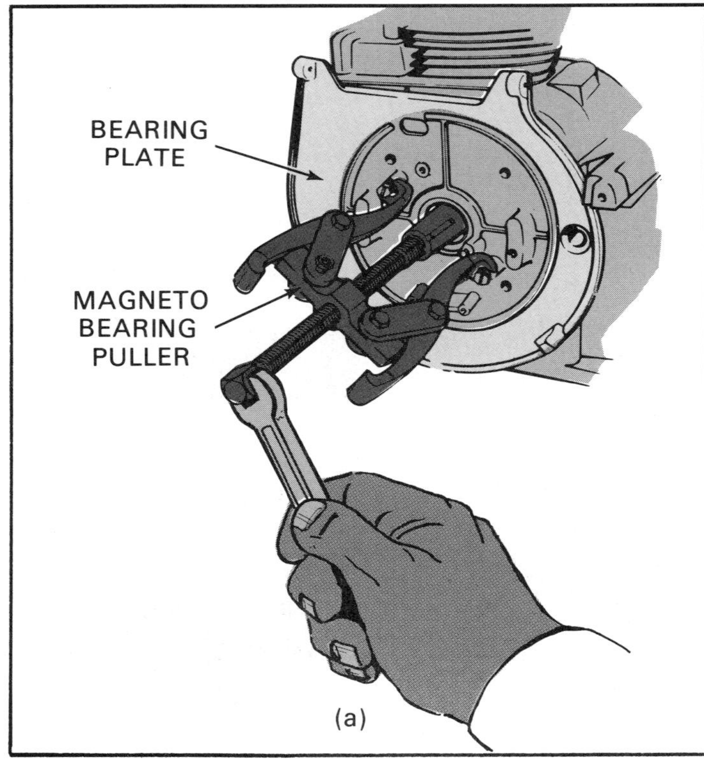

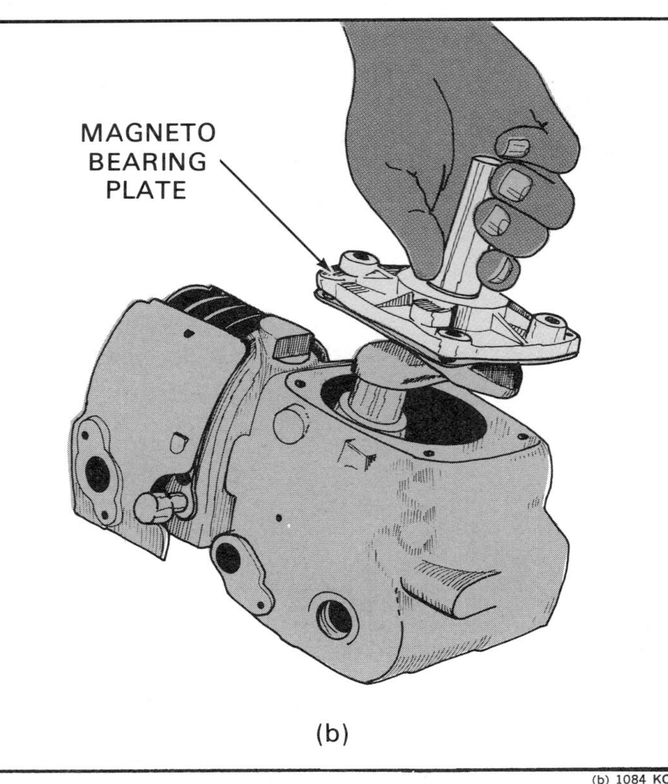

FIGURE 302. (a) Removing the bearing plate with a puller. (b) Removing the PTO-bearing support along with the crankshaft after the opposite bearing plate is removed.

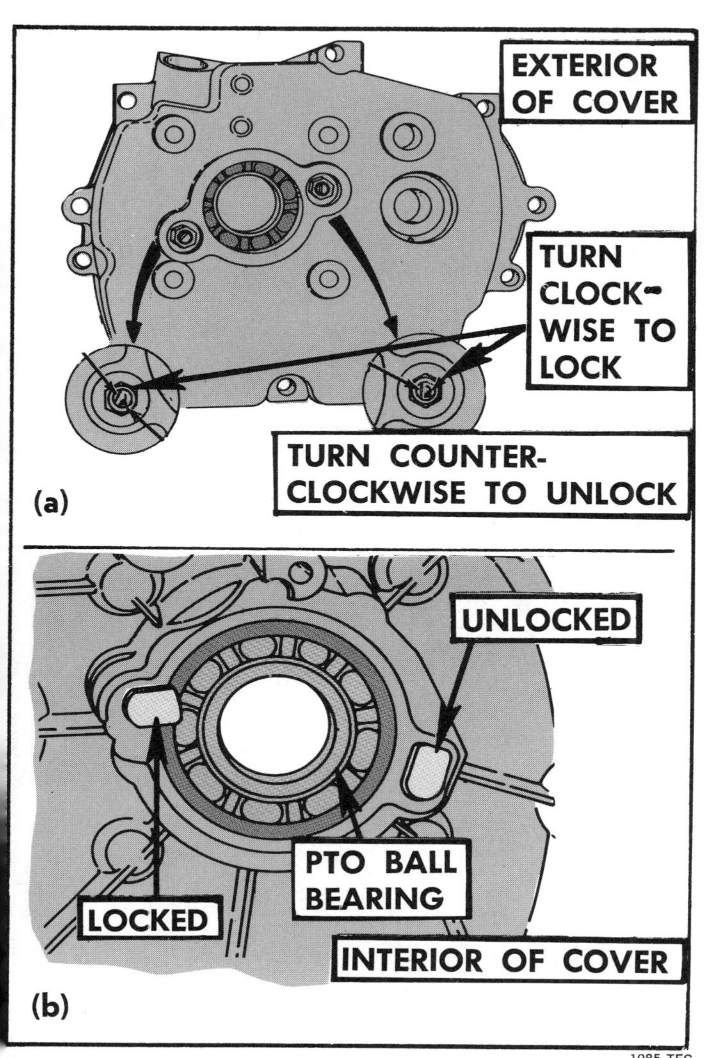

FIGURE 303. Cam locks for ball bearings. (a) Exterior view. Cam slots for turning with a screwdriver. (b) Interior view with one cam locked and one cam unlocked.

8. *Disconnect the connecting rod from the crankshaft.*

   Refer to "Removing the Piston-and-Rod Assembly," page 218.

   If the engine has **tapered-roller bearings** (Figure 304), remove the crankshaft by pulling gently by hand.

   If the engine has **ball or plain bearings,** tap (or press) the crankshaft out (Figure 305).

9. *Remove oil seals if not already removed.*

   Use a screwdriver or a special tool (Figure 306).

FIGURE 304. Removing a crankshaft which has tapered-roller bearings.

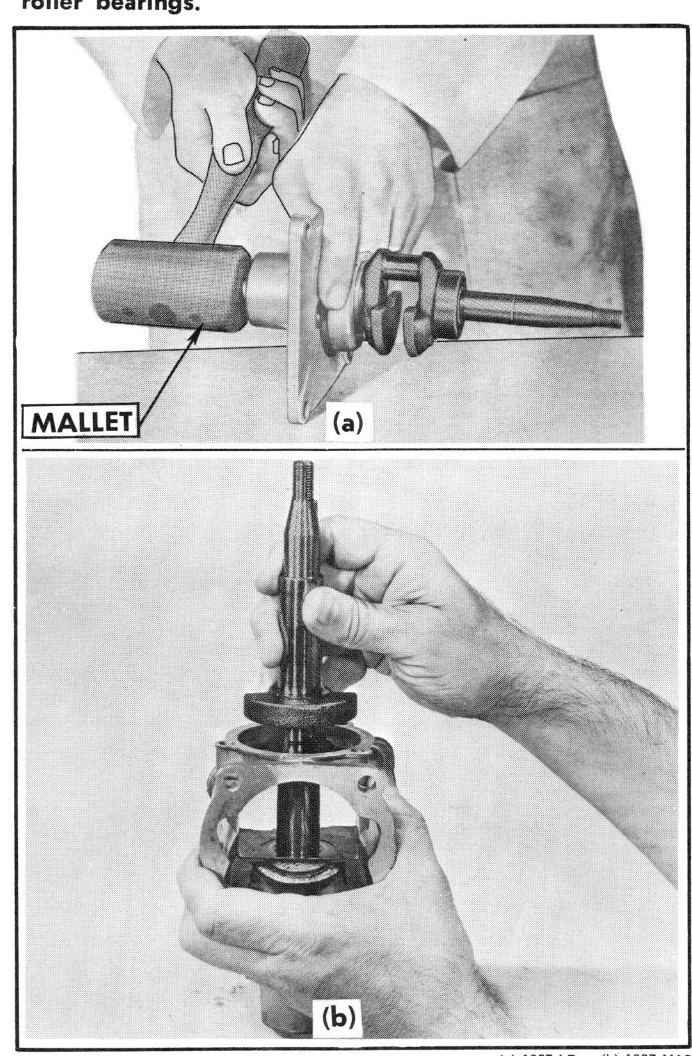

FIGURE 305. Removing a crankshaft equipped with ball or plain bearings from: (a) a removable bearing plate, and (b) the crankcase.

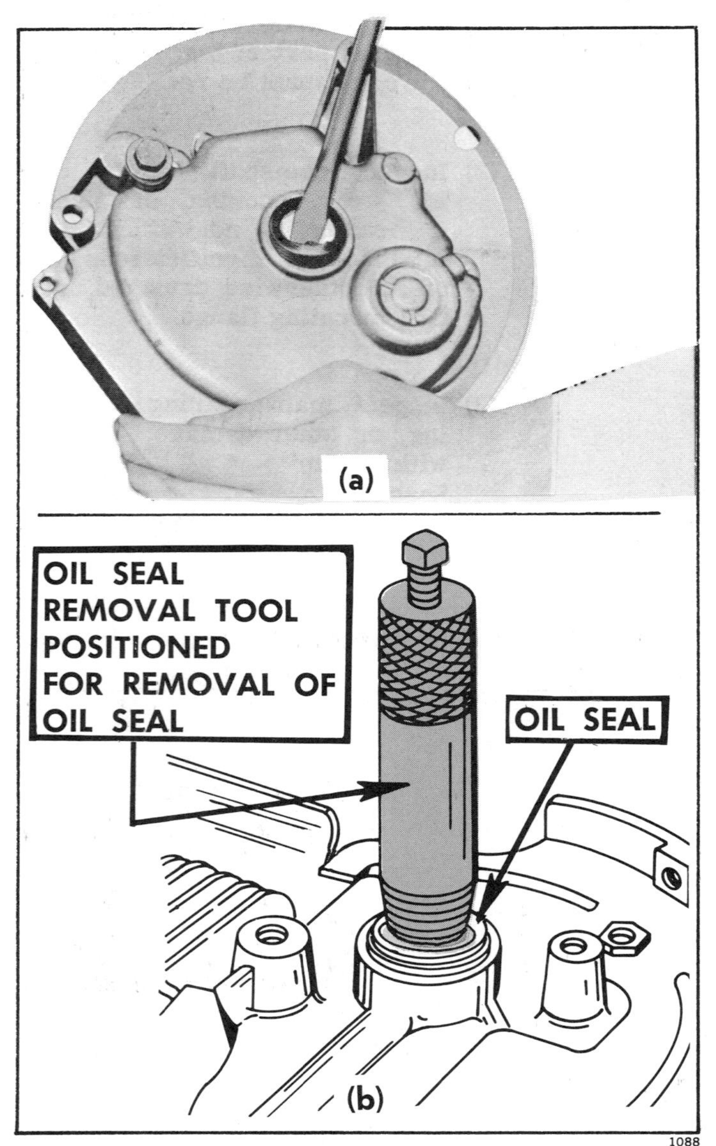

FIGURE 306. Removing the oil seal with (a) screwdriver, or (b) special tool for pulling the seal.

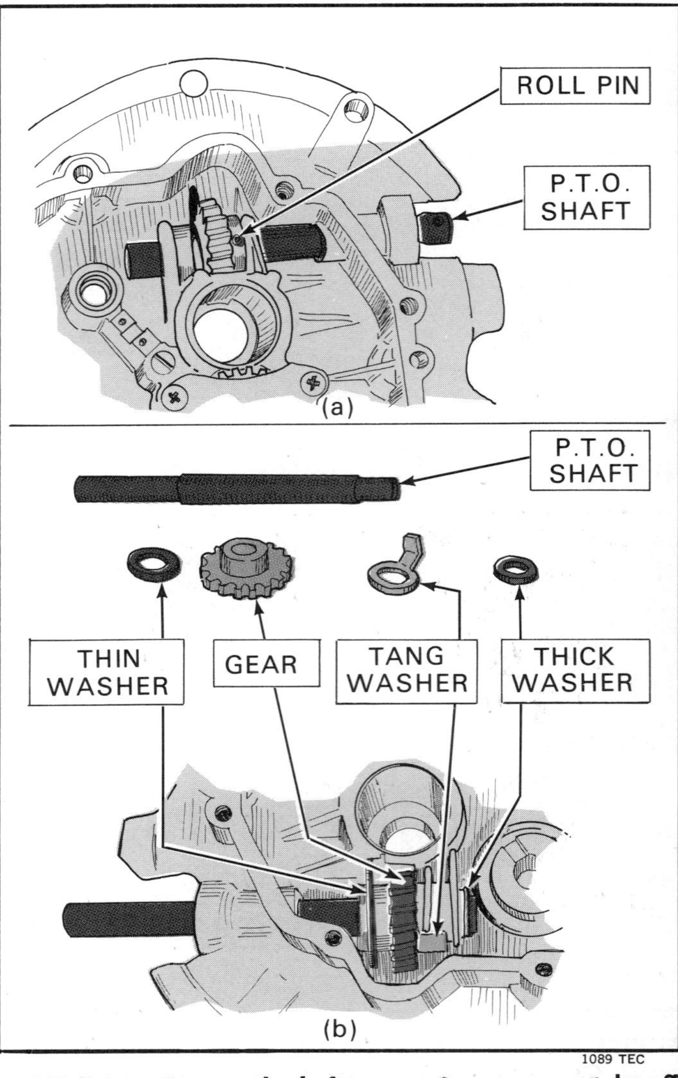

FIGURE 307. Two methods for removing a power take-off shaft are (a) punching out a roll pin, and (b) removing a shaft retainer.

10. *Remove the power take-off shaft, if installed (Figure 307).*

    Some engines have a secondary shaft for supplying power at slower speed. It is not the camshaft, as used on some engines. If you wish to service it, remove and check it while the crankshaft is removed.

CLEANING AND INSPECTING THE CRANKSHAFT

1. *Clean with petroleum solvent.*
2. *Check for wear (Figure 308).*

   Measure crankshaft journals for **out-of-roundness.** Replace crankshaft if it is out-of-round more than .025 mm (.001 in). The journals are the bearing surface of the crankshaft.

   Check **gear teeth** for chips and wear.

   Check **threads.** Threads can be reconditioned with a die.

3. *Check the crankshaft for straightness (Figure 309).*
4. *Check for score marks or roughness at bearing points.*

   Light scoring or roughness can be smoothed with an emery cloth.

5. *Check the crankshaft taper, if tapered.*

   If rust is present on the tapered portion which fits into the flywheel, this indicates

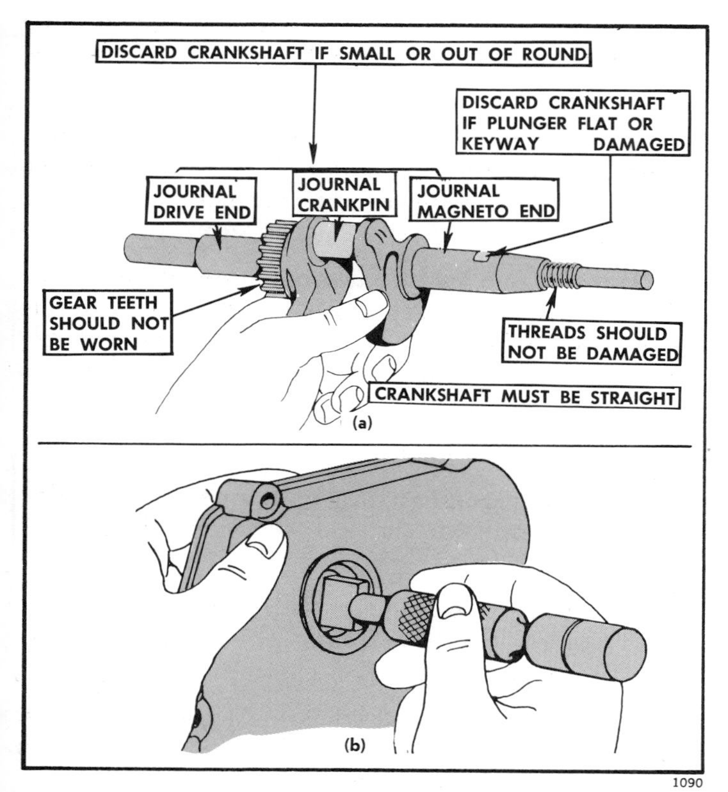

FIGURE 308. (a) Measure the diameter of the journals. Compare them with tolerances given in your repair manual. (b) Checking bearing with "go, no-go" gage.

the engine has been operating with a loose flywheel.

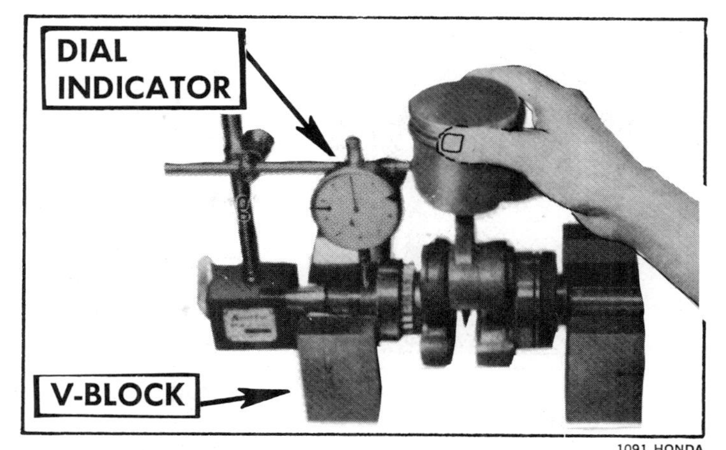

FIGURE 309. Checking the crankshaft for straightness with a dial indicator.

6. *Check the keyway.*

   If the keyway is worn excessively, replace the crankshaft. This worn keyway will cause the ignition timing to be off.

7. *Check the breaker-point plunger for wear, if installed (Figure 101).*

   This step is for engines that have the breaker points activated by a cam or trip on the crankshaft and/or camshaft. The plunger is not a part of the crankshaft, but the crankshaft must be removed in order for you to check it.

CLEANING AND INSPECTING THE CRANKSHAFT BEARINGS

Do not remove bearings from the crankcase unless you already know they are defective or unless it is necessary to remove them with the crankshaft.

1. *Clean the bearings in petroleum solvent.*

   Do not direct compressed air onto ball or roller bearings so they will spin. Spinning a dry bearing causes rapid wear and may ruin both the balls, or rollers, and the raceways.

2. *Check bearings for damage and wear.*

   Rotate **ball or roller bearings** by hand to check their condition. They should run smoothly. Balls and rollers should be true, with no flat sides or abrasions.

Caged bearings should remain in the cage and not fall out.

Check to see if the outer race is turning in the crankcase housing. If it is, peen the housing to tighten it.

Replace ball, roller or needle bearings if damaged.

Check **plain bearings** (Figure 310) for wear, scratches and scoring. Use an inside gage and an outside micrometer to measure the diameter of plain bearings, or use a special go, no-go gage (Figure 308b). Rebore and replace plain-bearing inserts when worn or damaged.

REMOVING AND REPLACING BEARINGS

Do not remove bearings unless there are indications that they are worn or damaged.

The removal method you use depends on the type of bearing and the material from which the crank-

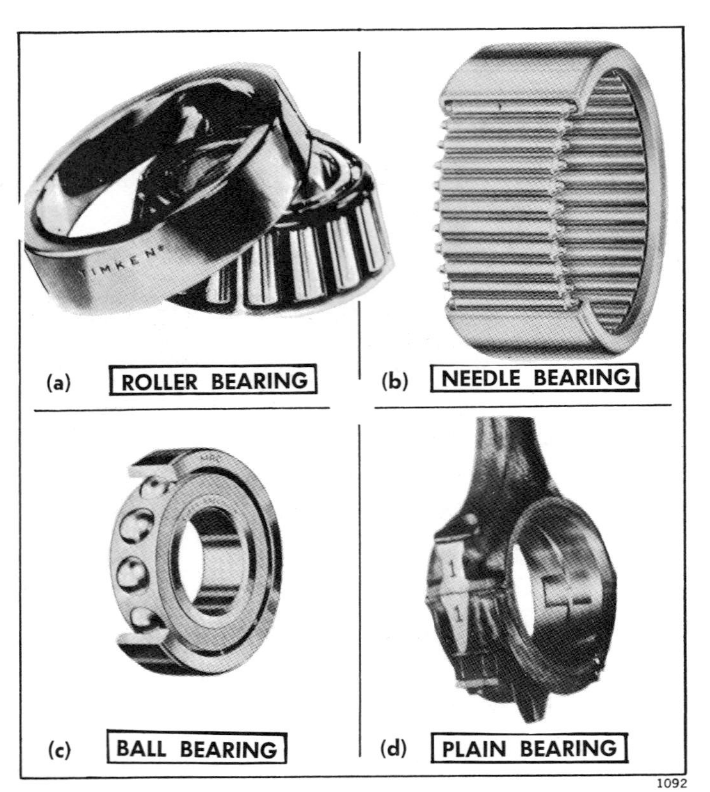

FIGURE 310. Types of bearings used on crankshaft assemblies. (a) Tapered-roller bearings withstand high speeds as well as heavy radial and thrust loads. They are used on heavy duty engines and engines of high horsepower. (b) Needle bearings take a heavy radial load. They are used in 2-cycle, high-speed engines. (c) Ball bearings withstand high speeds, moderate radial loads and moderate thrust. You will find ball bearings in all sizes of small engines. (d) Plain bearings are used in low-horsepower engines.

case is made. If the engine has **ball, needle or roller bearings** (Figure 310a, b and c), remove and replace them.

If the engine has a **cast-iron crankcase with plain-bearing inserts** (Figure 310d), they can also be removed and replaced; but the new inserts must be reamed.

If the engine has an **aluminum crankcase with plain bearings** which consists of a hole bored in the crankcase housing (no inserts), it may be possible to rebore and install an insert. The insert must be reamed to the shaft diameter.

Proceed as follows:

a. If the crankshaft has **ball, needle or roller bearings,** similar to the ones shown in Figure 310, proceed as follows:

1. *Install a bearing puller (Figure 311a), or place the crankshaft in an arbor press (Figure 311b).*

   Do not tighten the bearing separator against the crankshaft (Figure 311a). It will scratch the crankshaft bearing surface.

2. *Turn the removal screw with a wrench (Figure 311a).*

   Strike with a mallet occasionally.

3. *Install a new bearing on the crankshaft.*

   Ball bearings are press fitted. Cool the

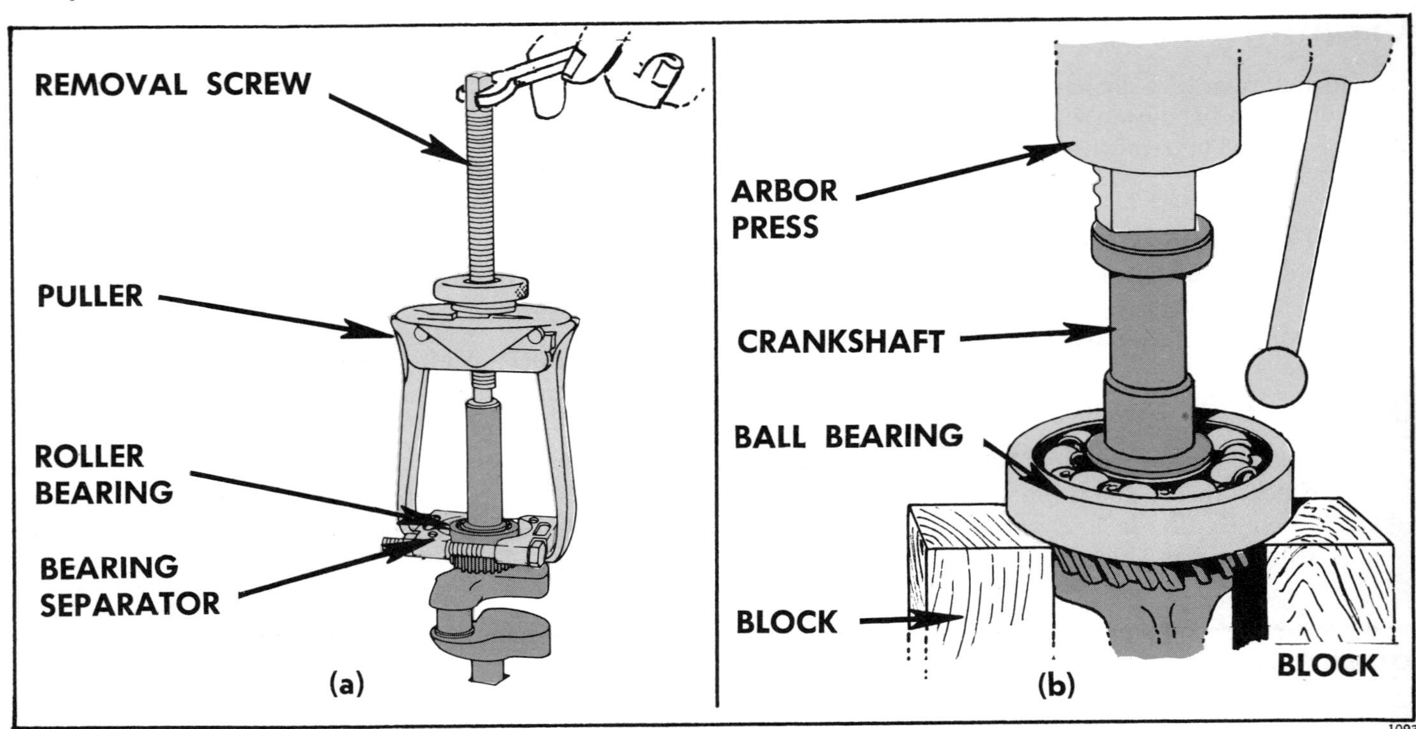

FIGURE 311. Removing ball bearings from the crankshaft with (a) a bearing puller, or (b) an arbor press.

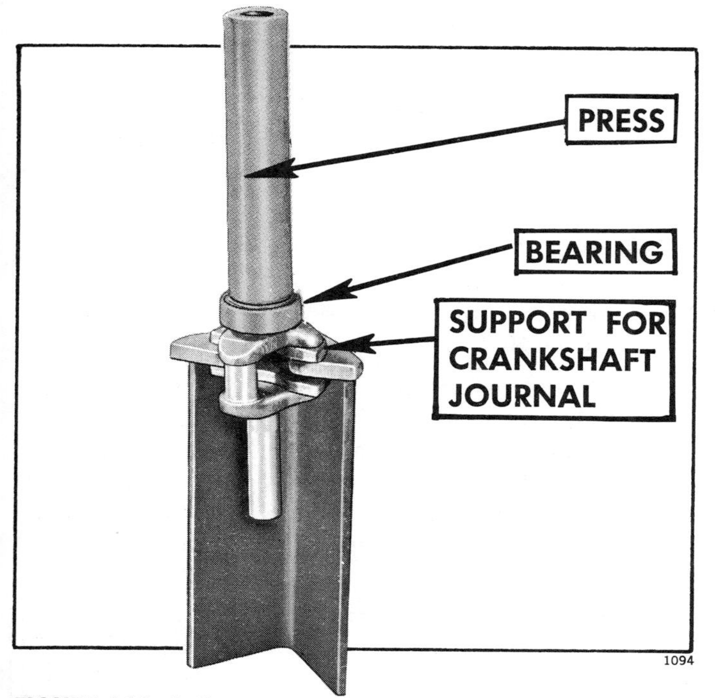

FIGURE 312. A jig supporting the crankshaft throw while bearing is fitted.

crankshaft and heat the bearing in hot oil—160°C (325°F) maximum. Do not direct flame against the bearings as it may distort it. Press the bearing all the way to the shoulder of the crankshaft (Figure 312).

The ball bearing must be sweat fitted onto the shaft; then sweat fitted to the housing. Do not press or drive bearings onto the crankshaft without supporting the throw (Figure 312).

Install outer races for new **tapered-roller bearings** in the housing (Figure 313a).

This is done by cooling the race, heating the housing and pressing the bearings into the housing.

Place needle bearings in the outer race by hand. Pack with multi-purpose grease to hold them in place until the shaft is installed.

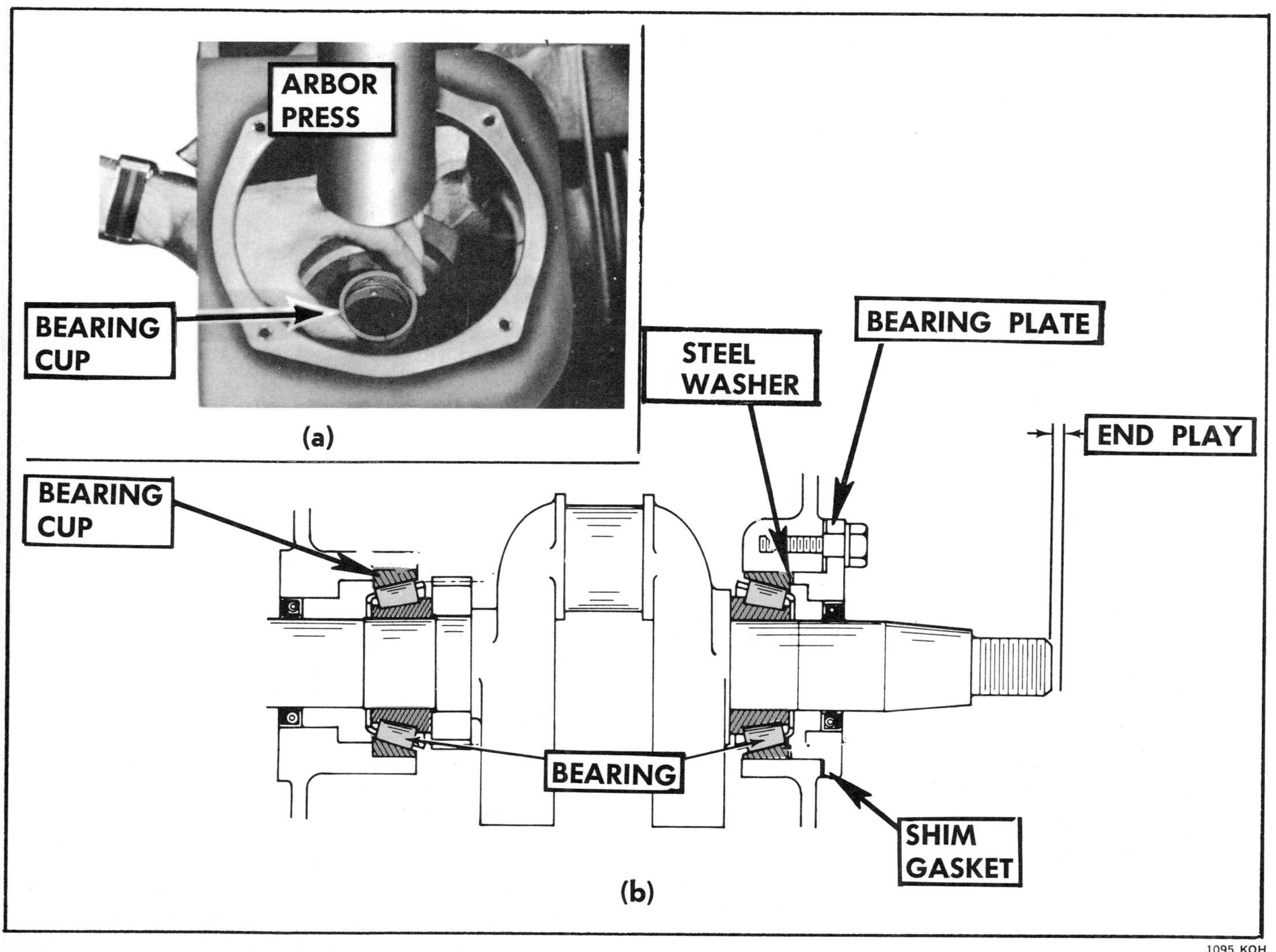

FIGURE 313. (a) Installing roller bearings race (bearing cup) in the crankcase. (b) Tapered roller-bearing assembly after pressing into position.

b. If the engine has a **cast-iron crankcase with plain-bearing inserts,** proceed as follows:

1. *Remove old bearings (Figure 314).*

   Remove bearing on PTO-end first. In these procedures you will remove, replace and ream one bearing before removing the opposite bearing. In this way you use one as a guide for reaming the other. When you are removing the bearing insert, always press it toward the inside of the crankcase. Use an arbor press.

2. *Press in a new bearing.*

   Install bearing toward the outside. Be sure you align the oil hole in the bearing with the oil hole in the crankcase.

   Make the outer end of the bearing flush with the outer end of the bearing hole in the crankcase to allow room for the oil seal.

3. *Select the proper reaming tools.*

   Special reaming tools are provided by each manufacturer. The size, design and procedures for their use vary; so check your service manual.

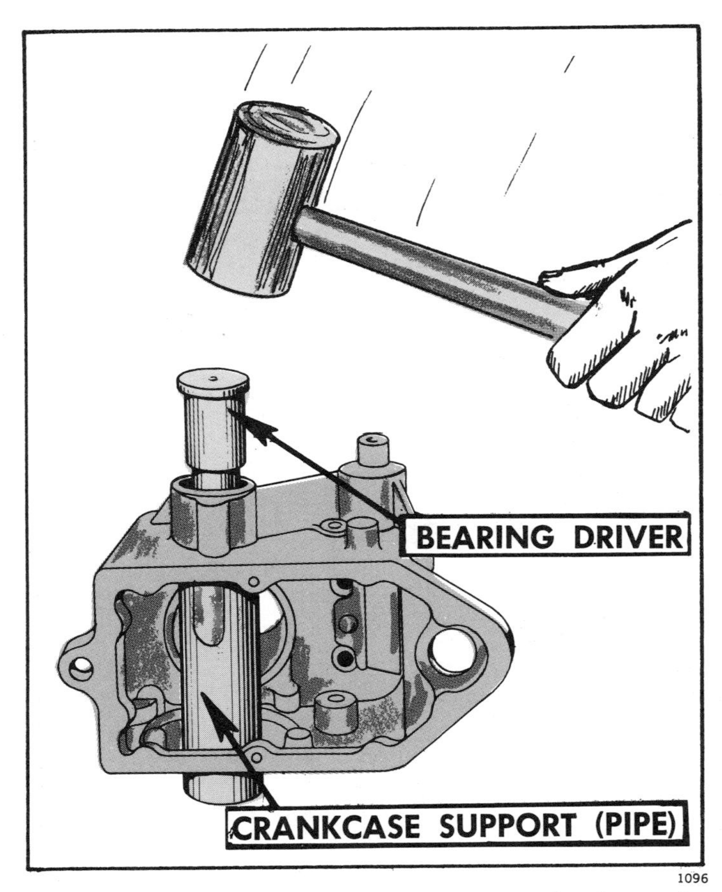

FIGURE 314. Removing a plain bearing. Crankcase is supported by large pipe in bearing area.

4. *Insert the reamer and pilot.*

   Procedures for two types of reamers are given here (Figure 315).

   If the reamer has a **pilot and guide bushing** (Figure 315a), place the guide bushing in the opposite bearing with the flange toward the inside; then reassemble the crankcase if separated.

   If the reamer is a **two-piece unit with the shank used as a pilot** (Figure 315b), assemble the crankcase — if disassembled — for alignment before you start. Then insert the reamer shank through the opposite bearing. With this type, the opposite bearing acts as the guide.

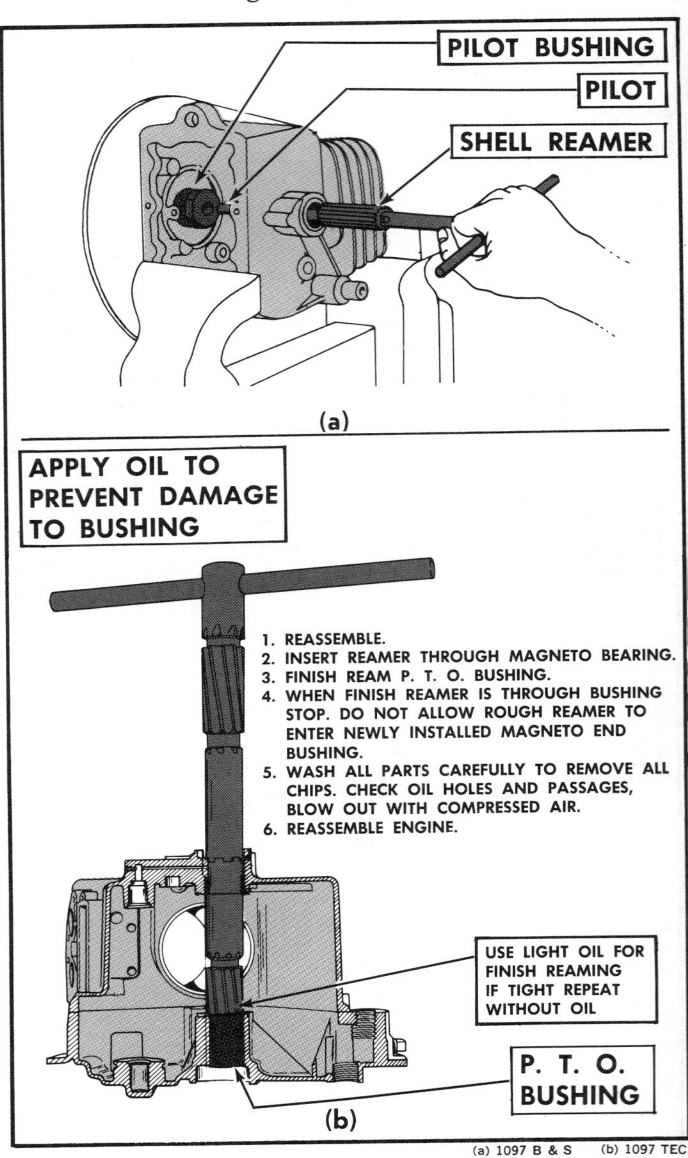

FIGURE 315. Reaming a bearing insert with (a) a reamer which has a separate pilot and guide bushing or (b) a reamer with the reamer shank and opposite bearing used as a guide.

5. *Ream bearing to size (Figure 315).*

   Turn reamer clockwise with a steady, even pressure until it is completely through the bearing. Follow instructions in your repair manual as to whether to use cutting oil or not. Cutting oil, kerosene or fuel oil will cause the reamer to cut a smaller hole than when it is cutting it dry. Some manufacturers recommend cutting wet; some, dry.

6. *Remove the magneto-bearing plate before attempting to remove the reamer.*

   This removal is to prevent you from having to pull the reamer back through the bearing insert. Each time the reamer passes through the bearing, it is enlarged. It is well to make one pass through, if possible. On some engines one pass through is not possible. The reamer must be removed from the top. The reamer size is adjusted for this.

7. *Remove the reamer from the opposite end of the crankcase.*

   Do not remove the reamer back through the bushing unless necessary. This removal will cut the bearing larger. Remove guide bushing (Figure 315a) if installed.

8. *Clean away all metal chips.*

   Make certain metal chips do not get into the oil passageways.

   CAUTION! Wear protective goggles when using compressed air to blow away metal chips.

9. *Check new bearing for size.*

10. *Remove and replace the other bearing in the same manner.*

    Always replace both bearings — not just one. This will give you proper crankshaft bearing alignment.

c. If the engine has an **aluminum crankcase with plain bearings** (not inserts), proceed as follows:

1. *Determine if the bearing is reparable.*

   It is possible on some aluminum engines to counterbore the aluminum crankcase and install a bearing insert. If yours is reparable, proceed to step 2.

   Some engines are not designed for having the bearing repaired, and no special tools or instructions are provided by the manufacturers. Some are designed for repairing only one of the bearings. The instructions in your service manual will provide this information.

2. *Select the proper counterboring tools.*
3. *Counterbore the crankcase for a new bearing insert.*

   If you do not know the difference between "counterboring" and "reaming," it is important that you understand the difference. The tools look the same and are used in the same manner, but the results are different. **Counterboring** consists of enlarging an existing hole with a rough cut. **Reaming** consists of sizing a hole to an exact dimension and finishing the surface with a fine finish.

   Follow steps 4 through 8 under "b" preceding, except use the counterbore cutter instead of the finish reamer (Figure 316). When using the counterboring tool with the pilot-guide bushing, install a steel guide bushing in the oil-seal recess at the bearing which you are counterboring (Figure 315a). The guide bushing centers the reamer so that the

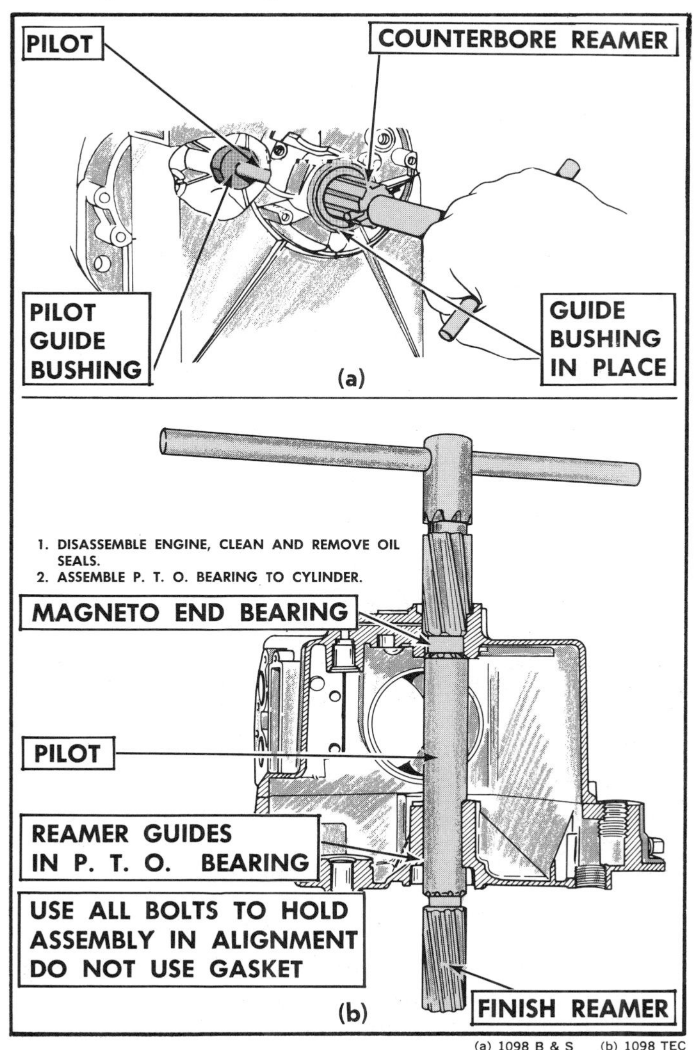

FIGURE 316. Counterboring (a) with a counterboring tool equipped with a pilot and guide, and (b) with a tool which uses the shank and opposite bearing as a guide.

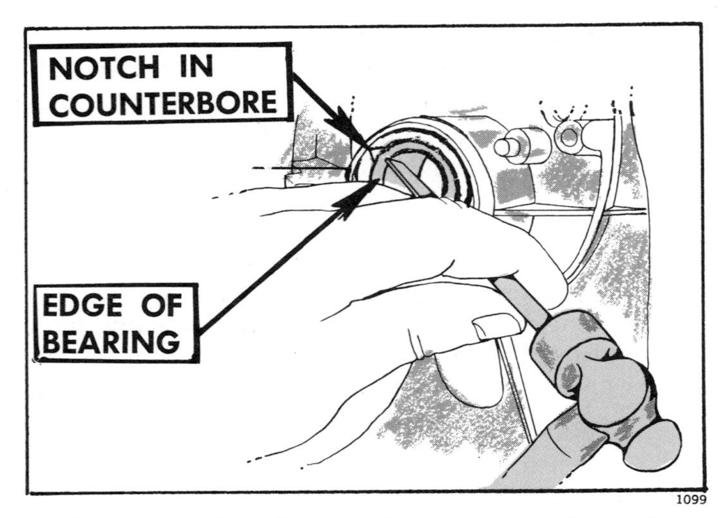

FIGURE 317. Staking (keying) the counterbore for anchoring the new bearing insert, in an aluminum crankcase.

enlarged hole is in line with the original bearing.

4. *Hold the new bearing insert against the outer end of the counterbore with the oil grooves aligned.*

5. *Mark the hub at a point opposite the split in the bearing insert.*

6. *Make a notch on the outer edge of a counterbore at the point marked (Figure 317).*

   One manufacturer recommends using a cold chisel to notch it at a 45° angle to the bearing surface. This notching is to stake (key) the bearing insert to the crankcase so that it will not turn during operation.

7. *Press in the new bearing.*

   Make sure oil passages are aligned. Make the bearing flush with the housing at the outer end.

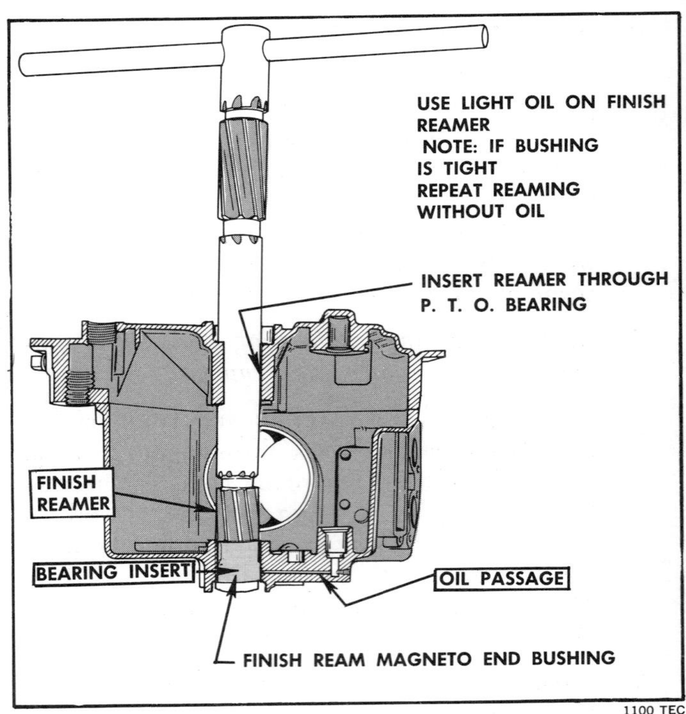

FIGURE 318. Reaming the front bearing.

8. *Stake the new bearing into the notch provided for in step 6, if recommended by your manufacturer.*

   Use a blunt chisel.

9. *Ream the new bearing insert (Figure 318).*

   Follow procedures in steps 3 through 10 under "b" preceding for reaming the new bearing.

10. *Check bearing for size.*

    Note procedures under "Cleaning and Inspecting Crankshaft Bearings," page 257.

11. *Repeat these procedures for the opposite crankshaft bearing if applicable.*

INSTALLING THE CRANKSHAFT

Before installing your crankshaft be sure all parts are clean and well oiled. Do not handle steel parts with your fingers unless absolutely necessary. The acid from your fingers will cause them to corrode. Use a rag or gloves for handling.

1. *Install the oil seals (Figure 319).*

   Oil seals may be installed in the bearing plate before the plate is attached, or they may be installed after the crankshaft is installed. In either case, be careful not to damage the oil seals. If you install the seal before bearing plate is attached, proceed as follows:

   (1) *Apply gasket cement to the outside of the seal.*

   As the oil seal bottoms when installed, the gasket cement will bond the outside of the seal to the crankcase.

   (2) *Seat the seal flush with the receptacle (Figure 319b).*

   (3) *Check to see that no gasket paste gets into the oil passageways.*

   (4) *Lubricate the oil seal with crankcase oil.*

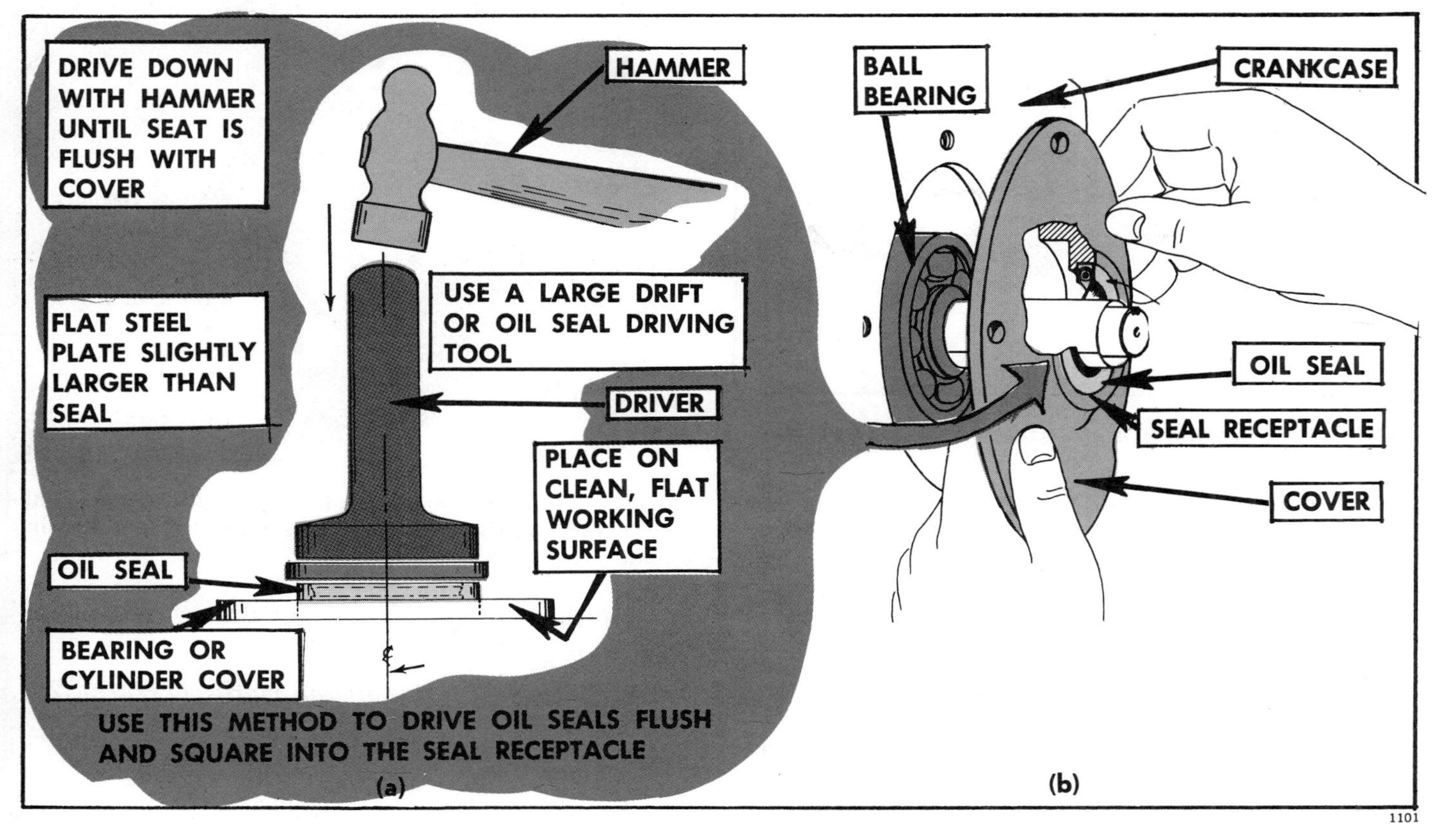

FIGURE 319. (a) Installing the oil seal before installing the crankshaft. (b) Installing the bearing plate and oil seal on the crankshaft.

2. *Install the camshaft assembly first, if it is the type that rotates on a pin (Figure 320a).*

   For procedures refer to "Installing the Camshaft Assembly," page 173. Unless you install the camshaft assembly first, you will not be able to get the camshaft and the valve tappets past the crankshaft after it is installed.

   The camshaft assembly should be installed loosely (not tightened securely) until the crankshaft can be installed on some engines. Lubricate all parts before installing them.

3. *Install the crankshaft in the crankcase.*

   If the crankshaft has a thrust washer, be sure it is installed.

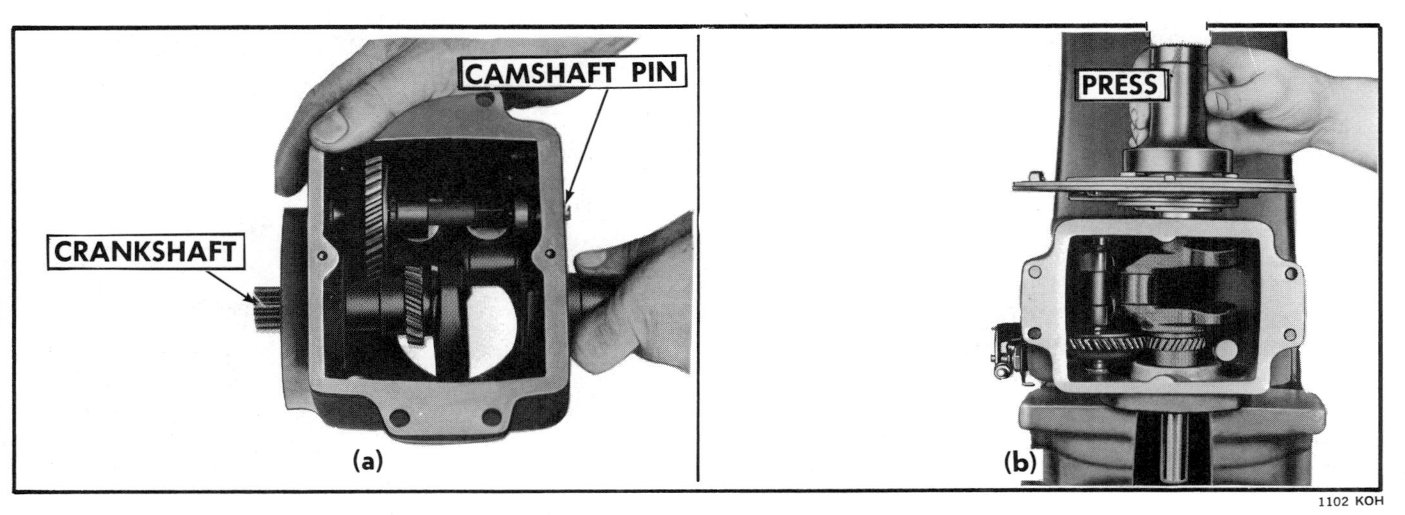

FIGURE 320. (a) Installing a crankshaft with ball bearings. (b) Pressing the bearings and crankshaft in place.

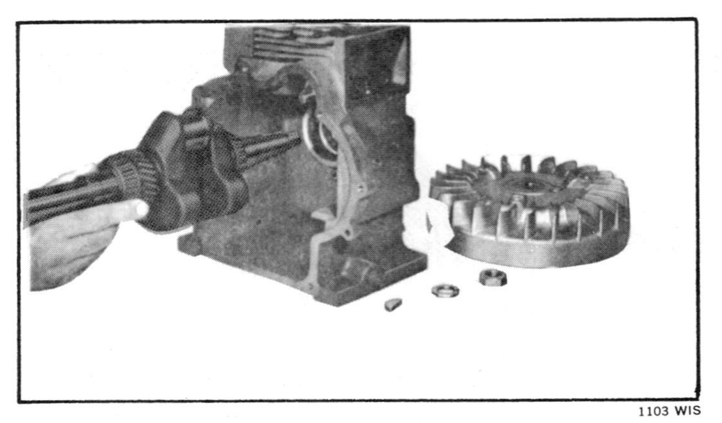
1103 WIS

FIGURE 321. Installing a crankshaft with tapered-roller bearings.

Watch for the breaker-point plunger, or the breaker-point actuating arm or the governor drive arm. These must be aligned properly. There is a danger of breaking them if you force the crankshaft into position.

If your engine has **ball bearings,** press the shaft and bearings in place (Figure 320).

If your engine has **tapered-roller, needle or plain bearings,** place the crankshaft in by hand (Figure 321).

4. *Connect the piston rod.*

   For procedures refer to "Installing the Piston-and-Rod Assembly," page 229.

5. *Install the camshaft assembly, if it has not already been installed (Figure 322).*

1105 KOH

FIGURE 323. Align the valve-timing marks on the crankshaft and camshaft gears before replacing the bearing plate.

   Refer to "Installing the Camshaft Assembly," page 247.

6. *Align the valve timing marks (Figure 323).*

   Refer to "Installing the Camshaft Assembly," page 247.

7. *Press in the camshaft pin if yours is the type that rotates on a pin (Figure 324).*

8. *Install the governor assembly if one is located inside the crankcase (Figure 322).*

9. *Install oil line from oil pump to upper main bearing if used (Figure 325).*

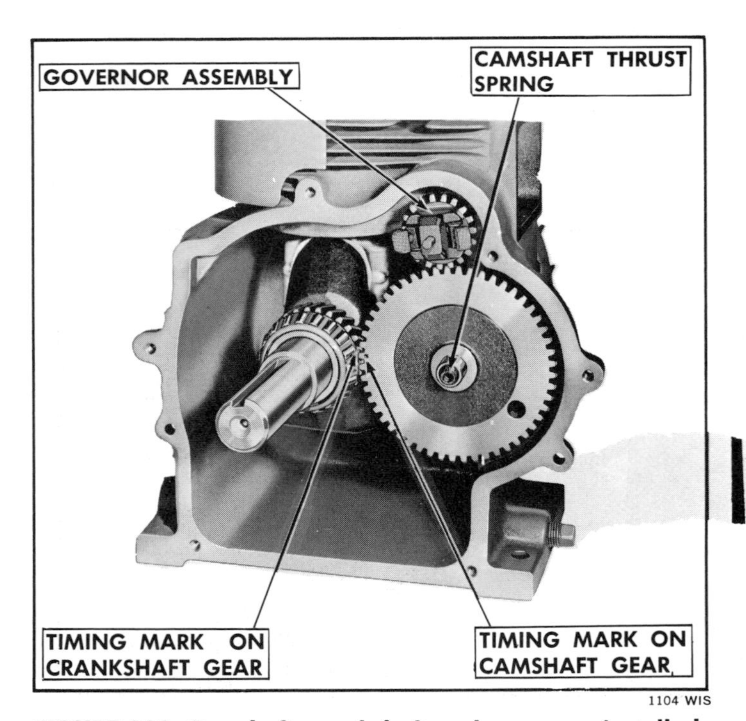

1104 WIS

FIGURE 322. Camshaft, crankshaft and governor installed.

1106 KOH

FIGURE 324. Pressing in the camshaft pin.

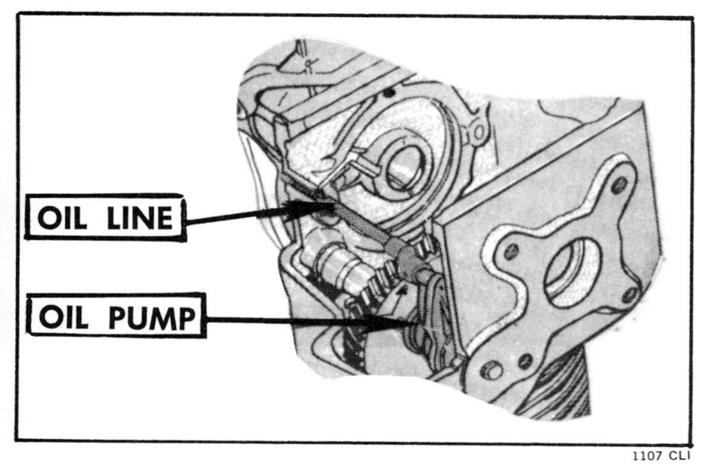

FIGURE 325. Some vertical-crankshaft engines have an oil line from the pump to the upper-main bearing.

10. *Install gasket(s) (Figure 326a).*

    On some engines the crankshaft endplay is controlled by the number of gaskets which are installed on the bearing plate, or bearing support.

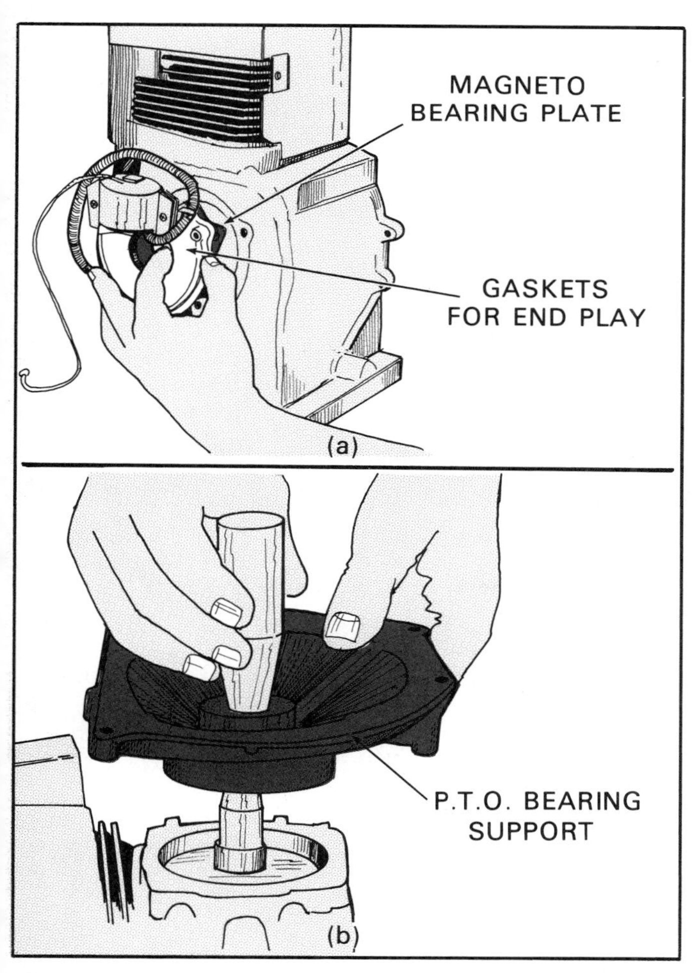

FIGURE 326. (a) Installing the magneto-bearing plate. (b) Installing the PTO-bearing support.

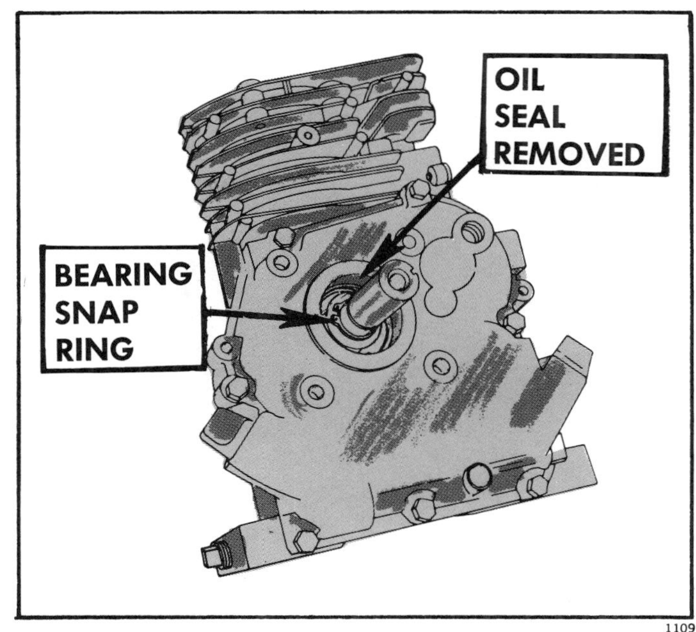

FIGURE 327. A crankshaft-bearing lock ring is installed on some engines.

11. *Install the magneto-bearing plate, or the PTO-bearing support (Figure 326).*

    Either of these plates contains the crankshaft bearing or provides for installing a bearing.

12. *Tighten magneto-bearing plate, or PTO-bearing support, with a torque wrench.*

    Torque recommendations vary from 82 N·m to 110 N·m (75 lb f·in to 100 lb f·in).

13. *Install crankshaft-bearing lock ring, if applicable (Figure 327).*

14. *Install oil seals if they are not already installed.*

    Refer to step 1.

15. *Check crankshaft endplay, if it is adjustable (Figure 328).*

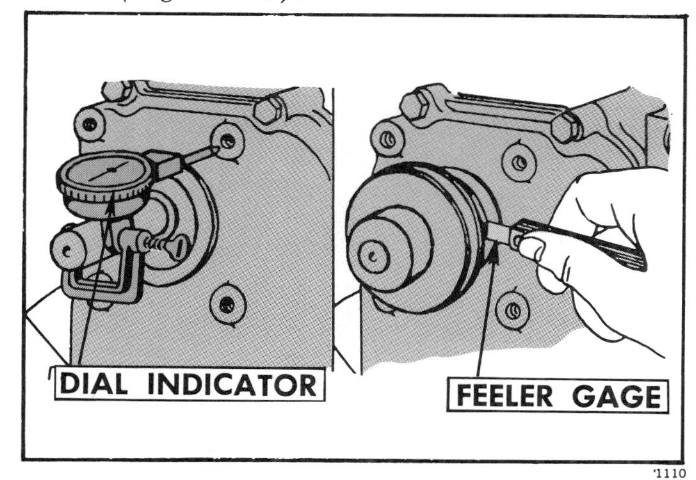

FIGURE 328. Checking the crankshaft endplay with a dial indicator.

OWNER'S ENGINE-INFORMATION FORM

GENERAL INFORMATION: ______________________________

NAME OF EQUIPMENT (ON WHICH ENGINE IS MOUNTED) ______________________________

NAME AND ADDRESS OF EQUIPMENT MANUFACTURER ______________________________

____

NAME AND ADDRESS OF ENGINE MANUFACTURER ______________________________

____

OPERATING POSITION OF CRANKSHAFT: VERTICAL □, HORIZONTAL □, MULTI-POSITION □.

ENGINE CYCLE: 2-Cycle □, 4-Cycle □.

MODEL NUMBER, OR NAME ______________________________

SERIAL NUMBER ______________________________

SPECIFICATION NUMBER ______________________________

TYPE NUMBER ______________________________

HORSEPOWER ______________________________

TYPES OF ACCESSORIES AND MAJOR UNITS:

CARBURETOR AIR CLEANER: Oil bath□, Oiled filter□, Dry filter□.

FUEL STRAINER: Combination screen and sediment bowl□, Screen inside the fuel tank□.

CRANKCASE BREATHER: Reed valve□, Floating disk valve□.

STARTER: Rope-wind□, Rope-rewind□, Wind up□, Electric, AC□, Electric, DC□.

IGNITION SYSTEM: Flywheel magneto□, External magnet□, Battery□.

FUEL PUMP: Mechanically driven□, Differential pressure driven□.

CARBURETOR: Float□, Suction lift□, Diaphragm□.

GOVERNOR: Air vane□, Centrifugal□.

SERVICE AND MAINTENANCE SPECIFICATIONS:

OIL: SAE 5W □ (add) 5W – 20 □, 5W – 30 □, 10W – 30 □, 10W – 40 □, classification: SC □, or SD □, SE □, CC □, 2-cycle □.

Solid state □

TYPE OF SPARK PLUG: ______________________________

Gap setting: 0.5 mm (.020 in), o.6 mm (.025 in)

IGNITION BREAKER-POINT GAP: 0.3 mm (.012 in) □, 0.4 mm (.015 in) □, Other ______________.

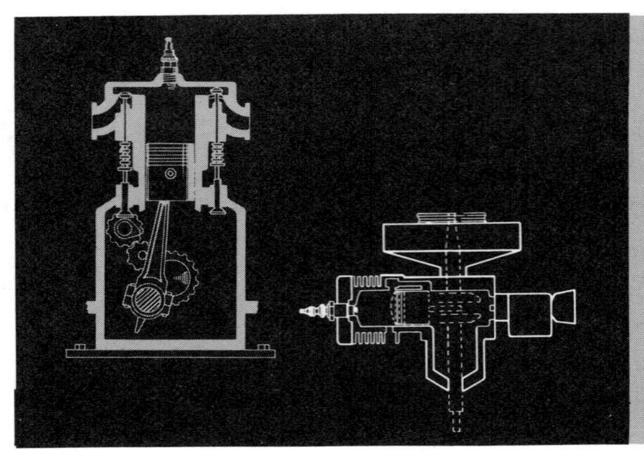

References

SPECIFIC REFERENCES

1. *Long Life Through Service Maintenance;* Johnson, Charles, Briggs & Stratton Engine Corporation; Publication No. MS-4976-114.

2. *Moving Parts Last Longer in Clean Engines;* Shop Service, McQuay-Norris Manufacturing Company, St. Louis, Missouri.

3. *The Internal-Combustion Engine*; Taylor, C. F. and Taylor, E. S., International Textbooks in Mechanical Engineering, Second Edition, August.

4. *Effect of Oil Change Interval on Low-Temperature Wear and Deposit-Forming Characteristics of Crankcase Oils*; Gulf Research and Development Company.

5. *Maintenance Inspections of Sixty Farm Tractors*; Weber, J. A., University of Illinois, Agricultural Experiment Station, Bulletin 624.

6. *Instructions for Starting and Operating;* Wisconsin Motor Corporate; Instruction Book and Parts List, No. MM-300-B.

7. *Research Report;* Briggs and Stratton Engine Corporation.

8. *General Theories of Operation;* Briggs & Stratton Corporation; No. MS-3553-24.

9. *Factors Affecting Piston Ring Life and Performance in Small Two and Four Cycle Engines*; Paul, D. A. and Radtke, J. B., Society of Automotive Engineers, Paper No. 601A.

GENERAL REFERENCES

(Partial List)

Teaching Aid for Small Gas Engines; Rogers, C. J., Agricultural Engineering Department, University of Florida.

A Proposed Course of Study on Small Engine Repair; Shawen, Robert, Department of Agricultural Education, University of Idaho.

Add Life to Your Lawnmower; Bear, W. F., Iowa Farm Science, Vol. 15, No. 11, College of Agriculture, Experiment Station, Iowa State University.

Small Tractor and Equipment Standardization—The Manufacturer Views Standardization; Bryant, William R., ASAE Paper June 1965, Paper No. 65-118.

General Principles of Internal Combustion Engines, Agriculture Education Department, Mississippi State University.

Power Mechanics A Curriculum Guide; Office of Curriculum, Louisville Public Schools, Louisville, Kentucky.

Internal Combustion Engines; by Weston, Curtis R., Lyndon Bays, Glen C. Shinn, Richard Linhardt; Department of Agricultural Education, University of Missouri, Columbia, Missouri, State Department of Education, Jefferson City, Missouri.

The Gasoline Engine Electrical System, Cameron, Walter A., Instructor, Agricultural Mechanics Education, Agricultural Education—Agricultural Engineering Department, Virginia Polytechnic Institute, Blacksburg, Virginia.

Micrometers and Related Measuring Tools; College of Agriculture, Vocational Agriculture Service, VAS 3023, University of Illinois, Urbana, Illinois.

4-H Small Engine Project I, II, III; Cooperative Extension Service, Iowa State University, Ames, Iowa.

Small Gasoline Engines; Curriculum Laboratory, Dept. of Community Colleges, St. Board of Education, Raleigh, North Carolina.

Tractor & Farm Implement Serviceman; U. S. Dept. of Health, Education and Welfare; Office of Educ. Washington, D. C.

Guide for Teaching, Operating and Maintaining Small Four-Stroke Cycle Air-Cooled Engines; Dept. of Agric. Educ., College of Education, Univ. of Georgia, August, 1965.

Formats for Instructional Materials in Trade and Industrial Education; Dept. of Trade and Industrial Education, College of Education, Univ. of Georgia, Athens, Georgia.

Small Gas Engine A Noisy Miracle for Teaching; The Agricultural Education Magazine, Vol. 39, Number 1, July, 1966.

Small Gasoline Engines; Cooperative Extension Service, University of Connecticut.

Small Engines—Repair and Overhaul; College of Agriculture, Vocational Agriculture Service, University of Illinois, Urbana, Illinois.

Small Engines—Principles of Operation, Trouble Shooting and Tune-Up, College of Agriculture, Vocational Agric. Serv., University of Illinois, Urbana, Illinois.

All About Small Gas Engines; Purvis, J., Goodheart-Wilcox Co. Inc., Homewood, Illinois.

Small Gasoline Engines; Stephenson, G. E., Delmar Publishers, Albany, New York.

Selecting and Storing Tractor Fuels and Lubricants; American Association for Vocational Instructional Materials.

Tractor Maintenance, Principles and Procedures; American Association for Vocational Instructional Materials.

The Internal-Combustion Engine; Taylor, C. Fayette and Taylor, Edward S.; International Textbooks in Mechanical Engineering, Scranton, Pennsylvania, Second Edition, August, 1962.

The Internal-Combustion Engine in Theory and Practice; Taylor, Charles F., Massachusetts Institute of Technology, Volume 1, 1960.

Carburetion; Volume 1, Fisher, Charles H., Fourth Edition, 1963.

Motor Oils and Engine Lubrication; Georgi, Carl W., Quaker State Oil Refining Corp.

Some New Designs for Internal Combustion Engines; Long, E. Melvin, Implement & Tractor.

Small Engines Repair Manual; Technical Publications, Inc., Kansas City, Missouri.

Automotive Electrical Systems; Engine Service Institute, Automotive Electric Association, Detroit, Michigan, 1954.

Automotive Fuel Systems; Engine Service Institute, Automotive Electric Association, Detroit, Michigan.

Symposium on Lubricants for Automotive Equipment; American Society for Testing and Materials, ASTM Special Technical Publication No. 334.

Chemistry and Prevention of Piston-Ring Sticking; Denison, G. H., Jr., and Clayton, J. O., California Research Corp., Subsidiary of Standard Oil Co. of Calif., SAE Journal (Transactions), Vol. 53, No. 5.

Engine Varnish and Sludge; Gibson, Harold J., SAE National Fuels and Lubricants Meeting, Society of Automotive Engineers, Inc.

Combustion Chamber Deposition and Power Loss; Gibson, H. J., Hall, C. A. and Huffman, A. E.; Society of Automotive Engineers, Inc.

Deposition and Wear in Light-Duty Automotive Service; Kendall, Norman and Greenshields, R. J., SAE Quarterly Transactions Vol. 2, No. 3.

A Symposium on Varnish in Engines; Kishline, F. F., SAE Journal Transactions.

Crankcase Ventilation; Moir, H. L. and Hemmingway, H. L., Pure Oil Co., SAE Paper National Fuels and Lubricants Meeting, July, 1947, Vol. 1, No. 3.

The Where and Why of Engine Deposits; Spindt, R. S., and Wolfe, C. L., SAE National West Coast Meeting, Society of Automotive Engineers.

Flame Photographs of Autoignition Induced by Combustion Chamber Deposits; Withrow, L. L. and Bowditch, F. W., SAE Journal.

Small Tractor Service Manual; Technical Publications, Inc., Kansas City, Missouri.

Air Cleaner Test Code; Society of Automotive Engineers, SAE J 726a.

Piston and Piston Ring Nomenclature; Society of Automotive Engineers, SAE J 612.

Voltage Drop for Starting-Motor Cable; Society of Automotive Engineers, SAE J 541.

Crankcase and Transmission Dipstick Marking; Society of Automotive Engineers, SAE J 611.

Spark-Plug Installation—Torque Requirements and Gasket Dimensions; Society of Automotive Engineers, SAE J 550a.

Minimum Identification Markings for Small Air-Cooled Engines; Society of Automotive Engineers, SAE J 608.

Test Code for Small Stock Air-Cooled 2-Stroke and 4-Stroke Cycle Gasoline Engines (6 HP and Below); Society of Automotive Engineers, SAE J 607.

Small Gasoline Engines; Ted Pipe, Training Manual, Photofact Publication GEP-1 Howard W. Sams & Co., Inc., The Bobbs-Merrill Company, Inc.

Bearings Can Reveal the Cause of Damage; Service Engineering, Magic Circle, Dana Corp., Toledo, Ohio.

Rx Doctor of Motors, Prescription For Better Gasoline Engine Overhauls; Perfect Circle, Technical Service Department, Dana Parts Company, Hagerstown, Indiana.

Rx Doctor of Motors, Prescription for Preventing Oil Loss Through Valve Guides; Perfect Circle, Dana Parts Co., Hagerstown, Indiana.

Service Manual for the Doctor of Motors; Fifth Edition, Dana Parts Company, Hagerstown, Indiana.

Battery Service Manual; The Assoc. of American Battery Manufacturers, Inc., Sixth Edition.

Principles of Operation, Service, Maintenance—Air Cooled Engines; Engine Service Institute. Automotive Electric Association, Detroit, Michigan.

American Standard Safety Specifications for Power Lawn Mowers; American Standards Association, Inc., N. Y., N. Y.

Chain Saw Safety; National Safety Council, 425 N. Michigan Ave., Chicago, Illinois.

Designing the Rotary Power Lawn Mower for Safety; Knapp, L. W. and McConnell, W. H., Institute of Agricultural Medicine, University of Iowa, Iowa City, Iowa, ASAE Paper 65-652.

Small Gasoline Engines, Operation and Maintenance; Hobar Publications, St. Paul, Minnesota 55112.

CONTRIBUTORS OF LITERATURE AND INFORMATION

MANUFACTURERS OF SMALL ENGINES:

Briggs and Stratton Corporation, 3300 N. 124th St., Milwaukee, Wisconsin 53201.

Clinton Engine Corporation, Maquoketa, Iowa, 52060.

Detroit Engine Corporation, 285 Piquette Street, Detroit, Michigan 48200.

Cushman Motors, Lincoln, Nebraska 68500 (Division of Outboard Marine Corporation).

Gravely Tractors, Division of Studebaker Corporation, 4400 Gravely Lane, Dunbar, West Virginia 25064.

Honda Motor Company, Ltd., Tokyo, Japan.

Jacobsen Manufacturing Company, 1721 Packard Avenue, Racine, Wisconsin 53400.

Kohler Company, Kohler, Wisconsin 53044.

Lawn Boy, Division Outboard Marine, Galesburg, Illinois 61401.

McCulloch Corporation, 6101 West Century Boulevard, Los Angeles, California 90000.

Onan, 1400 73rd Ave., N.E, Minneapolis, MN 55432.

Tecumseh Products Company, Lauson-Power Products Engine Division, Grafton, Wisconsin 54436.

West Bend Company, Hartford Division of Chrysler Outboard Corporation, Hartford, Wisconsin 53027.

Teledyne Wisconsin Motor, 19 South 53rd St., Milwaukee, WI 53219

O & R Engines, Inc., 3340 Emery Street, Los Angeles, California 90023.

Beaird-Poulan, Inc., Shreveport, Louisiana 71100.

MANUFACTURERS OF TOOLS AND EQUIPMENT:

Ammco Tools, Inc., North Chicago, Illinois 60600.

Electro-Tech, Inc., Hapeville, Georgia 30054.

Merc-O-Tronic Instruments Corp., 215 Branch St., Almont, Michigan 48003.

Neway Sales, Inc., Corunna, Michigan 48817.

Wood's Powr-Grip Mfg. Co., Wolf Point, Montana 59201.

Zim Manufacturing Co., Cambridge, Massachusetts 01922.

Graham Research, Inc., Minneapolis, Minnesota 55418

Sunnen Products Company, St. Louis, Missouri 63143

MANUFACTURERS OF SMALL ENGINE PARTS AND ACCESSORIES:

AC Spark Plug, Division of General Motors, Flint, Michigan 48502.

Champion Spark Plug Company, Toledo, Ohio 43600.

Gould Engine Parts Division (Pistons, Rings, Valves, Etc.), Gould-National Batteries, Inc., St. Paul, Minnesota 55101.

McQuay-Norris Manufacturing Co. (Pistons, Rings, Valves, Etc.), St. Louis, Missouri 61300.

Perfect Circle (Piston Rings), Dana Parts Corporation, Toledo, Ohio, 43601.

Ramsey Corporation (Piston Rings, Etc.) St. Louis, Missouri 63166.

The Bendix Corporation (Electrical Accessories and Carburetors), South Bend, Indiana 46620.

Warner Electric Brake & Clutch Company (Electrical Accessories), Beloit, Wisconsin 53512.

Colt Industries (Electrical Accessories), Fairbanks Morse Engine Accessories Operation, Beloit, Wisconsin 53512.

Scintilla Division (Electrical Acessories), Bendix Aviation Corp., Sidney, New York 13838.

Tillotson Manufacturing Company (Carburetors), Parts & Service Division, Toledo, Ohio 43600.

WICO Division (Electrical Accessories), Globe-Union Inc., West Springfield, Massachusetts.

Koppers Company (Pistons, Rings, Etc.), Metal Products Division, Baltimore, Maryland 21203.

Owatonna Tool Company, Owatonna, Minnesota 55060.

Stewart-Warner Corporation, Chicago, Illinois 60614.

Snap-On Tools Corporation, Kenosha, Wisconsin 53140.

S-K Wayne Tool Company, Chicago, Illinois 60604.

P & C Tool Company, Portland, Oregon 97222.

Herbrand, Kelsey Hayes Tool Division, Orangeburg, South Carolina 29115.

Proto Tool Company, Los Angeles, California 90054.

Products of Cedar Rapids Engineering Company, Cedar Rapids, Iowa 52400.

Index

Index

NOTES

NOTES

NOTES

NOTES

NOTES

NOTES

NOTES

Other Bestsellers From TAB